NATO ASI Series
Advanced Science Institutes Series

A series presenting the results of activities sponsored by the NATO Science Committee, which aims at the dissemination of advanced scientific and technological knowledge, with a view to strengthening links between scientific communities.

The series is published by an international board of publishers in conjunction with the NATO Scientific Affairs Division

A	Life Sciences	Plenum Publishing Corporation
B	Physics	London and New York
C	Mathematical and Physical Sciences	D. Reidel Publishing Company Dordrecht, Boston and Lancaster
D	Behavioural and Social Sciences	Martinus Nijhoff Publishers
E	Engineering and Materials Sciences	The Hague, Boston and Lancaster
F	Computer and Systems Sciences	Springer-Verlag
G	Ecological Sciences	Berlin, Heidelberg, New York and Tokyo

Series C: Mathematical and Physical Sciences Vol. 157

Molecular Astrophysics
State of the Art and Future Directions

Molecular Astrophysics

State of the Art and Future Directions

edited by

G. H. F. Diercksen

Max-Planck-Institut für Physik und Astrophysik,
Institut für Astrophysik, Garching-bei-München, F.R.G.

W. F. Huebner

Theoretical Division, Los Alamos National Laboratory,
Los Alamos, New Mexico, U.S.A.

and

P. W. Langhoff

Department of Chemistry, Indian University,
Bloomington, Indiana, U.S.A.

D. Reidel Publishing Company

Dordrecht / Boston / Lancaster

Published in cooperation with NATO Scientific Affairs Division

Proceedings of the NATO Advanced Research Workshop on
Molecular Astrophysics - State of the Art and Future Directions
Bad Windsheim, F.R.G.
8-14 July, 1984

Library of Congress Cataloging in Publication Data

NATO Advanced Research Workshop on Molecular Astrophysics—State of the Art and
    Future Directions (1984 : Bad Windsheim, Germany)
    Molecular astrophysics.

    (NATO ASI series. Series C, Mathematical and physical sciences ; vol. 157)
    "Published in cooperation with NATO Scientific Affairs Division."
    Bibliography: p.
    Includes indexes.
    1. Molecular astrophysics—Congresses. 2. Interstellar molecules—Congresses.
3. Astronomical spectroscopy—Congresses. I. Diercksen, G. H. F. II. Huebner, W. F.
(Walter F.), 1928-   . III. Langhoff, P. W. (Peter W.), 1937-   . IV. Title.
V. Series: NATO ASI series. Series C, Mathematical and physical sciences ; no. 157)
QB462.6.N37   1984   523.01'96   85-14198
ISBN 90-277-2081-9

Published by D. Reidel Publishing Company
P.O. Box 17, 3300 AA Dordrecht, Holland

Sold and distributed in the U.S.A. and Canada
by Kluwer Academic Publishers,
190 Old Derby Street, Hingham, MA 02043, U.S.A.

In all other countries, sold and distributed
by Kluwer Academic Publishers Group,
P.O. Box 322, 3300 AH Dordrecht, Holland

D. Reidel Publishing Company is a member of the Kluwer Academic Publishers Group

---

All Rights Reserved
© 1985 by D. Reidel Publishing Company, Dordrecht, Holland.
No part of the material protected by this copyright notice may be reproduced or utilized
in any form or by any means, electronic or mechanical, including photocopying, recording
or by any information storage and retrieval system, without written permission from the
copyright owner.

Printed in The Netherlands.

NATO ADVANCED RESEARCH WORKSHOP ON
MOLECULAR ASTROPHYSICS —
STATE OF THE ART AND FUTURE DIRECTIONS
BAD WINDSHEIM (GERMANY)
8. JULY - 14. JULY 1984

TABLE OF CONTENTS

PREFACE, G.H.F. Diercksen, W.F. Huebner, P.W. Langhoff

PART I. ASTROPHYSICAL OBSERVATIONS AND MODELS

A. Dalgarno
MOLECULAR ASTROPHYSICS   3

M. Guelin
CHEMICAL COMPOSITION AND MOLECULAR ABUNDANCES OF
    MOLECULAR CLOUDS   23

O.E.H. Rydbeck and A. Hjalmarson
RADIO OBSERVATIONS OF INTERSTELLAR MOLECULES   45

T.L. Wilson and C.M. Walmsley
SUB-MILLIMETER ASTRONOMY AND ASTROPHYSICS   177

N.Z. Scoville
INFRARED SPECTROSCOPY OF INTERSTELLAR MOLECULES   201

J.H. Black
VISIBLE AND ULTRAVIOLET STUDIES OF INTERSTELLAR
    MOLECULES   215

E. Herbst
THE PRODUCTION OF COMPLEX MOLECULES IN DENSE
    INTERSTELLAR CLOUDS   237

R.M. Crutcher and W.D. Watson
CHEMICAL REACTIONS IN DIFFUSE CLOUDS AND THE
    CHEMISTRY OF ISOTOPE FRACTIONATION IN THE
    INTERSTELLAR GAS   255

A. Dalgarno
THE CHEMISTRY OF SHOCKED REGIONS OF THE INTER-
    STELLAR GAS   281

B.T. Draine
THEORETICAL MODELS OF SHOCK WAVES IN MOLECULAR
    CLOUDS   295

W.F. Huebner
COMETARY COMAE   311

C. de Bergh
MOLECULAR SPECTRA OF THE OUTER PLANETS AND
    SATELLITES   331

PART II. THEORETICAL AND EXPERIMENTAL LABORATORY STUDIES

S. Leach
MOLECULAR IONS IN ASTROPHYSICS AND IN THE
    LABORATORY   353

M. Winnewisser, B.P. Winnewisser and G. Winnewisser
MILLIMETER AND SUB-MILLIMETER WAVE SPECTROSCOPY
    IN THE LABORATORY AND IN THE INTERSTELLAR MEDIUM   375

R.J. Saykally
STUDIES OF ASTROPHYSICALLY IMPORTANT MOLECULAR IONS
    WITH ULTRASENSITIVE INFRARED LASER TECHNIQUES   403

P. Erman
TIME RESOLVED PROPERTIES OF SMALL ASTROPHYSICAL
    MOLECULES   421

U. Buck
ROTATIONAL EXCITATION IN MOLECULAR BEAM EXPERIMENTS   433

D. Smith and N.G. Adams
ION-MOLECULE REACTIONS AT LOW TEMPERATURES   453

E.E. Ferguson
ION-MOLECULE REACTIONS IN INTERSTELLAR MOLECULAR
    CLOUDS   471

P.R. Bunker
THEORETICAL PREDICTIONS OF THE INFRARED AND MICRO-
    WAVE SPECTRA OF SMALL MOLECULES   491

J. Schaefer
RECENT THEORETICAL RESULTS OF ROTATIONAL-TRANSLATIONAL
    ENERGY TRANSFER   497

G.D. Billing
ROTATIONAL AND VIBRATIONAL ENERGY TRANSFER IN
    DIATOMIC AND POLYATOMIC MOLECULES   517

J. Oddershede
CALCULATION OF RADIATIVE LIFETIMES OF ALLOWED AND
    FORBIDDEN TRANSITIONS   533

P. W. Langhoff
MOLECULAR PHOTOIONIZATION PROCESSES OF ASTROPHYSICAL
    AND AERONOMICAL INTEREST   551

W. P. Kraemer and A. U. Hazi
DISSOCIATIVE RECOMBINATION OF INTERSTELLAR IONS:
    ELECTRONIC STRUCTURE CALCULATIONS FOR HCO^+ 575

CONTRIBUTED PAPERS

L. M. Ziurys and B. E. Turner
DETECTION OF INTERSTELLAR ROTATIONALLY EXCITED CH   591

S. S. Prasad
INTERSTELLAR CLOUDS: FROM A DYNAMICAL PERSPECTIVE ON
    THEIR CHEMISTRY   603

T. J. Millar
MODELS OF GALACTIC MOLECULAR SOURCES   613

J. L. Destombes
MILLIMETER WAVE SPECTROSCOPY OF TRANSIENT SPECIES
    IN RF AND DC DISCHARGES   621

M. M. Graff
PHOTOFRAGMENT SPECTROSCOPY OF INTERSTELLAR
    MOLECULES   625

B. R. Rowe, J. B. Marquette, and G. Dupeyrat
MEASUREMENTS OF ION-MOLECULE REACTION RATE
    COEFFICIENTS BETWEEN 8 AND 160 K BY THE CRESU
    TECHNIQUE   631

H. Böhringer and F. Arnold
MEASUREMENTS OF ION-MOLECULE REACTION RATE
    COEFFICIENTS WITH AN ION-DRIFT-TUBE METHOD AT
    TEMPERATURES FROM 18 TO 420 K   639

W. Federer, H. Villinger, P. Tosi, D. Bassi, E. Ferguson,
and W. Lindinger
LABORATORY STUDIES OF ION REACTIONS WITH ATOMIC
    HYDROGEN   649

N. G. Adams and D. Smith
DISSOCIATIVE RECOMBINATION OF H_3^+, HCO^+, N_2H^+,
    AND CH_5^+ 657

C. M. Sharp
MOLECULAR DISSOCIATION FUNCTIONS OBTAINED FROM
    THERMODYNAMIC AND SPECTROSCOPIC DATA   661

W. H. Kegel, S. Chandra, and D. A. Varshalovich
A THEORETICAL STUDY OF H_2O MASERS   673

E. F. van Dishoeck
THE EXCITATION OF INTERSTELLAR C_2 681

E. F. van Dishoeck
PHOTODISSOCIATION PROCESSES IN THE OH AND HCℓ
    MOLECULES   683

F. Palla
LOW TEMPERATURE ROSSELAND MEAN OPACITIES FOR
    ZERO-METAL GAS MIXTURES   687

CONTRIBUTED ORAL PRESENTATIONS

G. A. Blake
A HIGH-SENSITIVITY SPECTRAL SURVEY OF ORION A IN THE
    1.3 MM REGION   697

S. Federman
THE IMPORTANCE OF NEUTRAL-NEUTRAL REACTIONS IN THE
    CHEMISTRY OF CN IN DIFFUSE CLOUDS   699

H. W. Kroto
LONG CHAIN MOLECULES IN INTERSTELLAR CLOUDS   700

A. Leger, M. Jura, and A. Omont
DESORPTION OF MOLECULES FROM INTERSTELLAR GRAINS   701

D. A. Williams
MOLECULAR MANTLES ON INTERSTELLAR GRAINS   702

R. McCarroll
ELECTRON CAPTURE BY MULTIPLY CHARGED IONS:
    POPULATION OF EXCITED STATES   703

F. Rostas
PHOTOABSORPTION AND PHOTODISSOCIATION OF CO IN THE
    900-1200 Å REGION   704

P. Andresen
PHOTODISSOCIATION OF H_2O IN THE FIRST ABSORPTION BAND
IN THE VUV AS THE PUMPMECHANISM OF THE
ASTRONOMICAL OH-MASER   706

C. R. Vidal
LASER SPECTROSCOPY ON SMALL MOLECULES OF ASTRO-
PHYSICAL INTEREST IN THE VACUUM UV   709

J. B. A. Mitchell
EXPERIMENTAL MEASUREMENTS OF THE BRANCHING RATIO
FOR DISSOCIATIVE RECOMBINATION PROCESSES   711

T. G. Heil
ASTROPHYSICALLY IMPORTANT CHARGE TRANSFER REACTIONS:
RECENT THEORETICAL RESULTS   712

E. Roueff
POTENTIAL SURFACE OF C_2-H_2 714

A. J. Sadlej
THEORETICAL PREDICTIONS OF ATOMIC AND MOLECULAR
PROPERTIES FOR THE CONSTRUCTION OF EMPIRICAL
SCATTERING POTENTIALS   715

D. R. Flower
ROVIBRATIONAL EXCITATION OF $^{12}C^{16}O$ BY H_2 717

LIST OF PARTICIPANTS   719

SUBJECT INDEX   729

MOLECULE INDEX   737

"...molecules are unique diagnostic probes of the physical conditions in which they are found... they influence in critical ways the evolution of the astronomical systems of which they are a part."

A. Dalgarno
Bad Windsheim, July 1984

PREFACE

Electromagnetic radiation has long provided the major source of information on astronomical objects. Although early observations were largely restricted to the visible region of the spectrum associated with radiation from stars and other hot objects, significant recent advances in technology within the last three decades have led to observations of extraterrestrial electromagnetic radiation also in the radio, infrared, and ultraviolet portions of the spectrum. Such studies have resulted in the detection in both absorption and emission of a large and growing number of molecular species in various astrophysical circumstances, including extragalactic objects, circumstellar shells and HII regions, stellar and planetary atmospheres, comets and planetary satellites, and, particularly, in the form of dense interstellar clouds thought to be the sites of star formation. It has become clear in recent decades that molecules, in addition to providing useful probes of astrophysical conditions, can play a central role in the condensation processes required to initiate the formation of massive objects. As a consequence, the field of molecular astrophysics is now an increasingly active and well-defined discipline that spans a significant range of topics and interests in astronomy, physics, and chemistry.

Molecules and their spectra have long been of astronomical interest in connection with stellar atmospheres and structure. Progress in molecular aspects of this area are summarized, for example, in the well-known Liege International Astrophysical Symposia. The importance of molecules in other astrophysical connections has developed more recently, in accordance with the broadening scope of observations. Thus, the early International Astronomical Union (IAU) Symposia devoted to interstellar clouds involved largely aerodynamical aspects of these objects [1-3], with significant attention devoted only comparatively recently to interstellar grains and to the growing numers of molecules detected therein [4,5]. The Les Houches school on the physics of interstellar matter, held approximately ten years ago, perhaps heralded the establishment of the discipline of molecular astrophysics [6]. Many of the rapid developments since then are summarized in the proceedings of the 1977 Liege Symposium [7], and in the IAU Symposium of 1979 devoted to interstellar molecules [8]. Excellent relevant monographs [9,10], related timely proceedings [11], and recently published elementary textbooks [12,13] further help to define the pedagogical scope of molecular astrophysics.

A significant financial investment has been made in the establishment of ground- and satellite-based observational facilities for molecular astrophysical studies. In the coming years, a wealth of experimental data is bound to accumulate, in which connection close interactions between observers, astrophysical modellers, and molecular physicists and chemists can play a helpful role in analysis and interpretation. In view of the increasing pace of activity in the field of molecular astrophysics, and in the apparent absence of relevant international meetings since the Liege 1977 and IAU 1979 Symposia, it was deemed appropriate and timely by the organizers to hold a workshop in 1984. Consequently, the NATO Advanced Research Workshop, "Molecular Astrophysics - State of the Art and Future Directions", was organized and held at Bad Windsheim, West Germany, from 8 to 14 July 1984.

The choice of speakers and subject matter of the Workshop was largely subjective, but designed to include most of the generally accepted areas of molecular astrophysical study. Workers from the fields of radio, infrared, and uv-optical observations, astrophysical modelling, laboratory spectroscopy, reaction chemistry, collision physics, and theoretical molecular physics and chemistry, were invited to present survey lectures in their areas of speciality. In addition, contributed papers were accepted from other senior workers in these fields in order to broaden the base of subject matter. Virtually all aspects of molecular astrophysics were touched upon in the invited and contributed lectures, some in more depth than others. In order to acquaint the largest possible community of workers in molecular astrophysics with activities at the Workshop, the present Proceedings includes all invited papers and all contributed papers for which manuscripts were received. In addition, abstracts are included of all contributed oral presentations, and comprehensive subject and molecule indices are provided. In the following paragraphs, brief descriptive accounts are provided of highlights of the reports presented at the Workshop. These remarks are included to furnish a modest overview of the scope of the meeting, and also as a convenience to the general reader who may not yet be familiar with the discipline of molecular astrophysics or with its dramatis personae.

The Workshop was divided into two equal parts, the first involving astrophysical observations and models, the second theoretical and experimental laboratory studies. Professor A.

Dalgarno (Harvard) kindly provided a general overview of the discipline of molecular astrophysics in opening remarks, in addition to a more specialized report (see below). His account of the role of molecules as actors as well as diagnostic spectators in the astronomical play ranged from descriptions of pregalactic and interstellar clouds to the stars, comets, and planets, providing a concise but comprehensive introduction to the Workshop. Chemical compositions of molecular clouds, difficulties that can arise in determinations of column densities from electromagnetic signals, and connections with stellar abundances were described by M. Guelin (IRAM, Granada), whose written report tabulated abundances in typical dense clouds. It emerged in discussion that the depletion of elemental abundances in the interstellar medium relative to solar values continues to be an interesting and controversial issue.

Aspects of radio observations were surveyed by P. Solomon (SUNY, Stony Brook), A. Hjalmarson (OSO, Chalmers), and T. L. Wilson (MPI, Bonn). Solomon reported on an extensive survey involving $\sim 4 \times 10^4$ individual 2.6 mm CO emission spectra taken over a five year period of giant molecular clouds, largely in the galactic plane. He presented estimates of their size, shape, density, temperature, and velocity distributions, and indicated that their large average mass ($\sim 10^5$ $M_\odot$) and their large numbers may affect the motion of the galactic disk. A general survey of radio observations of interstellar molecules was provided by Hjalmarson, who reviewed historical aspects, discusses aspects of abundance determinations from measured antenna temperatures, and described circumstances of molecular detection in some specific cases (CN, OH, CH, SiH, H_2CO, H_2O). In related contributed reports, L. M. Ziurys (Berkeley) described the first radio observations of rotationally excited CH through a Λ-doubling transition, and G. A. Blake (Cal Tech) presented results of a high-spatial-resolution survey of the Orion molecular cloud in the 1.3 mm region detecting over 500 emission lines, which were assigned to 28 individual molecules. Recent developments in sub-millimeter astronomy were reported by Wilson, with particular reference to determinations of physical conditions in molecular clouds and circumstellar envelopes, and to their spatial structures. Wilson described briefly aspects of continuum emission from dust grains, but focused mainly on the information complementary to longer wavelength radio observations provided by measurements of sub-millimeter emission lines. The higher spatial resolution provided by the shorter wave length measurements was emphasized. An account of the design characteristics of the new 10 meter MPI-Arizona submillimeter telescope was also provided.

Aspects of infrared and uv-visible observations were surveyed by N. Z. Scoville (Cal Tech) and J. H. Black (Arizona), respectively. Scoville emphasized the importance of infrared spectroscopy in relatively hot objects, such as shocks and HII regions, which can give rise to observable H^+ recombination lines, high-J pure rotational transitions (CO), quadrupole rotational transitions (H_2), and various vibrational transitions. Illustrations were given in the cases of the Orion molecular cloud and the Becklin-Neugebauer object. Possible reasons for the failure to observe to date a protostar were indicated, and future prospects in infrared observations employing the space telescopes (ISO, SIRTF) to be launched during the next decade were reported. Black reviewed recent developments in optical (uv-visible) observations of molecules in absorption, largely against background OB stars, emphasizing their historical importance and the very small angular width provided along the line of sight by such measurements. Particular molecules in clouds, nebulae, and quasi-stellar objects were discussed, and the possibility of observations of highly obscured stars with Space Telescope were indicated. The need for better understanding of photodissociation, particularly in CO, and of collisional excitation mechanisms, was stressed.

Chemical reactions in dense and in diffuse clouds were reviewed by E. Herbst (Duke) and W.D. Watson (Illinois). Herbst described the synthesis of complex molecules in dense clouds largely on basis of ion-molecule reaction schemes initiated by radiative association and terminated by electon-ion recombination mechanisms. He cautioned against complacency with present cloud models in view of continuing discrepancies between observed abundances and known physical conditions in clouds, against use of unrealistic physical conditions in model studies, and urged continued examination of the role of grains in this connection. A useful account of the various chemical models employed recently for synthesis of more complex molecules in dense clouds was provided. Watson reviewed the chemistry of diffuse clouds, emphasizing the importance of these relatively well understood objects as testing grounds of astrochemical models, and also described studies of isotope fractionation in both diffuse and dense clouds. The importance of grain surface reactions in H_2 formation, radiative association reactions for other molecules, ion-molecule and dissociative recombination reactions, and self-shielding in the Lyman and Werner bands of H_2 to limit its photodissociation were emphasized. In Watson's view there is as yet no demonstrated mechanism sufficiently effective under general conditions for the ejection of molecules other than H_2 from grains. The relevance of $^{13}CO/^{12}CO$ isotopic ratios in placing constraints on physical conditions in diffuse clouds, and deuterium enhancement in dense clouds by charge exchange with H_3^+, were indicated in reviewing recent aspects of isotope fractionation. Dissociative recombination of H_3^+ with electrons was noted as important in the latter regard. A number of contributed reports were of interest in connection with cloud chemistry. Specifically, S.S. Prasad (JPL) and T.J. Millar (Manchester) described dynamical studies of cloud chemistry, S. Federman (JPL) reported on the importance of neutral reactions in the chemistry of diffuse-cloud CN, H.W. Kroto (Brighton) described experimental and observational aspects of long-chain molecules, and A. Omont (Saint-Martin) and D.A. Williams (Manchester) reported on aspects of molecular desorption mechanisms from grains.

The chemistry and physics of shocked regions of the interstellar gas, produced by supernovae, expanding ionized regions, stellar winds, and other mechanisms, were reviewed by Dalgarno and B.T. Draine (Princeton). Distinction was drawn by Dalgarno between chemistry in dissociative and nondissociative shocks. In the former case, hydrogen reactions similar to those in pregalactic clouds are important, as is precursor recombination radiation which dissociates molecules ahead of the shock. Detailed models have yet to be constructed in this case. Endothermic reactions are enhanced in non-dissociative shocks, generating elaborate oxygen, sulfur, silicon, carbon, and nitrogen chemistries, and leading to vibrational and rotational excitation. This mechanism possibly accounts for observed molecular enhancements and for maser action. The cooling effects of molecular radiation couple the chemical and thermal structures of the shock, particularly for heteronuclear molecules. Draine reviewed physical processes in shock waves, which are largely of so-called C-type in molecular clouds, in which the thermodynamic variables are continuous across the shock front. Under conditions of low fractional ionization, in the presence of magnetic fields, ions can be accelerated prior to the neutral gas, resulting in heating by collisions with streaming ions in the region of the shock. Draine reported on the principal coolants in such shocks, on the collisional and reaction mechanisms involved, and presented the results of detailed calculations of H_2 and CO line emission in models of regions of the Orion molecular cloud. Non-magnetic shocks apparently cannot reproduce the observed line intensities in this case.

Concluding the first part of the workshop, W.F. Huebner (Los Alamos) reviewed aspects of cometary comae, and C. de Bergh (Meudon) described spectral observations and analyses of the outer planets and their satellites. Huebner described the atmosphere, or coma, of a comet that develops when it is within ∼ 2 AU from the sun. A comet's emission spectrum should provide knowledge of its pristine composition, possibly related to the earliest history of the solar system. However, a complex photochemistry involving dissociation and ionization, as well as chemical reactions, complicates the interpretation of measured data, requiring detailed modelling and a considerable body of photochemical and physical data, much of it known only in approximate form. Such models are required, in particular, in support of the 1986 Giotto mission to Halley's comet. Our knowledge of the atmospheres of planets and satellites of the outer solar system, based on remote-site observations in the continuing absence of space probes, was reviewed by de Bergh. Molecular spectroscopy plays a central role in such observations, as illustrated by Voyager-IRIS spectra which provide molecular abundances, temperatures, and cloud structures for Jupiter and Saturn. Both planetary atmospheres are ~ 90-95% H_2, and mixing ratios of many trace elements are now known. Studies of Titan indicate a ~ 95% N_2 atmosphere, but little is known of the atmospheres of Uranus, Neptune, and Pluto. The launch of Space Telescope and of the Infrared Space Observatory, as well as visits by Voyager 2 to Uranus in 1986 and to Neptune in 1989, and by Galileo to Jupiter in 1989, should contribute greatly to our knowledge of the outer planets.

Introductory remarks to the second part of the Workshop were provided by G. Herzberg (Ottawa), who reviewed aspects of the early interactions between laboratory work and astronomical observations, and described spectral studies of molecular ions done in Ottawa and elsewhere. By previous agreement, Herzberg did not submit a written manuscript in light of recent publication of a related report [14]. Aspects of molecular ion spectroscopy were surveyed by S. Leach (Orsay and Meudon), M. Winnewisser (Giessen), and R.J. Saykally (Berkeley). Leach provided a comprehensive overview of laboratory aspects of the formation of molecular ions, of their spectroscopy, and of their dynamical attributes, including nonradiative transitions. He reported on a number of ions studied at Orsay, particularly sulfur compounds, and indicated the possible importance of doubly charged ions in astrophysical situations. Winnewisser reviewed developments in millimeter and sub-millimeter molecular spectroscopy in the laboratory and also in the interstellar medium, focusing attention on long-chain molecules and on molecules with inversion. He emphasized the importance of interaction between laboratory developments and interstellar measurements, and reported on recent measurements with the new Cologne 3- meter radiotelescope. A review of infrared spectroscopic studies of molecular ions was given by Saykally, focusing particular attention on far-infrared laser magnetic resonance (LMR) and velocity-modulation infrared studies, the former technique providing sensitivity, the latter a means for separating ion from neutral lines. A useful summary table of molecular ions studied to date with high-resolution infrared techniques was given, specific astrophysically relevant ions were discussed, and typical LMR spectra were reported. In related contributed remarks

J. L. Destombes (Lille) reported on millimeter spectra of CO^+, CH, and ^{13}CN in a cooled RF discharge, and R. McCarroll (Bordeaux) described theoretical aspects of electron capture by multiply charged ions.

Experimental and aspects of theoretical procedures for determining radiative molecular lifetimes or transition probabilities were surveyed by P. Erman (Stockholm). He indicated that accurate f numbers and equivalent widths can provide reliable molecular abundances from optical observations of clouds. Erman described supersonic jets employed in time-resolved spectroscopic measurements of optical lifetimes, cited useful compilations of molecular lifetime measurements, urged care in lifetime-to-f number conversions, and described aspects of radiationless transitions and collisional de-excitation processes of importance in radiative lifetime measurements. In related contributed remarks, photodissociation measurements on CH^+ and CH were reported by M. M. Graff (Harvard), comprehensive experimental photoabsorption and dissociation studies of CO were described by F. Rostas (Meudon), and laser-induced-fluorescence studies of photodissociation of H_2O as an OH pump mechanism were reported by P. Andresen (MPI Göttingen), who emphasized that collisional pumping theories give anti-inversion for the Λ-doublet states in OH. Laser spectroscopic studies of NO and CO in the vacuum uv were described by C. R. Vidal (MPI Garching).

A survey of rotational excitation processes as studied by molecular beam experiments was given by U. Buck (Göttingen), who compared the difficulties of ab-initio calculations of the necessary anisotropic potential surfaces with those of experimental determinations of state-to-state differential cross sections. Complementary experimental-theoretical approaches involving anisotropy-sensitive differential energy loss measurements and approximate potential surfaces were described, and results reported for astrophysically important collision partners (H_2, D_2, CO, OH).

Experimental aspects of ion-molecular reactions were reviewed by D. Smith (Birmingham) and E. E. Ferguson (Boulder). Smith cautioned against uncritical general use of room-temperature rate coefficients under low-temperature cloud conditions. Fortuitously, a large class of astrophysically important reactions are collision rate limited, in which case such extrapolation is valid. Low temperature measurement procedures, including the Birmingham variable temperature selected ion flow tube method, were described and astrophysical examples, including isotope exchange reactions, were presented. Ferguson reviewed recent low-temperature data, and proposed empirical guidelines for rate estimates based largely on intermediate-complex, radiative-association, and vibrational-relaxation considerations. Recent developments in methane chemistry, spin-violated reactions, endothermic trapping studies, and atomic hydrogen reactions were reviewed. A number of contributed papers reported on additional recent studies in ion chemistry. Aspects of measurements of ion-molecule reaction rate coefficients at low temperatures using a supersonic beam were described by B.R. Rowe (CNRS, Meudon), and by H. Böhringer and F. Arnold (MPI, Heidelberg) using an ion drift tube, in a paper not reported at the Workshop but included in these Proceedings. W. Lindinger (Innsbruck) reported on rate coefficients for ion reactions with hydrogen atoms obtained with a drift tube apparatus. Aspects of dissociative recombination measurements on astrophysically important ions employing a flowing afterglow apparatus were described by N.G. Adams (Birmingham), and measurements of recombination cross sections for specific exit channels were reported in a contributed abstract by J.B.A. Mitchell (Western Ontario), who was unable to attend the Workshop. Theoretical aspects of charge-transfer reactions for low energy collisions of multiply charged ions with hydrogen and helium atoms were reported by T.G. Heil (Athens, Georgia).

G.H.F. Diercksen (MPI, Garching) reviewed ab-initio methods for calculations of the potential surfaces required in structure determinations, chemical reactions, and molecular collisions. The critically important basis-set problem was indicated, particularly in connection with weak van der Waal's interactions, and representative potential surfaces were reported. Ab-initio calculations of the potential surface for the H_2-C_2 system were presented by E. Roueff (Meudon) in contributed remarks, and A.J. Sadlej (Lund) described calculations of multipole moments and polarizabilities for use in construction of potential surfaces in the asymptotic region.

P.R. Bunker (Ottawa) described corresponding ab-inito studies of rotational-vibrational states and related quantities on given potential energy surfaces, with particular reference to quasi-linear and quasi-planar molecules, for which harmonic-oscillator, rigid-rotor approximations are poor. Applications to various molecules (CH_2, CH_3, SiH_3), and ions (H_3O^+) were reviewed, and the importance of calculations in guiding laboratory spectral searches indicated. In contributed remarks on studies of molecular partition functions in stellar atmospheres, C.M. Sharp (MPI, Garching) noted the inadequacy of the harmonic approximation to bending-mode contributions in C_3. Further use of potential energy surfaces, in rotational-translational energy transfer studies, were reviewed by J. Schaefer (MPI, Garching). Accurate potential surfaces for collisions of simple molecules (H, He, H_2) provide the basis for computational studies by coupled-rotational-channel methods of rotational relaxation and angle-differential scattering cross sections, and of other properties, in good accordance with measured values. Aspects of studies in progress of the H_2–CO system were indicated. Computational procedures for study of rovibrational transitions during molecular collisions were reviewed by G. D. Billing (Copenhagen). He contrasted and compared aspects of quantum and semiclassical treatments of rovibrational excitation in H_2–^{4}He collisions, and in rotational excitation in ^{4}He–NH_3 collisions and described semiclassical predictions of rovibrational energy transfer in larger systems (N_2, CO_2, HF, CO). In related contributed remarks, W. H. Kegel (Frankfurt) reported on collisional pumping of H_2O by H_2 as a maser mechanism, and D. L. Flower (Durham) described rovibrational excitation calculations of CO by H_2 collisions.

Calculations of radiative lifetimes in diatomic molecules were surveyed by J. Oddershedde (Odense), largley in the context of approximate treatments of the polarization propagator. The history of calculated transitions in CH^+ was reviewed, and work in progress on polyatomic compounds (C_3, SiC_2, Si_2C, Si_3) indicated. Theoretical studies of photoexcitation and dissociation in C_2, OH, and HCℓ were reported in contributed remarks by E. F. van Dishoeck (Leiden) employing quantum mechanical calculations, and astrophysical applications were discussed. Additionally, F. Palla (Florence) reported on calculations of opacities for primordial gas mixtures. Recent experimental and theoretical photoionization studies, including higher-lying channels, were reviewed by P. W. Langhoff (Bloomington). Particular attention was drawn to resonance features in measured and calculated cross sections, which are common in polyatomic molecules. Heavy-atom molecular cross sections are large at thresholds, possibly leading to enhanced ionization in strong uv regions.

Finally, W. P. Kraemer (MPI, Garching) surveyed experimental and theoretical aspects of dissociative recombination processes. Generally, poor agreement obtains between measured and calculated recombination cross sections for light diatomic ions (NO^+, N_2^+, O_2^+), although reasonable results are obtained for H_2^+ ions. The theoretical values are very sensitive to the shapes and relative positions of the ionic and neutral-complex potential surfaces, as illustrated by the results reported of recent detailed calculations on HCO^+.

The diversity of subject matter evident in the preceding descriptive remarks emphasizes the breath and vitality of molecular astrophysical studies. As the field expands, timely workshops will be increasingly important for maintaining communication among workers in the various disciplines. There was a general feeling among the participants of the 1984 Workshop that a similar meeting would be appropriate in 1987. Additionally, considerable interest was expressed in the organization of a related summer school on molecular astrophysics for the benefit of students and other beginners in the field. These matters will be pursued subsequently by an appropriate committee.

Many of the organizational aspects of the Workshop benefitted from the competence and good sense of N.E. Gruener and J. Steuerwald (MPI, Garching), to whom we are grateful. Expert secretarial assistance was kindly provided by B. Joseph and G. Kratschmann. Finally, on behalf of all the participants at the Molecular Astrophysics Workshop, the organizers thank NATO and the International Programs Division of the U.S. National Science Foundation for their kind financial assistance.

G. H. F. Diercksen
W. F. Huebner
P. W. Langhoff

Munich, April 1985

REFERENCES

1. *Problems of Cosmical Aerodynamics*, J. M. Burgers and H. C. van de Hulst, Editors (Air Force Documents Office, Dayton, Ohio, 1951).
2. *Gas Dynamics of Cosmic Clouds*, J. M. Burgers and H. C. van de Hulst, Editors (North Holland, Amsterdam, 1955).
3. *Cosmical Gas Dynamics*, J. M. Burgers and R. N. Thomas, Editors, Rev. Mod. Phys. $\underline{30}$, 905 (1958).
4. *Interstellar Gas Dynamics*, H. J. Habing, Editor (Reidel, Dordrecht, 1970).
5. *Interstellar Dust and Related Topics*, J. M. Greenberg and H. C. van de Hulst, Editors (Reidel, Dordrecht, 1973).
6. *Atomic and Molecular Physics and the Interstellar Matter*, R. Balian, P. Encrenaz, and J. Lequeux, Editors (North Holland, Amsterdam, 1975).
7. *Laboratory and Astrophysical Spectroscopy of Small Molecular Species*, I. Dubois, Editor (Universite de Liege, Institut d'Astrophysique, 1977).
8. *Interstellar Molecules*, B. H. Andrew, Editor (Reidel, Dordrecht, 1980).
9. *Diffuse Matter in Space*, L. Spitzer, Jr. (Wiley, New York, 1968).
10. *Physical Processes in the Interstellar Medium*, L. Spitzer, Jr. (Wiley, New York, 1978).
11. *The Photochemistry of Atmospheres*, J. S. Levine, Editor (Academic, New York, 1985).
12. *The Physics of the Interstellar Medium*, J. E. Dyson and D. A. Williams (Manchester University Press, 1980).
13. *Interstellar Chemistry*, W. W. Duley and D. A. Williams (Academic, London, 1984).
14. G. Herzberg, in *Molecular Ions: Spectroscopy, Structure and Chemistry*, T. A. Miller and V. E. Bondybey (North Holland, Amsterdam, 1983), pp. 1-9.

PART I

ASTROPHYSICAL OBSERVATIONS AND MODELS

MOLECULAR ASTROPHYSICS

A. Dalgarno

Harvard-Smithsonian Center for Astrophysics
Cambridge, Massachusetts, U.S.A.

Molecules are remarkably durable systems which are able to exist in significant abundance in environments that are inhospitable to their formation and hostile to their survival. They occur in a diversity of astronomical surroundings ranging from the recombination era of the early Universe to the circumstellar shells of red giants. Because of the complexity of their energy level structure and of their emission and absorption spectra, molecules are unique diagnostic probes of the physical conditions in which they are found. Molecular observations have been used to study the interstellar gas in external galaxies, molecular clouds and galactic structure, masers and star formation, circumstellar shells and mass loss, protostars, HII regions, planetary nebulae, Herbig-Haro objects, sun spots, comets and the atmospheres of the planets and their satellites. Molecules remain to be discovered elsewhere in intergalactic space, active galactic nuclei and quasars.

Molecules are much more than probes of exotic conditions. They often control the thermal and ionization structure and their presence can enhance or suppress dynamical instabilities so that they influence in critical ways the evolution of the astronomical systems of which they are a part.

Molecules are detected by observations in the radio, submillimetre, infrared, visible and ultraviolet regions of the electromagnetic spectrum. The interpretation of the observations is usually complicated and demands a deep understanding of the extraordinary array of processes that occur when a cosmic molecular gas is subjected to radiation. An extensive data base containing reliable values of the parameters which characterise molecular processes is an essential component of quantitative theories about the nature of the entities in which the molecules are created and destroyed.

In my introductory lecture, I will make an arbitrary selection of topics, beginning at the beginning. According to the standard big bang cosmology, the Universe expanded from an initial singularity. In the early stages, Thompson scattering of electrons and photons maintained matter and radiation in thermal equilibrium. As the expansion continued, the densities decreased and the heating of matter by Thompson scattering failed to overcome the cooling by adiabatic expansion. After about a million years, the temperature of matter had diminished to about 4000K, cool enough for recombination to occur (and with it molecular astrophysics).

With recombination, matter changed rapidly from a fully-ionized state to a largely neutral condition. Thermal contact between matter and radiation was largely lost and matter and radiation evolved independently. During the recombination era, the Universe was composed mostly of neutral hydrogen atoms and photons with a few electrons, protons, and light nuclei.

As illustrated in Fig. 1, molecular hydrogen was formed then by two reaction sequences, each necessarily initiated by a radiative process. Radiative attachment

$$H + e \rightarrow H^- + h\nu \tag{1}$$

formed negative hydrogen ions and was followed by associative detachment

$$H^- + H \rightarrow H_2 + e. \tag{2}$$

The second sequence was radiative association of protons and hydrogen atoms,

$$H^+ + H \rightarrow H_2^+ + h\nu \tag{3}$$

followed by charge transfer or ion-atom interchange

$$H_2^+ + H \rightarrow H_2 + H^+. \tag{4}$$

The destruction processes of photodetachment

$$H^- + h\nu \rightarrow H + e \tag{5}$$

and photodissociation

$$H_2^+ + h\nu \rightarrow H + H^+ \tag{6}$$

by the background radiation field of the universe restricted the abundance of molecular hydrogen that was formed in the early stages. Fig. 2 shows the fractional abundances of H, H^+ and H_2 as functions of the red shift z (1). The late enhancement in H_2 at z = 100 arises from the contribution from reactions (1) and (2).

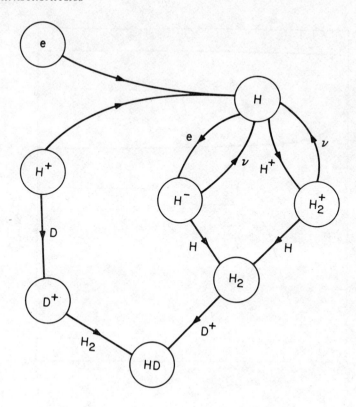

Fig. 1. The chemistry of the early Universe.

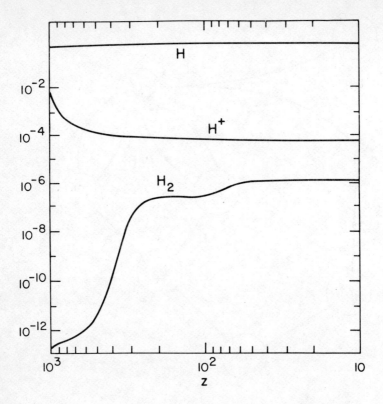

Fig. 2. The abundance of H_2 in the early Universe as a function of red shift (adopted from ref. 1).

The molecular formation processes affected the relict fractional ionization remaining as the expansion continued. The limiting value of $n_e/n(H)$ was about 2×10^{-6}. Because of the coupling between matter and radiation maintained by the electrons, the fractional ionization is a crucial parameter in the determination of the size of the first pre-galactic objects (2), which in an uncertain scenario must have formed in response to fluctuations in density.

The molecule HD was formed in the recombination era by the deuterium analogues of reactions (1) to (4) but also and more rapidly by the sequence

$$H^+ + D \rightarrow H + D^+ \tag{7}$$

$$D^+ + H_2 \rightarrow HD + H^+ . \tag{8}$$

Because of reaction (8), a large fractionation occurred and the limiting value of the ratio $n(HD)/n(H_2)$ was probably enhanced by a factor greater than 100.

The importance of H_2 molecules in the collapse of pregalactic clouds has been emphasized repeatedly. Because of the molecular cooling by vibrational and rotational excitation, the temperature rise during the collapse is slowed and the critical Jeans mass required for collapse is greatly reduced. In the initial stages of the collapse, cooling is dominated by collisional excitation of HD.

In a collapsing pre-galactic cloud, three-body recombinations

$$H + H + H_2 \rightarrow H_2 + H \tag{9}$$

$$H + H + H_2 \rightarrow H_2 + H_2 \tag{10}$$

become the major formation mechanisms of H_2 molecules as the density increases (3). The presence of H_2 (and HD) may lead to thermal instabilities (4) which affect cloud fragmentation and the initial mass function of the first stars.

Molecular astrophysics increased greatly in complexity after the heavy elements were formed by this first generation of stars. Because of the heavy elements, present in the gas phase and in grains, interstellar molecular clouds are very different objects from pre-galactic clouds. They differ also because they are subjected to cosmic rays. The cosmic rays are the source of an ionization-driven chemistry that produces an extensive collection of molecules which are shielded from the destructive action of the interstellar radiation field by the dust grains.

Table 1 is a list of the molecules that have been detected in interstellar clouds. They are many unattributed spectral lines suggesting the presence of molecular candidates still to be identified. Recent addition to the list are methyl cyanoacetylene CH_3C_3N (5) and methyl diacetylene CH_3C_4H (6,7,8).

Table 1

Interstellar Molecules

H_2	Hydrogen	CH	Methylidyne
CH^+	Methylidyne Ion	OH	Hydroxyl
C_2	Carbon	CN	Cyanogen
CO	Carbon Monoxide	NO	Nitric Oxide
CS	Carbon Monosulphide	SiO	Silicon Monoxide
SO	Sulphur Monoxide	NS	Nitrogen Sulfide
SiS	Silicon Sulphide	C_2H	Ethynyl
H_2O	Water	HNC	Hydrogen Isocyanide
HCN	Hydrogen Cyanide	HCO^+	Formyl ion
HCO	Formyl	H_2S	Hydrogen Sulphide
N_2H^+	Protonated Nitrogen	OCS	Carbonyl Sulphide
HCS^+	Thioformyl ion		
SO_2	Sulphur Dioxide	H_2CO	Formaldehyde
NH_3	Ammonia	H_2CS	Thioformaldehyde
HNCO	Isocyanic Acid	HNCS	Isothiocyanic Acid
C_3N	Cyanoethynyl	CH_2CO	Ketene
$HOCO^+$	Protonated Carbon Dioxide	HCOOH	Formic Acid
CH_2NH	Methanimine	HC_3N	Cyanoacetylene
NH_2CH	Cyanamide	CH_3CH	Methyl Cyanide
C_4H	Butadinyl	CH_3SH	Methyl Mercaptan
CH_3OH	Methyl Alcohol	CH_3C_2H	Methyl Acetylene
NH_2CHO	Formamide	CH_2CHCN	Vinyl Cyanide
CH_3NH_2	Methylamine	$HCOOCH_3$	Methyl Formate

CH_3CHO	Acetaldehyde	CH_3C_4H	Methyl Diacetylene
HC_5N	Cyanodiacetylene	$(CH_3)_2O$	Dimethyl Ether
CH_3C_3N	Methyl Cyanoacetylene	CH_3CH_2CN	Ethyl Cyanide
CH_3CH_2OH	Ethyl Alcohol	HC_9N	Cyano-octatetra-yne
HC_7N	Cyanohexatriyne		

The diatomic molecule carbon monoxide is the most abundant of the interstellar molecules which emit in the radio region of the spectrum. It is an ubiquitous molecule, which coexists with molecular hydrogen, the major component of dense interstellar clouds, and emission from CO is used to map the molecular regions in galaxies and to elucidate the varying rates of star formation (9).

Emission and absorption by CO and by other interstellar molecules can be interpreted to yield the temperature and density of the ambient medium. The derived densities depend upon the cross sections for excitation of molecular energy levels in collisions with ortho and para-hydrogen and with helium. An important question is the relationship between the measured abundance of CO and H_2. Molecular hydrogen has no radio frequency spectrum and is not observable in dense molecular clouds.

Several chemical sequences have been postulated which lead to the formation of CO. One such is illustrated in Fig. 3 which is initiated by the radiative association of C^+ and H_2. The sequence leads to CH_3^+ and on to the formation of the complex molecules by an extensive array of reactions, few of which have been studied by experimental or theoretical methods. Recent work by Herbst (10), Suzuki (11) and Millar and Freeman (12) has revealed promising pathways for the formation of the complex species but only empirical values derived from diffuse cloud models of CH are available for the rate coefficient of the initiating reaction

$$C^+ + H_2 \rightarrow CH_2^+ + h\nu \tag{11}$$

at cloud temperatures. The role of CH_3^+ is critical in molecular cloud chemistry. It may undergo dissociative recombination

$$CH_3^+ + e \rightarrow CH_2^+ + H$$
$$\rightarrow CH + H_2 \tag{12}$$

if the fractional ionization is high or radiative association

$$CH_3^+ + H_2 \rightarrow CH_5^+ + h\nu \tag{13}$$

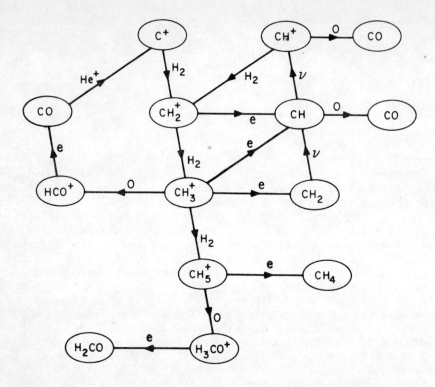

Fig. 3. The carbon chemistry in cold clouds.

if it is not. In a remarkable experiment, Barlow, Dunn and Schauer (13) have measured the rate coefficient of (13) at 13K. Following (13), methane CH_4 and formaldehyde H_2CO are readily produced. Methane is difficult to detect but H_2CO is widespread suggesting that the fractional ionization is generally low.

Despite the uncertainties in the chemistry of complex molecules, Stahler (14) has argued that the observed regular decline in the abundance of the cyanopolyenes $HC_{2k+1}N$ as a function of length implies a sequential formation process. The age of the cloud is given by the time it takes to form the longest chain that is present.

The sequence leading to CH_3^+ which consists of abstraction reactions is characteristic of interstellar ion-molecule chemistry. Fig. 4 illustrates a similar sequence for oxygen-bearing molecules, which begin with the cosmic ray ionization of H_2 and H.

Isotopic variants of many of the molecular species have been formed which provide information about nucleosynthesis in the galaxy. The isotopic molecular abundance effects must be clarified before the elemental isotopic abundance ratios can be derived. Of special utility are the deuterated molecules which offer the potential of yielding a value for the cosmic abundance ratio $[D]/[H]$. The $[D]/[H]$ ratio is a measure of the original matter density in the Universe. The deuterated molecules undergo large enhancements through exothermic reactions such as

$$H_3^+ + HD \rightarrow H_2D^+ + H_2 \qquad (14)$$

(15) and

$$HX^+ + D \rightarrow DX^+ + H \qquad (15)$$

(16), where X is some neutral atomic or molecular constituent. Because the fractionation can be interrupted by dissociative recombination of the deuterated ions and by their reactions with neutral material the enhancements depend upon the fractional ionizations and the heavy element depletions as well as the temperatures in the clouds. Crucial to the interpretation of the observational data is the rate coefficient for dissociative recombination of H_3^+,

$$H_3^+ + e \rightarrow H_2 + H \qquad (16)$$
$$\rightarrow H + H + H .$$

Some recent experimental (17) and theoretical studies (18) but not all (19) argue that for H_3^+ in its lowest vibrational state the rate coefficient is two orders of magnitude less than the value usually adopted. The upper limits derived for the fractional ionization are correspondingly less severe.

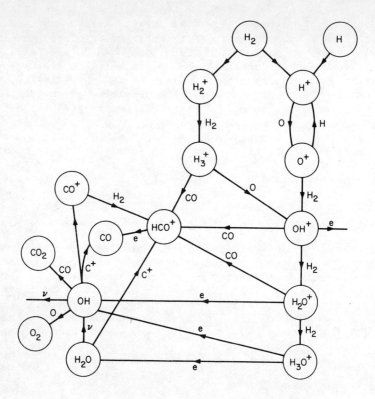

Fig. 4. The oxygen chemistry in cold clouds.

The first discoveries of interstellar molecules were made by optical absorption measurements looking towards stars. The diatomic species CH, CN and CH^+ were also detected. Fig 5 reproduces the spectrum seen towards ζ Oph (20).

The relative intensities of the absorption lines are consistent with an excitation temperature of 2.7K and provide a measure of the temperature of the cosmic microwave background radiation at 2.6 mm and 1.32 mm. Woody and Richards (21) had found evidence of spectral deviations from a black body. It appears they do not originate in the cosmic background.

In diffuse clouds into which photons penetrate, the carbon ions C^+ are produced mostly by ultraviolet photoionization of neutral carbon atoms rather than, as in dense clouds, by cosmic ray ionization of helium and the reaction of He^+ with CO.

The OH radical is produced in diffuse and dense clouds by the sequence shown in fig. 4. Cosmic ray ionization drives the chemistry and the abundance of OH reflects the magnitude of the cosmic ray flux. For ζ Per the derived cosmic ray ionizing frequency lies between 1 and $3 \times 10^{-16} s^{-1}$ and for ζ Oph it is at least $4 \times 10^{-16} s^{-1}$ (22). These estimates are an order of magnitude larger than those found with less accurate cloud models and are large enough that cosmic ray heating is not negligible. The measured rate for high energy cosmic rays is $10^{-17} s^{-1}$. There may be some contribution to OH towards ζ Oph by formation in a shock heated gas (23).

Cosmic rays also produce HD by the sequence of reactions (7) and (8) so that given the cosmic ray flux [D]/[H] can be derived from the measured HD abundances. Values near 5×10^{-6} appear to be indicated, close to the lower limit found by Vidal-Madjar et al. (24) from absorption measurements of atomic deuterium in the interstellar gas.

The theoretical models of molecular abundances in diffuse clouds are reasonably satisfactory though many uncertainties remain in the data on the rates of the chemical reactions and the photodissociation and photoionization processes. For dense clouds, the models have met with qualitative success but there are serious quantitative discrepancies with measured abundances. Some are surely due to the inadequacy of our chemical understanding but many arise from the simplicity of the assumed cloud conditions. Thus the observation of substantial amounts of neutral atomic carbon in dense clouds is difficult to reproduce without invoking the exposure of carbon monoxide to ultraviolet photons (25). Time-dependent models may suffice (26). It is not clear that either steady state or time-dependent uniform models can explain the complex hydrocarbons. It is interesting to note that the generation of complex species increases rapidly with the ionization rate (27). Heated gas, whether by x-rays (28) or by shocks (29), may also enhance their formation. In diffuse clouds, the chemistry of fig. 3, fails to reproduce the measured abundances of CH^+. Elitizur and Watson (23) have proposed that an additional source

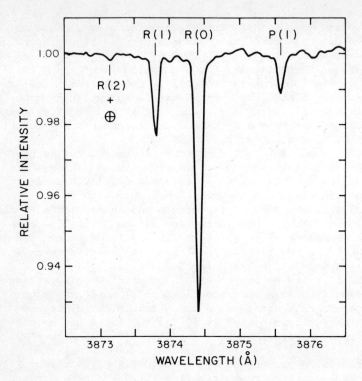

Fig. 5. The absorption spectra of CN (from ref. 20).

of CH^+ is produced in shocked regions of the interstellar gas through the endothermic reaction

$$C^+ + H_2 \to CH^+ + H \ . \tag{17}$$

The production of OH, H_2O and HD will also be increased.

Shocked gas is clearly evident in many astronomical objects, particularly in association with highly supersonic motions seen in CO emission which appears to be the result of outflow from very young stellar objects. The shocked gas is revealed by thermal emission from low-lying excited vibrational and rotational levels of molecular hydrogen, which indicate temperatures of the order to 2000K. Infrared emission lines from the heated gas provide detailed insight into its structure (29), the major sources being the oxygen fine-structure transitions and vibrational and rotational transitions of CO, H_2O, H_2 and OH.

Magnetic fields exert a large influence (31,32). The chemistry is controlled by reactions with H_2, both exothermic and endothermic, and the molecular composition is quite different from that found in cold clouds (30). In fast shocks, the molecules are fully dissociated (33) and then formed in new arrangements in the cooling post-shock gas. Precursor radiation may have a significant effect on the ultimate composition of the gas.

Vibrationally excited H_2 is also produced by ultraviolet pumping. The resulting fluorescence appears in the ultraviolet. It was seen first in sunspots and subsequently in Herbig-Haro objects (34). Most of it arises following the absorption of Lyman alpha by vibrationally excited H_2. There occur other line coincidences. Lyman beta lies close to the 6-0 P(1) line of the Lyman system of H_2 and produces fluorescence at 160.7nm. The fluorescence may be a sensitive means of detecting HD which has a different spectral response from H_2.

Vibrationally excited H_2 has been observed in many planetary nebulae (35,36). Some contribution to the population of the excited levels may arise from reaction (2). A tentative detection of H_2^+ has been claimed which if correct appears to require a dynamic model of an ionization front expanding more rapidly than the molecular envelope (37).

Optical, infrared and radio observations of molecules in the solar atmosphere and in stellar atmospheres have been extensively utilised to obtain element abundances. The chemistry is near equilibrium and accurate partition functions are needed to interprete the data (38). More recently observations of molecules in circumstellar shells have provided much information on the advanced stages of stellar evolution, nucleosynthesis and mass loss (39). A list of molecules observed in the bright infrared carbon star IRC +10216 (CW Leo) is given in Table 2 (40,41).

Table 2

Molecules in IRC +10216

CO	CS	SiO
SiS	CN	
HCN	HNC	C_2H
SiC_2	C_3N	NH_3
C_3H	C_2H_2	CH_4
C_4H	HC_3N	C_2H_4
HC_5N	C_2H_6	HC_7N
HC_9N	SiH_4	CH_3CN

The chemistry is very different from the low density chemistries of interstellar space. Deep inside in the hot dense layers where conditions are similar to those prevailing in the photospheres of cold stars, thermodynamic chemical equilibrium prevails. In the outer regions, the temperatures and densities diminish and the chemical reaction times become larger than the envelope expansion time scale. Depending on the specific reaction times, the abundances of the molecular constituents may be frozen out at their equilibrium values. The rate of collisions with grains is often faster than the expansion rate so that chemical reactions will take place on grain surfaces. In the outermost layers photodissociation tends to destroy the molecules and the study of the spatial distributions of different molecules may give insight into photodissociation mechanisms. For H_2 and CO, just as in interstellar clouds, self-shielding occurs (42,43).

Molecular masers OH, H_2O, SiO and SiS are found in circumstellar envelopes associated particularly with late-type stars. The SiO molecule mases in the v=0, 1 and 2 vibrational levels. Interstellar masers have been observed in OH, H_2O, CH_3OH and SiO and weakly in CH, HC_5N and H_2CO. The widely distributed OH and H_2O masers are indications of early stages in the formation of stars (44).

For maser action, population inversion must be achieved. Population inversion in interstellar masers results from energy transfer collisions, chemical pumping and ultraviolet and infrared pumping. The analysis is made difficult by complicated radiative transfer effects (45). Many data on collision and radiative processes are needed to work out the pump cycle (46).

Comets provide another arena whose molecular observations

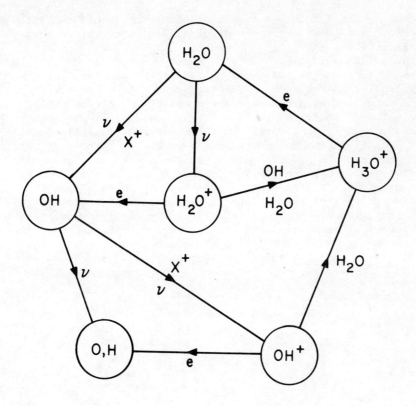

Fig. 6. The chemistry of a water-dominated cometary coma.

provide diagnostic information. From them, it should be possible to determine whether comets originate in interstellar space or within the solar system.

Table 3 is a list of molecules that have been identified in the atmospheres and tails of comets (47). The chemistry of cometary molecules shares some of the features of the chemistry of diffuse interstellar clouds and the chemistry of circumstellar shells. Ion-molecule reactions occupy a central role and photodissociation and photoionization, for comets, by the solar radiation field, limit the molecular abundances. The main constituent is probably H_2O and H_2O replaces H_2 as the major participant in the chemistry. For some comets, CO_2 or CH_4 may be the dominant parent molecule. As the parent molecules stream away from the cometary nuclei they are ionized and dissociated by solar ultraviolet radiation and by the interaction of the solar wind. Lists of the chemical reactions have been presented by Huntress et al. (48), Heubner and Giguerre (49) and Biermann, Giguerre and Heubner (50). Fig. 6 illustrates the chemistry of a water-dominated coma.

Table 3

Cometary molecules
Coma

CN	C_2	CH
NH	OH	CO
C_2	CS	S_2
C_3	NH_2	H_2O
HCN	CH_3CN	

Tail

CO^+	CH^+	N_2^+
OH^+	CN^+	CS^+
SH^+	CO_2^+	H_2O^+
H_2S^+		

In comets the molecules are usually detected in emission. The interpretation of the intensities is rarely straightforward and involves an analysis of the several processes that populate the excited states. Atomic emissions result from molecular photodissociation and dissociative recombination. Thus the red line of atomic oxygen is produced by

$$H_2O + h\nu \rightarrow H_2 + O(^1D) \tag{18}$$

and by

$$OH + h\nu \rightarrow H + O(^1D) \tag{19}$$

and the 193.1nm line of atomic carbon is produced by

$$CO^+ + e \rightarrow O + C(^1D). \tag{20}$$

Solar radiation also dominates the molecular processes that occur in the atmospheres of the planets, as they rotate in orbit around the Sun. Planetary atmospheres constitute perhaps the most sophisticated and most developed of those areas of research that belong to molecular astrophysics. In response to the quality and extent of the observational data, gathered by ground-based and balloon, rocket and space-craft instrumentation, and supported by an extensive laboratory and theoretical program of basic studies, refined quantitative theories have been developed of the properties of planetary atmospheres and of their evolution. With the Earth as a convenient object for detailed local investigations, it has been possible to construct critical tests of the theories. Extra-terrestrial ion-molecule chemistry can be traced back to the work of Bates and Massey (51,52) who envisaged the terrestrial sequence

$$h\nu + O \rightarrow O^+ + e \tag{21}$$

$$O^+ + N_2 \rightarrow NO^+ + N. \tag{22}$$

On Mars and Venus where carbon dioxide is the principal neutral component the sequence

$$h\nu + CO_2 \rightarrow CO_2^+ + e \tag{23}$$

$$CO_2^+ + O \rightarrow O_2^+ + CO \tag{24}$$

or

$$CO_2^+ + O \rightarrow CO_2 + O^+ \tag{25}$$

and

$$O^+ + CO_2 \rightarrow O_2^+ + CO \tag{26}$$

establishes O_2^+ as the major ionic constituent.

Closer to interstellar chemistry is the chemistry of the Jovian planets where H_2 and He are the major neutral gases.

The H_2^+ ions arising from photoionization and cosmic ray ionization react to form H_3^+ ions. At the higher densities, three-body reactions may be faster than two-body reactions and H_3^+ may be converted to H_5^+ before reacting with minor constituents such as CH_4, NH_3 and C_2H_6. Vibrational excitation can effect the chemistry. Complex hydrocarbons can be produced readily. Propane which has been detected in the atmospheres of Saturn and Titan is produced by a series of reactions terminating with

$$C_3H_9^+ + e \to C_3H_8 + H. \qquad (27)$$

The study of molecules in astrophysical environments, of the processes which form them and which destroy them, the collisions which transfer energy between them and the radiative interactions through which molecules scatter, absorb and emit photons is an integral component of astrophysics. The study links experimenters, theorists and observers in a common enterprise as we seek a unified description of the evolution of the Universe, the collapse of clouds, the formation of galaxies, stars and planets, and the development of pre-biotic compounds in an environment that can initiate and support life.

ACKNOWLEDGMENT

This work was supported in part by the National Science Foundation under Grant AST-81-14718.

REFERENCES

1. Shull, M.J. and Lepp, S. 1984, Ap. J. 280, pp. 465-469.
2. Carr, B.J. and Rees, M.J. 1984, Mon. Not. Roy. Astron. Soc. 206, pp. 315-325.
3. Palla, F., Salpeter, F.E. and Stahler, S.W. 1983, Ap. J. 271, pp. 632-641.
4. Silk, J. 1983, Mon. Not. Roy. Astron. Soc. 205, pp. 705-718.
5. Broten, N.W., MacLeod, J.M., Avery, L.W., Irwine, W.M., Hoglund, B., Briberg, P. and A. Hjalmarson 1984, Ap. J. Lett. 276, pp. L25-L29.
6. MacLeod, J.M., Avery, L.W. and Broten, N.W. 1984, Ap. J. Lett. in press.
7. Walmsley, C.M., Jewell, P.R., Snyder, L.E. and Winnewisser, G. 1984, Astr. Ap. 134, pp. L11-L15.
8. Loren, R.B., Wooten, A. and Broten, N.W. 1984, Ap. J. Lett. in press.
9. Scoville, N.Z. 1984, Proceedings of a Workshop on Star Formation, Occ. Rep. Roy. Obs. Edinburgh (Ed. R. Wolstencroft) to be published.
10. Herbst, E. 1983, Ap. J. Suppl. 53, pp. 41-53.
11. Suzuki, H. 1983, Ap. J. 272, pp. 579-590.
12. Millar, T.J. and Freeman, A. 1984, Mon. Not. Roy. Astron. Soc. 207, pp. 405-423.
13. Barlow, S.E., Dunn, G.H. and Schauer, M. 1984, Phys. Rev. Lett. 52, pp. 902-905.
14. Stahler, S.W. 1984, Ap. J. in press.
15. Watson, W.D. 1976, Rev. Mod. Phys. 48, pp. 513-552.
16. Dalgarno, A. and Lepp, S. 1984, Ap. J. Lett. submitted.
17. Smith, D. and Adams, N.G. 1984, Apl J. Lett. in press.
18. Michels, H.H. and Hobbs, R.H. 1984, Ap. J. Lett. in press.
19. MacDonald, J.A., Biondi, M.A. and Johnsen, R. 1984, Planet. Spa. Sci. 32, pp. 651-654.
20. Meyer, D.M. and Jura, M. 1984, Ap. J. Lett. 276, pp. L1-L4.
21. Woody, D.P. and Richards, P.L. 1979, Phys. Rev. Lett. 42, pp. 925-929.
22. van Dishoeck, E.F. 1984, Thesis: University of Leiden.
23. Elitzur, M. and Watson, W.D. 1980, Ap. J. 236, pp. 172-181.
24. Vidal-Madjar, A., Laurent, C., Gry, C., Bruston, P., Ferlet, R. and York, D.G. 1983, Astr. Ap. 120, pp. 58-62.
25. Boland, W. and de Jong, T. 1982, Ap. J. 261, pp. 110-114.
26. Langer, W.D., Graedel, T.E., Frerking, M.A. and Armentrout, P.B. 1984, Ap. J. 277, pp. 581-604.
27. Krolik, J.H. and Kallman, T.R. 1983, Ap. J. 267, pp. 610-624.
28. Lepp, S. and McCray, R.A. 1983, Ap. J. 269, pp. 560-567.
29. McKee, C.F., Chernoff, D.F. and Hollenbach, D.J. 1984, in Galactic and Extragalactic Infrared Spectroscopy, Ed.

M.F. Kessler and J.P. Phillips (Reidel) pp. 103-131.
30. Mitchell, G.F. 1983, Mon. Not. Roy. Astron. Soc. 205, pp. 765-772.
31. Draine, B.T. 1980, Ap. J. 241, pp. 1021-1038.
32. Draine, B.T., Roberge, W.G. and Dalgarno, A. 1983, Ap. J. 264, pp. 485-507.
33. Hollenbach, D.J. and McKee, C.F. 1979, Ap. J. Suppl. 41, pp. 555-592.
34. Schwartz, R.D. 1983, Ap. J. Lett. 268, pp. L37-L40.
35. Isaacman, R. 1984, Astron. Ap. 130, pp. 151-156.
36. Storey, J.W.V. 1984, Mon. Not. Roy. Astron. Soc. 206, pp. 521-527.
37. Black, J.H. 1983, in Planetary Nebulae, I.A.U. Symp. 103 (Reidel) pp 91-102.
38. Rossi, S.C. and Maciel, W.J. 1983, Astrophys. Spa. Sci. 96, pp. 205-212.
39. Zuckerman, B. 1980, Ann. Rev. Astron. Ap. 18, pp. 263-288.
40. Johannson, L.E.B., Andersson, C., Ellder, J., Friberg, P., Hjalmarson, Hoglund, B., Irvine, W.M. Olofsson, H. and Rydbeck, G. 1984, Astron. Ap. 130, pp. 227-256.
41. Thaddeus, P., Cummins, S.E. and Linke, R.A. 1984, Ap. J. Lett. in press.
42. Glassgold, A.E. and Huggins, P.J. 1983, Mon. Not. Roy. Astron. Soc. 203, pp. 517-532.
43. Morris, M. and Jura, M.A. 1983, Ap. J. 264, pp. 545-553.
44. Reid, M. and Moran, J. 1981, Ann. Rev. Astron. Ap. 19, pp. 231-276.
45. Langer, S.H. and Watson, W.D. 1984, Ap. J. in press.
46. Elitzur, M. 1982, Rev. Mod. Phys. 54, pp. 1225-1260.
47. Lüst, R. 1981, Topics in Current Chemistry 99, pp. 73-98.
48. Huntress, W.T., McEwan, M.J., Karpes, Z. and Avicich, V.G. 1980, Ap. J. Suppl. 44, pp. 481-488.
49. Heubner, W.F. and Giguerre, P.T. 1980, Ap. J. 238, pp. 753-762.
50. Biermann, L., Giguerre, P.T. and Huebner, W.F. 1982, Astron. Ap. 108, pp. 221-226.
51. Bates, D.R. and Massey, H.S.W. 1946, Proc. Roy. Soc. A187, pp. 261-296.
52. Bates, D.R. and Massey, H.S.W. 1947, Proc. Roy Soc. A192, pp. 1-16.

CHEMICAL COMPOSITION AND MOLECULAR
ABUNDANCES OF MOLECULAR CLOUDS

M. Guelin

I.R.A.M., Avenida Divina Pastora, Granada,
Spain

I- Chemical composition of the clouds

The study of the chemical composition of the
interstellar medium (ISM) (i.e. essentially the determination
of the abundance of its trace elements which altogether
represent only 1 percent in mass) is one fundamental topic of
astrophysics. The trace element abundance not only controls
the IS gas chemistry, but also the cooling of this gas and
its degree of ionization (hence indirectly the evolution of
the IS clouds, the rate of star formation, and the IMF), as
well as the chemical composition of the stars (hence their
evolution).
 The ISM chemical composition is partly inherited from
the protogalactic gas and partly the result of the recycling
of the ISM through stars; for some elements (Li,Be,B) it may
also result in part from nuclear reactions triggered by
cosmic ray bombardment (see e.g. Reeves 1974). The time
scales for astration and spallation are small with respect to
the age of the Galaxy, so one can expect that the chemical
composition of the ISM has evolved during the history of the
Galaxy and that it differs from one region to another.
Whether it differs between two clouds located in the same
region of the Galaxy depends of the age of the clouds.
 What exactly is the age of molecular clouds is a matter
of controversy. Uncertainties on the velocity and magnetic
fields in these clouds, as well as on the density
distribution, prevent accurate theoretical estimates. At the
most can it be said that the free-fall time of collapse of a
compact cloudlet ($t \approx 10^5$ y) should be a lower limit to this
age. For the giant molecular clouds associated with HII

regions, the age of the HII region, or the age of the embedded stars (usually, $t \approx 10^6$ y) yields more significant lower limits to the cloud age.

Additional information on the age of the giant molecular clouds comes from the CO J=1-0 maps of our galaxy. Whether CO emission, and hence the molecular gas, is or is not confined into the spiral arms implies that the clouds are young or old relative to the time needed for the gas to cross a spiral arm (few $\times 10^7$ y). Unfortunately, the arm/interarm contrast is difficult to measure, owing to the difficulty of converting radial velocities into distances and owing to excitation temperature effects. Cohen et al. (1980) see in the Columbia CO survey evidence that this contrast is high (up to 10/1), from which they derive a mean cloud age $< 10^8$ y, while Scoville et al. (1984) find that the average contrast is low and argue for significantly larger cloud ages.

Clouds with ages as large as few $\times 10^9$ year could well have developed peculiar element abundances, while massive clouds with shorter ages must have abundances similar to those of the surrounding ISM.

Chemical composition of the ISM in the solar neighborhood

This composition can be derived from studies of the optical and UV spectrum of stars, from the observation of IS absorption lines in the line of sight to nearby bright stars, and from that of emission lines in nearby HII regions.

Stellar abundances
....

Early-type main sequence stars have recently condensed out of the ISM, so their original material is very similar to the present day IS material. Nucleosynthetic processes since then have altered their light material content (D, Li, Be, B), but must not have changed the abundance of the heavy elements. Indeed, the synthesis of C, N, O and a fortiori of heavier elements, and their appearance at the surface of the star, occurs at temperatures only reached in the red giant stage (or even in later stages of stellar evolution). The heavy element content of these stars is thus expected to be the same for all the stars and to faithfully reflect the present ISM content. Table 1 lists the abundances of the principal elements observed in a few young stars and in the sun: within the errors (50 per cent in the best cases), these abundances are the same for all the young stars and are very similar to the solar ones. For this reason, the solar abundances (or more generally the solar system ones), which are accurately derived, are often referred to as "cosmic abundances".

TABLE 1: Element abundances in stars and in the solar system
logN(X)-logN(H)

	10Lac (O9)	Sco (B0)	Her (B3)	Sun	C1 chondrites
H	0	0	0	0	0
He	-0.8	-1.0	-0.8	-0.7	-1.10
O	-3.2	-3.3	-3.6	-3.17	-3.13
C	-3.6	-3.9	-3.9	-3.47	-3.35
Ne	-3.3	-3.4	-3.4		-3.86
N	-3.6	-3.7	-4.3	-4.09	-4.04
Mg	-3.8	-4.5	-4.7	-4.43	-4.40
Si	-4.3	-4.4	-4.9	-4.35	-4.43
Fe		-4.7	-4.6	-4.4	-4.48
S		-4.8	-4.9	-4.8	-4.72
Ar					-5.42
Al		-5.8	-5.9	-5.6	-5.51
Ca			-5.6	-5.64	-5.65
Na				-5.7	-5.68
Ni					-5.74
P				-6.6	-6.42

Abundances in stars are taken from Unsold and Baschek (1983) and references therein, chondrites abundances are from Anders and Ebihara (1982).

Gas chemical composition
....

Table 1 gives the chemical composition of the present day IS matter in all its forms: gas and dust. Although the dust represents only some 0.6 per cent of the mass of the IS matter (Jenkins and Savage, 1974), it contents a large fraction of the heavy elements. Infrared stellar spectra show absorption bands, due to dust, which can be identified with C, N, O and Si-bearing radicals or molecules (see e. g. Savage and Mathis 1979, Lacy et al. 1984). Direct evidence of the gas depletion in heavy elements is given by the Copernicus satellite observations of IS abbsorption lines in the direction of bright nearby stars (see e. g. Spitzer and Jenkins 1975). Si and the metals (Mg, Al, Fe...), which have a high condensation temperature are found underabundant by factors of 40 and more, C, N, O and S by a factor of 1.5 - 4.

(In their recent analysis of neutral carbon lines Jenkins et al. 1983 found that toward some stars carbon may be not depleted at all). Note that these results apply to the "diffuse" clouds, i. e to low density, relatively warm clouds (T= 40 to 80 K) transparent to UV radiation.

The diffuse cloud abundances are more uncertain than those of Table 1, because cloud overlap along the line of sight and the impossibility to observe all the states of ionization of each studied atom, make them model dependant. Determinations of the main element abundances accurate within a factor of 2-4 are available for only three stars.

It is interesting to note (see the discussion of Spitzer and Jenkins 1975) that while the degree of depletion of most elements shows a good correlation with their condensation temperature, indicating that they probably condensed on grains at high temperature in the outer atmosphere of stars, C, N, and O seem to show some "extra" depletion probably resulting from collisions with dust grains in the IS medium. The collision time of one oxygen atom with a dust grain is few x 10^7 year in a diffuse cloud, but about 50 times shorter in a dense cloud (n= 10^4 cm-3). Depletion of C, N an O, via atom/grain or molecule/grain collisions can be there an order of magnitude higher if the grain temperature is low enough (for example, less than 17 K, Leger 1983). Evidence for CO molecules frozen on grains comes from the IS 4.6 micron absorption feature (Lacy et al. 1984).

One isotope whose abundance can be better derived from IS absorption measurements than from any other data is deuterium. Deuterium is efficiently destroyed in the interior of the stars (T > 10^6 K) and, because of this, has never been observed in any star, even the sun (Ferlet et al. 1983a). The knowledge of the deuterium abundance is important for cosmological studies as well as for the determination of the factor of enhancement of this isotope in molecules. Measurements of the interstellar D/H abundance ratio have been made in the direction of several nearby stars with Copernicus. The derived ratios differed from star to star by up to a factor of 20 (Laurent et al. 1979), a result difficult to understand which made doubt that a local IS value of the D/H ratio could be defined. A recent reanalysis of the Copernicus data (Vidal-Madjar et al. 1983) has fortunately cleared up this problem (which was due to a blending of the deuterium line by a high velocity hydrogen component) and leads to a single value of D/H over the line of sight to all the stars: D/H 5×10^{-6}.

Abundances in HII regions
....

Optical, radio and infrared observations of atomic emission lines from HII regions allow an estimate of the chemical composition of the giant clouds associated with them. Relative abundance determinations involving intensity ratios of allowed transitions can be very accurate in "normal" HII regions (i. e. regions excited by OB stars): typically a factor of 2 and even 1.1 in the case of He/H.

The chemical composition of the nearby HII regions shows little difference from region to region, as well as with that of the young main sequence stars. O and N seem underabundant in Orion by a factor of 2, with respect to their cosmic abundances. There is no other evidence for differences in nucleosynthetic evolution for the nearby giant molecular clouds.

Abundances outside the solar neighborhood

Emission line studies in HII regions provide the only straightforward way to determine the ISM chemical composition outside a radius of 3 kpc from the sun. Such studies can be carried out not only in remote parts of our galaxy, but in other galaxies as well (Peimbert 1975, Dennefeld and Statinska 1983).

Investigations of HII regions in our galaxy indicate the existence of a galactocentric gradient of the abundance of the heavy elements. This gradient is fairly similar for N, O, S and the metals (15/100 per kpc in the direction of the galactic centre). Similar or stronger N, O and S gradients are often found in external galaxies: as in our galaxy, O/H increases in spiral galaxies from the outermost HII regions to the centre, while S/O remains constant and about equal to its solar value. N/O remains constant or increases toward the centre. Observations of planetary nebula give similar results (see e. g. Pagel and Edmunds 1981).

The abundances derived from HII regions pertain only to the gas. These gradients are however too large to be explained by depletion on grains. X-ray absorption measurements in our galaxy are consistent with an overabundance of heavy elements in the inner ISM (gas+dust) (Ryter et al. 1975).

The question of deuterium and helium abundance gradients in our galaxy is an important one from the point of view of cosmology. The deuterium abundance determinations from the D and H Lyman lines, quoted above, pertain only to the solar neighborhood. Attempts have been made to derive the IS D/H abundance ratio from molecular observations (Penzias 1979). The abundance of the deuterated molecules is however enhanced, with respect to that of their hydrogenated counterparts, by factors of the order of 1000 through fractionation (these molecules just wouldn' t be observable otherwise). Entangling an abundance gradient effect from a fractionation effect is therefore very difficult. Radio

recombination line studies show a clear decrease of the He/H
line intensity ratio toward the galactic centre. This
decrease, which originally has been interpreted in terms of
an He/H abundance gradient, seems to be only a consequence of
the O/H abundance gradient noted above: due to their higher
abundance in heavy elements, the stars in the central region
of the galaxy emit less energetic UV photons capable of
ionizing helium atoms than the stars of the outer regions
(Thum et al. 1980).

Summary of the results obtained from atomic line observations

Except for a few light elements, the chemical
composition of the ISM (gas + dust) in the solar neighborhood
is very similar to that in the solar system and can for most
elements be considered as relatively well known. Si and the
metals (Mg, Fe...) are mostly condensed on grains with about
or less than 1 per cent in the gas phase. C, N, O and S are
little or not condensed on grains in the diffuse
clouds(<factor of 4), but could be more heavily depleated in
the gas at the center of the dense clouds (see below). Few
isotopic abundance determinations can be made from atomic
line observations. Recent reanalysis of the H and D Lyman
lines have lowered the D/H abundance ratio by a factor of 4
with respect to the previously admitted value.
   The gas plus dust chemical composition does not seem to
change significantly from one cloud to another in the solar
neighborhood. It does vary, however, between the solar
neighborhood and the galactic centre region: the abundances
of N, O, S (and probably C), relative to H, increase by at
least a factor of 2 from the outer parts of the galaxy to the
inner parts; the N/O ratio increases in the same direction
(confirming that N is at least partly a secondary element),
while the S/O ratio seems everywhere constant and about equal
to its solar value (which means that S and O are probably
formed in the same stars). The O/H (primary element to H) and
N/O (secondary to primary element) abundance ratios vary from
galaxy to galaxy and show some correlation with the Hubble
type.

II- Molecular abundances

Fine structure, rotational, vibrational and electronic
spectra of molecules are now observed at radio, IR, optical
and UV wavelengths. These spectra withold information on the
structure, dynamics and physical properties of the IS clouds,

on the UV and X-ray interstellar radiation fields and, of course, on the gas molecular content. The extraction of this information, even in the case of properly calibrated data, is however far from being straightforward and supposes that the following conditions are fulfilled:
1) the studied lines should be unambiguously identified and should not be blended with other lines;
2) the molecular excitation parameters (collisional excitation and radiative de-excitation rates) should be well known;
3) the observed cloud(s) should have a simple structure and should be easily modelizable; enough lines should be observed to determine its physical parameters.
4) when line intensity ratios are considered, the lines should arise from the same region(s) of the cloud(s);
5) the lines should not be optically very thick.

Identification of molecular lines

The identification of spectral lines is usually easier in the radio frequency range where are observed most of the molecular lines (the pure rotational lines fall essentially in the millimeter and submillimeter ranges). Rotational lines arising in cold and quiescent clouds, such as the dark clouds HC12 and L1544, can be exceedingly narrow ($\Delta \nu/\nu \approx 5 \times 10^{-7}$) and their frequency can be measured with a very good accuracy; the identification of such lines is quite secure when laboratory frequency measurements of comparable accuracy are available. Even when one has good laboratory data, the identification of a line in the millimeter wave spectum of the giant molecular clouds Orion A and Sgr B2 is more difficult. This spectrum is so rich in lines (lines which are not anymore so narrow: $\Delta \nu/\nu = 3 \times 10^{-5}$ to 3×10^{-4}) that over large frequency intervals it is often not possible to reach the continuum. Superposition of unrelated lines is there a common feature and the chances of misidentification of a weak line relatively high. The situation is even worse for submillimeter and infrared lines as the laboratory data at these wavelengths are scarce and as the spectra become more complex. It is therefore not surprising that some of the 50 or so molecules reported in these sources have been wrongly identified from lines belonging to other species. Among these misidentified species we can quote CH_4 (the lines observed in Orion by Fox and Jennings 1978 belong in fact to methyl formate, HCO_2CH_3-- Ellder et al. 1980), O_3 (Phillips et al. 1980), HNO (Ulich et al. 1977, Hollis et al. 1981), CO^+ (the line reported by Erickson et al. 1981 has now been identified as arising from ^{13}C methanol --see G. A. Blake, this conference), and CS^+ (an unsuccessful search for an intense band at 6840 A has infirmed the tentative identification of

Ferlet et al. 1983b--E. Roueff, private communication). The detections of HOC^+ (Woods et al. 1983) and NaOH (Hollis and Rhodes, 1982) are also far from secure, although accurate laboratory frequencies are available. Chance superposition of lines cannot be ruled off even in the relatively poor millimeter-wave spectrum of the envelope of the carbon star IRC+10216: for example, the line at 114.22 GHz, which was attributed by Scoville and Solomon (1978) to vibrationally excited CO is nothing else than the J=23/2-21/2 component of the N=12-11 rotational transition of C_4H, a radical for which no frequency data was available at that time (Guélin, Green and Thaddeus 1978). Secure identification of a molecular species in an astronomical source requires usually the observation of more than one line and a good prediction of the spectrum emitted by all other astrophysical molecules. This spectrum is unsufficiently known below 1mm, where the stretching of fast rotating molecules disturbs in an complicated way the line frequencies.

The extreme physical conditions encountered in the IS clouds allow the survival of reactive molecular species which would be instantly destroyed in a terrestrial environment. The study of these species presents a great interest from the point of view of IS chemistry as well as from that of physical chemistry. The identification of such species, for which little or no laboratory data are available, is understandably challenging. It is quite impressive that half a score of radicals and ions have been detected and identified in space before being observed in a spectroscopic laboratory. These are N_2H^+, C_2H, C_4N, C_4H, HC_7N, HC_9N and HCS^+, which have been identified on the basis of a characteristical line pattern (fine structure, etc...), of simple minded "astrochemical" considerations and of quantum mechanical ab initio calculations (see e. g. Guélin and Thaddeus 1977, Wilson and Green 1977, Guélin, Friberg and Mezaoui 1982). More remarkably, HCO^+ and HNC have each been identified from the observation of one single unresolved line, on the basis of chemical considerations and quantum mechanical calculations (e. g. Klemperer 1970, Kraemer and Diercksen 1976). Confirmation of these identifications (except for HC_9N) has come later from laboratory work (see e. g. Gottlieb et al. 1983). The latest reactive species (and fourth molecular ion) discovered in the IS gas is HCO_2^+, protonated carbon dioxyde. These species have been proposed with HOCN by Thaddeus, Guélin and Linke (1981) as one of the two most likely carriers of three millimetric lines observed in SgrB2. Since then, accurate quantum mechanical calculations (DeFrees et al. 1982) have ruled out HOCN on the basis of its too small rotational constant and confirmed that HCO_2^+ could well be the sought molecule. Definitive confirmation of the correctness of this identification has just been reported by Bogey et al. (1984) who succeeded in measuring the frequencies of six rotational lines of HCO_2^+ in the lab. Millimetric lines in IRC+10216 and Orion A have recently been identified with $HC_{11}N$ (Bell et al. 1982), SiCC (Thaddeus, Cummins and Linke 1984), C_3H (Johansson et al. 1984) and HCl (see Blake, this conference). These latter identifications have yet to be confirmed.

Molecular excitation parameters

The population of the lowest rotational levels of a molecule can be at thermal equilibrium when this molecule is immersed in a dense region (for example the core of the Orion A cloud) and/or if its permanent dipole moment is very small (for ex. CO). In this case, the molecule column density can be calculated from the intensity of a single optically thin emission line, provided the energy of the levels, the dipole moment and the equilibrium temperature are known.

When the populations are not thermalized, the statistical equilibrium equations have to be solved, which requires the knowledge of the radiative and collisional excitation rates. Rotational transition radiative rates are easily derived from the molecule inertia moments and from its permanent dipole moment; this latter has been measured or calculated with a good accuracy for all astrophysical molecules. The derivation of the collisional rates, on the other hand, requires lengthy and delicate calculations, which have been performed for only a few molecules. Excitation rates for collisions with H_2 molecules have been calculated for CO, CH, OH and calculations are in progress for NH_3 (G. H. F. Diercksen, P. Valiron, private communication). Rates for collisions with He (which are often scaled to estimate the collisional rates with H_2) have been calculated for HCN, N_2H^+, CS, HC_3N, H_2CO, CO, CH, OH and NH_3; the use of empirical interaction potentials make most of these rates rather unaccurate. When no such calculations are available, the collisional rates are estimated from those of structuraly similar molecules (for ex. HCO^+ from N_2H^+) or simply guessed. The derived column densities can then easily be wrong by factors of 2 or more.

Problems arising from optical thickness of the lines

In dense clouds, line opacity, limited angular resolution and inhomogeneities along the line of sight limit obviously the meaning of the derived "beam averaged" molecular abundances. These limitations are less severe in the diffuse clouds where absorption measurements against stars can be made; there, the main problem is usually the **weakness** of the lines. Even in the best cases (i. e. the conjunction of a bright star and of a cloud of reasonably large gas column density, as for the star ζ Oph), only few molecules can be observed (see Table 2).

It has been thought that molecular abundance studies would be easier in the dark clouds (which are cold and dense clouds located within a few hundred parsecs from the sun), particularly in the Bok globules: the shape of these objects on optical plates strongly suggests spherical symmetry and clear cut boundaries; their angular size is large enough that they can be resolved with medium size radio telescopes and their molecular column densities are large. Detailed studies have however revealed that these objects are quite complex. They usually exhibit a clumpy structure and are surrounded by a halo of gas. This halo, although tenuous, can efficiently absorb the radiation emitted by the cloud core, often making abundance determinations unsecure. For example, in the Bok globule L183, the intensity of the J=1-0 line of the ^{13}C isotope of HCO^+ is a factor of 2 stronger than that of the main ^{12}C isotope, although this latter is a factor of 90 more abundant! (see Guélin, Langer and Wilson, 1982). For HNC, the $^{13}C/^{12}C$ intensity ratio is about 1. Similar results are found in the dark cloud HCl2. In such extreme cases, radiative transfer calculations would certainly not allow to directly estimate the ^{12}C isotope abundance. Reliable abundance determinations require the use of optically thin lines --possibly by turning to very rare isotopic species and assuming that the rare to abundant isotopic ratio is known.

Interstellar isotopic abundance ratios

The elemental isotopic abundances in the interstellar gas, just as the chemical composition of this gas, are essentially the result of nuclear synthesis in the interior of stars. The solar system isotopic abundances are expected to reflect mainly the ISM abundances at the time of formation of the sun, some 5 billion years ago. Since that time, the ISM has continued to evolve and so should have its isotopic abundances. A comparison with the elemental abundance variations discussed above suggests, however, that this evolution has been slow and has not much affected most isotopic abundance ratios, at least outside the innermost region of the Galaxy.

Except for deuterium and helium, the ISM isotopic abundance ratios have to be derived from molecular line observations: optical absorption lines for the nearby diffuse clouds, centimetric, millimetric or infrared lines for the dense clouds. Here again, the optical data are the more accurate, but are restricted to a few lines of sight and to only one ratio: $^{12}CH/^{13}CH$. This ratio is found to be very close to the solar system $^{12}C/^{13}C$ ratio, 89.

In the dense clouds, the derivation of isotopic abundance ratios is in principle much easier than that of absolute molecular abundances, since one compare two species with very similar excitation and spectroscopic parameters and with similar spatial distributions. The $C^{18}O/C^{17}O$ ratio, for example, has been accurately measured in several molecular clouds (3.2±0.2, Penzias 1981) and is found equal to the $HC^{18}O^+/HC^{17}O^+$ (Guélin, Cernicharo and Linke 1982) and $^{18}OH/^{17}OH$ (Bujarrabal, Cernicharo and Guélin 1983) ratios, so that one can speak with some confidence of an interstellar $^{18}O/^{17}O$ abundance ratio (this ratio is 1.5 times smaller than the terrestrial $^{18}O/^{17}O$ ratio). However, In the case of isotopes of very different abundances the lines of the main isotopic species are often optically thick when those of the rarer species become detectable. This is particularly the cases for the $^{12}C/^{13}C$ and $^{16}O/^{18}O$ ratios for which the interpretation of the observational data is far from being straightforward. Earlier attempts to derive these ratios (using the wings of the lines to overcome saturation problems and/or assuming equal excitation for the two isotopic species) have led to the belief that their local ISM values were at least a factor of two smaller than the terrestrial ones. Recent observations of the very rare $^{13}C^{18}O$ and $^{13}C^{34}S$ isotopic species (Penzias 1983, Guélin, Linke and Cernicharo 1984) and of IR ^{13}CO and ^{12}CO lines (Scoville et al. 1983) have established that the $^{12}C/^{13}C$ IS ratio is close to the terrestrial ratio except perhaps toward the galactic centre sources. The $^{16}O/^{18}O$ ISM ratio, as derived from OH and H_2CO after correction for radiation trapping effects in the lines of the ^{16}O species, are found also to be terrestrial to better than a factor of 2. A critical review of the existing data leads to the conclusion that the common isotopic ratios in the ISM are within a factor of two equal to the terrestrial ones, except for D/H (an effect of isotopic fractionation) and for $^{14}N/^{15}N$ in the galactic centre; the ^{13}C abundance in molecules can be enhanced by factors of 4-6 by isotopic fractionation at the edge of the clouds.

Assuming terrestrial isotopic abundance ratios, when deriving the abundance of a molecule from that of one of its rare isotopes, introduces thus an error which in most cases is small in front of the uncertainties on the molecular column density determinations.

The abundance of the H_2 molecule

In molecular clouds, hydrogen is essentially neutral, in the form of atoms (HI) or molecules (H_2). In diffuse clouds, the HI and H_2 abundances can be directly measured from UV absorption studies in the line of sight to bright stars. In the dense and opaque clouds, or further away from the sun, one has to rely on 21-cm line measurements for HI and on IR measurements for H_2. 21-cm line measurements are often plagued by insufficient angular resolution, confusion along the line of sight and line opacity. Except in small regions close to embedded stars or behind shock waves, the gas in molecular clouds is too cold to excite collisionally H_2 in any but its lowest rotational level (the J=2 level lies already 510 K above the J=0 level); IR H_2 emission measurements (to date, only vibration-rotation lines are detected) provide thus little information about the H_2 abundance in common dense clouds.

An analysis of the UV spectra observed with the Copernicus satellite towards 100 stars (Bohlin, Savage and Drake, 1978) yields for the local 1kpc region an average total hydrogen density, $n(HI +H_2)$, of about 1 at. cm-3, some 25 percent of which is in the molecular form. The percentage of molecular hydrogen is not constant and ranges from ≈ 0, toward stars with low reddening such as ζ Pup, to 0.3 toward ζ Per (E(B-V)= 0.6). This variation was expected and reflects the self-shielding of H_2 for sufficient gas column densities ($N(HI+H_2)$) > few x 10^{20}, according to Federman, Glassgold and Kwan, 1979). For clouds with large gas column densities and particularly for the dense clouds, one expects nearly all hydrogen in the form of H_2.

One of the most important results of these UV studies is the very good correlation between reddening (hence the amount of dust in the clouds) and $N(HI+H_2)$. Bohlin et al. (1978) find $N(HI+H_2)$ =5.8x10^{21} xE(B-V) cm-2 for E(B-V) between 0.01 and 0.6. The reddening, E(B-V), is simply related in most clouds to the visual extinction, Av, a quantity which can be derived for the nearby dark clouds from star counts: Av≈ 3 E(B-V). In the dark clouds, the extinction Av is too large to allow the observation of the UV H_2 lines. The extrapolation of the above relation for E(B-V) ≳0.6 (i. e. Av ≥ 2) provides an simple way to estimate the H_2 column density from star counts. This extrapolation, which is currently applied in dark cloud studies, supposes that the dust grains are the same in the diffuse and in the dense clouds and that the gas to dust ratio does not depend on the gas density or temperature -- two assumptions may well not be fulfilled.

Another way of estimating the hydrogen column density in a dense cloud is to determine the gas density from observed line intensity ratios and molecule excitation calculations. If the cloud shape is simple enough that its thickness can be estimated from its extent in the plane of the sky, one can roughly derive $N(H_2)$. Column densities derived from star counts and by this method seldom agree to better than a factor of 2 --presumably, because of cloud clumpy structure.

Abundance of CO

The CO column density can be derived in diffuse clouds from UV absorption or microwave emission measurements. The CO UV measurements can be linked to HI and H_2 measurements towards the same stars, directly yielding the CO fractional abundance. This latter (see Table 2 and Dickman et al. 1983) is typically 10^{-6} and does not show any significant variation with E(B-V).

In the dark clouds, one has to rely on microwave emission observations and on star counts to estimate CO/H_2. The rotational lines of the main CO isotope, $^{12}C^{16}O$, are usually completely saturated, so the CO column density must be derived from the observation of rarer isotopes.

The first CO/Av studies in dark clouds were made by Encrenaz et al. (1976) and Dickman (1978). Due to the poor sensitivity of the instrumentation, they were mainly restricted to $^{13}C^{16}O$ J=1-0 line measurements and yield ^{13}CO $\approx 2.5 \times 10^{15}$ Av cm-2. The $^{13}C^{16}O$ J=1-0 line, however, is easily saturated at the center of the dark clouds and the $^{13}CO/^{12}CO$ abundance ratio at the cloud edges is affected by ^{13}C fractionation; ^{13}CO, therefore, is not a good tracer of the ^{12}CO abundance. More accurate studies require the detection of rarer isotopes, such as $^{12}C^{18}O$.

By far, the most complete study of the $C^{18}O/Av$ ratio is that of Cernicharo and co-workers (Cernicharo and Guélin 1984, Bachiller and Cernicharo 1984, Duvert, Cernicharo and Baudry 1984). This study is based on 300 high sensitivity J=1-0 $C^{18}O$ spectra observed in the dark clouds HC12 and L1495 and in the Perseus dark cloud. It was carried out with the Bordeaux telescope which has an angular resolution of 4.5'. In the area covered by one Bordeaux beam, on average 70 background stars are observed on low obscuration sections of the Palomar Sky Survey print; star counts over such an area provide accurate extinction values up to Av =4-5 magnitudes. The thermalization of the J=1-0 line was checked at places by observing the J=2-1 line with the University of Texas millimeter-wave telescope. In both clouds one derive, for Av between 1.5 and 6 mag, a remarkably clean linear relation:
$$C^{18}O = 2.3 \times 10^{14} (Av-1.5) \text{ cm-2}.$$
In directions with less than 1.5 mag of extinction, $C^{18}O$ is not detected, probably because it is photodissociated (a result which fits with the very low CO/Av ratio in diffuse clouds). In HC12, Av$\approx$ 3 E(B-V), and the extrapolation of the $N(HI+H_2)/E(B-V)$ relation to large Av and the use of a terrestrial $^{16}O/^{18}O$ ratio ($\approx$500) yields: $CO/H_2 \approx 10^{-4}$.

Star counts do not allow to measure visual extinctions larger than 6 mag. These can only be inferred from IR reddening measurements in the line of sight to background stars. The extinction values then refer to a much smaller **solid** angle than the CO data and can be affected by dust

associated with the star-- two efffects which lead to
underestimate the CO/Av ratio. By using this method,
Frerking, Langer and Wilson (1982) estimate $C^{18}O/Av$ at very
large extinctions. For Av=4-21 mag, they find
$C^{18}O/Av=1.7\times10^{14}$ cm-2 mag-1, a value only marginally
different than that found for lower extinctions, in view of
the biais noted above. We conclude, therefore, that the
fractional abundance of CO, CO/H_2, is close to 10^{-4} in the
dark clouds.

Crude estimations of CO/H_2 in the giant molecular clouds
(see Table 2) lead to about the same value.

Abundance of other molecules

As stressed above, only indicative values of the
molecular abundances can be given in dense clouds. These
values are nevertheless of interest as they reveal large
differences in cloud molecular composition. In Table 2, we
have listed the abundances in 4 dense clouds of 39 of the
most abundant IS molecules. These abundances, which result
from studies of various authors, have been as much as
possible homogenized:
-by selecting good quality observations with similar
  beamwidths (1-2', unless stated-- the data are taken from
  references preceded by an asterisk),
-by assuming a simple physical model for each cloud,
-by using the lines of rare isotopes to estimate the
  abundance
of a species, when the lines of the main isotope are
saturated.

They are compared in Table 2 to the abundances derived
in the best studied diffuse cloud, the cloud intercepted by
the line of sight toward the star ζ Oph (see Snow 1980 and
references therein).

The dense molecular clouds are Orion A(KL), SgrB2(OH),
HCL2(TMC1) and L183.

At least four different gas components are seen by a 1'
beam in the direction of the KL nebula in Orion A (see e. g.
Johansson et al. 1984): an extended envelope, observed mainly
in CO; a narrow, elongated ridge, only 20' broad in the east-
west direction, observed in CS, CN,...; a dense, compact (
10'x15') core, from where arise most molecular lines observed
in Orion A; finally, a small (30'x30') region of shocked gas,
with a large velocity dispersion, usually referred to as the
"plateau" component. The distance to Orion A is about 500 pc.
Average densities and temperatures in the ridge are 10^4-10^5
cm^{-3} and 40-60 K, in the core 10^6-10^8 cm^{-3} and 90-200 K, and
in the plateau 10^6 cm^{-3} and 100-150 K. Note that these can be
very different when derived from different molecules or from
transitions arising from levels of very different energies

(which indicates, as does the presence of a velocity
structure, that each component itself is complex). Because of
the core small size and clumpy stucture, "beam averaged"
abundances depend much on the size of the beam. The emission
from molecules like NH_3 arises essentially from tiny clumps,
in which NH_3/H_2, as observed with a 3' resolution, is as high
as 10^{-5}; averaged over a 40' beam, this abundance drops by a
factor of 20 (Johansson et al. 1984).

The giant molecular cloud associated with the SgrB2 HII
region is located near the center of our galaxy. Despite its
distance and owing to its large size (20 pc), this cloud is
one of the strongest molecular sources of the sky; it is
detected in the lines of nearly all known interstellar
molecules. Line emission is observed to be the most intense
in the direction of the Sgr B2 OH maser. In that direction,
the line of sight crosses four or five gas components, with
densities and temperatures ranging from 10^2 to 10^6 cm-3 and
20 K to 200 K (Scoville et al. 1975, Linke et al. 1981,
Wilson et al. 1982, Vanden Bout et al. 1983). The most
massive component is probably the 60 kms-1 core, which has a
mean density of about 10^5 cm-3 and a mean temperature of
20-50 K. As in Orion A, the molecules are not evenly
distributed, the rare isotopes of HCO^+, HCN, SiO and CS being
mostly concentrated in a cloud with a lower velocity (Guélin
and Thaddeus 1979).

HCL2(TMC1) and L183(L134N) are two cloudlets belonging
to the nearby Taurus and L134 dark cloud complexes. These
cloudlets have been extensively studied (see e. g. Toelle et
al. 1981, Guélin et al. 1982b) and their physical parameters
are relatively well known: they both consist of dense clumps
≈ 0.1 pc in size (with densities $\approx fewx10^4$ cm-3 and
temperatures ≈ 10 K), surrounded by an extended, low density
envelope (see above). Both exhibit strong emission from a
variety of interstellar molecules, but, although they look
physically quite similar, have a different molecular content
(see Table 2).

The molecules in Table 2 are listed by order of
decreasing abundance in Orion A. At first glance, after
allowance is made for elemental abundances (i. e. the low
abundance of S, Si,...), one can say that this order is that
of increasing complexity for the stable species: the list
starts with H_2, CO, H_2O, and ends with HC_5N, the heaviest
molecule detected in Orion. Sulfur and silicon-bearing
molecules (CS, H_2S, SiO,...) are two or more orders of
magnitude less abundant than their oxygen or carbon homologs;
in Orion, they are mostly concentrated in the "plateau"
component. Radicals (CH, OH, CN, C_2H,...) and ions also have
low abundances; in Orion, they are mostly found in the ridge
(Turner and Thaddeus 1977, Thaddeus et al. 1981).

TABLE 2: H_2 column density and molecular fractional abundances ($logN(X)-logN(H_2)$) in five molecular clouds

	ORIONA(KL)	SGRB2(OH)	HCL2(TMC1)	L183	ζ OPH
H_2 (CM-2)	2×10^{23}	2×10^{23}	10^{22}	1.3×10^{22}	4.5×10^{20}
CO (/H_2)	-4i	-4i	-4i	-4i	-5.6
H_2O	-5u				<-7.2
CH				-7.7a	-7.3
NH_3	-6.6c			-7	
SO	-6.6c(P	-8.7	-7.3	7.8i	
CH_3OH	-6.6	-6.7i			
SO_2	-6.8 (PL)	-8.5	<-8.7		
HCN	-7.1i		-7.7i		
OH	-7.2a	-6.5a	-7a		-6.9
H_2CO(ortho)	-7.5i	-7.0i	-8.0i		
OCS	-7.9 (PL)	-7.8i	<-8.8		
SiO	-7.9 (PL)		<-9.4		
CH_3OCH_3	-7.9	-8.5			
$HCOOCH_3$	-8u	-9.			
CN	-8.0i	-7.6i	-8u	<-9	-8.0
H_2S	-8.3u(PL)	-9u			
C_2H	-8.3			-8.1	-9.0
HCO^+	-8.3	-7.7i		-8.1i	-8.2
CH_3CH_2CN	-8.5	-9.9		<-9	
CH_3C_2H	-8.6	-7.8		-8.2	
HNCO	-8.6	-8.0i		-9.8	
HC_2O	-8.9	-8.8			
CS	-9.0 (PL)	-7.8u		-8.4i	-8.8i
HNC	-9.0			-7.5i	
CH_3CN	-9.2	-9.2		-9.2	
HC_3N	-9.3	-8.3i		-8.2	-9.8
N_2H^+	-9.3			-9i,u	-8.8i,u
CH_2CHCN	-9.4	-9.8		-9.5	<-9.6
HCS^+	-10.3	-9.7		-10u	
HC_5N	-10.6	-9.0		-8.7	
NO	<-7	-7.0			
C_4H	<-9.7	<-8.7		-7.5	<-9
CH_3C_4H				-8.6	
HC_7N	<-10	<-9.4		-8.9	
HC_9N				-9.3	
CH_3C_3N				-9.3	
C_3N	<-10.3	<-9.8		-9.0	<-9.8
HCO_2^+ (d)	<-10	-8.6			

Notes to Table 2.
a: from low resolution data
c: clumpy distribution; $x(NH_3)$ observed with few arc sec resolution is -5, $x(SO)$ is -6
d: assuming a permanent dipole moment of 2D, Green et al. (1976).
i: derived from observation of rare isotopes, assuming terrestrial isotopic abundance ratio
u: very uncertain
PL: mainly in the "plateau" component.

   A closer look at col 2 of Table 2 shows that not all stable molecules follow the abundance/complexity relation. Ethyl cyanide, CH_3CH_2CN, is a factor of 10 more abundant than its unsaturated homologs CH_2CHCN and HC_3N; quite unexpectedly, it is also a factor ≈ 5 more abundant than methyl cyanide, CH_3CN. It is true that ethyl cyanide is not observed in most clouds and that in Orion, its emission is mainly restricted to the low velocity component of the core (Johnson et al. 1977). Dimethyl ether, CH_3OCH_3, is much more abundant in Orion than its barely detected isomer ethyl alcool, and acetic acid is not seen, while its isomer methyl formate, $HCOOCH_3$, is quite abundant. Ring molecules have not been detected so far, except for the SiCC radical in the envelope of the IR star IRC+10216 (Thaddeus et al. 1984).
   Large abundance variations are observed from cloud to cloud, not only between the diffuse and the dense clouds, but also between these latters. In HCl2, the abundance of oxygen-bearing molecules (with the exception of CO) is much smaller than in Orion, whereas that of the long, unsaturated carbon chains (HC_3N, HC_5N, HC_7N, HC_9N, and the radicals C_3N and C_4H) is astonishingly large. The HC_3N/HCN and HC_5N/HC_3N abundance ratios in Orion are equal to $\approx 1/100$ and $1/20$; in HCl2 they are $1/6$-$1/2$ and the HC_7N/HC_5N, HC_9N/HC_7N, CH_3C_3N/CH_3CN, CH_3C_4H/CH_3C_2H and C_3N/CN ratios are similarly large. C_4H, in HCl2, appears even more abundant than C_2H. This wealth of HCl2 in carbon chains (which is more accentuated in some fragments of this cloud-- see e. g. Toelle et al. 1981, Cernicharo, Guélin, and Askne 1984) is not shared by all the dark clouds, as shown by the low abundances of HC_3N and HC_5N in L183 (Table 2 and Ungerechts et al. 1980).
   Another source with a peculiar chemical composition is the envelope of the star IRC+10216. Like HCl2, this envelope appears exceptionally rich in radicals and carbon chains and poor in oxygen compounds. It is the only source where C_3H, SiCC (see above), C_2H_2 and SiH_4 have been detected (the **latter two** in absorption near 10 microns--Goldhaber and Betz

1984). A detailed comparison between the molecular abundances
in IRC+10216 and in the other sources is however difficult,
because part of the molecular lines are observed in
absorption against the tiny (<1') IR source while others are
seen in emission in the extended (1-5') envelope. The
excitation of the microwave lines, which are seen in
emission, strongly varies with the distance to the source.

C_2H_2 and SiH_4 are detected in IRC+10216 not necessarily
because they are more abundant in this source, but mainly
because of the strength of its 10 micron emission. These
species, which have no permanent electic dipole moment, can
be easily observed only in absorption in the infrared. Other
non-polar molecules likely to be abundant in molecular
clouds, but not observed at radio wavelengths, are H_2, C_2,
N_2, O_2, CO_2, C_3, C_4H_4 and CH_4. H_2 and C_2 are seen in diffuse
clouds in absorption at optical wavelengths and are known to
be abundant (C_2 is seen in a diffuse part of HC12 and has
there an abundance comparable to those of CH and H_2CO--Hobbs,
Black and van Dishoeck 1983). The others are thought to be
abundant because related species such as N_2H^+, SO, OCS,
HCO_2^+, SiCC, CH_3OH,... are observed (Herbst et al. 1977). The
most surprising result on the abundance of non-polar
molecules in dense interstellar clouds, is probably that on
N_2 in the source DR21(OH). Using the rare isotopes of N_2H^+
Linke, Guélin and Langer (1983) found that N_2 is about as
abundant as CO in this source; this result raises interesting
questions about the depletion on grain of these two species.

References

Anders, E. and Ebihara, M. 1982, Geochim. et Cosmochim. Acta,
    46, p. 2363.
Bell, M. B., Feldman, P. A., Kwok, S. and Matthews, H. E.
    1982, Nature, 295, p. 389.
Bachiller, R., Cernicharo, J. 1984, Astr. and Ap., in press.
*Bieging, J. H. 1976, Astr. and Ap., 51, p. 289.
Bogey, M., Demuynck, C. and Destombes, J. L. 1984, Astr. and
    Ap., in press
Bohlin, R. C., Savage, B. D. and Drake, J. F. 1978, Ap. J.,
    224, p. 132.
*Broten, N. W., MaLeod, J. M., Avery, L. W., Irvine, W. M.,
    Hoglund, B., Friberg, P. and Hjalmarson, A. 1984, Ap. J.,
    276, p. L25.
*Bujarrabal, V., Guélin, M., Morris, M. and Thaddeus, P.
    1981, Astr. and Ap., 99, p. 239.
Cernicharo, J. and Bachiller, R. 1984, in preparation.
Cernicharo, J., Guélin, M. and Askne, J. 1984, Astr. and Ap.,
    138, p. 371.

Cernicharo, J and Guélin, M. 1984, Astr. and Ap., in press.
Cohen, R.S., Cong, H., Dame, T.H. and Thaddeus, P. 1980, Ap. J. (Letters), 239, p.L53.
DeFrees, D.J., Loew, G.H. and McLean, A.D. 1982, Ap. J., 254, p.405.
Dennefeld, M. and Statinska, G. 1983, Astr.and Ap.,118, p.234.
Dickman, R.L. 1978, Ap. J. Supplement, 37, p.407.
Dickman, R.L., Somerville, W.B.,Wittet, D.C.B., McNally, D. and Blades, J. 1983, Ap. J. Supplement, 53, p.55.
Duvert, G., Cernicharo, J. and Baudry, A. 1984, in preparation.
Ellder, J., Friberg, P., Hjalmarson, a., Hoglund, B., Irvine, W.M., Johansson, L.E.B., Olofsson, H., Rydbeck, G., Rydbeck, O.E.H. and Guélin, M. 1980, Ap. J. (Letters), 242, p.L93.
Federman, S.R., Glassgold, A.E. and Kwan, J. 1979, Ap. J., 227, p.466.
Ferlet, R., Dennefeld, M., Spite, M. 1983a, Astr. and Ap., 124, p.172.
Ferlet, R., Roueff, E., Horani, M. and Rostas, J. 1983b, Astr. and Ap.,125, L5.
Fox, K. and Jennings, D.E., 1978, Ap.J. (Letters), 226, p.L43.
Frerking, M.A., Langer, W.D. and Wilson, R. W. 1982, Ap. J., 262, 590.
Goldhaber, D.M. and Betz, A.L., 1984, Ap.J. (Letters), 279, p.L55.
Gottlieb, C.A., Gottlieb, E.W., Thaddeus, P. and Kawamura, H. 1983, Ap. J., 275, p.916.
Guélin, M. and Thaddeus, P. 1977,Ap.J. (Letters), 212,p.L81.
Guélin, M., Green, S. and Thaddeus, P. 1978, Ap. J. (Letters), 224, p.L27.
Guélin, M. and Thaddeus, P. 1979, Ap.J.(Letters), 227, L139.
*Guélin, M., Friberg, P. and Mezaoui, A. 1982a, Astr. and Ap., 1982, 109, p.23.
*Guélin, M., Langer, W.D. and Wilson, R.W. 1982b, Astr. and Ap., 107, p.107.
Guélin, M., Cernicharo, J. and Linke, R.A. 1982, Ap. J. (Letters), 263, p.L89.
Guélin, M., Linke R.A. and Cernicharo, J. 1984, in preparation.
Herbst, E., Green, S., Thaddeus, P. and Klemperer, W. 1977, Ap. J., 215, p.503.
Hobbs, L.M., Black, J.H and van Dishoeck, E.F. 1983, Ap. J. (Letters), 271, p.L95.
Hollis, J.M., Snyder, L.E., Blake, D.H., Lovas, F.J., Suenram, R.D. and Ulich, B.L. 1981, Ap. J., 251, p.541.
Hollis, J.M. and Rhodes, P.J. 1982, Ap. J. (Letters), 262, p.L1.

Jenkins, E.B. and Savage, B.D. 1974, Ap.J., 187, p.243.
*Johansson, L.E.B., Andertsson, C., Ellder, J., Friberg, P., Hjalmarson, A., Hoglund, B., Irvine, W.M., Olofsson, H. and Rydbeck, G. 1984, Astr. and Ap., 130, p.227.
*Johnson, D.R., Lovas, F.S., Gottlieb, C.A., Gottlieb, E.W., Litvak, M.M., Guélin, M. and Thaddeus, P. 1977, Ap. J., 218, p.370.
Klemperer, W. 1970, Nature, 227, p.1230.
Kraemer, W.P. and Diercksen, G.H.F. 1976, Ap. J. (Letters), 205, p.L97.
Lacy, J.H., Baas, F., Allamandola, L.J., Persson, S.E., McGregor, P.J., Lonsdale, C.J. Geballe, T.R. and van de Bult, C.E.P. 1984, Ap. J., 276, p.533.
*Lafont, S. 1982, These de troisieme cycle, Universite de Grenoble.
Laurent, C., Vidal-Madjar, A., York, D.G. 1979, Ap.J., 229, p.923.
Leger, A. 1983, Astr. and Ap., 123, p.271.
*Linke, R.A., Cummins, S.E., Green, S. and Thaddeus, P. 1981, Proc. Symp. on Neutral Clouds near HII Regions, Penticton.

*Linke, R.A., Guélin, M. and Langer, W.D. 1983, Ap. J. (Letters), 271, p.L85.
*MacLeod, J.M., Avery, L.W. and Broten, N.W. 1984, Ap.J. (Letters), Feb 2
Pagel, B.E.G. and Edmunds, M.G. 1981, Annual Rev. of Astr. and Ap., 19, p.77.
Peimbert, M. 1975, Annual Rev. of Astr. and Ap., 13, p.113.
Penzias, A.A. 1979, Ap. J., 228, p.430.
Penzias, A.A. 1981, Ap. J., 249, p.518.
Penzias, A.A. 1983, Ap. J., 275, p.916.
Phillips, T.G. and Knapp, G.R. 1980, BAAS, 12, p.440.
Reeves, H. 1974, Annual Rev. of Astr. and Ap.,12, p.437.
Ryter, C., Cesarsky, C. and Audouze, J. 1975, Ap. J., 198, p.103.
Savage, B.D. and Mathis, J.S. 1979, Annual Rev. of Astr. and Ap., 17, p.73.
Scoville, N.Z., Solomon, P.M. and Penzias, Astr. and Ap. 1975, ApJ., 201, p.352.
Scoville, N.Z. and Solomon, P.M. 1978, Ap. J. (Letters), 220, p.L103.
Scoville, N.Z., Kleinman, S.G., Hall, D.N.B. and Ridgway, S.T. 1983, Ap. J., 275, p.201.
Snow, T.P.Jr. 1980, IAU Symp. 87 Interstellar Molecules, ed. B. Andrews, Reidel
Spitzer, L.Jr. and Jenkins, E.B. 1975, Annual Rev. of Astr. and Ap.,13, p.133.
*Toelle, F., Ungerechts, H., Walmsley, C.M., Winnewisser, G. and Churchwell, E. 1981, Astr. and Ap., 95, p.143.
Thaddeus, P., Guélin, M. and Linke, R.A., 1981, Ap. J.

(Letters), 246, p.L41.
Thaddeus, P., Cummins, S.E. and Linke, R.A. 1984, Ap.J. (Letters) in press.
Thum, C., Mezger, P.G. and Pankonin, V. 1980, Astr. and Ap., 87, p.269.
Ulich, B.L., Hollis, J.M. and Snyder, L.E. 1977, Ap. J. (Letters), 217, p.L105.
Turner, B.E. and Thaddeus, P. 1977, Ap. J., 211, p.755.
*Ungerechts, H., Walmsley, C.M. and Winnewisser, G. 1980, Astr. and Ap., 88, p.259.
Unsold, A., and Baschek, B. 1983, "The New Cosmos", Springer-Verlag.
Vanden Bout, P.A., Loren, R.B., Snell, R.L. and Wootten, A. 1983, Ap.J.,271, p.161.
Vidal-Madjar, A., Laurent, C., Gry, C., Bruston, P., Ferlet, R. and York, D.G. 1983, Astr. and Ap., 120, p.58.
Wilson, S. and Green, S. 1977, Ap. J. (Letters), 212, p.L87.
Wilson, T.L., Ruf, K., Walmsley, C.M., Martin, R.N., Pauls, T.A. and Batrla, W. 1982, Astr. and Ap., 115, p.185.
Woods, R.C., Gudeman, C.S., Dickman, R.L., Goldsmith, P.F., Hugenin, G.R., Irvine, W.M., Hjalmarson, A., Nyman, L.A. and Olofsson, H. 1983, Ap.J., 270,p.583.

RADIO OBSERVATIONS OF INTERSTELLAR MOLECULES,
of their behaviour, and of their physics

O.E.H. Rydbeck and Å. Hjalmarson

Onsala Space Observatory, OSO,
Chalmers University of Technology
S-439 00   ONSALA, Sweden

ABSTRACT The radio observation of interstellar molecules has grown into an important, exciting, and challenging field since the OH radical (emitting in its Λ-doublet state at $\lambda \sim 18$ cm) was detected in 1963. Searches for OH, CH, and SiH had been proposed by Shklovsky (USSR), in 1948, and by Townes (at the Washington Conference on Radio Astronomy) in 1954. Unsuccessful OH searches were performed by Barret and Lilley in 1956. With a good telescope, the Millstone 25.6 m instrument, more advanced technology (including Weinreb's autocorrelator), and with more accurate main-line frequency data (obtained by Townes and associates at Columbia) the detection of OH finally became possible.

CH was detected much later, in 1973, at OSO. But this discovery required a maser pre-amplifier, and a pattern recognition method, because the signals were weak and the CH line frequencies not known (CH is a regular doublet with only one main-line; this triplet has never been observed in the laboratory).

Thus far SiH has not been detected, and its frequencies are not known. It would be at least as difficult to detect SiH as CH.

With the discoveries in 1968/69 of NH_3, H_2O (a strong maser in the detected state), and of formaldehyde, H_2CO (seen in absorption against the cosmic radiation), the field became even more exciting, but theoretical difficulties remain (as do the experimental). The hyperfine structure of the masing H_2O transition have not been observed, for example, nor is there a really satisfactory theory for the H_2O and OH masers (the latter does not always mase), for example in the HII-regions.

These discoveries were immensely stimulating and led to a very rapid development of the millimeter wave radio astronomy. SiO was found to be a strong maser at about 43 and 86 GHz, but it had to vibrate (which H_2O does not), and a CO, J=3-2 rotational transition ($\lambda \sim 0.87$ mm) was found at the entrance to the submillimeter wave range.

64 molecules (including the optically radiating H_2) have been detected to date (1984), and, of course, various isotopic variants. HDO was found in Orion A and, quite recently, also in the HII region W51M. With the linear molecule $HC_{11}N$, the number of "molecular atoms" has increased to 13. A complex molecule like methyl formate, $HCOOCH_3$ ("methylated" formic acid), for example, abounds in various transitions in Orion A, but its long sought for isomer, acetic acid, CH_3COOH, has not yet been found. Nor has the simplest of the amino acids, glycine, NH_2CH_2COOH (note: methylaneimine, H_2CNH, was detected in 1972) been observed.

CH is not the only interstellar molecule that is "terrestrially" unstable; HCO^+ and N_2H^+, are similar examples. Observations, therefore, of cold clouds can yield spectroscopic information that would be almost impossible to obtain in the laboratory.

One method of detecting such molecules, or completely unexpected ones for that matter, is to "scan" substantial portions of (hopefully by octaves in the future) the millimeter wave spectrum with a high sensitivity, high stability telescope. This has recently, after some efforts, been done at Onsala, towards Ori A and the evolved, carbon-rich star IRC+10216, through the range 72-91 GHz. Results of these scans will be presented. Quite generally it can be said that many new molecular transitions are found but no unexpected molecules and radicals, except perhaps for C_3H. Would improved (and very expensive) sensitivities of the order 0.01 K reveal any new, complicated molecules or are they, if present, locked onto grains? Or would the spectral density increase be ten-fold or more, and would organic molecules like urea, $CO(NH_2)_2$ rise above a confusion free detection treshold?

Even though the possibilities to detect new molecules are viewed with limited optimism, one has reason to be more optimistic about various types of radicals, for example those containing carbon chains - or fragments of them, which may be difficult to detect and accurately observe in the terrestrial laboratory.

Very long baseline interferometer (VLBI) observations of OH (18 cm) polarized maser hotspots were performed already in 1968 between OSO (using a rutile travelling wave maser as a preamplifier) and telescopes from east to west in the US. Diameters as small as 10 AU (i.e., of solar system size) and spot brightness as high as 10^{12} K were actually found. Hot spot magnetic fields of typically 5 m Gauss were observed and several years later confirmed by extra high resolution VLBI.

Extension of telescope frequencies to the 22 GHz range, controlled by hydrogen line masers, has made it possible to trace the proper motions of interstellar H_2O masers by VLBI methods, which for the first time, could yield rotation-model independent, direct distance determinations in the Galaxy. With future short millimeter wave arrays it will perhaps (using hydrogen frequency standards of *extreme* accuracy) be possible to study the structure also of (circumstellar) SiO masers and their peculiar transient properties.

Interstellar molecules can, with some advantage, also be used to trace gross molecular cloud properties of other galaxies. With the OSO 20 m millimeter wave telescope, using cooled diode mixers, CO (J=1-0, at about 115 GHz), and ^{13}CO, have been observed, and fairly well resolved, in remote galaxies, such as M51 (about 9 Mpc distant), and in M82 ($\sim$ 3 Mpc distant). This marks the beginning of a new, technically and observationally extremely difficult era.

The fascinating, and rapid development of the astrophysics of interstellar molecules is a good example of the statement, that new discoveries generally are made with (or require) new, advanced (and frequently more expensive) instruments. The difficult observations of extragalactic molecules is a typical confirmation of this assertion.

The following table of contents demonstrates how the authors have lived through, and with, the years of expanding knowledge of astro-molecular theory, of higher quality telescope building, and or relentless improvements of the quantum electronical devices.

CONTENTS

1. INTRODUCTION

2. THE EARLY DISCOVERIES

3. FURTHER RADIO OBSERVATIONS OF CN

4. OH - THE THIRD INTERSTELLAR RADICAL

5. WORK AT THE ONSALA SPACE OBSERVATORY (OSO) IN MOLECULAR RADIO ASTRONOMY, WITH MASER RADIOMETERS

6. OH – VERY LONG BASELINE INTERFEROMETRY (VLBI)
   Early results and later astrophysical interpretations

7. FURTHER OBSERVATIONS OF ROTATIONALLY EXCITED OH IN THE W3 REGION

8. MOLECULAR LINE INVERSION, ESPECIALLY OF OH, IN THE INTERSTELLAR MEDIUM

9. H_2CO – THE EARLY PROBE OF INTERSTELLAR MEDIUM PROPERTIES

10. H_2O – THE MOST POWERFUL MASER

11. HDO – THE ELUSIVE, IMPORTANT MOLECULE

12. SiO – THE THIRD INTENSE MASER, A CIRCUMSTELLAR RADIATION SOURCE

13. CH – THE WEAK UBIQUITOUS MASER

14. SOME SAMPLES OF OBSERVATIONAL RESULTS OBTAINED WITH MM TW MASER RADIOMETER SYSTEMS

15. ON SEARCHES FOR NEW INTERSTELLAR MOLECULES

16. THE ONSALA SPECTRAL SCAN OF ORION A AND IRC+10216
    a) New lines
    b) Line identifications
    c) New species detected
    d) Molecular excitation in different regions
    e) Chemical composition
    f) Isotopic abundance ratios
    g) The helium abundance
    h) Non-detections, and why?

17. LINE STATISTICS – TOWARDS THE CONFUSION LIMIT

18. MOLECULES IDENTIFIED IN DIFFERENT REGIONS

19. SURVEY OF MOLECULAR ABUNDANCES

20. EXTERNAL GALAXIES

21. SUMMARY AND FUTURE PROSPECTS

1. INTRODUCTION

Many molecules, organic and inorganic, have been discovered in the interstellar medium during the past years. Of the 64 molecules found, some organic species are quite complex, with as many as 13 atoms (1983). Some are chemically active radicals, with a short terrestrial life time, such as OH and CH. Of these molecules, 45 rare isotopic variants have been discovered to date. A few of the molecules have been found in the Magellanic clouds and in 50 other galaxies, all of them within a range of about 70 million ly. Nine of the 64 molecules have been detected in these remote objects, and two rare isotopic variants.

Since almost all of the molecules have been discovered by their radio emissions, in the microwave and millimeter wave regions, it is the advanced ultra-low-noise electronic detection schemes and, in many cases, also radio spectroscopy, in the laboratory as well as in space, that has made these discoveries at all possible. To substantially extend our molecular recording range in space, even more sophisticated detectors, approaching the uncertainty limitations set by quantum theory, are required, perhaps to be flown above our atmosphere.

Molecules play important roles in many astrophysical processes. Our Galaxy, for example, is believed to contain as much interstellar matter in molecular form, as in atomic, or ionized, plasma forms, probably amounting to some 10^9 M_o. Since this type of matter serves as "building materia" for stars, the giant molecular clouds of our Galaxy are its dominant star (formation) factories. The interstellar molecules, moreover, appear to be the main cooling agent of the denser interstellar clouds, efficiently converting their kinetic energy into electromagnetic radiation, and hence making gravitational collapse towards star formation at all possible. Most of these clouds contain regions, where a local thermodynamic equilibrium (LTE) does not exist. Intensities, shapes and shifts of the molecular spectral lines depend upon many local parameters. Therefore, the analysis of the molecular emissions has become a most important tool to determine the chemical and isotopic abundances, as well as the dynamics, the masses, densities, and temperatures of the clouds. This has lead to a rapid development of an essentially new scientific field, *molecular astrophysics*.

2. THE EARLY DISCOVERIES

In 1926, Eddington (1) recognized that the stationary Ca^+ (detected by J.F. Hartmann already in 1904; "die ruhenden Calcium Linien"), and Na absorption lines, observed towards distant stars, must be due to the presence of these atoms, in a very tenuous gas, in interstellar space. The further detections of other atoms, and of the first molecules, were made possible by the subsequent development of high-dispersion spectrographs, especially the Coudé spectrograph at Mount Wilson. In 1937 Dunham, and Adams (2,3) with this, high quality instrument observed three extremely narrow lines at 4300.31, 4232.58, and 3957.74 Å, which they could not identify. It fell to Swings and Rosenfeld that same year, in their communication *"Considerations regarding interstellar molecules"* (4), to identify the first one of these lines. It coincided with the $R_2(1/2)$ line of the well known CH band at 4315 Å. This line is the only one in the band, that connects with the lowest rotational level of CH, the $J=1/2$, $^2\pi_{1/2}$ state. The weak triplet *radio* lines of its ground state Λ-doublet, a regular one, at about 9 cm, were detected with the Onsala 25.6 m telescope 36 years later, in 1973 by Rydbeck et al. (5), using one of two different, ultra-low-noise travelling wave masers especially built for the purpose (note: the rest frequencies of these micro-wave lines have not, as yet, 1983, been verified in the laboratory, Rydbeck (6), although a number of higher frequency transitions have recently been measured). Our further CH-investigations, which also covered many giant molecular clouds, and dark clouds, will be dealt with somewhat more thoroughly in a later section.

Swings' and Rosenfeld's tentative identification of CH was safely confirmed, when three additional lines of it, in the 3870 Å band, in 1940 were predicted by McKellar (7) to arise from the lowest rotational state, and were thereafter observed by Adams in the spectrum of ζ Ophiuchi. His results appeared in a memorable paper, *Some results with a Coudé spectrograph of the Mount Wilson Observatory* (8).

The remaining lines, observed by Dunham, and Adams in 1937, and two additional ones, at 3745.33 and 3579.04 Å, found by Adams in 1940, were subsequently, in laboratory experiments, performed by Douglas and Herzberg 1941, 1942 (9,10) shown to be due to CH^+. Not unexpectedly, all three are R(0) lines, and come from the lowest rotational level of the ground state.

Furthermore, four lines of interstellar CN, belonging to 3883 Å band, were predicted by McKellar and observed, "in the blue", by Adams (*), at 3874.6 Å, R(0); 3874.0, R(1); 3875.8, P(1), and 3873.8, R(2), see Fig. 1.

(*) *Walter Sidney Adams was George Ellery Hale's oldest, and most trusted associate. He succeed him as director of the Mount Wilson Observatory. Adams was 64 years old when he detected the interstellar CN-lines.*

On the basis of the line intensity ratios McKellar, in a now

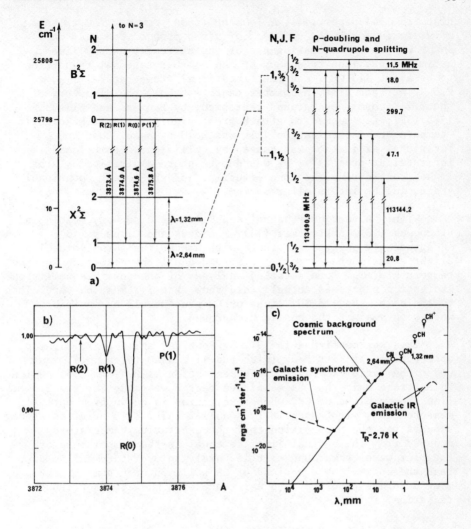

Figure 1: The CN radical, one of the most important astrophysical molecules and one of the most extensively studied spectroscopic species.
a) Energy levels of CN, indicating the transitions that correspond to the optical absorption lines.
b) Optical spectrum of CN, discovered by W.S. Adams at Mt. Wilson, 1940, against the reddened star ζ Ophiuci, $\sim 10^6$ years old, well known for its strong, visible interstellar absorption lines.
c) Spectrum of the microwave background radiation, indicating rotational temperatures of the N=1-0 & 2-1 transitions.

historic 1941 paper (11), *Molecular lines from the lowest state of diatomic molecules composed of atoms probably present in interstellar space*, making use of the extremely narrow lines, estimated the line frequency of the N=1-0 rotational transition to be about 113 GHz, and the "rotational" temperature, T_{10}, to 2.3 K. At that time, a quarter century before the discovery of the 2.7 K cosmic background radiation, by Penzias and Wilson, 1965 (12), few seemed to attach great significance to McKellar's "temperature determination". It was held by some, that the rotational excitation was caused by collisions ("some local interstellar process"), or was due to photo-excitation by starlight, considering the long lifetime, of the lowest rotational state, in the interstellar medium.

Within a year of Penzias' and Wilson's discovery, Field and Hitchcock (13), Shklovskii (14), Thaddeus and Clauser (15), independently pointed out that, within McKellar's (unavoidable) slight experimental error, T_{10} should equal the cosmic background temperature, T_R. More detailed analyses of the possible collisional excitation processes actually did show that T_{10} must be very close to the same. The 113 GHz lines of CN thus should be invisible, when seen against the cosmic background.

This was confirmed in 1972 by Penzias' and Jeffert's famous observations at Kitt Peak (with the 11 m NRAO telescope), reported under the title *Interstellar CN excitation at 2.64 mm* (16), which in Abstract said that "A sensitive search was made for the 2.64 mm line from a cloud of CN, whose excitation is known from optical measurements, with essentially a null result. This provides strong support for the proposition that the excitation temperature deduced from the optical lines is equal to the temperature of the microwave background". It is hardly an overstatement to say, that the optical observations of CN still furnish some of the most accurate proofs of the existence of a mm-wave background radiation.

When this had been written, and Fig. 1 had been finished, there appeared in January, 1984, a remarkable paper, by Meyer and Jura (17a) on "the microwave background temperature at 2.64 and 1.32 millimeters", in which the authors present extremely high signal-to-noise observations of the 3874 Å band of interstellar CN towards ζ Oph. These observations were performed with the Lick Observatory 1872 element Reticon photodiode detector on the Coudé spectrograph, of the 3 m telescope. The author's resulting spectrum is the most sensitive observation of interstellar CN to date, with a signal-to-noise ratio of about 2000; a most remarkable improvement since the days of W.S. Adams. The authors' R(2) observations are also the most accurate to date (see Fig. 1 in (17a)).

The $R(1)/R(0)$ line strength ratio yields an excitation temperature (T_{10}) of 2.83 ± 0.04 K for the $J=1 \to 0$ transition at 2.64 mm. Since the $R(0)$ line is saturated to some extent, this value must, as the authors point out, be an upper limit on T_{10}. Using the linewidth observed towards ζ Oph, by Hegyi et al. 1972 (17b), for CN, the authors were able to correct for the saturation of the $R(0)$ line and ultimately obtained $T_{10} = 2.73 \pm 0.04$ K. Furthermore the $R(2)/R(1)$ ratio corresponds to an excitation temperature (T_{21}) of 2.8 ± 0.3 K for the $J=2 \to 1$ transition at 1.32 mm, "right on top of the black body maximum", see Fig. 1.

The presence of local CN excitations towards ζ Oph should, however, increase T_{10} and T_{21} slightly above T_R. The observed values thus are upper limits. It was pointed out by Thaddeus in 1972 (18a), that collisions with electrons is the only, local process likely to influence the CN excitation towards ζ Oph. Recent estimates, however, of the electron density, in the interstellar ζ Oph cloud, yield a collisional contribution, T_c, to $T_{01} < 0.1$ K. Therefore, the authors' determination of $T_{01} = 2.73 \pm 0.04$ K is consistent with a T_R-value of 2.7 K at $\lambda = 2.64$ mm. This is close to the long-wavelength measurements, $T_R = 2.74 \pm 0.087$ K, as reviewed 1980 by Weiss (18b). A most interesting fact, since an eventual departure of the cosmic microwave background from a Planck spectrum would be very important for our understanding of the physics of the early Universe.

Finally it should be mentioned that earlier Fabry-Perot type observations of CH, and CH^+, whose $R(1)$ lines end at the rotational levels $\lambda \sim 0.56; 0.36$ mm, led to higher temperature limits and uncertainties, as provisionally indicated in Fig. 1. Bortolot and Thaddeus, in 1969, observing with the Coudé spectrograph at the Lick 120-inch telescope, towards ζ Oph, found a new interstellar line at 4232.08 Å (by adding twenty-five spectra on a computer), which they attributed to $^{13}CH^+$. For an equivalent linewidth of 0.68 mÅ, they estimated a $^{12}C/^{13}C$ abundance ratio = 82(+55, -15), suggesting that the ^{13}C-rich carbon stars contributed little to the interstellar material in the direction of ζ Oph.

The ρ-doubling of the N=1-level of CN, which cannot be resolved in the optical spectra, experiences hyperfine, nitrogen quadrupole splittings as shown in Fig. 1. By means of optically determined rotation and ρ-doubling constants did it become possible to determine the main radio line's ($|1, 3/2, 5/2> \to |0, 1/2, 3/2>$) frequency with acceptable, preliminary accuracy. This line was detected, 1970 at Kitt Peak, in emission towards Orion A, at a tentative frequency of 113490 MHz. The antenna temperature was 1.5 K, i.e., considerably higher than the normal value for CH at Onsala, whose radio emission was to be discovered three years later (at a much lower frequency). At that time it had not been possible to determine the quadrupole splittings by radio spectrographic methods in the laboratory on account of the short (terrestrial) lifetime of the CN-radical, a case almost analogous to that of CH.

3. FURTHER RADIO OBSERVATIONS OF CN

CN:s main line radio frequency, upon which the hyperfine transitions had to be based, was determined in a round-about type of way, much as in the case of CH, viz., by the use of the line rest frequency for ^{13}CO, 110.2 GHz (i.e. not far from the CN lines), which had been determined with great accuracy in the laboratory. Assuming that the CN and CO molecules were well mixed in the cloud, then they would have the same Doppler-shifts, and one simply had to measure the direct frequency difference between their respective lines. This led in 1974 to a CN main line frequency of 113491 ± 0.2 MHz, with hyperfine components as shown in Fig. 1, with errors of ± 500 kHz. This would correspond to about ± 15 kHz for the CH transitions, or five times the frequency uncertainty estimated by us.

Once the CN frequencies were determined, one could, through an analysis of the system-Hamiltonian, determine the radicals rotational and ρ-doubling "constants", and the quadrupole moment with considerable accuracy. This demonstrates, as was later found to be the case for CH, that the interstellar medium is an excellent "laboratory" for the instrumentally well equipped radio "astrophysicist". This was, moreover, amplified and proved by the recent detection (1982) of five lines, belonging to the N=2-1 group, at the Owens Valley Observatory (OVRO), by Wootten et al. (19), using the number one 10.4 m OVRO mm wave telescope ($\sim$ 30" resolution). This time, however, it was possible to pre-compute the line frequencies, determining the spin-rotation, the magnetic fine structure, and the quadrupole coupling constants from the CN(N=1-0) laboratory spectrum measured by Dixon, and Woods 1977 (20), and with the use of their rotation and centrifugal distortion constants. This led to *17 lines in all*, covering a range from 226886.27 to 226287.51 MHz, i.e., 598.76 MHz wide. The main line, $|2, 5/2, 7/2> \rightarrow |1, 3/2, 5/2>$ has a frequency of 226874.78 MHz. With a relative intensity of 200 for the same, the two nearest "satellites" $|2, 5/2, 5/2> \rightarrow |1, 3/2, 3/2> = 226974.29$ MHz, $|2, 3/2, 5/2> \rightarrow |1, 1/2, 3/2> = 226659.46$ MHz get the relative intensities 126, and 125. In this difficult experiment Dixon and Woods relied heavily on the knowledge of the "astronomical" frequencies, just as in their earlier high precision measurements to secure identifications of the astronomically detected species HCO^+, HNC and HN_2^+.

Detected were, besides the main line, the nearest upper

satellite, the highest frequency transition (strength 24), and the following ones, $|2, 5/2, 5/2\rangle \rightarrow |1, 3/2, 5/2\rangle$ = 226892.94 (strength 24), $|2, 5/2, 3/2\rangle \rightarrow |1, 3/2, 1/2\rangle$ = 226876.05 MHz (strength 75). (*)

(*) *The N=2-1; 3-2; 4-3 transitions of CN, in its ground as well as lower vibrational states, very recently have been measured by Skatrud et al.: 1983, J. Mol. Spectrosc. 99, 35.*

These safe detections were done towards OMC-1 (Orion Molecular Cloud One), centered on the so called BN/KL-region with an antenna temperature, T_A of about 8 K, towards OMC-2, and the molecular envelope of the carbon rich, late type star IRC+10216, $T_A \sim 2$ K.

Since one has not seen any infrared vibration-rotation absorption lines, and since the CN(N=2-1) envelope of IRC+10216 is observed to be comparatively extended, it is suggested that the source of CN radicals is located in the outer, circumstellar shell. Analysis of the chemical, expansion and photodissociation time scales in the outer shell seems to indicate, that CN is produced through photodissociation of HCN by the interstellar radiation field. Such a process would explain the very large observed CN abundance compared to that predicted by chemical equilibrium models of the shell, and should confirm the importance of photochemistry in such circumstellar shells.

The N=2-1 lines, just discussed were detected already in December 1979 with an InSb hot-electron bolometer mixer receiver, operating at the Cassegrain focus, yielding a 350 K single sideband system noise temperature, at an instantaneous bandwidth of 1 MHz. It should be possible to detect also several of the lower satellites; this has perhaps been done at our time of writing (with advanced low noise, multi-channel receivers).

That the (cold) interstellar medium really is a remarkable laboratory has also been shown by the discoveries, at Onsala Space Observatory (OSO), of CH, by the detection of some hyperfine components in the ortho formaldehyde $CH_2(O)$ transition $1_{10} \rightarrow 1_{11}$, at 4829.64 MHz, and, of several others in the $NH_3(1,1)$ para transition (at about 23.69 GHz). Later in the text, we will return to these characteristic examples of what can really be done and found by high quality low noise, electronic detection schemes.

4. OH - THE THIRD INTERSTELLAR RADICAL

Interests in the important OH radical and its bands are quite old, see Jack, 1927 (21). Oldenberg of Harvard, 1935 (22), was one of the early investigators. Studying OH concentrations

in flames, he noticed that "in emission the rotational-intensity
distribution may be quite abnormal under certain conditions";
see also Dieke and Crosswhite, 1962 (22).

Shklovskii in 1949 (23) expressed interest in the radio
emissions of the OH (which, in contrast to CH, and CN had not
been detected optically in interstellar space) and CH radicals,
estimating the wavelength of the former λ_{OH}, to 9.52 cm (3.15
GHz; not much less than the lowest satellite line of the CH
ground state doublet $^2\pi_{1/2}$, J=1/2). In a later Liège communication (24), Shklovskii recalls, that he 1948 pointed out that "in
the radio spectrum in the Galaxy, along with the hydrogen line
λ = 21 cm, the presence of a number of lines, caused by the
interstellar molecules may be expected. These lines must occur
at the permitted transitions between the components of Λ doubling
of the main rotational state of the OH, CH, and other molecules".
Using the formulae of Van Vleck, 1929 (25), Mulliken and Christy,
1931 (26), he computed and obtained for OH, λ_{OH} = 18.3 cm (J=3/2),
for CH, λ_{CH} = 9.45 cm (J=1/2), values not too far from the actual
ones, and for SiH (which has not yet been detected), λ_{SiH} = 12.5
cm (J=1/2). Of these three radicals, CH & SiH are regular ones,
for which $^2\pi_{1/2}$, J=1/2, is the ground state.

Shklovskii (in the Liège communication) prophetically added:
"Of special significance will be the attempt to discover these
lines in some dark nebulae, where a considerable concentration
of interstellar molecules may be expected. Other radio-lines
might also be expected, in particular the lines caused by the
isotopes of the OH, CH, and SiH molecules, which open important
prospects for the investigation of the interstellar diffuse
medium". OH had not been observed optically at the time, "probably
explained by the small value of the oscillator strength for the
resonance transition of that molecule", and it would take many
years until Snow (27) in 1975 with the *Copernicus* telescope-
spectrometer, at about 1222 Å, reported a possible detection of
OH towards o Per (note: previously identified were CO, CN, CH,
and CH$^+$), and, in 1981, Storey et al. (28), recorded the $^2\pi_{3/2}$,
J=5/2 - 3/2 rotational transition, at about 120 μm (to which
we will return in a later section).

At the Washington Conference on Radio Astronomy 1954, it
was said that "finally Townes' calculations give strong support
to Shklovsky's suggestion[10] that we should search for evidence
of OH in the vicinity of 1665 Mc/sec". ([10] is our ref. 23).
Already in 1952, Sanders et al. (Townes' group), (30) had
searched for the J=9/2, $^2\pi_{3/2}$, ΔF = 0 transitions of OH (23826.90,
& 23818.16 MHz), which 27 years later were observed in absorption
against W3(OH), by Winnberg et al., using the Bonn 100 m telescope (31). In August 1955 (32), Townes reported to the IAU-
Symposium at Jodrell Bank, that "there is a limited number of

radio frequency resonance for which immediate search with radio telescopes appears justified. Of these, the Λ-doubling of OH, the hyperfine structure of H^2 and perhaps of the He^3 II are outstanding. A number of additional resonance can probably be found if their frequencies can first be accurately determined in the laboratory. These include the Λ-doubling of CH and the hyperfine structure of ionized N^{14}". Townes gave the OH ground state, $J=3/2$, $\Delta F = 0$, main lines the preliminary frequencies 1667.0, and 1665 MHz. For CH, Townes estimated $\sim$ 1 GHz, for $^2\pi_{1/2}$, $J=1/2$ (the ground state), as well as for $^2\pi_{3/2}$, $J=3/2$ (the first, rotationally excited state).

In 1956, Barrett & Lilley (32), used the 50 ft telescope at NRL with the "comparison type" radiometer built by Hagen et al. in 1955 (converted from 21 to 18 cm wavelength) to search for OH. "A very thorough search was devoted to the spectrum near 1667 Mc/sec with Cassiopeia A in the beam. The entire interval covered in the Cassiopeia A search was approximately 25 Mc/sec, centered about 1667 Mc/sec. No evidence was found for either emission or absorption features in the Cassiopeia A direction", Barrett and Lilley noted. Earlier in their paper, however, they wrote that "in this experiment a value of 20^0 K is chosen as the minimum detectable temperature change, although in some portions of the frequency search a somewhat lower value might be adopted". Remembering, that Ehrenstein, Townes, and Stevenson in 1959 (34) observed the OH ground state main lines ($\Delta F = 0$), in the laboratory with remarkable accuracy, viz., $F=2-2 = 1667.34 \pm 0.03$; $F=1-1 = 1665.46 \pm 0.10$ MHz, Barrett and Lilley thus were observing in the right frequency range, but, unfortunately towards Cas A, where OH is seen in moderately strong absorption. But a strong OH maser emission could, in principle, have been detected (and probably much surprised them). Ehrenstein et al. wrote: "One of the molecules whose presence in interstellar space may be detectable by means of its radio-frequency spectrum is the OH radical. Attempts at observing it with radio telescopes have been unsuccessful thus far. To make future searches more fruitful, frequencies of the appropriate absorption lines were measured in the laboratory, with a Zeeman-modulated spectrometer".

In October, 1963, there appeared, a now historic paper, *Radio Observations of OH in the Interstellar Medium*, by Weinreb et al. (35). Using the 84 ft parabolic Millstone Hill antenna (a D.S. Kennedy dish), and a spectral line, autocorrelation radiometer [described by Weinreb in his Ph.D.-Thesis, *A Digital Spectral Analysis Technique and its Application to Radio Astronomy* - originally built for the purpose of detecting the deuterium spin-flip line at about 310 MHz, which corresponds to the 21 cm line of hydrogen - Technical Report No. 412, Research Laboratory of Electronics, MIT, August 30, 1963], Weinreb and colleagues detected the two ground state, OH main lines ($F=2 \rightarrow 2$, ~ 1667 MHz;

F=1 → 1, ∿ 1665 MHz) in absorption, against the radio source
Cas A; their historic 1667 MHz spectrum is depicted in Fig. 2a.
The initial frequency errors were remarkably small, about 20 kHz
for the 2-2 line, and about 60 kHz for the other, whose LTE
intensity is 5/9 of the formers. This discovery truly opened up
a new, fruitful and immensely important era of research, radio
molecular astrophysics. Weinreb's technology also marked an
important step forward in the techniques of observational radio
astronomy. Bigger correlators were built, for larger and larger
bandwidths with oversampling, the originator remembering J.H.
Vleck's classical work, 1943, on *"The spectrum of clipped noise"*
(36). Modern integrated circuit technology makes high sampling
rates possible, through small system pulse delays.

With Weinreb's work, digital techniques and digital recordings made their appearance on the radio astronomical scene,
probably never to disappear. Modern very long baseline interferometry (VLBI), across continents and oceans, would not have been
possible without this technique and good atomic clocks.

In 1964, Barrett & Rogers (37), interested in the possibilities of detecting OH lines, other than the main transitions, in
the interstellar medium, presented a remarkably accurate computation of the transition frequencies for the $^2\pi_{3/2}$ rotational
ladder, yielding 1720.56 MHz (F=$2^+ \to 1^-$); 1667.357 ($2^+ \to 2^-$);
1665.402 ($1^+ \to 1^-$), and 1612.20 ($1^+ \to 2^-$), which was of great help
to all radio astronomers concerned. More accurate frequency determinations were late in coming. But, in 1972, Ter Meulen and
Dymanus (38) made amazingly accurate, OH measurements, in the
lowest rotational state, and found: 1720.52998; 1667.35903;
1665.40184; 1612.23101 MHz. Such great precision (5σ ∿ 100 Hz)
is possible because of the very narrow lines that molecular beam
methods allow (∿ 2.5 kHz widths). Barrett's & Roger's errors in
the main lines (to be mentioned for comparison) were as small as
2 kHz, or less. 500 Hz corresponds to a linewidth of about 0.1
km/sec, i.e., for a very cool cloud with very little turbulence.

In October 2, 1965, Weaver et al. (39) reported *"observations
of a strong unidentified microwave line ("mysterium") and of
emission from the OH molecule"* in the direction of the W3(OH)
source (more detailed properties of this very interesting OH
region, will be discussed in a later section). "Our observations",
the Weaver group wrote, concerned about the strong deviation
from the LTE line ratios, 1:9:5:1 (counted from 1720 to 1612),
"indicate that 'mysterium' is found in strong HII regions. We
found no 'mysterium' in the Rosette Nebula, M8, M16, or M17".
What they had detected was the OH, 1665 maser line; see Fig. 2b.
Weaver et al., using the Hat Creek (northern California) 85 ft
antenna, employed a tunable paramp, with a bandwidth of about
20 MHz, followed by a 100 channel (contiguous, 10 or 2 kHz wide)

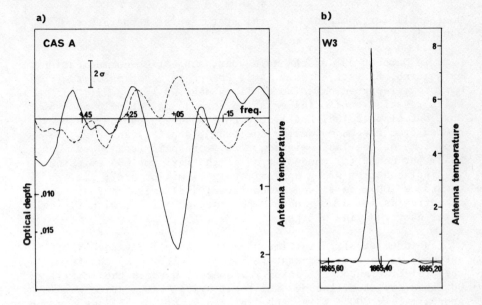

Figure 2, a): The *first radio observation* of an interstellar molecule, the OH radical. OH ground state, main line, $^2\pi_{3/2}$ (3/2), F=$2^+ \to 2^-$, ~1667 MHz ($\lambda \sim 18$ cm) discovered by S. Weinreb, et al. (USA) in October, 1963, seen in absorption against the strong radio source Cassiopeia A, ~ 3 kpc distant, believed to be a Type II supernova remnant. A novel correlator-radiometer was used in the 84 ft Millstone Hill antenna, Westford, Mass. The discovery marked the beginning of a new era in the natural sciences, molecular astrophysics. The solid line shows 8000 sec of data with the antenna beam directed towards Cas A, the dashed line 6000 sec of data with the beam pointed slightly off source.

Figure 2, b): The first observation of an interstellar maser line. OH ground state, main line, $^2\pi_{3/2}$ (3/2), F=$1^+ \to 1^-$, ~1665 MHz, detected by H. Weaver et al. in 1965 (in the direction of the HII region W3), with a tunable parametric amplifier in the 85 ft Hat Creek antenna, in California. The discovery of the narrow OH maser line, originally called "mysterium", is one of the greatest in the history of molecular astrophysics. A serendipitous discovery, followed by two other ones, viz., of the H_2O and SiO masers.

post amplifier, obtained a total system noise temperature of about 130 K.

In October 30, of the same year, another OH-communication, by Weinreb, Barrett, et al. (40), *Observations of polarized OH emission*, reported emission of lines near W3 (i.e., W3(OH)), which revealed that: "(a) emission is present not only at 1665 Mcs, but also at 1667 and 1720 with frequency spacings as predicted from the OH molecular spectrum"; see (37) and Fig. 12; "(b) some of the 1665 emission is linearly polarized by as much as 37 per cent; (c) the position of the maximum 1665 emission is displaced from W3 by approximately 14 min of arc"; i.e., towards W3(OH), as it was named later; "(d) emission features as narrow as 1-1.5 kc/s have been observed"; i.e., with typical maser narrowed line widths.

Weinreb et al. this time used the (Neroc) Haystack 120 ft antenna, and a room temperature paramp, followed by the former's spectral-line autocorrelation radiometer, used in the initial OH discovery at Millstone. The total system temperature was approximately 200 K. "As yet", Weinreb et al. wrote, "we have not observed other regions. In our opinion 'mysterium' is anomalously excited OH".

The authors furthermore added, that "a large anomaly exist in the intensity ratios of the lines", that "the 1612 Mc/s line was not detected during our limited search", and that "the existence of strongly polarized OH emission is both surprising and difficult to explain. Possible polarization mechanisms are: (1) Zeeman effect; (2) Stark effect; (3) resonance scattering; (4) amplification, through a maser-type population inversion, of polarized background radiation". To our knowledge, this was the first time that the maser-concept appeared in radio astronomy; it rapidly became unanimously accepted. (A few months later C.H. Townes, N.G. Basov, and A.M. Prokhorov shared the physics Nobel Prize for their maser-laser work). The authors finally noted that "the Zeeman effect is unique among the four mechanisms listed above because it is the only mechanism which will give rise to circular polarization. Therefore, observations of this region with circular polarization are extremely important". This was very stimulating news for everyone interested in molecular radio astronomy (astrophysics).

5. WORK AT THE ONSALA SPACE OBSERVATORY (OSO) IN MOLECULAR RADIO ASTRONOMY, WITH MASER RADIOMETERS

The 84 ft (25.6 m) Onsala telescope, using a D.S. Kennedy dish, with punched aluminium plate panels, erected in 1963, was designed with one basic purpose in mind, viz., to detect OH, CH, and H_2CO (hence the punched plate surface, and not the mesh that had been used earlier - "good for 21 cm use") with the new Onsala travelling wave masers, which would make a dish of modest size highly competitive. The senior author (OR) witnessed Townes', Giordmaine's and Mayer's early work (1958) with an experimental ruby cavity maser, in the NRL 50 ft "gun-mount telescope", recording planetary emissions (at about 10 GHz) with amazing stability, and became convinced that a travelling wave maser (with short gain-length, for example with rutile as active material) would be the best, most sensitive preamplifiers for the Onsala telescopes. In the early communication (41) *Rutile traveling-wave maser system for the Onsala 84-foot radio telescope*, O.E.H. Rydbeck and E. Kollberg (1968) reported on the basic system, frequency range 1.3 - 3.4 GHz (to include also CH). The initial total (zenith) system noise, in the ground state OH band, became $\leqslant 30$ K, a value that ultimately could be lowered. Subsequently, tw-masers were built for the 5 GHz band (primarily for the H_2CO ground state, $1_{10} - 1_{11}$, and for the OH excited state $^2\pi_{1/2}(1/2)$), for the 6 GHz band (primarily for OH, $^2\pi_{3/2}(5/2)$), and thereafter two different masers for CH, whose $^2\pi_{1/2}(1/2)$ ground state was not well known (a rebuilt version of one of them has later been used at Parkes, Australia, for CH observations in the southern hemisphere, cf., (42). Later, technically different rutile tw-masers were built for the 66 ft (20 m), mm wave telescope, covering several bands in the range 22 - 35 GHz. Likewise, the 20 m telescope became very competitive in these frequency bands. For the higher frequency ranges, however, Schottky barrier and SIS-element mixers are used. It is intrinsically difficult to build good tw-masers for frequency levels as high as 100 GHz.

Outstanding OH ground state spectra were obtained with the new maser radiometer in 1967 (41). Meanwhile OSO was grant bound to let the Scandinavian Telesatellite Committee use the 25.6 m telescope (in cooperation with OSO) for extensive multi channel telephone experiments, which were carried out during the period 1964-67. These experiments guided the Scandinavian countries to establish a common ground station (in Sweden), instead of building vulnerable microwave, multi channel links to the mid-European ground stations. Gradually, however, it became possible for OSO to use the telescope full time for radio astronomy work.

In due course, the 5 and 6 GHz masers, using liquid helium cryostats for one week's uninterrupted operation, were installed in the telescope. Fig. 3 a and 3 b, shows early low noise recordings, obtained towards W3(OH) during the period 1968-69, of interstellar OH-excited states, $^2\pi_{3/2}(5/2)$ (6035 MHz, F=$3^- \to 3^+$; 6030; F=6031, F=$2^- \to 2^+$, and $^2\pi_{1/2}(1/2)$ 4765, F=$1^- \to 0^+$); see also Fig. 12 for the energy levels and their parities. The three lines were all seen in maser emission, Rydbeck et al., 1970, 1972 (43,44).

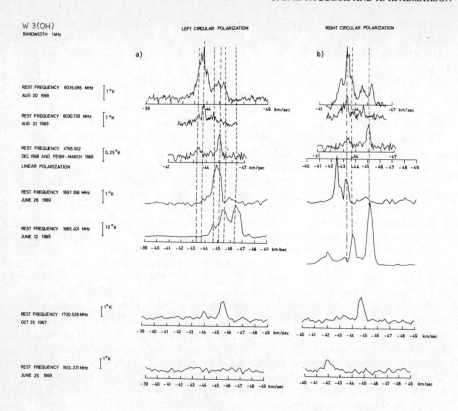

Figure 3, a) & b): Early low noise Onsala Space Observatory (OSO) recordings of emissions from rotationally excited OH states, $^2\pi_{3/2}(5/2)$, $F=3^- \to 3^+$, ~ 6035 MHz, $F=2^- \to 2^+$, ~ 6031 MHz, and $^2\pi_{1/2}(1/2)$, $F=1^- \to 0^+$, ~ 4765 MHz, in the direction W3(OH). Observations of the left and right circularly polarized 6 GHz emissions, in W3(OH) as well as in W75N, indicate the presence, across the emitting regions, of 5-6 milli Gauss magnetic fields. As expected, the $^2\pi_{1/2}$ - state showed no signs of polarized emission. For the reader's additional information, and comparison, complete OH ground state spectra, at 10 kHz resolution, have been included in the figures.

The 6035, 6030 features were roughly homologously redshifted for right circular polarization, (RCP), and blueshifted for LCP; the mean shifts being about 0.35, and 0.50 km/s. This corresponds to about equal longitudinal magnetic fields for both transitions of about 6.3 milli Gauss, directed into the source (Rydbeck et al., 1970). No circularly polarized feed was built for the $^2\pi_{1/2}$, J=1/2 state, at the time, because its g factor is very small (in fact equal to zero for a pure $^2\pi_{1/2}$ state, Hund's case b)). On a later occasion, using a circular polarizer for this state, no CP was detected.

Fig. 4 shows details of the 6 GHz center-features, indicating, see b), that several maser components with fields as strong as 7-8 mG, are contained in the main envelope. Of the ground state spectra, the $2^+ \to 1^-$ transition (the 1720 MHz line) displays the simplest Zeeman pattern, corresponding to about 6.6 mG (Rydbeck, et al. 1970, (43,44)).

The Zeeman splitting of the OH, $^2\pi_{3/2}$, F=2 → 1 transition produces three triplet sets, corresponding to Δm_F = 1, 0, -1. This would make it difficult to separate the RCP and LCP lines from the rest of the Zeeman pattern, unless its absence (a partial depolarization) could be explained. For a saturated maser, about 10^{15} cm long, with a magnetic field of ~ 5 mG, a free electron density ~ 5 cm^{-3} (not difficult to obtain in the vicinity of HII regions) is needed for a sufficient depolarization, as Lo et al. pointed out in 1975 (45). In a saturated maser, a trapped far IR radiation redistributes the upper sublevel populations, see Figs. 12 & 13, and suppresses the slow-rate transitions, if crossrelaxations are greater than stimulated emissions. Furthermore, if the angle between the magnetic field and the direction of the maser radiation is less than about sixty degrees, the π-components will be more or less suppressed, and the σ-components, with the largest transition rate, will be fully circularly polarized. The emerging Zeeman pattern thus will be composed of the m_F = 2 → 1, and -2 → -1 lines, which corresponds to 0.65 kHz, or 0.11 km/s, per mG. For the 2 → 1 transition, seen towards NGC6334A, Rydbeck et al. 1970 (44) deduced a field value of 6.8 mG, i.e., about the same as for W3(OH), but now directed *out of the source*.

Fig. 5a and 5b displays another set of OH records, obtained with a maser radiometer by OSO, towards W75B (now N) in December, 1969. It is an unusual source, since it exhibits temporal variations, of the time scale one year, in its 1667 MHz emissions. These variations are in part anticorrelated with those of the 6035 emissions (also found towards W75B by OSO), Rydbeck et al. 1970 (44). In this remarkable source, Elldér 1971 (45) detected a RCP 1667 MHz maser-flare (duration a couple of months), but at a velocity as high as 27 km/s. A similar maser flare appeared in

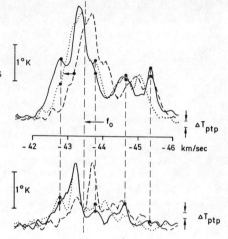
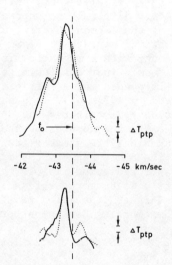

Figure 4: Early details of 6035 & 6030 GHz spectra, indicating a fairly homogeneous Zeeman splitting within and across the multi-spot main line features.

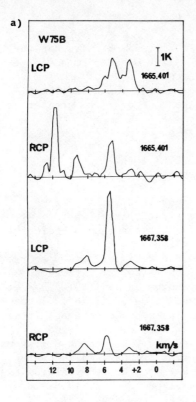
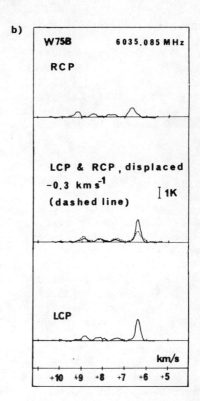

Figure 5, a): First OH ground state spectra, obtained towards the W75B region (now called W75N), discovered by OSO as an OH radio source, in 1969.
b): OH, 6035 MHz spectra, right and left circularly polarized (RCP & LCP), obtained towards W75B, at 3 kHz resolution. The Zeeman splitting, about 5.4 milli Gauss, for the four spectral components, indicates a fairly homogenous magnetic field across the masing region. As in the case of W3(OH), the magnetic field (as seen from the earth) is directed into the source. In the case of NGC6334A the magnetic field is directed out of the source, which in this case may have an accreting disk of opposite symmetry (opposite bi-polar type flow) to that of the W3(OH) source and its associated HII region (see Fig. 7b).

April-May 1972, at +28.4 km/s.

Two interesting facts emerge from Fig. 5b. First, that the Zeeman-type splitting for the 6035 MHz line was 0.3 km/s (corresponding to ~ 5.4 mG); second, four features (probably coming from different regions) had about the same shifts, which indicates that the magnetic field might be homogeneous over the greater part of the emitting region.

6. OH - VERY LONG BASELINE INTERFEROMETRY (VLBI)
Early results and later astrophysical interpretations

That the OH maser condensations - or "hot spots" - were small indeed, was demonstrated by the first transatlantic, Very Long Baseline OH Interferometry (VLBI), between the following three radio telescopes in the USA, viz., Haystack (120 ft), Westford, Mass., NRAO (140 ft), Green Bank, W. Va., and Hat Creek (85 ft), California, and with Onsala, OSO, (84 ft) in Sweden, on Jan. 29, 1968, Moran et al. (47a), Burke et al. (48). The four telescopes simultaneously observed W3(OH), in the $1 \to 1$ transition (1665.4 MHz) at RCP & LCP, forming a phase coherent interferometer, with six baselines. The data were digitized and recorded at each site on magnetic tape, and the local oscillators were locked to atomic frequency standards (today usually hydrogen masers). All possible pairs of tapes were correlated yielding the fringe visibility at the projected baselines of 4.7, 19.5, 22.4, 30.6, 34.6, and 42.6×10^6 (OSO - Hat Creek) wavelengths (at 18 cm). Previous to these measurements, it was believed that the only unresolved feature in W3(OH) was the "hot spot" at about -43.7 km/s, which was estimated to be smaller than 0.005 sec. of arc. Fig. 6a shows the single antenna spectrum, in RCP, and the fringe amplitude for the Haystack - NRAO baseline. The -43.7 km/s "hot spot" appears totally resolved, whereas at least two maser spots are hidden in the strong -45.1 km/s feature. Fig. 6b depicts the visibility as a function of baseline for the -43.7 km/s feature. The projected distance from OSO to Hat Creek, 7.700 km, about 43 million wavelengths, is barely sufficient to resolve the same.

It is also of historical interest, in the present context, to mention that almost simultaneously with the molecular measurements, VLBI observations in the continuum (at 1670, and 5010 MHz) were performed between NRAO (140 ft) and OSO. The baselines were 35, and 105 million wavelengths long, K. Kellerman et al., O. Rydbeck et al., M.H. Cohen, and D.L. Jauncey (49). These continuum observations did show, for the first time, that a number of radio galaxies and quasi-stellar sources contain intense components, with angular dimensions of the order of 0".001 or less.

Fig. 6c depicts a 1665 W3(OH) map based on the early US

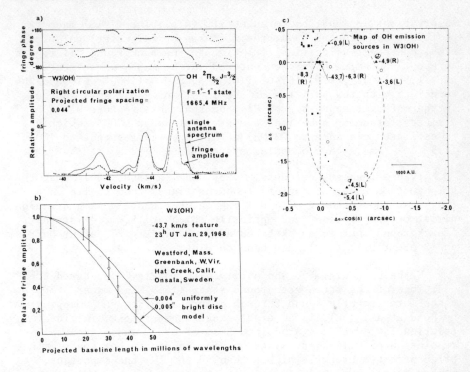

Figure 6, a): Single antenna OH, $F=1^+ \to 1^-$ (1665.4 MHz), W3(OH) maser spectrum, fringe amplitude and fringe phase for the Haystack (Mass.) - NRAO (W. Virginia), Very Long Baseline Interferometer (VLBI) leg, January, 1968.
b): Relative fringe amplitude, at right circular polarization, as function of base line length, in millions of wavelengths, for the first, transatlantic, OH-VLBI experiment (transition $F=1^+ \to 1^-$), Jan., 1968. Atomic clocks were used at all stations and the data were digitized. The distance Hat Creek - Onsala was needed to resolve the diameter of the smallest maser hot spot (at -43.7 km/s), about 10 AU. This was the first time that a maser spot (actually its most brilliant center region) was shown to be of solar system size.
c): W3(OH), "hot spot" map based on the early US observation, in 1968, at 1665.4 MHz (Moran et al., (47b). The 1968 maser positions are denoted by ▲, later ones (determined by the Moran group in 1969) have been entered by circles (for brightness temperatures $\gtrsim 6 \times 10^{11}$ K), otherwise by circular spots. 6035 GHz ($F=3^- \to 3^+$) maser positions, determined by the same group have been marked by ■, in rough proportion to the brightness temperature.

observations in 1968, by Moran et al. (47b). The (1968) maser
positions are denoted by ▲, their senses of (circular) polarization, and their velocities, relative to the background HII region,
at about -50 km/s, have been entered in the maps, whose center,
at α (1950) = 02^h, 23^m, 16.46^s ($\pm 0.01^s$), and δ (1950) = $61°38'57''.64$,
coincides with the position of the important -43.7 km/s source.
Furthermore, it appears from Fig. 6b that this feature has a
diameter (assuming spherical symmetry) of about $0''.004$, which at
an estimated distance to W3(OH) of 2.2 kpc, corresponds to 10 AU
(1.5×10^{14} cm), i.e., of solar system size.

It has been suggested (e.g., by Cook in 1968, (50)) that the
brightest OH maser components (condensations) tend to lie in a
ring near the boundary of a HII region. Fortunately such a region,
almost of proper size, and optically thick at 18 cm, exists as
a background source, as sketched at 5 GHz in Fig. 7, after Harris,
and Scott (1976, (51)), and denoted HIIA.

We have in Fig. 6c, and partly also in Fig. 7, entered the
positions of the strongest ground state (1665 MHz) maser components more recently found through a more accurate VLBI mapping
by Reid et al. (52), in an eight station experiment, in linear
polarization, with a synthesized beam of $0''.01$, and a spectral
resolution of 0.14 km/s. Rings in Fig. 6c denote hot spots with
brightness temperatures higher than 6×10^{11} K; the important,
weaker ones are marked by black spots. It is, in this context,
interesting to notice, how one could take the view, that the
1665 MHz maser spots in Fig. 6c were lying on an ellipse, perhaps
a projected circle, whose normal angle to the line of sight is
$\leqslant 60°$. The circle diameter (or major axis), $\leqslant 2.4''$, roughly equals
8×10^{16} cm. It describes a $120°$ cone, with its apex located at
the (stellar) center of HIIA, about as shown in Fig. 7, very
approximately 2.3×10^{16} cm from the maser circle plane.

For gravitational, free fall collapse, of the remnant accreting material, containing the OH condensations, about as outlined
by Reid et al. (52), the conic angle should be $2 \times \tan^{-1}(1/\sqrt{2}) \sim$
$\sim 70°$, ($= 2 \times \theta m$) instead of $120°$. To obtain such a sharp cone,
the distance to the central star would have to be increased to
5.6×10^{16} cm (which might be possible within the experimental
errors). The interpretation of the maser circle's (or the shell's)
velocity structure is made difficult on account of the considerable, main Zeeman shifts, 0.59 km/s per milli Gauss for the 1665
MHz transition. Moreover, it is difficult safely to identify any
ground state Zeeman pairs on account of the inherent gain competition between the LCP & RCP maser modes at the very high amplification in question.

Since the electron temperature of the HIIA region is of the
order 10^4 K, and maser-spot brightness temperatures of 10^{12} K are

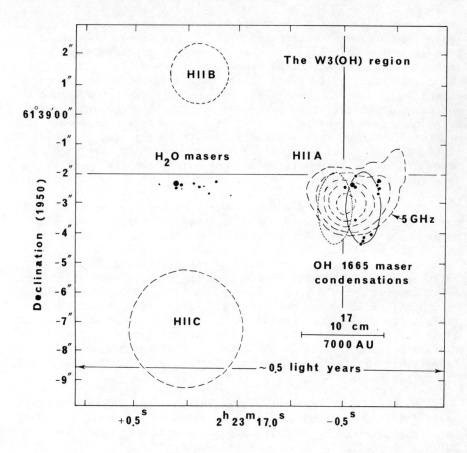

Figure 7: 5 GHz continuum contour map of the W3(OH) HIIA region. The OH ground state maser spots appear to lie along the periphery of an (obliquely inclined) accreting disk. A point-drawn mirror disk, perhaps located behind the HIIA region (as a bi-polar symmetry partner), could contain (invisible) masers. Note the characteristic property, of regions like W3(OH) and NGC7538 (see Fig. 19), viz. seemingly isolated islands of H_2O masers.

quite common (about 1.5×10^{12} K for the historic -43.7 km/s feature), OH maser amplifications of 10^8 are frequently required. This situation must hold for most OH masers associated with HII regions (i.e., the so called Type I, or *interstellar masers*). Indeed, as Reid et al. write (52), the reason these masers are always found associated with HII regions, may be that a bright source of input radiation is needed for the maser to achieve a high brightness temperature ($> 10^{10}$ K). In contrast, it should be mentioned, that OH masers associated with long-period variable stars (i.e., Type II, or *stellar masers*) generally have much lower brightness temperatures and are mainly amplifying their own spontaneous emission.

For the unsaturated W3(OH) maser sections, radiation along ray paths, that emanate from the HIIA region, must be several orders of magnitude brighter than those which do not intersect the same. Therefore, a typical, initial maser emission will be "beamed" into a solid angle equal to that subtended by the emitting part of the HIIA region from the maser condensation. This "beaming" should remain fairly constant until the maser saturates, and many ray parcels begin to complete for amplification, i.e., to induce OH transitions. Maximum gain probably occurs along rays with small line of sight velocity gradients. As the radiation grows it will induce almost all of the available OH transitions and thereby suppresses further growth along diverging, nearby ray paths.

The brightness temperature, at which a homogeneous, spherical maser (as the OH condensations are assumed to be) would saturate, is simply obtained by equating the stimulated emission rate to one half of the collisional decay rate, Γ. For $\Gamma \sim 0.03$ sec^{-1}, mainly caused by collisions of OH with H_2 - in this case at a temperature of say 100 K and a density of 10^8 cm^{-3}, the saturation begins at a maser depth, τ_s, of about 10, at which level the brightness temperature becomes $\sim 10^4 \times \exp(\tau_s) \simeq 2 \times 10^8$. Therefore the gain of the masers *after saturation* must be comparable to the unsaturated gain in order to generate the intense 10^{12} maser spots.

Integrating the equation of transfer through the saturated regime, one can show, assuming an unsaturated population inversion, ΔN_o, of 0.1 cm^{-3}, that the brightness temperature is 10^{12} K of a 0".005 hot spot (2.2 kpc distant), and that the molecular hydrogen density, N_{H_2}, equals 10^8 cm^{-3}, that an *amplification length* of (very roughly) 2×10^{15} cm $\simeq$ 130 AU really is required (that the intense spot size is less than one tenth of this value is not surprising, however). It is interesting also to note, that this amplification length is close to the size of maser condensations, and to coherence lengths, for a gravitationally collapsing, accreting envelope deduced by Reid et al. (52). Furthermore, it should be added, that the estimated maximum diameter, 2".4, of the maser disk corresponds to about 8×10^{16} cm, or to 40 approximate maser lengths. Finally, since the collisional rate, Γ, is proportional to $\Delta N_o \times N_{H_2}$, a determination of for example N_{H_2} would yield the OH $(1^+ \to 1^-)^2$ population inversion.

The physical conditions of OH masers have recently been discussed and reviewed by Guilloteau (53). The general conclusions are that the kinetic temperature (not surprisingly) must be > 60 K, that the mean density should be 2×10^7 cm^{-3} within a factor 3, and that the OH abundance should be at least 10^{-5}. Guilloteau's conditions must, of course, vary somewhat between the maser groups. The kinetic temperature is likely to be higher, at least up to 150 K, in regions which form the strong excited state masers. The strongest 6035 GHz masers, VLBI-mapped by Moran et al. across W3(OH) in 1978 (54), and marked by squares on the W3(OH) map, Fig. 6c, are found in the north east corner of the same, generally outside the 1665 MHz disk, and probably closer to the HIIA region. On the other hand, 1665 MHz masers can rather easily be inverted at lower kinetic temperatures, ~ 60 K, in the presence of strong IR-radiation.

The fact that one mainly observed OH maser condensations in front of, or at the edge of the HIIA region, does not rule out the presence of OH, perhaps also in clumps, in the general molecular, CO-rich cloud (velocity range -41 to -52 km/sec) through which HII might be moving. Recently Guilloteau et al.(55), using the Very Large Array (VLA), mapped the $NH_3(2,2)$ (a para state) inversion line, 23723 MHz, which does not differ much in frequency from the OH $^2\pi_{3/2}$ (9/2) main lines, $5^- \to 5^+ = 23827$ MHz, and $4^- \to 4^+ = 23817$ MHz, earlier found by Winnberg et al. (56), *in absorption* against the ultracompact (nucleus) of HIIA. Their spatial resolution was 0''.26 ($\sim 8 \times 10^{15}$ cm, or about 4 OH ground state amplification lengths).

It is most interesting that the observations of Guilloteau et al. led to the suggestion, that the absorbing medium should be an accreting disk, inclined about 45° to the line of sight (for the OH 1665 MHz disk we found - or estimated - roughly 60°), in which the masers are "embedded". The temperature in the disk is estimated to be about 60 K, and the density to 10^7 cm^{-3} (cf. 10^8 cm^{-3} used by Reid et al. in estimating the maser length), which corresponds to a mass of about 1 M_o. Overall, the NH_3 and OH velocities seem to agree.

The situation appears to be somewhat similar to that of W75N, where Haschick et al. 1981 (57), after 1665 MHz OH VLBI-mapping found an edge-on-like disk, traced by OH masers, suggesting that such disks represent a normal region of star formation. The general morphology of the W75N area, with three compact HII regions, and a large molecular cloud (rich in HCN, for example),

with the OH masers on edge in HIIB, and the H_2O masers located
slightly to the east, but north of a 53 µm IR source (see Fig. 10
in (57)), reminds us very much of the W3(OH) region. The lack of
strong 10, and 20 µm IR emissions, near the OH and H_2O masers,
might indicate a higher extinction and a high dust density in
their positions. Lo et al. (58) also suggested a "scenario" for
star formation, in which the H_2O masers, with no adjacent HII
regions, are indicators of the *earliest* stage of stellar evolution. It is held that OH masers, with associated ultracompact HII
regions, represent a slightly later stage, and finally that HII
regions, possibly with near IR emission, represent even later
ones. Thus, in W75N there probably are stars in the earlier
stages of formation, indicated by the H_2O and OH masers, in
regions of highest gas and dust density, while the older, more
evolved stars are located at the edge of the molecular cloud.

Radio continuum maps of the ultracompact HII regions, at
higher frequencies, indicate a complex structure, as shown in
Fig. 8, basically from Guilloteau et al. (55), of the 23.7 GHz
W3(OH) HIIA continuum structure with superimposed $NH_3(2,2)$
absorption depths, at -43.2 km/s, and a number of important,
1665 MHz maser positions. As emphasized by Guilloteau et al.,
the velocities of OH masers projected beyond the "edge" of HII
might imply a slight rotation of the disk, on the order of 1.5
km/s, a value close to the maximum allowed for the disk to remain
stable against fragmentation (cf., Ostriker, and Bodenheimer,
1973 (61)). Also Haschick et al. (57) maintain that the velocity
gradient, and the alignment of the OH maser features, suggests
that these masers are contained in a rotating disk of accreting
material. Because collapsing, rotating disks are unstable to the
formation of a binary system, also a W3(OH) model might favour
a very young one. A rear 1665 MHz maser disk, as shown dashed
in Fig. 7, is also something one would have to consider in this
context, and eventually look for, in the form of intense, blue
shifted (with respect to HIIA) outer edge, east-background
masers.

When the present manuscript had been written, there appeared
a most interesting preprint, *Detection of a New Type of Methanol
Maser*, by T.L. Wilson, C.M. Walmsley, L.E. Snyder & P.R. Jewell
(Max-Planck-Institut, Bonn, Preprint No. 178, March 1984). The
authors report discovery of emission and absorption at 23121 MHz,
attributed to the $9_2 - 10_1$ A^+ transition of methanol (CH_3OH).
The emission lines come from W3(OH), Orion-KL, and NGC7538-IRS1
(see its 4.8 GHz map in Fig. 19), whereas absorption was found
toward the compact HII region in W31. It is interesting *in the
present context* to note that the emission from W3(OH) is caused
by maser amplication of the background continuum source, whereas
the Orion-KL results appear to be consistent with a LTE population. The emission from NGC7538 is probably of maser type. The

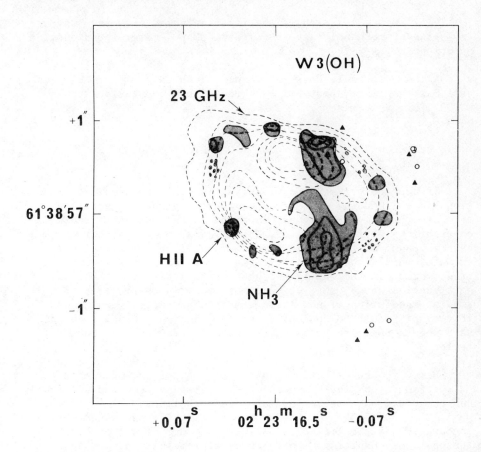

Figure 8: 23.7 GHz contour maps of the W3(OH) HIIA region, with superimposed NH_3 (2,2) absorption depths at -43.27 km/s (after Guilloteau et al. (55) and some 1665 MHz OH maser spots for comparison.

$9_2 - 10_1$ A$^+$ maser emission is the first detected from the A symmetry state of methanol, and the first methanol maser found outside of Orion-KL.

Since a molecule does not require much inversion to be seen as a maser against a hot HII background, one should look toward W3(OH) for other masers.

7. FURTHER OBSERVATIONS OF ROTATIONALLY EXCITED OH IN THE W3 REGION

It appears from the previous section that physical conditions in the neutral gas surrounding HII regions, like HIIA, are rather poorly understood, in spite of the extensive radio observations performed so far. This is due to several circumstances. Many of the atomic and molecular line emissions, for example, which cool the neutral gas, *just beyond* the ionization front, occur in the far IR and have, with a few exceptions, see next section No. 8, not yet been observed. Furthermore, poor angular resolution of non-masing, molecular radio lines makes it very difficult to spatially resolve the (compressed) neutral gas, unless they are seen against the background of compact HII regions.

However, the detection, in absorption, of the OH, J=9/2, $^2\pi_{3/2}$ transitions (by Winnberg et al., already referred to, (56)), and the detailed observations of the J=7/2 lines, $4^+ \to 4^-$ (13441 MHz), $3^+ \to 3^-$ (13435), by Baudry et al. (1981), have created encouraging possibilities to use OH as an efficient "probe" of the physical conditions in the immediate vicinity of compact HII regions. An interesting part of the spectra of Baudry et al., is produced in Fig. 9, showing both masing and (a weak) absorption in the $4^+ \to 4^-$ mode. The results were obtained with the Effelsberg 100 m telescope, using a two-stage, uncooled parametric amplifier, which gave a system temperature of about 240 K. The strongest J=7/2, and 9/2, main line features were observed at about -42.5, -45.1 (cf., Fig. 6a). Whilst the maximum antenna temperature of the $4^+ \to 4^-$ maser was about 17 K, and 0.34 K for the adjacent, absorption feature, the line temperatures of the J=9/2, $5^- \to 5^+$, and $4^- \to 4^+$ modes, only seen in absorption, were -0.13, and -0.11K. Thus, even for the 100 m telescope, an extremely good maser amplifier would be needed to detect the (desireable) J=11/2, 37 GHz main lines.

Excited OH observations of the types just described sample the gas in front of the HII region, and thus do not suffer from the spatial resolution-lack of many other molecular observations. At the same time, the analysis is not quite simple, since OH frequently tends to have non-LTE population distribution. These

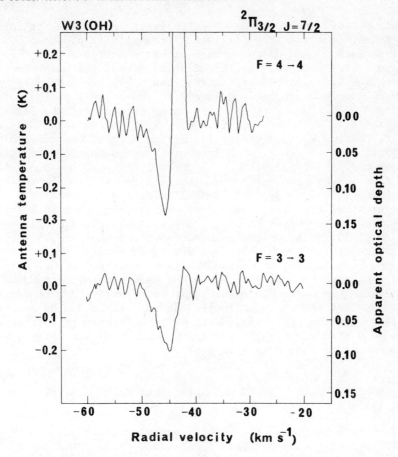

Figure 9: An interesting, and characteristic recording of the OH, $^2\Pi_{3/2}$, J=7/2 main lines (at about 13.4 GHz), towards W3(OH), obtained by Baudry et al. in 1981 (62). Clipping the peak of the F=4 - 4 main line feature, an adjacent absorption line appears, at about the same radial velocity as the F=3 - 3 absorption. For J=9/2, ~23.8 GHz, both main lines, F=5 - 5, and 4 - 4, detected by Winnberg et al. in 1978 (56), are seen in absorption. Main line inversion is no longer possible (see also Fig. 13).

highly excited states of OH decay to lower levels via far infrared transitions (see Fig. 12) with probabilities of about 0.5 to 1 per second and hence require either densities of $\gtrsim 10^8$ cm^{-3}, or/and a very intense, far IR radiation field. The existence of absorption lines in OH excited states now appear to be very common in molecular clouds adjacent to compact HII regions. The 6035 and 6030 lines were observed in absorption for the first time in 1973, by Rydbeck et al. (63), with the Onsala 6 GHz maser, towards W3C1, see Fig. 10, during a joint experimental project, with the Univ. of Massachusetts, and the Neroc Haystack Observatory (equipped with a ruby travelling wave maser - probably the first experiment performed with a maser at 22 GHz), to look for faint H_2O, $6_{16} - 5_{23}$ (22.23 GHz) emissions (weak masers) and also eventual traces of the H_2O-hyperfine components {Kukolich, 1969, (64)}, Yngvesson et al., Rydbeck et al., 1975, (65), see also Fig. 22. No hyperfine components were found. Fig. 9 shows what sensitivity can be obtained with good masers (the maximum antenna temperature, of the 6030 line in absorption, was only -0.025 K), and how the 6035, 6030 MHz maser lines, from W3(OH), were picked up on the antenna side lobe. Finally, one finds from Fig. 10, that the ratio of the 6030/6035 optical depths generally is less than the LTE ratio, 7/10. It is difficult to determine with good accuracy, how it varies with the radial velocity, and from source to source (HII regions).

8. MOLECULAR LINE INVERSION, ESPECIALLY OF OH, IN THE INTERSTELLAR MEDIUM

In 1976, Bertojo et al. (66) suggested, that OH, and CH could be population inverted, in the interstellar medium, by collisions with H, H_2 or He. Such collisions, considering the anisotropy of the OH- (para) H_2 interactions, have recently been analyzed by Dewangan, and Flower (67). Para-H_2 has J=0 in its ground rotational state. It is also indicated that collisions with ortho-H_2, in its rotational ground state, will be qualitatively similar. The author's results seem to confirm that rotational excitation by H_2, followed by efficient radiative decay, can lead to population inversion in the OH ground state Λ-doublet.

Collisional transitions within the doublets, however, seem to thermalize the population, at a rate that remains large even at low temperatures, where the rotational excitation rate coefficients are decreasing rather rapidly. Since the rates of these processes are comparable for $T \sim 60$ K, transitions within the doublet tend to dominate at lower temperatures. Therefore, if collisional excitation by para-H_2 is the main process leading to population inversion, kinetic temperatures in the maser regions must be relatively high.

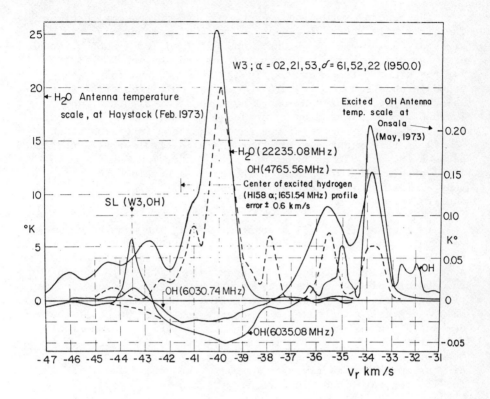

Figure 10: The *first observation* of OH, $^2\pi_{3/2}$, J=5/2 seen in *absorption*, by Rydbeck et al. in 1973 (63). The 6 GHz spectra were obtained as by-products of the authors' search, at Haystack (using a ruby 22 GHz travelling wave maser, designed by K.S. Yngvesson), for hyperfine splittings of weak H_2O ($6_{16} - 5_{23}$) masers. Note that masing, 6 GHz main lines, from W3(OH), are picked up by the antenna, and that these lines are much more narrow than the absorption lines, recorded in the main beam; as illuminating comparison.

The general situation, neglecting hyperfine complications, is fairly well described even by the simpler rotational excitations of CS (J=1→0, ∿ 49 GHz), and SiO (J=1→0, ∿43 GHz), as demonstrated in Fig. 11 (Varshalovich, and Khersonskii, 1978 (68)) for the kinetic temperatures (T_k) 50 & 100 K, and for a (background) radiation temperature T_B ∿ 2.7 K. Collision cross sections, calculated by "the strong coupling method" of S. Green, and P. Thaddeus, approximated by T. de Jong, S. Chu, and A. Dalgarno, were used in the analysis. One notes that the inversion range ($\delta > 1$) is fairly narrow, and that there is little real difference between CS, and SiO as functions of N_{H_2}. For low densities the level populations are close to the Boltzmann values for a T_x (the excitation temperature) = T_B, and for $T_x = T_k$, when the density is high.

Furthermore, Fig. 11 demonstrates that the population inversion of the first adjacent levels makes the system quantitatively behave like a *three level maser*. Hyperfine splittings, in a different kind of molecule (for example HCN), would not change the overall picture much, but, perhaps, generate a modest population redistribution among the hyperfine levels (in the J=1 state). When δ begins to increase, with decreasing N_{H_2}, the collisional de-excitation becomes smaller than the radiative, which finally dominates, and $T_x \rightarrow T_B$. The inversion, or pumping of OH is, however, more complex than that of the simpler diatomics, CS & SiS, as appears directly from Fig. 12, the energy levels of OH, for its two rotational ladders, $^2\pi_{3/2}$, and $^2\pi_{1/2}$. $^2\pi_{3/2}$(3/2) is the lowest OH state, and not $^2\pi_{1/2}$(1/2) as for CH, which is a regular doublet. The OH radicals entire, dramatic interstellar behaviour appears to be due to its irregularity.

That the complicated, and asymmetric IR-transitions of OH make it difficult to analyze the optical (or IR) pumping, and possible final inversion, is evident. We have, to somewhat facilitate its understanding, in Fig. 13 plotted the relative IR absorption coefficients for OH (based on the tables of Brown et al., 1982, (69)), for an assumed (high) effective temperature ∿ 500 K, which takes turbulence of the gas into account. According to observations v_{therm} ∿ 0.2 km/s, and v_{turb} ∿ 0.7 km/s.

A very simple example basically illustrates what could happen, for example by IR pumping within the $^2\pi_{3/2}$ ladder only. Absorption of far IR photons (λ ∿ 120 μm) through line (transitions) 9, see Fig. 12, can transfer molecules from level 2^+ to level 2^-. These can decay radiatively to level 1^+ as well as to level 2^+, i.e., lines 2 (1665 MHz) and 3 (1612 MHz) tend to be inverted. Continuum photons will, however, also excite the "image mechanism", viz. by absorption through line 6, with a net transfer from 2^- to 1^-, reinforcing the 1612 inversion and weakening, or even destroying the 1665 inversion. This effect would, of course,

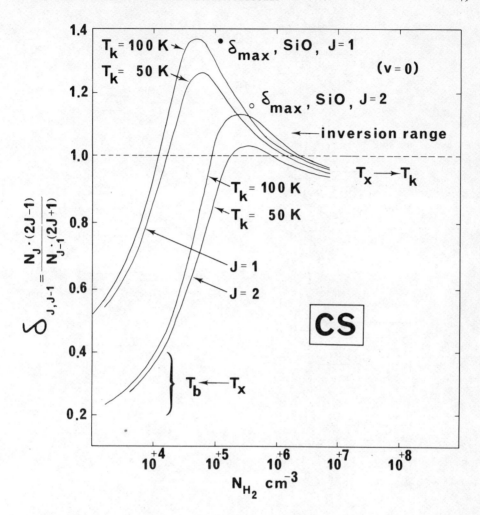

Figure 11: Collisional inversion of CS (and of the much similar SiO) as function of the particle density. The occurrence of inverted population levels can be qualitatively explained as follows. During collisions the system is excited not only to the adjacent level, $\Delta J = 1$, but also to overlying levels, $\Delta J \geqslant 2$. The total probability of all the $\Delta J \geqslant 2$ collisional transitions must be sufficiently high (see for example Varshalovich & Kersonskii, 1978, (68)); but the system can (at the same time) quite well be understood in terms of a three-level maser. Non-equilibrium collisional pumping can therefore lead to population inversion in astronomical masers. In principle, the detection of a population of any pair of rotational levels, *for a linear molecule*, can serve as a rather sensitive indicator of the $N(\delta_{max})$ density of the medium.

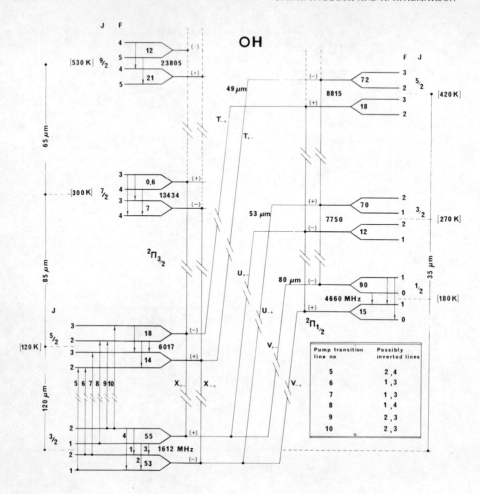

Figure 12: Energy level structure for OH, an irregular Λ-doublet. Routes of probable far-IR pump transitions are shown in the Figure, as well as a table of possibly inverted ground state lines.

hardly occur if the exciting photons, absorbed in line 9, came in a properly Doppler shifted 10 line (see Fig. 13), from an approaching OH pump cloud. In principle, therefore, a net inversion at 1665, or 1612 may be produced. One can also construct basic situations in which the 1667, or the 1720 MHz would be inverted.

It should be emphasized in this context, that the hyperfine structure of the X_{+-}, and X_{-+} components, of the 120 μm transitions in Fig. 13, are almost identical, i.e. the "image mechanisms" would nearly cancel out.

One of the more efficient mechanisms to invert the OH main line masers would be to *make use of the $\Delta J = 0$ transitions*, at 49 μm, and 53 μm in Fig. 13. They are the only lines that connect the upper and lower Λ-doublet levels and show remarkable *IR line overlaps* already at moderate temperatures.

Let us first look at the 53 μm transition. Assume that level 2^+ in the $^2\pi_{1/2}(3/2)$ state has an initial overpopulation. A strong transfer of photons from line 23 (now from $2^+ \to 2^-$, see Fig. 13) to line 24 (direction $1^+ \to 1^-$) would pump the molecules from 1 to 2. For intermediate OH column densities, this mechanism would lead to an inversion of the 1665 MHz line, but perhaps also to an annihillation of the 1667 MHz line.

A similar, but principally even more pronounced effect, could be due to the 49 μm transition, be means of lines 32 and 31. This might, to first order, increase the eventual inversion of a 6030 GHz transition.

The enhanced asymmetries induced in the infrared rates by overlaps of the far IR components, were proposed several years ago (1969) by Litvak (70) as sources of main line inversions, but the complexity of the OH level structure (and of the involved radiative transfers) for years prevented the development of more detailed models based on his suggestions.

More recently, however, interesting work has been done on models for the pumping of OH main line masers in HII/OH regions by the effects of overlaps, due to differential Doppler effects, in different parts of an OH region (see for example Lucas, 1980, (71)).

In order to simplify the complicated analysis somewhat, but still maintain a physically realistic picture, we may consider a quiescent maser cloud, submitted to the far IR OH lines emitted by a large, dense and warm external OH pumping cloud, of a given radial velocity, V_p, (with respect to the maser cloud) lying within the velocity range shown in Fig. 13. If one looks at the transitions (absorption coefficients) 25, 24, 14, 15, & 5 on the one hand and 22, 23, 16, 7, & 6 on the other, and considers the Doppler shifts of the external pump lines, one gets, to first order, the impression that the 1612 MHz line could be inverted (fairly uniformly), when V_p lies in the $-1 \to -4.5$ km/s region, and for the 1720 MHz in the $+1 \to +4.5$ km/s (collapse) region. There could hardly be any uniform correlation with the 1665 & 1667 MHz lines, whose upper levels would be exposed to more irregular and variable population "streams". These general, first order assessments, of an (in reality) extremely complicated overlapping scheme, seem to agree with numerical results obtained by Guilloteau et al. (1981, (72) - Fig. 13 a-c). The authors assume a pumping OH cloud, with $T_p \sim 150$ K (the rotational excitation temperature), $N_{OHP} \sim 10^{17}$ cm^{-2}, a linewidth due to thermal broadening at 200 K, and a maser cloud with $T_k = 200$, $n_{H_2} \sim 10^6$ cm^{-3}, and $N_{OH} \sim 10^{16}$ cm^{-2}. An important point is, that N_{OHP} must be large (as indicated) in order to generate sufficient amounts of pump photons, streaming into the quiescent cloud.

It appears from what has just been said, that the radiative excitation of OH certainly is complicated by the fact that far-IR continuum radiation from warm dust clouds (whose grain absorption efficiency may vary as λ^{-2} in the 35-120 μm range), associated with the OH, could cause preferential transitions between the various IR rotation levels. Direct, and extended observations of the same would facilitate the understanding of the pumping mechanism of many OH maser lines.

Since high densities ($n_{H_2} \sim 10^9$ cm^{-3}) are required to thermalize the rotational transitions of OH, it will be difficult to detect far-IR emission lines from quiescent molecular clouds. Assuming an average OH/H$_2$ ratio of 10^{-7}, one finds that the dust clouds, even for n_{H_2} as low as (say) 10^6 cm^{-3}, may be sufficiently thick, in the far-IR, to mask the OH line emissions. Furthermore, typical molecular cloud temperatures are somewhat low to populate even the J=5/2 state (E $\sim$ 120 K, see Fig. 12) well enough. Moreover, the fractional abundance of OH (not to mention CH) decreases in very high density regions. IR-OH will therefore, as Storey, Watson et al. point out (1981, (73)), be most apparent in the diffuse, outer regions of clouds, where it would show up in absorption against the central, hot, background component.

The passage of a shock wave, on the other hand, through a molecular cloud is exactly the type of process that could throw the molecular gas out of equilibrium with the dust, to such an extent, that the far IR, OH emission lines would be detectable. It is interesting to note, (see Watson, Storey et al. (1980, (74)), that high-J-CO lines (J=21 $\to$ 20, 124.2 μm; and 22 $\to$ 21, 118.6 μm) have been detected from such regions.

It was not surprising that Storey et al., using their tandem Fabry-Perot spectrometer (with a Ge:Sb photoconductor, a 24 Hz chopper and a 40" field of view) in NASA's Kuiper Airborne Observatory at 12.5 km altitude, detected the (important) $^2\pi_{3/2}$ (J=5$^-$/2 → 3$^+$/2, 119.233 μm; J=5$^+$/2 → 3$^-$/2, 119.331 μm) OH far-IR transitions (see also Figs. 12 and 13), towards Sgr B2. The historic spectrum, depicted in Fig. 14, represents eight 30 sec scans, divided by a scan of the calibration background. The expected $^2\pi_{3/2}$, J=7/2 - 5/2 transition of CH was not seen, however.

The limited spectral resolution, probably unavoidable in these author's demanding experiments, did not make it possible to resolve the structure of the OH absorption lines (half power widths about 250 km/s), which appeared to be centered at about 0 → 25 km/s, and, finally, to trace them to the OH ground state (18 cm) Sgr B2-cloud velocities, +62 km/s, and -90 km/s (the weaker absorption). It is significant, in this context to point out, that while the 18 cm lines are seen in absorption against the free-free continuum of the HII regions, the far-IR, OH lines should be seen against thermal emission from the dust.

Spectra were also taken (by Storey et al.) in the direction of Orion, towards the Kleinmann-Low Nebula, but "no significant" emission was detected. The authors point out that, given the strength of the J=5/2 → 3/2 emission, it is possible to predict an upper limit for the intensities of the J=5/2 Λ-doublet transition, at about 6 GHz, assuming that all lines are optically thin. Even in the most favourable case, the OH 6 GHz antenna temperature would hardly be more than about 1 mK. A direct comparison (or correlation) with the intensities of the 6 GHz OH "hot spots", recorded with ground based radio telescopes, appears to be impossible at present. But one thing is evident: The Townes' group has initiated work in a new and important research area, IR-radio-molecular astrophysics. It is well worth remembering that C.H. Townes gave his first talk on OH at the Washington Conference on Radio Astronomy - 1954, recommending a search for the molecule in interstellar space.

Developments have been tremendous, and overwhelming, after the discovery of the OH interstellar masers about 20 years ago. Storey et al. write, for example, that the hydroxyl radical "has become one of the best studied denizens of molecular clouds". In his book, *Microwave Spectroscopy of Free Radicals*, Academic Press, 1974, Alan Carrington writes in the same spirit (on p. 191) that "OH certainly qualifies as the free radical most studied in the laboratory and is a good example of the way in which a molecular wave function can be probed by high resolution spectroscopy", and he adds that "OH is certainly keeping many (radio) telescopes and even more brains quite busy".

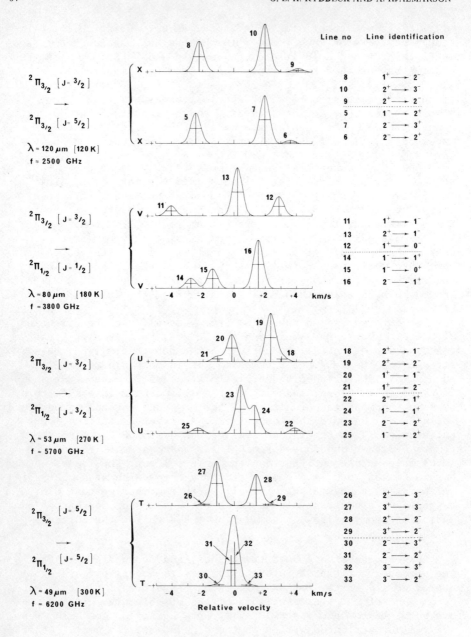

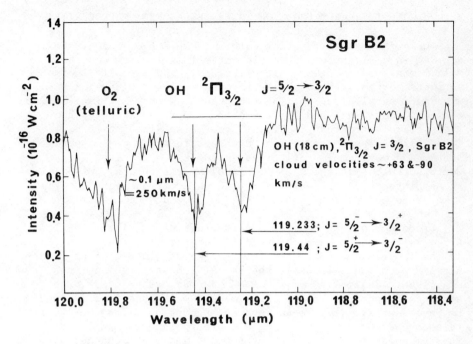

Figure 14: Historic, far-IR, OH absorption spectrum, obtained by Storey, Watson & Townes, 1981, (73), with a tandem Fabry-Perot spectrometer in the 91.4 cm telescope of NASA's Kuiper Airborne Observatory.

←

Figure 13: The hyperfine absorption line structure (relative intensities in LTE) of the far-IR transitions, X_{+-} to $T_{\pm\mp}$ (see also Fig. 12), linking the OH ground state to other rotational states of the $^2\pi_{3/2}$, and $^2\pi_{1/2}$ ladders. The horizontal scale is drawn in relative, radial, velocity, km/s. The linewidths are, for the purpose of demonstration, shown for an efficient temperature, $T_{eff} \sim 500$ K, which takes into account turbulent motion of the gas in the sub-source. According to observations $V_{therm} \sim 0.2$ km/s, and $V_{turb} \sim 0.7$ km/s (see for comparison also V.V. Burdyuzha and D.A. Varshalovich, Sov. Astron., 17, No. 3, Nov.-Dec. 1973).

9. H$_2$CO - THE EARLY PROBE OF INTERSTELLAR MEDIUM PROPERTIES

Like the faster rotating water molecule H$_2$O, H$_2$CO (formaldehyde) is a slightly asymmetric rotor (small difference between main moments of inertia). Also like H$_2$O, H$_2$CO is split in two variants, ortho with parallel hydrogen nuclear spins, and para with anti-parallel spins. The sub quantum number k_{-1} is even for the para modes, and odd for the ortho modes, as shown by the energy level structure in Fig. 15 (see also Kirchoff et al., 1972, (75)). There are no line doublets in the $k_{-1} = 0$, para state, and very small splittings in the $k_{-1} = 2$ state. It is worth noting, that the $k_{-1} = 1$, ortho variant rotates, with J=1 at absolute zero, whereas the $k_{-1} = 0$ para variant does not (just like ortho and para H$_2$ & H$_2$O; see Fig. 20 for the latter).

To our knowledge only two para transitions, $1_{01}^- \rightarrow 0_{00}^+$, & $2_{02}^+ \rightarrow 1_{01}^-$, have been observed in interstellar space, but many in the laboratory (75). Most interesting of all the other interstellar H$_2$CO radio lines is the one of lowest frequency (see Oka, 1960, (76)), $1_{10}^+ \rightarrow 1_{11}^-$ (4829.66 MHz, at F=$2^+ \rightarrow 2^-$, the strongest hyperfine level; $\lambda \sim 6$ cm). It was discovered in 1969 by Snyder et al. (77), seen in deep absorption, with the 140 ft telescope at Green Bank, using a cooled parametric amplifier (system noise temperature about 100 K), and a 400 channel autocorrelation receiver. The mean rotational energy of this transition corresponds to a frequency of about 315 GHz ($\lambda \sim 0.95$ mm), or 15 K. The ground state of the first ortho chain (in Fig. 15) thus is relatively easily excited, even in cold clouds. Quite remarkable, and very important, was the finding that the 6 cm line is seen in absorption even against the cosmic background radiation. Thus, its excitation temperature must be lower than 2.7 K, and considerably lower than the kinetic temperature in clouds of moderate densities. It is, under these circumstances, a stronger absorber than it should be under thermal equilibrium, originally named a Daser (De-Amplification by Stimulated Emission of Radiation) by the Townes group - also called an anti-maser.

Townes and Cheung in 1969 (78), with characteristic intuition, remembering H$_2$CO:s almost equal main moments of inertia, noted that preferential collisions could get the states out of equilibrium, but in an opposite sense compared to OH. For this molecule (assumed to be in either of the two lower Λ-doublet levels) an approaching H atom can, if it comes close enough, form a temporary chemical bond with the O atom, in a configuration that roughly resembles the water molecule. As the H atom then continues past the OH molecule, the bond is broken (the hyperfine states are usually left unperturbed) and the unpaired electron is left in the upper + orbital of the J=5/2 state (the efficiency has to be evaluated over all directions of collisions). If the density is not too high, radiative deexcitation (fluorescence) takes

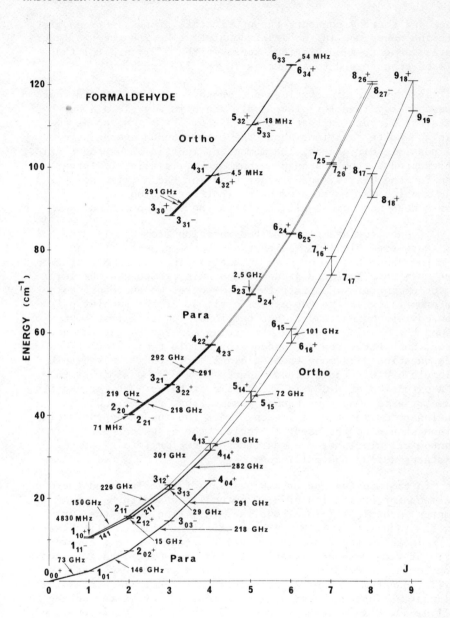

Figure 15: Energy level structure of para and ortho formaldehyde (H_2CO).

place and thereby connects the upper state of the J=5/2 doublet with the upper one in the J=3/2 doublet; likewise the two lower states connect. A population excess in the 5/2 + orbital thus will be transferred to an excess in the 3/2 + orbital.

In the case of formaldehyde, whose configuration and four lowest energy (doublet) levels are shown in Fig. 16 a & b, the classical picture is that the two H atoms *stick out* so as to help to excite rotations about the c-axis, upon averaging over all possible directions. The moment of inertia about the c-axis, represented by the sub quantum number k_+, is slightly larger than for the b-axis, much larger than for the a-axis (see Fig. 16 a), which is represented by k_- (note: rotation about the b-axis does not come explicitly into the state notation, $J_{k_-k_+}$).

If one, for example, excites rotation from the 1_{11} to the 2_{11} state (see Fig 16 b, neither of the components of angular momentum about the a- or c-axis changes. The change of rotation thus must have appeared around the b-axis, which has the lower moment of inertia. Contrary to the OH case, we therefore do not preferentially excite the upper level, 4, of the doublet pair (now corresponding to J=2) but the lower level, 3 (rotation around the c-axis). Instead of getting a maser we get the anti-maser. Such early, correspondence-principle type calculations were questioned by many theoreticians. It actually takes an H_2 molecule of 10 K (cold cloud) energy $\sim 6 \times 10^{-13}$ s to cross the ~ 2 Å extension of H_2CO, in which time it turns $\sim 40°$ around its symmetry axis. One thus cannot assume, that the molecule is stationary during the collision, i.e., the sudden approximation method should not be used. To include the classical effect in quantum calculations, one must thus *allow for (close) coupling* of states during the collision.

In an important paper, 1975, Evans (79) calculated the cross sections for excitation of H_2CO in collisions with H_2, in a quantum mechanical "close coupling" form (including the four lowest states of ortho-H_2CO). The effect of this coupling really is to increase the cross sections for excitation to the lower state of *each* doublet, relative to the cross sections for excitation to the upper state. This fact, Evans wrote, "suggests that earlier quantum mechanical calculations differed from classical calculations because they neglected coupling of states during the collision". Evan's results lend credence to the collisional model for explaining the observations of anomalous absorption in H_2CO at 6 and 2 cm wave lengths.

At about the same time as Evans, Garrison et al. (80), 1975, verified the cooling of the 6 & 2 cm transitions of interstellar (ortho) formaldehyde by a collisional pump, through a series of more accurate quantum mechanical calculations, using He as a

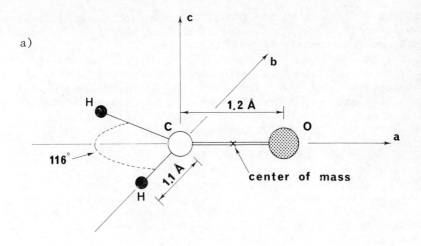
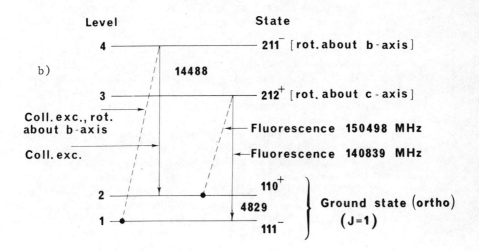

Figure 16: H$_2$CO collision configurations; a) rotation projection, & b) the four lowest ortho levels.

scatterer instead of H_2 (to reduce the scope of the computation). It is interesting, but not surprising, that the J=3, ortho-doublet (28975 MHz) is found to be an integral part of the collisional pumping scheme. By varying the number of states, used in the equations of statistical equilibrium, the effect of different J doublets on the cooling were assessed. Neglect of the J=4 levels, for example, caused less than 0.2 K changes in the effective temperatures for He concentrations at which cooling occurs. Omission of the J=3 levels, however, resulted in no cooling.

In a later paper by the same authors, {Green et al., 1978, (81)}, their previous rates of excitation, by collision with He, have been extended to higher rotational levels and kinetic temperatures. Rates for para-H_2CO have also been computed. It is suggested that excitation by low-temperature para-H_2 (i.e., all molecules in their lowest J=0 level) is faster by about 50% than excitation by He. Furthermore, excitation by ortho-H_2 (J=1) is expected to lead to significant enhancement of dipole allowed transitions of H_2CO.

The theoretical studies just summarized certainly underline the tremendous importance of high quality observational radio astronomy. If the anomalous absorption had not been found, who would have cared about detailed calculations of the collisional cross sections, and more seriously thought about the importance of allowing also for close coupling of states during collisions, etc.? Actually, there are many such examples of very important spin-offs from the radio observations. Main line OH masers, IR-pumping, molecular isotope ratios (& fractionation), deuterated molecules, excitation of complex molecules, terrestrial studies of detected radio molecules with unknown radio lines, such as the basic CH & CN for example. To our knowledge the radio lines of ground state CH have not as yet (10 years after their discovery) been verified in the laboratory, and only recently have the higher CN-lines been measured in the laboratory (the N=4 - 3; 3 - 2; 2 - 1 transitions). There are many other examples, e.g., the observations of the radicals C_3N, C_3H & C_4H (perhaps in the future also of C_5N, C_5H & C_6H). The list of such interplays between radio observations and advanced theory is long, and in its own right, interesting and thought provoking.

Although ground state H_2CO is not in equilibrium with the 2.7 K cosmic background radiation (or with other constituents of the dust clouds), the hyperfine populations should reach equilibrium among themselves, because of their rapid relaxation rates (see for example Kwan, 1974 (82)). The optical depth can therefore (in principle) be determined by measurements of the relative intensities of only two of the hyperfine transitions, shown in Fig. 17 (not to scale). Observations of this type require an ultra low noise receiver, such that the hyperfine components,

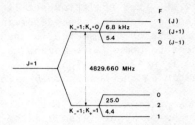

Figure 17: Hyperfine energy level structure (not drawn to scale) of the H_2CO, $1_{10} - 1_{11}$ (4.83 GHz) transition.

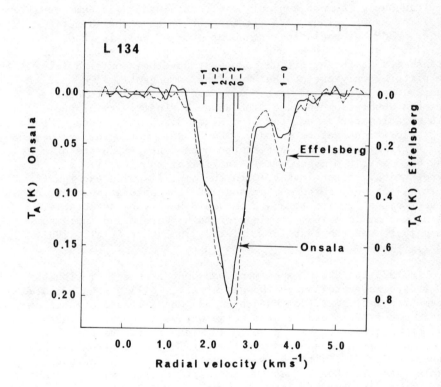

Figure 18: Formaldehyde absorption profile, obtained for the $1_{10} - 1_{11}$ transition by Sume et al. (83), using a travelling wave rutile maser in the Onsala 25.6 m telescope. Note the resolution of the F=1-0 hyperfine component. The spectrum (the first of its kind) is a good example of results that can be obtained with a good maser radiometer system, even in a telescope of moderate size.

only separated by a few kHz from the strongest line, $F=2^+ \to 2^-$ can be resolved. One can, of course, also determine the excitation temperature if the size of the source is known. But that could vary from component to component. Anyhow, a receiver of maser sensitivity class is needed, at least for a radio telescope of medium size, such as the Onsala 25.6 m instrument.

The preamplifier used at Onsala for this purpose was our 5 GHz rutile, travelling wave maser (also used for observations of the OH, $^2\pi_{1/2}$ (J=1/2) state), which gave a mean zenith system noise temperature of 29 K (1974). The half power beam width was 11' (at 4.8 GHz), and the beam efficiency 0.44. To secure the best possible line resolution, two 100 channel, post mixer, filter receivers, of channel widths 10 and 1 kHz, were used simultaneously. A most typical result obtained 1975 (Sume et al., (83)) for Lynds' cloud L134 (it lies well above the galactic plane, b=36°, at a distance of about 100-200 pc), is shown in Fig. 18, unattainable at the time. For comparison relative LTE intensities of the hyperfine component have been entered into the picture. Only one of the hyperfine components, $1^+ \to 0^-$ is clearly resolved, but the other ones are seen to de-symmetrize the main absorption feature. The unusual sensitivity of the 25.6 m radiometer is evident from Fig. 18; the amplitude of the 1-0 absorption feature is only 0.04 K. Remembering our unsuccessful, earlier attempts to detect the hyperfine components of a weak $6_{16} - 5_{23}$, H_2O maser, it is sufficient to visualize a (hypothetical) strong maser, with the amplification exp. $\{n x \alpha(f)\}$, where n is a very large number and $\alpha(f)$ the absorption profile, to see that a 1-0 type transition would hardly be detectable, unless one could clip a very narrow main feature and increase the gain to see the rest (cf., Fig. 2, in Baudry et al., 1981, (62)).

In 1976 the whole experiment was done over again, with the 100 m Effelsberg telescope (Downes et al., (84)), using a cooled parametric amplifier, followed by a 384 channel autocorrelation spectrometer. The total system noise was 95 K (about 3 times that at Onsala), the half power beam width as low as 2.6', and the beam efficiency about 0.65. The Effelsberg spectrum is shown dashed in Fig. 18, with the same spectral resolution as that Onsala used. It shows, that the intensity of the $F=1^+ \to 0^-$ component becomes stronger relative to that of the $2^+ \to 2^-$ component as a function of decreasing beamwidth. This result indicates that the ratios of the hyperfine components vary across the cloud. The optical depth of the $2^+ \to 2^-$, line for example, becomes greater than the Onsala mean value. Obviously, one should (try to) use telescopes with beam widths much smaller than the dimensions of the dust clouds.

Downes et al. (84) derived an excitation temperature for the

$1_{11} \to 1_{10}$ transition, T_{12}, of about 1.4 K, which (remarkably enough) agrees with theoretical values obtained from the early collisional pumping models (cf. Townes and Cheung, (78)). Garrison et al. (80) also predict excitation temperatures of this order, for interstellar clouds with kinetic temperatures, $T_k \sim 10$ to 20 K. In higher density regions, with $n_{H_2} > 10^5$ cm^{-3}, the excitation temperature of H$_2$CO may exceed 2.7 K and the 4.8 GHz transition would then appear in emission rather than in absorption, as some observations bear witness of (cf., the recent VLA mapping by Johnston et al.: 1983, Astrophys. J. 271, L89).

For n_{H_2} densities $< 10^5$ cm^{-3}, the rapid decay rates from levels 3 to 1 (see Fig. 16), and 4 to 2 should lead to an equilibrium with the background radiation, i.e., $T_{13} \sim 2.7$ K (see also Evans et al., (85)). The corresponding optical depth might be so large, however, that trapping of the 140.8 & 150 GHz lines (see Fig. 16 b) becomes important. This could increase T_{13} to, say, 3.5 K in the cloud center.

The elegance of determining formaldehyde cloud properties from cm as well as mm wave observations may now be in jeopardy. Langer et al., who 1979 detected the $2_{02} \to 1_{01}$ transition, at about 128.8. GHz, of deuterated formaldehyde, HDCO (86), find that the cm and mm wave lines of H$_2$CO must come from different regions within the dark clouds observed. Furthermore, these observations imply that formaldehyde probably is formed by gas-phase ion-molecule reactions, rather than on grains. Apparently as a by-product of their investigations, Langer et al. detected the para formaldehyde lines $1_{01}^- \to 0_{00}^+$, $2_{02}^+ \to 1_{01}^-$, and the ortho line $2_{12}^+ \to 1_{11}^-$ (see also Fig. 15). In this context it should be mentioned that the $k_- = 0$, & 1 states of HDCO are connected (by collisional and radiative transitions), as different from H$_2$CO. It is, of course, not obvious that the early excitation conditions for refrigerating formaldehyde (Townes and Cheung, (78)) would hold also for HDCO. From their mm wave observations Langer et al. furthermore find that the degree to which formaldehyde is deuterated (in L134N and CLD2) is *comparable* to that found for other molecules, for example DCO$^+$, H$_2$D$^+$ & DNC.

Interstellar HDCO was detected ten years after H$_2$CO, and at a frequency 27 times higher. The formation, excitation and fractionation of interstellar formaldehyde has almost become a science of its own. To which should be added that H$_2^{13}$CO has been found in the $1_{10} \to 1_{11}$, $2_{12} \to 1_{11}$, and $2_{11} \to 1_{10}$ transitions (the latter as high as ~ 147 GHz), and H$_2$C^{18}O in $1_{10} \to 1_{11}$ (at about 4.39 GHz), complete with six hyperfine components.

The formaldehyde maser – a late confirmation of an unexpected state. In 1974, the H$_2$CO ground state transition was detected in emission, associated with the ultracompact HII region, IRS1, near

NGC7538, by Downes and Wilson (87), using the 100 m Effelsberg telescope. Six years later, Forster et al. (88), observing with the Westerbork Radio Synthesis Telescope, found two $1_{10} \rightarrow 1_{11}$ emission spots, at radial velocities -57.6 and -59.8 km/s, near IRS1. The emission was partly unresolved (source size <4"), corresponding to a lower limit, 800 K, of the brightness temperature.

Not much later, Rots et al. (89), with the Very Large Array, at Socorro, New Mexico (VLA; twelve antennas were used, yielding 66 baselines) managed to resolve the H_2CO source in two components of about equal intensity, separated by about 0".1. The emission had angular sizes smaller than 0".15 (probably much smaller in reality), corresponding to linear dimensions of $\sim 10^{16}$ cm (at an assumed distance of 3.5 kpc). This implies higher brightness temperature limits, $\sim 5 \times 10^5$ K. The sizes are certainly small enough, and the brightness temperatures high enough to make the maser assumption plausible. Since the properties of the molecular cloud, in which NGC7538-IRS1 is embedded, probably do not differ much from those of other dark clouds, in which H_2CO is seen in anomalous absorption, the recently found maserlike emissions can hardly be explained by collisional pumping. It has been suggested by Boland and de Jong, in 1980, (90), that the $1_{10} - 1_{11}$ transition (of H_2CO) could be inverted by the free-free radio continuum radiation of a nearby compact HII region. It must be very compact, with emission measures of $10^8 - 10^{10}$ cm^{-6} pc, so that it is optically thick at 6 cm but rapidly becomes thin at millimeters. The masing H_2CO gas, located in front of the HII region, then should amplify its radio continuum radiation, just about as OH does in W3(OH) (in front of HIIA).

To demonstrate the approximate pump mechanism one has to include at least the four lowest rotational levels of (ortho-) H_2CO (see also Fig. 15) - as in the collisional case. Since we assume that the H_2CO maser is radiatively pumped by the radio continuum radiation only, the equations of statistical equilibrium become fairly simple and the lowest doublet (J=1) inversion ratio, n_2/n_1, easy to evaluate. One finds (see also Boland and de Jong), that inversion of the ground state doublet ($n_2 > n_1$) occurs, if the free-free continuum radiation field induces more downward transition from level 4 to level 2 (~ 150 GHz), than from level 3 to level 1, and that, at the same time, the 4 and 3 level populations are kept approximately equal by *rapid* radiative exchange. This means that the preferential downward transitions now take place via rotation about the b-axis, i.e., exactly the opposite of the collisional case.

To allow sufficiently rapid radiative pumping, mm wave photons must be able to escape from the cloud, which requires a large velocity gradient through the maser region. VLBI observations (at 6 cm) would probably reveal maser (correlation) lengths and spot sizes, similar to those deduced for OH in W3(OH).

A continuum, 4.8 GHz map (after Rots et al. (89)), Fig. 19, demonstrates the general similarity between the H_2CO-OH maser configurations in NGC7538 (IRS1) and the maser rings close to W3(OH), HIIA, shown in Figs. 6c and 7. An OH, 1665 MHz VLBI map, of IRS1, might perhaps also reveal a circular, or elliptical hot spot form, as in the case of W3(OH), and, at 6035 MHz, perhaps a displaced $^2\pi_{3/2}$ (5/2), spot circle, with a diameter corresponding to the distance between the unresolved H_2CO maser spots (at about 4.8 GHz). The relative situation of the remote H_2O (maser) concentrations may be characteristic for the regions. Figs. 7 and 19 are perhaps typical of space containing young HII regions, frequently surrounded by large, molecular clouds. CO, HCN, CN, HCO^+, and N_2H^+ have, for example, been observed towards NGC7538-IRS1, indicating that it moves through, or expands into, a molecular cloud with a velocity $\sim$ 10 km/s. A VLBI mapping of the maser regions, in NGC7538-IRS1, would be the obvious way to acquire more basic knowledge concerning the structure, and dynamics of the same and, if possible, minimize the parameter assumptions that presently have to be made.

10. H_2O - THE MOST POWERFUL MASER

Following the prediction, and suggestion by Snyder, and Buhl (1969, (91)), the 1.35 cm, $6_{16} - 5_{23}$ rotational line of (ortho) H_2O was detected in the interstellar medium by the Berkeley group (Cheung et al., 1969 (92)), initially in the sources Sgr B2, Orion & W49. Suitable electronic techniques were almost at hand, as spin-offs from the World War II radar at K-band, which, by chance it seems, almost coincided with the 1.35 H_2O line in the terrestrial atmosphere. Excess H_2O attenuation became an unexpected problem, and the asymmetrical shape of a strongly pressure broadened line one of immediate concern, to which Van Vleck's and Weisskopf's historic paper on the theoretical shape of pressure-broadened lines (93) bore witness.

The H_2O emission was detected with the twenty-foot radio telescope at the Hat Creek Observatory (where once the OH maser emission was discovered), employing the same technique that was used for the detection of interstellar NH_3 (94). Strong H_2O radiation, producing an antenna temperature of about 14 K, was observed from the Orion Nebula, and a temperature at least as high as 55 K for the radiation from W49. About their historic, and far-reaching discovery, Cheung et al. wrote as follows: "It is surprising that the transition is as strong as observed, since it involves levels of rotational energy 456 cm^{-1}, which can radiate to lower states in about 10 sec. Their excitation

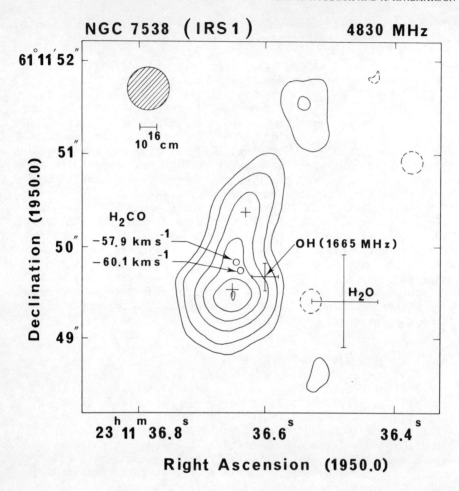

Figure 19: 4.83 GHz continuum map (Rots et al., 1981, (89)) of NGC7538, displaying H_2CO maser hot spots, amplifying the HII background, not located too far from the OH uncertainty cross (probably indicating the center of an OH, accreting disk, roughly believed to exist in W3(OH); see Fig. 7). As in W3(OH), the H_2O maser(s) form islands of their own; probably a characteristic, common feature.

requires moderately high temperatures, and also frequent excitations, either by collisions or by radiation. Rotational states of NH_3, which can similarly radiate, have not been found, indicating that NH_3, and H_2O have been detected in rather different regions. Presumably, the H_2O is present in rather special regions of higher-than-normal excitation. The matrix element for this H_2O line is appreciably less than that for the NH_3 inversion levels. Hence, if there is thermal equilibrium between $6_{16} - 5_{23}$ states, the population in these two levels, rather high above the ground state, must have a column density greater than what was found for NH_3, or about 10^{17} cm^{-2} in the Sgr B2 cloud". The authors continue: "The high intensity, very narrow lines suggest that perhaps thermal equilibrium does not occur and that there may even be maser action. -- If thermal equilibrium does in fact apply, this intense microwave radiation from H_2O, and the existence of strong HDO transitions in the radio region should allow an interesting measurement of the hydrogen-deuterium ratio".

What the Berkeley group had found, was the *second interstellar maser*, by far the strongest. Even compared with the third interstellar maser, SiO, which was detected and identified a few years later. Neither SiO, nor OH have such short and irregular, temporal variations. As Cheung et al. predicted, HDO was found, but a reliable determination of the D/H ratio has not yet been made. We will return to this important problem in the HDO section.

H_2O VLBI observations have revealed, as in the case of OH, that the masing sources (no thermal emission has been seen at the transition in question, but perhaps "superthermal" ones in more recently discovered states) consist of a number of hot spots, frequently in spherical, or shell like objects (see, for example, the OH ground state maser, hot spot configuration in Fig. 6 a) with typical dimensions of 10^{17} cm, on occasions much less, with spot sizes of a few AU, and spot brightness temperatures at least as high as 10^{13} K. One even obtains 10^{16}, or more, for the recently discovered H_2O maser flare source (radial velocity ~ 8 km/sec), in the Orion-KL region (L.I. Matveenko et al., 1980 (95)). Actually, the flare region, observed with the Simeiz (Crimea)-Puschino (Moscow) radio interferometer (using the 22 m, RT-22 telescopes - frontend maser equipped), also with the Effelsberg (Bonn) 100 m telescope, and with the Haystack-Green Bank baseline, has a complex spatial structure, with component brightness temperature as high as 10^{17} K, linewidths as small as ~ 6 kHz, and high intensity maser-length sections as short as 10^{13} cm (~ 1 AU). Antenna temperatures of *several thousand* degrees were recorded with the 37 m Haystack telescope by Abraham et al., 1981 (96). Furthermore, unexpected linear polarization up to as much as 50% was observed (Matveenko et al., 1983 (97)). The flare probably connotes "isolated" and energetic events in the neighbourhood of the Orion-KL region, which could be due to the action of an emerging protostar, or a region whose density and temperature favour enhanced masing (Abraham et al., (96)). Furthermore, the high degree of linear polarization could indicate a saturated maser. As different from the W3(OH) masers, the flare maser must amplify its own fluctuations. The flare maser mechanism is intrinsically coherent and coherent radiation possesses directivity. One could define the directivity of the radiation (see Matveenko et al. (95)) by the size of the area within which coherence is preserved. This would yield about 100 x 100 cm, such that the phase shift λ/D would be less than π, (it would be say 15 times more for OH), whilst the measured size of the flare emission region is $\lesssim 10^{13}$ cm. The source, therefore, must comprise a *large number of coherently radiating spots*, non-coherent among themselves, but distributed over an area of about 10^{13} cm in diameter. The great, temporal intensity variations observed by Abraham et al. (96) could (perhaps) be explained by pump intensity (or medium parameter) variations over a structure about one light month in extension, i.e., $\sim 10^{17}$ cm, or just the typical maser dimension (as mentioned in the introduction of this paragraph).

Of special, and important interest, are the VLBI proper motion observations of H_2O maser sources (Genzel et al., 1981, (98)). These observations, with relative positional accuracies of 0.1 milli-arcsec. (note: the historic Onsala-US, 5 GHz, continuum VLBI observations yielded ~ 1 milli-arcsecond accuracies already in 1968), showed that the proper motion (in the W51MAIN region) typically was about 1 milli-arcsec/year. This probably represents kinematic motions of the maser cloudlets (the kinematics of the maser hot spots in W51MAIN is more complex than of those in the Orion-KL region). Genzel et al. (98) note that the transverse motion of most maser features in W51MAIN appear to be random and can perhaps best be interpreted as turbulent motions created, when a strong stellar wind interacts with the (density) inhomogeneities in the surrounding molecular clouds. From a comparison of the transverse and radial-velocity dispersions, "a statistical parallax", the distance to W51 is determined to be 7 ± 1.5 kpc, which agrees with the "far" kinetic distance. This VLBI-technique to measure distances is a *direct* (and first) one, and it is, important enough, independent of the "hierarcy" of astronomical distance indicators and of the rotational models for the Galaxy.

The average angular diameters of the H_2O maser spots in W51MAIN appear to be 0.1 - 0.3 milliarcseconds. The true physical sizes of the maser clouds are probably greater (see, for comparison, the OH spots in W3(OH)), about 10^{14} cm. There is, of course, a large spread in brightness temperatures, e.g., from 10^{11} to 10^{14} K, low values compared with those of the H_2O **maser** flare in the Ori-KL region. It should be mentioned in this context, that a similar type of H_2O maser flare was observed (June, 1971) in the source W49 (distance about 14 kpc, or 28 times that of Ori-KL). The size was about 150 µ arcsec, and the brightness temperature $\sim 10^{16}$ K.

Maser lengths of 10^{16} cm were proposed very early, also neutral gas densities as high as 10^7 cm^{-3}, mostly hydrogen molecules (higher densities might thermalize the maser transitions), and (ortho) H_2O densities of 100 cm^{-3}, in the masing regions. In view of these high densities, long pathlengths, and of the strongest spontaneous transition rates, 0.1 sec^{-1} for H_2O, the optical depths in the strong, far-infrared lines must be extremely large, so that radiation trapping must be of great and decisive importance. This will be evident in what follows.

But first, let us remind our readers, that H_2O is believed to be one of the most abundant molecules in dense, interstellar clouds, $H_2O/H_2 \sim 10^{-5}$ (cf., Prasad and Huntress, 1980, (99)), a value predicted to increase under influence of shocks, by an order of magnitude or more. High H_2O production rates in stellar atmospheres of oxygen-rich chemical composition are another possibility (Scalo and Slavsky, 1980, (100)). The scant observational data (for Orion-KL only) are, however, in fairly good agreement with these predictions.

The high expected abundance, and the great number of strong transitions in the far-IR region, as shown in the energy structure, Fig. 20, for ortho and para H_2O (based on the H_2O-tables by Hall and Dowling, 1967, (101)) make water vapour of primary importance for the thermal balance of dense molecular clouds ($n_{H_2} \gtrsim 10^5$ cm^{-3}). Furthermore, considerable amounts of H_2O are assumed to be tied up in grain mantles. The abundance of H_2O will therefore also have to be considered, when one studies the evolution of interstellar dust particles, and their interaction with the gas (cf., Hagen et al., 1983, (102)).

It appears from Fig. 20, that it cannot be easy to determine the H_2O abundance through terrestrial observations. The multitude of important energy levels, E_ℓ, and the frequencies of transition between them (expressed in GHz when less than 1000 GHz, as shown in Fig. 20), range high, and the atmosphere becomes essentially opaque (for them) to the ground based observer. Although observations of OH maser regions indicate that they are moderately dense (with only the lowest rotational levels appreciably populated), observations of H_2O maser regions, on the other hand, indicate that these are denser, hot regions, since the transitions leading to maser emissions (usually - excellent tracers of star formation regions) occur in highly lying rotational levels.

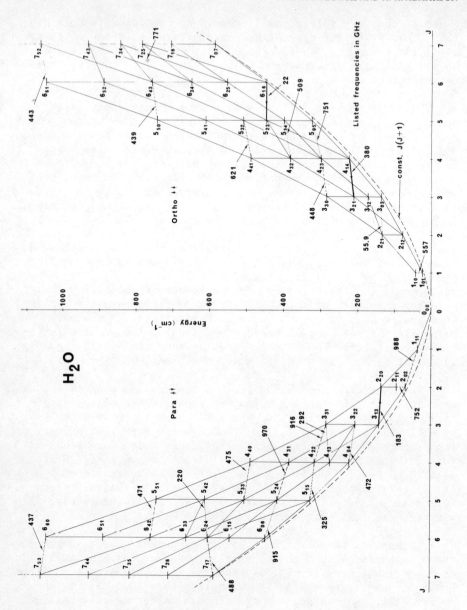

Figure 20: Energy levels of para and ortho H_2O, displaying many sub-mm wave transitions which are likely to be detected soon with airborne observatories. Detection of such transitions (the 183 GHz para line was detected by Waters et al., 1980, (103); the 380 GHz ortho line by Phillips et al., 1980, (105)) would greatly increase our understanding of the H_2O (masing) cloud physics.

The line strength of the maser transition, $6_{16} - 5_{23}$ (22235 MHz), is in itself quite weak, in spite of the high dipole moment; 1.88 Debye in the direction of the b axis. It is thanks to the water molecule being asymmetric, that the transition is allowed at all. The intrinsic lifetime of the 6_{16} level is much shorter than that of 5_{23}, and so will also be the effective lifetime of the 6_{16} level in thin clouds, where there will be no radiation trapping. For thick clouds, however, the effective lifetime of the 6_{16} level becomes comparable to that of the 5_{23} level, and, with sufficient external (volume) sources of pumping, population inversion may result, as sketched in Fig. 21, and commented upon as follows.

The molecules in the 6_{16} state fluorescence down via different branches (see Fig. 20), two of which are shown with their effective transition probabilities, $(A/Q)_{pq}$, where Q is the trapping factor (=1 for the thin cloud), and A is the Einstein coefficient. Q increases the lifetime of the 6_{16} state. We assume, for simplicity, that below a certain level (labelled 3 in Fig. 21; see also Fig. 20), characterized by the angular momentum J_m, we have reasonably good thermodynamic equilibrium between the rotational excitation temperature and the kinetic temperature, which therefore determines the population distribution below that level. Above this level, however, the fluorescence rate begins to *dominate over the collisions*, and the rotational temperature will be lower than the kinetic temperature of the particles. This is due to the important fact, that levels become more widely spaced the higher one comes up the rotational ladders, and that A is proportional to ν^3 (and, approximately, ν to J).

To be a little more specific, the condition for the rotational temperature to be less than the kinetic is, that the net collision rate down is smaller than the effective fluorescence rate between the two levels considered, or, according to the notations in Fig. 21, that $W \{1 - \exp(-h\nu/kT)\} < A/Q$, when $J > J_m$ (the difference in statistical weight is neglected).

From this basic inequality one finds that the conditions needed to yield
1) enough population *almost* in thermal equilibrium, up to a level fairly close to the maser levels, to provide for the reabsorption of the fluorescence radiation, and
2) enough optical depth to bring the lifetime of the 6_{16} level *up to that* of the 5_{23} (and give maser action), roughly are that
3) the H_2 density $\sim 10^7 - 10^8$ cm^{-3},
4) the ortho H_2O column density be 10^{18} cm^{-2} (or more), and finally that
5) the kinetic temperature be about 200 K (note: for $6_{16} - 5_{23}$, $E_\ell \sim 640$ K).

There could, of course, be other sources of pumping that the IR volume source actually assumed here. Collisional pumping only, followed by cascades down to the maser levels, would probably need higher densities, but then the problem of thermalizing, anti-inverting collisions across the maser transition would occur, and such collisions have large cross sections, as pointed out by Litvak (70).

Irrespective of the type of volume population source assumed, there should be an initial population enhancement at the "back bone" of the population ladders, for ortho as well as para H_2O, i.e., at the levels J_{1J}, and J_{0J} (for para states k_-k_+ are either even-even, or odd-odd), see Fig. 20.

Thus it was not surprising, that Waters et al., 1980 (103), discovered *para* H_2O by its 3_{13} (J_{1J})- 2_{20} transition, at about 183 GHz (see Fig. 20) in the Orion Nebula, using the NASA Airborne Observatory 91 cm telescope. The peak antenna temperature of the line was ~ 15 K, and its (LSR) velocity about 8 km/sec (the same as for the giant H_2O maser flare).

The velocity profile had characteristics similar to those for CO, a narrow (~ 4 km/s) "spike" centered at about 9.5 km/s, and a broad "plateau" with extended wings centered at 8 km/s. It is important to notice that the plateau emission appears to be enhanced above that expected for thermal excitation, "if it originates from the no greater than 1' region characteristic of plateau emission from all other observed molecules". Correcting the data of Waters et al. for beam filling (Olofsson, 1984,(104)), one finds that the brightness temperature is $\gtrsim 1500$ K (if the emission is disk dominated; $\lesssim 20" \times 20"$), or $\gtrsim 350$ K (if high velocity flow dominated). One should, in this context, note that $E_\ell \sim 200$ K.

Waters et al. (103) argue, that the spike emission is consistent with an optically thick source, of the approximate size of the well-known molecular ridge in Orion, having (para) H_2O in thermal equilibrium at $T \sim 50$ K. The column density then giving rise to the spike becomes $\sim 3 \times 10^{17}$ cm^{-2}, or about one third of that assumed for ortho water. The detection of para water is one of the more important discoveries in radio molecular astrophysics.

The Waters group used a Ga:As Schottky-barrier diode, developed by R.J. Mattauch of the University of Virginia, which was mounted on a quartz micro strip. The lowest double-sideband system temperature attained at 183 GHz was ~ 2000 K.

Using an In:Sb heterodyne bolometer receiver, with a diode harmonic generator for oscillator power (and a system temperature variable from 400 - 1500 K), in the same Kuiper Airborne Observa-

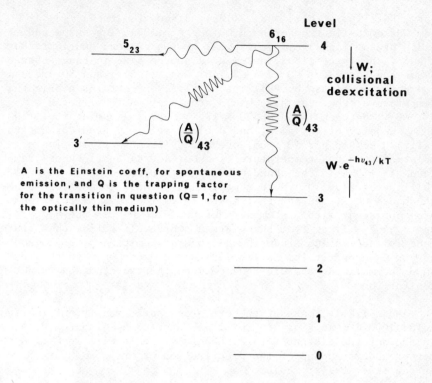

Figure 21: Sketch of the radiation trapping between the 4,3 levels, necessary to maintain the H_2O, $6_{16} - 5_{23}$, population inversion.

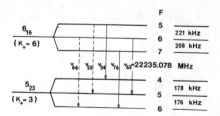

Figure 22: H_2O hyperfine, energy level structure (not drawn to scale). Thus far, hyperfine splittings have not been detected, even in the weakest masers.

tory, Phillips et al., 1981, (105), discovered the $4_{14}(J_{1J})-3_{21}$, 380 GHz emission from ortho H_2O (see Fig. 20). The shape and width of the line are more typical of the plateau source rather than of the large cloud. Assuming a source size in range between 50 arcsec found for CO (J=4-3, 461 GHz, (105)), and 30 arcsec found for SO_2 (also by Phillips et al.), one obtains a brightness temperature (the beam size was 3 arcmin at 380 GHz) for H_2O $4_{14}-3_{21}$) between 160 and 430 K ($E_\varrho \sim 320$ K). Assuming (see (105)) an H_2 density of 10^7 cm^{-3}, and a cloud velocity gradient of typically 20 km/sec per 10^{17} cm, one finds (from an excitation calculation at a gas temperature of about 100 K) a brightness temperature > 150 K in the abundance range $[H_2O]/[H_2] = (20-1) \times 10^{-6}$. Phillips et al. (105) note that the brightness temperature peaks at about 500 K, and add that the observations and calculations are consistent with peak maser action ($-\tau \geqslant 3$) for the $4_{14}-3_{21}$, and $3_{13}-2_{20}$ lines from the plateau source. If solid, this statement is interesting since these lines are emitted by ortho and para H_2O respectively. Earlier in this context we used the term superthermalized; the distance between the two concepts is not too far.

Most likely, many of the ortho and para transitions (marked below 1000 GHz in Fig. 20), will be detected by airborne detectors in the not too distant future. Hopefully, this will lead to a better understanding of the distribution of ortho and para H_2O in the interstellar medium.

Finally, a few words should be said about the old problem of the H_2O hyperfine structure. Fig. 22 demonstrates it for the $6_{16}-5_{23}$ transitions; see also Meeks et al. 1969, (106). The frequency difference, between the almost equally strong main lines, is quite small. In a very strong maser, like the narrow H_2O flare in Ori-KL, mode competition might lead to a survival of only the F=7-6 transition.

11. HDO - THE ELUSIVE, IMPORTANT MOLECULE

Already at the time of their detection of H_2O, the Berkeley group emphasized the importance of detecting HDO ($\mu_b = 1.73$ Debye). Initially not being fully aware of the thorough maser property of the $6_{16}-5_{23}$ transition, they thought that a determination of the (permanently sought for) D/H ratio would be possible by a discovery of an interstellar HDO transition.

The desired discovery was made, towards Orion, by Turner et al. in 1975 (107), of the $1_{10}-1_{11}$ transition at 80.6 GHz ($E_\varrho \sim 43$ K), using the 11 m Kitt Peak telescope; the single sideband receiver temperatures were ~ 750, and ~ 1200 K, for the two senses of linear polarization. The antenna temperature was about

0.3 K and the line was seen in Orion-KL only. Much later, Olofsson, 1983, (104), using the Onsala 20 m telescope, was able to find HDO also towards W51M, about 15 times more distant than Ori A. Olofsson used a cooled Schottky barrier diode mixer in his observations. The single sideband receiver temperature, measured at the horn, was 380 K. The mean peak antenna temperature was 1.8 K (i.e., about six times the Kitt Peak value; the telescope dish ratio is about 3.3).

Fig. 23 depicts a sketch of the lowest HDO energy level structure (based on tables by de Lucia et al., 1974, (108)), labelled after its $J_{k_-k_+}$ asymmetric quantum numbers. In this case, as different from H_2O, transitions between the para and ortho states are permitted.

To determine H_2O abundances from derived HDO abundances, one needs a reliable estimate of the possible deuterium fractionation that may occur in these molecules. In principle, neglecting the unknown density differences between interstellar ortho and para water, it should be possible to derive an [HDO]/[H_2O] - ratio in the Orion-KL region by comparing the Onsala data with the $3_{13} - 2_{20}$ results of Waters et al. (103). That such a comparison might make sense, in spite of the possible deviation from LTE, is apparent from Fig. 24, where we have plotted the H_2O, $3_{13} - 2_{20}$, and HDO, $1_{10} - 1_{11}$ line profiles as functions of the radial velocity.

Actually, a Gaussian decomposition of the Orion-KL HDO ground state feature reveals that it contains emissions from the so called ridge cloud (at 8 km/sec), from the hot core source, and from the weaker plateau source. In the decomposition of the map spectra (Olofsson, (104)), the weak plateau feature was suppressed, and the velocities of the two remaining components were left fixed at their map center values. As expected, the emission maximum, in the hot core component, is located somewhere in the region between the compact IR sources, IRc2 and IRc4, while the ridge feature mainly emanates in a small region close to IRc5, coincident with the 6 cm H_2CO emission peak, recently revealed by Johnston et al.: 1983, Astrophys. J. 271, L89 (see also the previous formaldehyde section, and Olofsson's Ori-KL map (104)). The HDO, $2_{11} - 2_{12}$ ($E_\ell \sim 84$ K, see Fig. 23) component, detected by Beckman et al. in 1982 (using an In:Sb bolometer, at the prime focus of the 5 meter millimeter wave telescope of the University of Texas, (109)) is clearly dominated by the hot core component. VLA maps of high-energy ammonia lines also indicate that this component consists of high temperature condensations of enhanced density ($10^7 - 10^8$ cm^{-3}) and temperature ($\sim 100 - 250$ K), located between IRc2 and IRc4 (Genzel et al., 1982, (110a)).

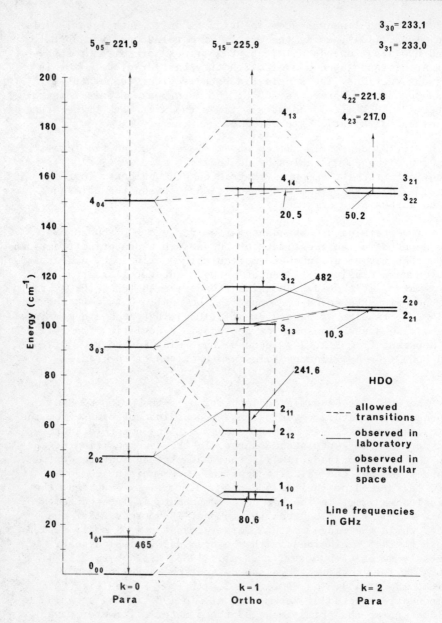

Figure 23: The lowest ground state levels of HDO. Each level is labelled after the $J_{k_- k_+}$, asymmetric quantum number. Transitions between para and ortho states now are permitted.

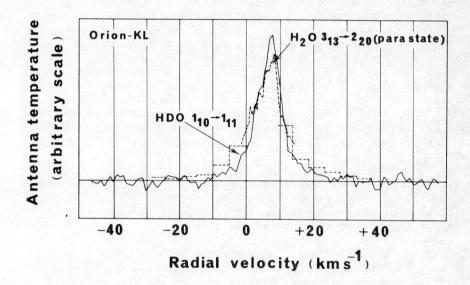

Figure 24: HDO, $1_{10} - 1_{11}$ spectrum, at 80.6 GHz, obtained with the OSO 20 m mm wave telescope (using a cooled OSO mixer) towards the Ori A - KL region. The para H_2O, $3_{13} - 2_{20}$ transition (Waters, et al., 1980), see Fig. 20, displays a roughly similar shape. HDO - the elusive molecule, has been detected also in W51M by the OSO group.

For the hot core region one finds (assuming LTE throughout) a total HDO column density of $(1-3) \times 10^{15}$ cm^{-2} for the rotational temperature range 100 - 200 K, and $\tau(1_{10} - 1_{11}) \sim 0.01$. Previous estimates of molecular [XD]/[XH] - ratios in Orion apply to the quiescent molecular gas (the ridge cloud) and range between 10^{-2} and 10^{-3} (see Table 1 in (109)). However, the regions of interest in the present context are considerably warmer, 100 - 200 K, than the cold dark clouds and the extended (~ 50 K) Orion ridge clouds. Watson and Walmsley, for example (1982, (110b)), seem to argue that, already at temperatures higher than ~ 40 K, the enhancement (over the D/H-value 10^{-5}, of York & Rogerson, 1976, (111)) should be less than an order of magnitude. Thus, $(HDO/H_2O)_{ortho} \lesssim 10^{-3}$ could be regarded as a trustworthy upper limit for the present. This yields $N(H_2O)_{ortho} \geqslant 10^{18}$ cm, and (in LTE), for a rotational temperature of about 150 K, $\tau \sim 3$ for the $3_{13} - 2_{20}$ transition (even for the spatially adjacent maser flare, $-\tau(6_{16} - 5_{23})$ presumably would be less than about $35 \sim 3.5 \times 10^3 \times \tau(HDO)$).

For the hot core source the beam filling actually would be so small, that the brightness temperature (of the $3_{13} - 2_{20}$ transition) might approach 3000 K, supporting the moderate maser concept proposed by Phillips et al. (105).

Since HDO is not a strong maser (if it is a maser at all), it should be mentioned that the HDO energy levels are hyperfine-split through the interaction of the D quadrupole moment with the molecular electric field gradient, and by the coupling of the magnetic moments of H & D to the magnetic field, generated by the molecular rotation. This hyperfine splitting, which was measured for the $2_{20} - 2_{21}$ transition (see Fig. 23) by Thaddeus et al. already in 1964 (112), is quite small, about 0.3 MHz and should be even smaller for the $1_{10} - 1_{11}$ transition, and probably of no immediate significance, since most of the energy is contained in the central hyperfine component.

The spectrum of W51M, shown together with that of Orion-KL, in Fig. 25, represents the *first detection of HDO outside the Orion region*. Except for Orion, it is also the only source searched by us at Onsala, in which we detected the CH_3OH ($7_2 - 8_1$, A^-, $E_\ell \sim 100$ K) line. Together with Onsala Spectral Scan data (Johánsson et al., 1984, (113)), it indicates that W51M contains regions with kinetic temperatures around 100 K. Direct evidence for such warm, and compact "condensations" has been provided by the VLA ammonia maps of Ho et al. (1983, (114)). They conclude that this area (and the similar ones associated with W51 S & N) is very similar in nature to the Orion-KL region (which is about three times smaller). An explanation in terms of a dense cloud "condensation", interacting with mass outflows from young stars is suggested.

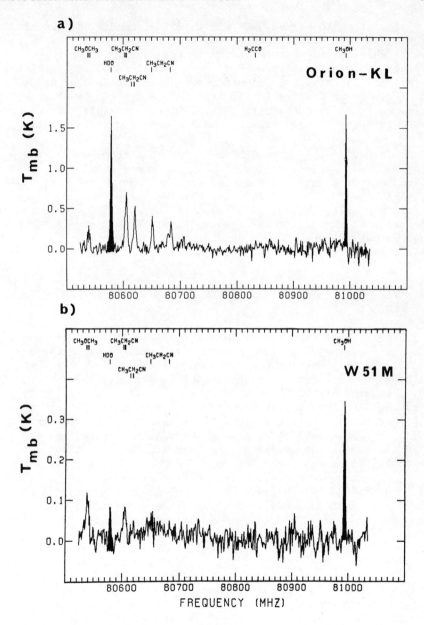

Figure 25, a): OSO spectral scan in the 80 GHz range, across the HDO line, *toward Orion-KL*, using a cooled OSO mixer. *b):* OSO scan *toward W51M*, also across HDO. CH_3OH (methyl alcohol) is strongly visible in both scan spectra.

The strength of the HDO line in W51M (see Fig. 25), scaled to the distance and size of Orion, approximately corresponds to the brightness temperature of the latter. Thus, on the rough average, we expect the same H_2O (ortho) and HDO column densities in W51M as in Orion. Our simultaneous detection also of CH_3CH_2CN, $(CH_3)_2O$, and CH_3OH in W51M (Olofsson, (104)) could be of some guidance in this context (see Fig. 25). Ethyl cyanide, however, has only been observed towards the hot core source in Orion.

Furthermore, it should be mentioned that the Onsala group searched for HDO in a region of possible shock-chemistry, IC443, in regions of presumably enhanced density and temperature, DR21(OH), NGC1333, NGC7538, and in W49, a strong H_2O maser source, but with negative results. W3(OH) is a remaining possibility. These results, and the weak HDO line towards W51M, indicate that other HDO sources outside Orion would be difficult to detect. One of the most interesting molecules is one of the most elusive.

12. SiO - THE THIRD INTENSE MASER, A CIRCUMSTELLAR RADIATION SOURCE

The discovery of the SiO maser was *serendipitous*, as was the discovery of the OH and H_2O masers. Furthermore, as in the case of OH, thermal emission had first been discovered from the molecule, in the SiO case from the transition $J=3-2$, $v=0$ (~ 130.25 GHz), towards Sgr B2, by R.W. Wilson et al. (1971, (115)), who used the Kitt Peak telescope with a forty-channel filter-bank superheterodyne receiver. About $2\frac{1}{2}$ years later (in December, 1973), Snyder & Buhl (116), using the same telescope, and the NRAO millimeter wave receiver, to look (in the direction of Orion-KL) for an unidentified molecular line in the upper sideband, detected, *in the lower one*, a "surprisingly strong group of unidentified molecular lines", at about 3.48 mm wavelength. The title of their paper, *Detection of possible maser emission near 3.48 millimeters from an unidentified molecular species in Orion*, indicated a mixture of expectation and uncertainty (the estimated line frequency was 86.245 GHz). The author's preliminary spatial map showed that the intensity fell off by 60 percent or more, only one beamwidth ($\sim 70"$) from the position of strongest emission. "Hence", the authors wrote, "the strongest emission arises from a compact area located in the direction of the central region of the Orion Nebula molecular cloud. Both the narrowness of the unidentified lines, and the limited spatial extent of the Orion emission region, suggest that we have detected a millimeter-wave maser".

Subsequently, F.J. Lovas & D.R. Johnson, of the US National Bureau of Standards, identified the narrow line radiation (that was shown to come both from the neighbourhood of Becklin's IR star in the Orion Nebula, and from a number of late type stars) as originating from the J=2-1 rotational transition of silicon monoxide *in its first excited (v=1) vibrational state*. The first millimeter wave maser (and the only strong one) had been discovered. Most remarkable was that it also had to vibrate in order to mase. In June, 1974, furthermore Buhl, Snyder, Lovas & Johnson, again observing at Kitt Peak, reported (117) the detection of maser emission from the J=1-0 rotational transition of the second vibrational state (v=2; ~ 42.8 GHz) in Orion as well as in a number of late M-type variable stars, of which many show OH & H_2O maser emission.

The SiO masers have simple energy levels, compared to those of OH & H_2O, as shown in Fig. 26. This simplicity gives us some hope, that the mechanism responsible for the population inversion will finally be understood. Maser emission in one or more rotational transitions, of the first three vibrationally excited states, for the SiO molecule have now been detected towards more than 60 (1983) giant and supergiant stars. Moreover, SiO offers the unique possibility of observing also thermal radiation in the same source. The fact that SiO energy levels as high as ~ 3500 K, and in some cases ~ 5300 K, exhibit population inversion of considerable optical depth, directly suggests that the SiO masers must be located very close to stars. VLBI experiments performed towards RCas, and VXSgr have shown (Moran et al., 1979, (118)) that the SiO masers are clustered within 4-6 stellar radii from the center of the stars. Studies of SiO maser properties, for example their temporal variations, will therefore be most helpful as tools in the analysis of the dynamical activities in regions within a few stellar radii from surfaces of red giants.

Fig. 27, based on observations performed with the Onsala 20 m millimeter wave telescope, demonstrates the remarkable temporal variations, in time scales less than a month, of the v=1 & 2, J=2-1 emissions from RLeo (an M giant Mira variable, mean period 312.5 days, about 150 pc distant; a water vapour source and one of the brightest Miras at infrared wavelengths), and from χ Cyg (a cool S-type variable), respectively. It has been very difficult to interpret the velocity structure variations of the SiO masers (without prompt access to VLBI observations). This is, most naturally, due to the complexity of the physical processes in a pulsating stellar atmosphere.

The pump mechanism responsible for the population inversion of the lower rotational levels of SiO is still a matter of debate. Elitzur (1980, (119)) argued, that only collisional excitation could meet the requirements imposed by the observational results then available, and that the masers had to be stellar atmospheric phenomena. But this, however, is not in accordance with VLBI results. Assuming that the SiO masers are located in

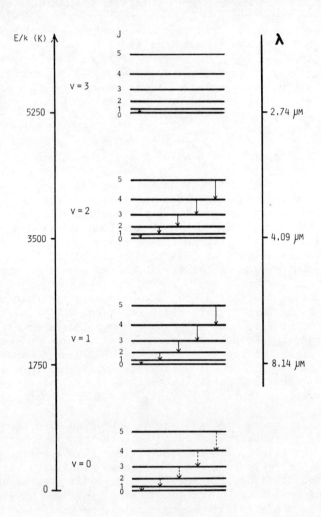

Figure 26: SiO energy levels - cf., Fig. 28.

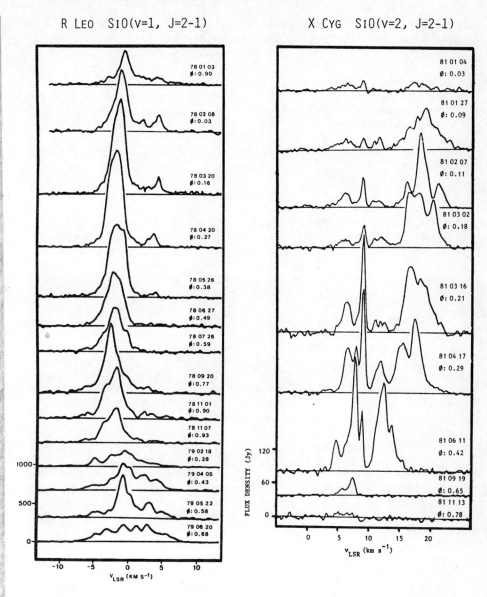

Figure 27: Temporal variations of SiO, v=1 & 2, J=2 − 1 maser emissions observed (at about 86 GHz) toward the red giants R Leo & χ Cyg, respectively (with the 20 m OSO telescope). The optical phase, φ, at each epoch is shown. φ = 0.0, & 1.0 correspond to maximum brightness in the visual region. The infrared peak normally occurs at φ ∼ 0.2. The *peculiar* v=2, χ Cyg transition was **discover**ed at OSO.

the inner part of the circumstellar shell, Bujarrabal and Rieu (120) arrive at the conclusion, that also radiative pumping at about 8 μm (see Fig. 26) could be a possible pumping mechanism.

The remarkable masing properties of vibrating SiO appears to have vitalized the general debate concerning masing conditions for simple molecules. Why are there only three strong masers? And why, for example, is CO, J=1-0, v=1 not a maser? Ultimately, many more weak molecular masers will probably be seen against various HII regions (amplifying the 10 000 K or so noise background of the same). But the fourth *strong* maser, where is it?

To make the reader familiar with the dimension, relative to the solar system, of the SiO emitting regions around highly evolved, red giant stars, we have in Fig. 28 reproduced an illuminating sketch prepared by Hans Olofsson of the Onsala group. It is self explanatory. Furthermore, we have in Fig. 29, as a second characteristic illustration of SiO emissions, reproduced isotope spectra, of the J=2-1 transition, observed towards the Ori-KL region by the Onsala group (Olofsson et al., 1981, (121)).

The v=0 spectra are roughly similar in shape, except for two maser-like spikes in the ^{29}SiO spectrum. These spikes roughly coincide in LSR velocity with the very strong maser spike of the vibrationally excited state (v=1). The irregular spike, seen at about +6 km/s, in the ^{30}SiO spectrum is not necessarily a ^{30}SiO maser feature, but perhaps mainly a superimposed methanol feature ($19_4 - 18_5$ E), "missing" in the chain of such features in the Onsala Spectral Scan. Such "confusion" phenomena are likely to show up, here and there, in future high sensitivity - high resolution scan spectroscopy.

The main velocity range, $\sim$-10 to +20 km/s of the ^{28}SiO (v=1), Orion maser spectrum is believed to define the inner part of an expanding gas shell, surrounding the infrared source IRc2 (see Genzel et al., 1980, (122)). This envelope is also interpreted to be the location of "shell type" OH and H_2O masers (Genzel et al., 1981, (123)). One interprets all SiO emissions in the $\sim$-10 to +20 km/s range in terms of an expanding shell, producing most of the detectable ^{29}SiO and ^{30}SiO emission, the major part of the wide ^{28}SiO emission profile, and the strong SiO masers.

A most likely explanation of the ^{29}SiO maser-like spikes seems to be that the major part of the ^{29}SiO shell is *optically thin*, making it possible to "sample" the inner part of the gas envelope. Here, the intense, infrared radiation field from the central star, or collisions in a suitable gas density range, may invert the rotational levels of the ground vibrational state.

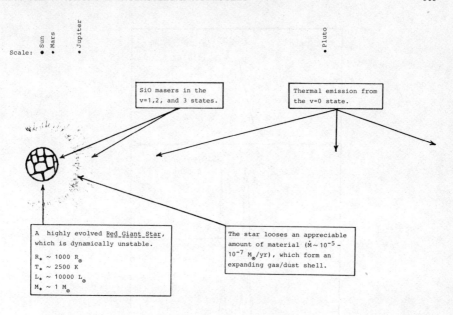

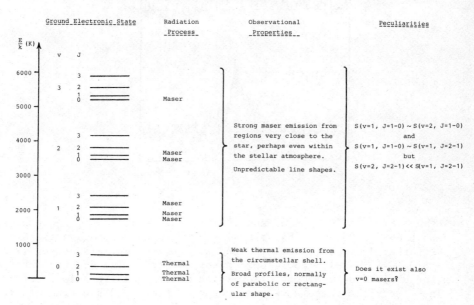

Figure 28: Circumstellar SiO maser and thermal emission sources, related to a typical, highly evolved *Red Giant Star*, sketched (for comparison) against a background, planetary distance scale.

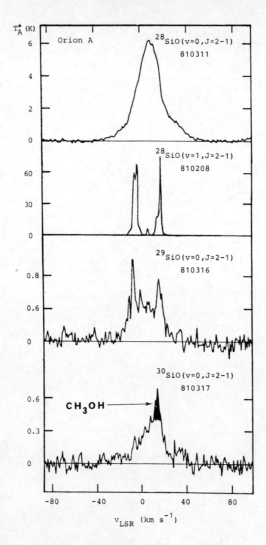

Figure 29: ^{28}SiO, ^{29}SiO, and ^{30}SiO; v=0, J=2-1 (rest frequencies about 86.85; 85.76, & 84.75 GHz) spectra obtained toward the Ori-KL infrared cluster, with the OSO 20 m telescope. Also shown is a ^{28}SiO (v=1, J=2-1; ∼86.24 GHz) maser spectrum. The spectral change (for v=0), from ^{28}SiO to ^{29}SiO (with maser anomalies) is dramatic. The maser-like feature, superimposed on the ^{30}SiO spectrum, on the other hand, probably stems from the CH$_3$OH, 19_4-18_5 E transition. Thus, no maser anomalies, like the time variable spike in ^{29}SiO, have been seen in the ^{28}SiO emission.

The region around the Orion BN/KL infrared cluster thus reveals, *mainly due to its proximity*, a number of interesting phenomena. The infrared objects, or at least some of them, seem to power a very energetic and apparently irregular gas flow. Wide velocity emissions ($\Delta v \gtrsim 100$ km/sec) from a number of molecules, including H_2, are present (see for example Rydbeck et al., 1981, (124)), as well as OH, H_2O, and SiO masers.

13. CH - THE WEAK UBIQUITOUS MASER

The radio astronomical detection and identification of the ground rotational state, $^2\pi_{1/2}$, J=1/2, of CH (Rydbeck et al., 1973, (5), the complete, regular Λ doublet; Turner and Zuckerman, 1974, (125), - the upper satellite), was made only after several unsuccessful attempts in preceding years (cf. Robinson, 1967, (126), and Rydbeck, 1974 - Radioastronomischer Nachweis von interstellaren CH-Radikalen, (6)).

Two previous Onsala searches for CH, with rutile travelling wave masers in the 25.6 m telescope (September, 1968 in the frequency range 3000 - 3135 MHz; October-November 1971 in the 3361 - 3385 range) gave negative results. Not surprising, because the CH lines do not lie in the bands searched. The final, rapid Onsala detection of the entire doublet (one main line and two satellites; cf., the irregular $^2\pi_{1/2}$, J=1/2 Λ doublet of OH) was facilitated by the use of a more sensitive maser (zenith system noise ≤ 35 K), with an appropriate tuning range, by precalculation of the expected hyperfine (satellite) splittings, and by a pattern recognition technique, through three spiral arm sections, against which the Cas A non-thermal radio source was seen. This (practical) method was necessary, since the CH line frequencies were not well known, and the signals, as expected, were very weak. What was finally found to be a trace of the main line, was seen as a doodle in all three spiral arm channels (Orion & Perseus). When this happened, the gradual change in local oscillator frequency was stopped, and the LO-frequency counted, while integration went on. The maximum CH antenna temperature, against Cas A, became low, as expected, about +0.22 K, whereas OH & H_2CO were seen in strong absorption, as shown in Fig. 30. This was the first indication that CH behaved as a weak maser, at least when seen against Cas A. Rapid searches toward most of the important HII regions, molecular and cool dust clouds revealed that CH was omni-present and was weakly masing. Its ubiquitous nature lead us to classify it as a main line, i.e. the transition F=1 - 1, for which the frequency became 3335.5 MHz ($=\nu_{11}$).

This turned out to be right. The satellite lines were detected after a few weeks of additional frequency-scan, pattern recognition searches, also toward Cas A. Like the main line, the

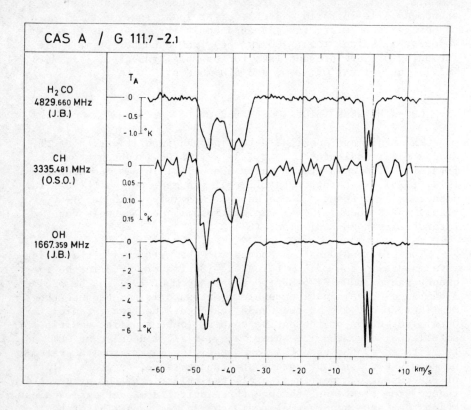

Figure 30: Historic CH, main line emission, discovered (by pattern recognition methods) toward the radio source Cas A, with the OSO 25.6 m telescope, using a rutile travelling wave maser, especially built for the "discover-CH"-project. The CH satellite lines were also seen in emission against Cas A. Interstellar CH was omnipresent, permanently with a net Λ-doublet inversion.
The OH & H_2CO Cas A absorption spectra (obtained by R. Davies at Jodrell Bank) were most useful, when the CH rest frequencies were determined.

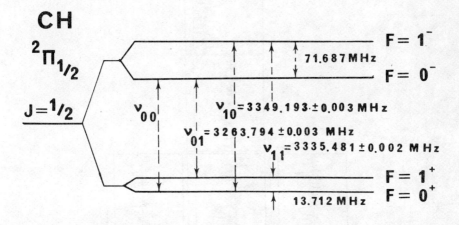

Figure 31: Energy levels of the CH ground state, a regular Λ-doublet, $^2\pi_{1/2}(J=1/2)$. By extensive, and careful observations (with the OSO, tw maser equipped 25.6 m telescope), the rest frequency of the main-line, $F=1^- - 0^+$, could be determined with an error of a mere ± 2 kHz (as compared to ± 3 kHz for the satellites; CH being a regular doublet has only one main-line in its ground state). Thus far, the ground state rest frequencies of CH (a chemically agressive radical) have not been determined in the laboratory, almost 11 years after their determination in interstellar space.

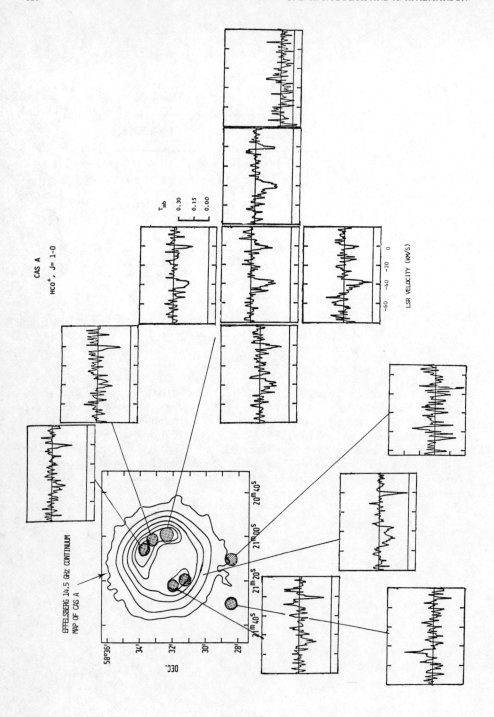

Figure 32: Absorption spectra of HCO^+, J=1-0 ($\sim$89.2 GHz), recently observed toward Cas A by L.-Å. Nyman (OSO staff member) with the OSO 20 m telescope. At such high frequencies (about 30 times those of ground state CH), the antenna beam (shown in the picture) is so much smaller than the angular size of the source, that rest frequency determinations, of the kind done for CH become impossible.

satellites were found to be masing. Only in sources with faint long wave, IR background radiation, did the satellites approach their LTE ratios, viz., 50% of the main line. The satellite lines were recorded at the frequencies 3349.2 MHz = ν_{10}, F=1 - 0, and 3263.8 MHz = ν_{01}, F=0 - 1. These results, presented November 9, 1973, at the annual meeting of the Swedish Royal Society in Uppsala (see "De celesta molekylernas radiostrålning", by Olof Rydbeck; Kungliga Vetenskaps-Societeten i Uppsala, Årsbok 1974, p. 37), are demonstrated in Fig. 31, with later higher order corrections of the main and satellite line frequencies.

In order to improve the frequency determination, the profile for the main line of CH, in the direction of Cas A, was compared with OH profiles taken by R.D. Davies and H.E. Matthews at Jodrell Bank, 1972, and with unpublished Onsala OH data. The presence of two components in the Orion arm feature at v ∼ -1 km/s, and the characteristic structure of the Perseus arm feature allows the CH and OH spectra to be compared with an accuracy of about ± 1 kHz (using spectra with higher resolution than those shown in Fig. 30). Another estimate (of $\delta\nu_{11}$, the main line error) was obtained by comparison of the main line spectrum of CH with that of H_2CO. A third, independent estimate of $\delta\nu_{11}$ was obtained by comparing the Onsala CH spectra with those of other molecules, observed towards four dark clouds. For details about the frequency determinations, see Rydbeck et al. 1974, (127). It is remarkable, that these ground state frequencies, have not been determined in the laboratory as yet, an indication of the chemical aggressiveness of the CH radical and of the difficulty of the experiment. Finally, it should be added that line frequency determinations, based on observations toward Cas A, are permissible only when the antenna beam is much greater than the angular size of Cas A. When this is not the case, one obtains position dependent spectra like those recently obtained by L-Å. Nyman (of the Onsala group) for HCO^+ (J=1 - 0; ∼ 89 GHz), with the 20 m telescope pointing across Cas A, see Fig. 32.

To demonstrate typical satellite to main line ratios for very different sources (for details, see also Rydbeck et al., 1974, (128)), we have in Figs. 33 & 34 reproduced spectra obtained towards Heiles Cloud 2 and W49. In the former case the line ratios approach the LTE (or statistical) values, in the latter, the upper satellite almost disappears in the 15 km/s velocity feature, associated with the HII, far-IR source. Not very much is left of the main line. On very rare occasions, such as toward M17, both upper lines are seen in light absorption through part of the spectrum (Rydbeck et al., Fig. 2 in (128)).

Fig. 35 demonstrates the CH energy level structure as presented by us in 1976, (129); for later upper state modifications, see M. Bogey et al., 1983, (130). The difference between the (regular) CH and the (irregular) OH states are noteworthy, and could be one of the reasons for CH always being such a weak maser. The enhancement of the lower satellite, over the statistical level (the overall, or net inversion of the ground state Λ-doublet is due to collisions, preferentially by atomic hydrogen, see (128), (129), and D. Bouloy et al., (131)), arises through the interaction of CH with far-IR dust induced radiation lines at about 150 μm ($\sim$96 K). It is of particular interest to study Fig. 35, since it indicates that the overlap of the pair of lines, connecting the upper halves of the Λ-doublets of $^2\pi_{3/2}$, J=3/2, and $^2\pi_{1/2}$, J=1/2 should be efficient in creating population inversion in the lower satellite of J=1/2 (as is frequently observed). However, both the far-IR field and the continuum background (sources) must be strong. This is in good agreement with the observations.

Furthermore, it should be noticed that also the pair of far-IR lines, which connect the lower Λ-doublet parts, should bring about an inversion, viz., of the upper satellite line. Such an enhancement has never been observed, however. Its permanent absence is probably due to the fact, that the hyperfine splitting of the lower J=1/2 level is as small as 14 MHz (corresponds to a velocity of about 1 km/s for the 150 μm line); see also (129).

Finally, it should be added that Bouloy et al., (131), have performed a systematic evaluation of the rate coefficients for collisions of CH with H and H_2. The coefficients for inelastic collisions are quite complicated, since they actually are based on the transfer of population from the four hyperfine levels of the rotational ground state ($^2\pi_{1/2}$, J=1/2), within the doublet itself, as well as to excited rotational states. To verify such results by experiments appears next to impossible for CH, whose ground state radio lines (ν_{11}, ν_{10}, ν_{01}) have not even been found in the laboratory.

In 1976 the Onsala group (Sume & Irvine) looked for the $^2\pi_{1/2}$, J=5/2 rotational state through the range 4677 - 4850 MHz, using a travelling wave maser in the 25.6 m telescope, towards a few selected objects. No line was detected to a peak-to-peak sensitivity of 0.1 K. Now that the main line frequencies have been determined in the laboratory, (Bogey et al., 1983, (130)), searches for the excited state should be resumed.

A preliminary survey of galactic CH, in the main line and in the galactic plane, has been performed by L.E.B. Johansson (1979, (132)). The large scale CH distribution is characterized by a maximum at a galactocentric distance of about 5.5 kpc. More than 90 percent of emission is found within the galactocentric distance interval 3 to 10 kpc. Observations perpendicular to the galactic plane at $\ell = 30°$, and in the W49 area, indicate a CH

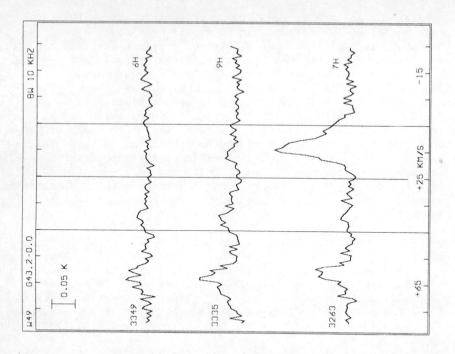

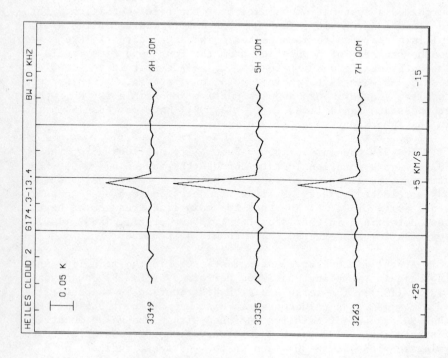

Figure 33: CH main and satellite lines (the ground state triplet) observed toward the cool dust cloud, Heiles 2, with the OSO 25.6 m telescope. The satellite lines are now seen almost in their LTE (statistical) ratio, 0.5 of the main line. One should note the extreme sensitivity of the maser radiometer system.

Figure 34: CH spectra obtained toward W49A, which demonstrate spiral arm features at about +62 and +41 km/s. The remarkable +15 km/s feature is associated with the HII, far-IR source, whose dust induced radiation lines (at about 150μm) probably over-inverts the $0^- \to 1^+$ transition (overpopulates the 0^- level; see also Fig. 35), i.e. it enhances the lower satellite line — a commonly observed phenomenon. Furthermore, it should be remarked, that both the upper satellite and the main line have, in extremely limited positions, been seen in faint absorption against a (strong) far-IR-background.

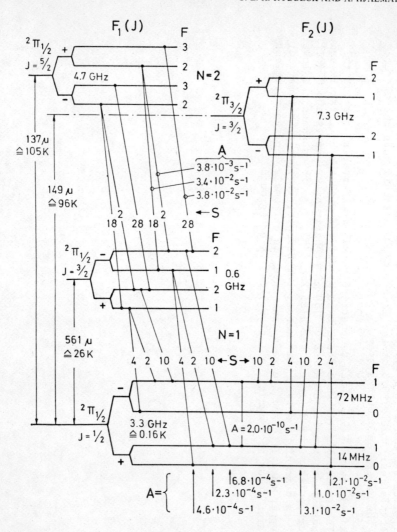

Figure 35: Approximate CH energy level structure, presented by Rydbeck et al. 1976 (129). A comparison with the OH structure (Fig. 12) demonstrates the great differences between the states of the regular and irregular Λ-doublets. CH is always inverted, but never a strong maser. OH is not always inverted, but frequently a strong maser.

layer thickness comparable to that found for CO, about 50 pc.

As has earlier been mentioned, one of the Onsala travelling wave CH masers was sent on loan to the Parkes 64 m telescope in Australia (operated by CSIRO) - and installed by one of the Onsala engineers, for a long term of observations towards the southern hemisphere (42).

The 3264-transition (the lower satellite line) of CH was detected in three galaxies, the Large Magellanic Cloud (LMC), NGC4945, and NGC5128. The line profiles were similar to those for H_2CO. The main line, the 3335 MHz transition, was detected in LMC and (probably) in NGC4945. With the exception of NGC5128, the line/continuum ratios were similar to those in our Galaxy. The results, furthermore, suggest an overabundance of CH relative to OH and H_2CO in a molecular cloud in NGC5128, which is probably located well in front of the nucleus. Generally, the observations indicate that the lower satellite line is a good tracer of regions with intense far-IR (150 μm) radiation.

As a by-product of the Parkes - Onsala observations, the $2_{11} - 2_{12}$ transition of H_2CS (thioformaldehyde), previously observed only towards Sgr B2 (1973), was detected in absorption against a total of 11 continuum sources. $[H_2CO]/[H_2CS]$ abundance ratios were also determined in the course of the sensitive investigations. The overall system noise temperature on cold sky was about 40 K.

A comparison of the results with profiles of the $1_{10} - 1_{11}$ transition of H_2CO, led to an improved rest frequency of 3139.391±0.002 MHz for the H_2CS transition (as accurate as the Onsala determinations of the CH rest frequencies).

* * *

Quite recently, Lovas et al. (1983, (133)) made a laboratory detection of the $4_{04} - 3_{13}$ rotational transition of CH_2, which has the following three fine structure components, J=5 - 4, at about 68.37 GHz; J=3 - 2, at 69.01; & J=4 - 3 at 70.68 GHz, all with hyperfine triplets.

Blint, Marshall & Watson, see (133), in 1978 concluded that a high formation rate of CH_2 is "unavoidable". Black, Hartquist, and Dalgarno (133), in the same year, modelled the ζ Persei cloud, predicting a CH_2 abundance larger than that observed for CH. Subsequently, Prasad and Huntress presented an Orion A cloud model for CH_2 (1980, (133)) and predicted substantially more CH_2 than C_2H (which has been observed in this source).

Unfortunately, the $4_{04} - 3_{13}$ transition is the only rotational **line of** CH_2 occurring below 400 GHz and arising from fairly low

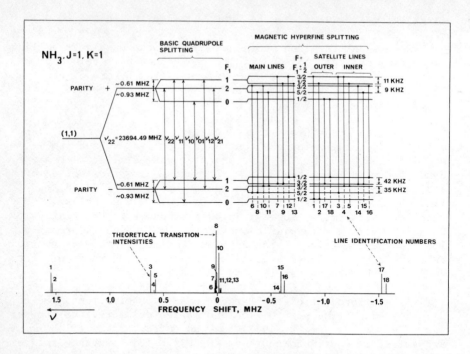
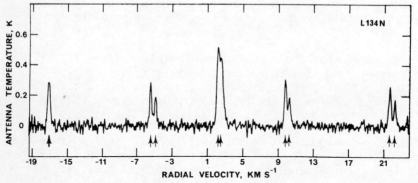

Figure 36: Sketch of the quadrupole and hyperfine energy levels for the (1,1) ammonia state, with a (LTE) spectral distribution profile. Note the many small components of the main-line group, and the conspicuously different splittings of the outer satellite groups.

Figure 37: 2 MHz, 512 channel, (1,1) ammonia spectrum from the dust cloud L134N. Arrows indicate the strongest hyperfine lines as given by S.G. Kukolich in 1967 (Phys. Rev. 156, 83). Since a kinetic temperature of 10 K yields linewidths greater than 0.16 km/s (those actually observed by the authors are about 0.3 km/s), it is impossible to resolve the highest outer satellite feature. There is a striking agreement between the theoretically and experimentally determined line group ratios, which strongly indicates that LTE excitation conditions prevail for the hyperfine lines (line groups) in the L134N dust cloud. The NH_3(1,1) spectrum shown (probably unmatched in quality), obtained with the OSO 20 m telescope, is one of several examples of what can be done with a good travelling wave maser in a good, moderate size telescope.

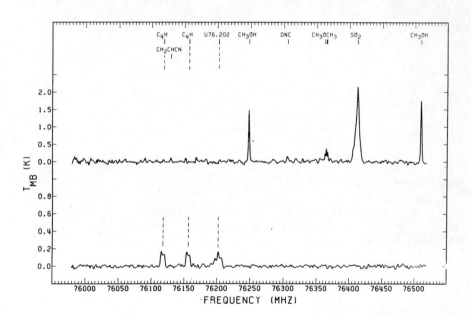

Figure 38: 76.0 - 76.5 GHz spectra towards the Orion A molecular cloud and the IRC+10216 circumstellar envelope - an illustration of differing chemical compositions. The peculiar line shape of U76.202 is ascribed to line overlap due to Λ-doubling of the J=7/2 and 5/2 energy levels of the $^2\pi_{1/2}$ state of the new interstellar radical C_3H (see § 16 c).

energy states, but still about 150 cm^{-1}. Actually, the next lowest transition, $2_{12} - 3_{03}$, lies near 444 GHz, and thus would be accessible only by the airborne observatory technique used by Phillips and Waters (already referred to) to detect sub-millimeter wave H_2O emissions.

There seems to be no possibility of detecting the next variants, CH_3 & CH_4, by radio methods. CH_3 appears to be plane, and CH_4 develops an appreciable electric moment first when it rotates very rapidly.

14. SOME SAMPLES OF OBSERVATIONAL RESULTS OBTAINED WITH MM TW MASER RADIOMETER SYSTEMS

The first OSO iron doped, rutile travelling wave maser (134) was designed for H_2O, as well as NH_3 high stability observations, in the 22 - 24 GHz range. The local oscillator was phase-locked to the observatory hydrogen maser frequency standards. The zenith system noise was less than 90 K in good weather. A scalar feed horn "illuminated" the movable Cassegrain subreflector. A 512 channel autocorrelation receiver was used with total bandwidths of 2 and 7.5 MHz, which yields resolutions of 7.8 & 29.3 kHz, or 0.10 and 0.37 km/sec at 24 GHz. The atmospheric, and radome attenuation was estimated from sky noise measurements at various elevations.

To resolve the quadrupole groups of the NH_3(1,1) para transition, as well as the stronger magnetic hyperfine lines of the former and of the main line feature, shown in the complicated energy level structure, Fig. 36 (Rydbeck et al., (135)), was the right challenge for the maser system advocates. Fig. 37 shows the result (obtained toward the dust cloud L134N), where arrows indicate the positions of the strongest hyperfine lines (135). It is, even today, one of the highest resolution, high sensitivity spectra obtained, and demonstrates the tremendous advantage of an ultra low-noise, *high stability* radiometer system.

It appears from Fig. 37, that interstellar ammonia must be an excellent tracer of interstellar cloud structure, density and temperature. The important and interesting properties of interstellar ammonia, in all its phases, have recently been reviewed by Ho, and by Townes (136), the far-seeing co-discoverer of NH_3.

As further examples of high resolution, high sensitivity spectroscopy in space, we may mention the astronomical detections and studies of the C_3N, and C_4H radicals in TMC-1 by Friberg et al. (1980, (137)), Irvine et al. (1981, (1938)), and Guélin et al. (1982, (139)). That these detections were at all possible was due to the maser preamplifiers, at $\sim$30 and 20 GHz, available at OSO, and NRAO (at Green Bank). These interesting carbon chain species have recently been studied in the laboratory by Gottlieb et al. (1983, (140)).

15. ON SEARCHES FOR NEW INTERSTELLAR MOLECULES

Such observations essentially have been performed in two opposite ways:
i) as *dedicated searches for specific molecules*, with laboratory measured, or otherwise estimated, transition frequencies. However, even a search for a specific species may require a very time consuming "spectral scan", if its frequency is uncertain. Perhaps the most difficult case of this kind is CH, which had to be sought, at observatories in Australia, Sweden, and USA, over a 15% range around its "nominal" frequency. Here a certain species is chosen, since it is simple enough, more complex than, or related to, already known compounds in a specific cloud, believed to be abundant, biologically important, or just because it is "interesting", or
ii) as time consuming, *unbiased spectral scans over large frequency ranges*, to a more or less uniform sensitivity level. Here attempts to identify new molecules, guided by laboratory spectroscopy data, harmonically related line frequencies, ab initio molecular quantum mechanics, etc., can only be performed after the "elimination" of all transitions due to already known interstellar species.

In Table 1 we list the molecules detected (as of spring 1984), including 45 isotopic variants. Of the 64 species entered roughly 25% are inorganic, 50% organic, and 25% are classified "unstable" (i.e., in terrestrial laboratories). 57 of these compounds are "radio molecules": HC_2H, $H_2C_2H_2$, CH_4, and SiH_4 have been detected in the envelope of the carbon star IRC+10216, in absorption against its infrared background, while H_2, C_2, and CH^+ have been observed in "diffuse" interstellar clouds (in absorption by UV or optical transitions). Although there is overwhelming *indirect* evidence that H_2 is the dominant species (collision partner) in dense gas clouds – where other molecules are only trace constituents, in total contributing to < 1% of the gas mass – the only *direct* evidence for H_2 in dense regions rests upon its (shock excited) rotation-vibration 2 μm, and rotation 12 μm lines. Although the majority of the interstellar molecules were detected in dedicated searches at specific rest frequencies, beginning with OH in 1963; NH_3 in 1968; H_2O and H_2CO in 1969; CO, CN, HCN, HC_3N, CH_3OH, and HCOOH in 1970, etc., it also very early became clear that the surprises of the interstellar medium not only consisted of anomalous excitation (maser and daser effects, as previously discussed), but also of *emission lines at unexpected frequencies*. The first such discovery, in 1970 by Buhl and Snyder (141),

INTERSTELLAR INORGANIC MOLECULES

DIATOMIC				TRIATOMIC				TETRAATOMIC			
H_2	[D]	IR, UV	(a)	H_2O	[D, ^{18}O(?)]		(a)	NH_3	[D, ^{15}N]	R, IR	(a)
	molecular hydrogen				water vapor				ammonia		
CC		IR, UV	(a)	H_2S			(a)				
	diatomic carbon				hydrogen sulphide						
CO	[^{13}C, ^{17}O, ^{18}O]	R, IR, UV	(a)	HNO			(a)				
	carbon monoxide				nitroxyl						
CS	[^{13}C, ^{33}S, ^{34}S]		(a)	NaOH(?)			(g)				
	carbon monosulphide				sodium hydroxide						
NO			(a)	OCS	[^{13}C, ^{34}S]		(a,n)				
	nitric oxide				carbonyl sulphide						
NS			(a)	SO_2	[^{34}S]		(a,j)				
	nitrogen sulphide				sulphur dioxide						
SiO	[^{29}Si, ^{30}Si]	vibr.	(a)								
	silicon monoxide										
SO	[^{34}S]		(a)								
	sulphur monoxide										
SiS	[^{29}Si, ^{30}Si, ^{34}S]		(a,n,p)								
	silicon sulphide										

INTERSTELLAR 'UNSTABLE' MOLECULES

RADICALS				IONS				ISOMERS			
CH		R, Opt.	(a)	CH^+	[^{13}C]	Opt.	(a,b)	HNC	[D, ^{13}C, ^{15}N]		(a)
	methylidyne				methylidyne ion				hydrogen isocyanide		
OH	[^{17}O, ^{18}O]	R, IR, UV	(a)	HN_2^+	[D, ^{15}N]		(a,m)				
	hydroxyl										
CN		R, Opt.	(a)	HCO^+	[D, ^{13}C, ^{17}O, ^{18}O]		(a,h)	HOC^+(?)			(1)
	cyanogen				formyl ion				isoformyl ion		
HCO			(a)	HCS^+			(d)				
	formyl				thioformyl ion						
C_2H			(a)	$HOCO^+$			(d)				
	ethynyl				(could be HOCN)						
C_3H(?)			(n)								
C_4H			(a)								
	butadinyl										
C_3N			(a)								
	cyanoethynyl										
C_3O			(q)								
SiC_2			(u)								

TABLE 1 Detected molecules (including recent references)

RADIO OBSERVATIONS OF INTERSTELLAR MOLECULES

INTERSTELLAR ORGANIC MOLECULES

(i) ALCOHOLES

CH_3OH [D, ^{13}C]   (a)
   methanol

CH_3CH_2OH (a)
   ethanol

(ii) AMIDS

NH_2CN (a)
   cyanamide

NH_2CHO (a)
   formamide

CH_2NH (a)
   methanimine

CH_3NH_2 [D]   (a)
   methylamine

(iii) CYANIDES

HCN   [D, ^{13}C, ^{15}N] R, IR (a)
   hydrogen cyanide

CH_3CN [$^{13}C(?)$]   vibr. (a,k)
   methyl cyanide

CH_2CHCN (a)
   vinyl cyanide

CH_3CH_2CN (a)
   ethyl cyanide

(i) ALDEHYDES & KETONES

H_2CO [D, ^{13}C, ^{18}O]   (a)
   formaldehyde

CH_3CHO (a)
   acetaldehyde

CH_2CO (a)
   ketene

(ii) ESTERS & ETHERS

$HCOOCH_3$ (a)
   methyl formate

$(CH_3)_2O$ (a)
   dimethylether

(iii) ACETYLENE DERIVATIVES

HC_2H IR (a)
   acetylene

$H_2C_2H_2$ IR (f)
   ethylene

CH_3C_4H (s)
   methyldiacetylene

CH_3C_3N (c)
   methylcyanoacetylene

HC_3N [D, ^{13}C]   (a,c)
   cyanoacetylene

HC_5N [D]   (a,e)
   cyanodiacetylene

HC_7N (a)
   cyanotriacetylene

HC_9N (a)
   cyanotetraacetylene

$HC_{11}N$ (i)
   cyanopentaacetylene

(i) ACIDS

HNCO   (a)
   isocyanic acid

HCOOH   (a)
   formic acid

(ii) SULPHUR COMPOUNDS

H_2CS (a)
   thioformaldehyde

CH_3SH (a)
   methylmercaptan

HNCS   (a)
   isothiocyanic acid

(iii) OTHERS

CH_4 IR (a)
   methane

SiH_4 IR (t)
   silane

TABLE 1 REFERENCES:

a. Cummins, A.P.C., and Williams, D.A.;
   Nature, 238, No. 5749, 721, (1980).
b. Vanden Bout, P.A., and Snell, R.L.;
   Ap. J., 236, 460, (1980).
c. Langer, W.D., Schloerb, F.P., Snell, R.L., and Young, J.S.;
   Ap. J. (Lett.), 239, L125, (1980).
d. Thaddeus, P., Guélin, M., and Linke, R.A.;
   Ap. J. (Lett.), 246, L41, (1981).
e. McLeod, J.M., Avery, L.W., and Broten, N.W.; Schloerb, F.P.,
   Snell, R.L., Langer, W.D., and Young, J.S.;
   Ap. J. (Lett.), 251, L33 & L37, (1981).
f. Betz, A.L.;
   Ap. J. (Lett.), 244, L103, (1981).
g. Hollis, J.M., and Rhodes, P.J.;
   Ap. J. (Lett.), 262, L1, (1982).
h. Guélin, M., Cernicharo, J., and Linke, R.A.;
   Ap. J. (Lett.), 263, L89, (1982).
i. Bell, M.B., Feldman, P.A., Kwok, S., and Matthews, H.E.;
   Nature, 295, No. 5848, 389, (1982).
j. Schloerb, F.P., Friberg, P., Hjalmarson, Å., Höglund, B.,
   and Irvine, W.M.;
   Ap. J., 264, 161, (1983).
k. Cummins, S.E., Green, S., Thaddeus, P., and Linke, R.A.;
   Ap. J., 266, 331, (1983).
l. Woods, R.C., Gudeman, C.S., Dickman, R.L., Goldsmith, P.F.,
   Huguenin, G.R., Irvine, W.M., Hjalmarson, Å., Nyman, L.-Å.,
   and Olofsson, H.;
   Ap. J., 270, 583, (1983).
m. Linke, R.A., Guélin, M., and Langer, W.D.;
   Ap. J. (Lett.), 271, L85, (1983).
n. Johansson, L.E.B., Andersson, C., Elldér, J., Friberg, P.,
   Hjalmarson, Å., Höglund, B., Irvine, W.M., Olofsson, H.,
   and Rydbeck, G.;
   A. & A., 130, 227, (1984).
o. Broten, N.W., McLeod, J.M., Avery, L.W., Irvine, W.M.,
   Höglund, B., Friberg, P., and Hjalmarson, Å.;
   Ap. J. (Lett.), 276, L25, (1984).
p. Ziurys, L.M., Clemens, D.P., Saykally, R.J., Colvin, M.,
   and Schaefer, H.F.;
   Ap. J., 281, 219, (1984).
q. Matthews, H.E., Irvine, W.M., Friberg, P., Brown, R.D., and
   Godfrey, P.D.;
   Nature, in press, (1984).
s. Walmsley, C.M., Jewell, P.R., Snyder, L.E., and Winnewisser,
   G.; Astron. Astrophys., 134, L11, (1984).
t. Goldhaber, D.M., and Betz, A.L.;
   Ap. J. (Lett.), 279, L55, (1984).
u. Thaddeus, P., Cummins, S.E., and Linke, R.A.; Preprint (1984).

(U89.2 GHz; U = unidentified), was a strong line (still one of the strongest!), widespread among the dense interstellar clouds. In the same year Klemperer (142) suggested that this line could be due to the lowest rotational transition of the *molecular ion* HCO^+. However, a secure identification had to await the successful laboratory synthesis and frequency measurement in 1975 by Woods et al. (143), although, almost simultaneously, Snyder et al. (144) had detected, in a number of interstellar clouds, a weak line at 86.754 GHz, close to the theoretically predicted $H^{13}CO^+$ frequency 86.720 GHz. Another important accidental discovery was that of the triplet U93.2 by Turner in 1974 (145), which was tentatively identified with another ion, HN_2^+, in an accompanying Letter by Green et al. (146). The laboratory verification was again done, in 1976, by Wood's group (147). *(*)* In the meanwhile "astrochemists" (Herbst and Klemperer, 1973 (149); Watson, 1974 (150)) had outlined the *ion-molecule reaction pathways*, which in more elaborate forms are still the common belief.

(*) A detailed account of these and other "interstellar detective stories" may be found in the excellent review chapter <u>Astrophysics of Interstellar Molecules</u>, by Winnewisser et al. (148).

In this context it should be remembered that even the extremely strong Orion SiO maser at 86.2 GHz, a rotational transition in the first vibrationally excited state (v=1, J=2 - 1), was an accidental detection in 1974 by Snyder and Buhl (151).

These and other chance detections of new lines formed a background to the present authors' early discussions of important millimeter wave projects for the new Onsala 20 m telescope. It thus seemed very desireable to "supplement" the narrow-band spectrometers, necessary for accurate observations of known interstellar species, with a wide-band spectrometer, by means of which "accidental detections" as well as "line searches over large frequency intervals" would be possible. "Besides", such a broad-band receiver was necessary for the foreseen observations of molecules in external galaxies (if the telescope system behaved as we hoped for). It may be interesting to mention that our considerations were further stimulated by the timely arrival of a most interesting preprint by Lóvas et al. (152). Inspired by their earlier (1974) discovery of $(CH_3)_2O$ (dimethyl ether) in Orion, they had searched, but with negative results, for the similarly complex species $(CH_2)_2O$ (ethylene oxide), $(CH_2)_2CO$ (cyclopropenone) and $(CH_3)_2CO$ (acetone). However, they "instead" - thanks to the 256 x 1 MHz filter bank then available at the NRAO Kitt Peak 11 m telescope - had discovered the new interstellar molecule SO_2, as well as a considerable number of lines from CH_3OH, CH_3CN, CH_3C_2H, $CH_3CHO(?)$, and ^{29}SiO.

The decision became a heavy (in the OSO reference frame) investment of manpower and money to build a stable 512 x 1 MHz filterbank spectrometer - covering the full bandwidth of the low noise 3 mm frontend also under development - combined with a 256 x 250 kHz "magnifier" to "tune in" interesting lines (The observatory already had in operation a 512 channel autocorrelator, with a flexible bandwidth between 0.5 and 15 MHz, for high resolution work.).

16. THE ONSALA SPECTRAL SCAN OF ORION A AND IRC+10216

This project was initiated in winter 1979, when a low noise 3.5 mm Schottky mixer was installed on the 20 m telescope. We decided to take advantage of the increased sensitivity towards small objects, provided by the narrow antenna beam (compared to earlier observations, mainly with the NRAO 11 m and Bell Laboratories 7 m telescopes), to survey the spectral range accessible to the mixer towards two, chemically different, but already well-studied sources: the young star formation region (Kleinmann-Low infrared cluster) in the Orion A molecular cloud and the carbon rich, mass-loss envelope around the evolved star IRC+10216. The differing chemical composition of the two objects may be illustrated by Figure 38.

Already during the very first days of observation a considerable number of "new" lines turned up in the Orion spectra. These first results were reported in our 1980 Letter *On Methyl Formate, Methane and Deuterated Ammonia in Orion A* (153), presenting the secure detection of CH_3OOCH in this cloud through the observation of some 25(!) lines clearly assigned to this species. Methyl formate transitions at 76.702, 76.711, 85.927, and 86.224 GHz corresponded to spectral lines previously attributed to CH_4, NH_2D, and $(CH_3)_2O$. This illustrates the problem of chance blend (of weaker lines), and clearly demonstrates that identification of a new species has to rest on more than one transition (or isotopic variant). Considering the observed CH_3OOCH signal strengths it appears that the detection of CH_4 in Orion must be considered very tentative. The NH_2D detection rests on a single line at $\sim$ 110 GHz, while the 85.927 GHz line may be a blend of $HCOOCH_3$ and NH_2D. Dimethyl ether, however, still was securely identified by a number of other lines and "its" 86.224 GHz feature, in fact, had been explained as an excitation anomaly (154). It is obvious that accurate laboratory spectroscopy data are crucial for this kind of identification work. In case of CH_3OOCH a detailed laboratory analysis by Bauder (155) was published in the valuable NBS Series on Molecules of Astrophysical Interest during the course of our work, and to secure our assignment of the two 76.7 GHz lines as mainly due to methyl formate Destombes and Lemoine of Université de Lille kindly measured their rest frequencies with high precision.

These first results gave us great hope for detecting new lines, and perhaps also new interstellar molecules. The results of this OSO spectral scan of the frequency range 72.2 to 91.1 GHz have now been published by Johansson et al. (156), and we will here briefly summarize, and expand on, some of the more interesting findings from this unbiased search to a sensitivity level (in terms of main beam brightness temperature) normally below 0.1 K:

a. New lines

In total $\sim$ 240 lines were observed, more than *half of which were astronomically new detections*. Approximately 170 lines, from 24 already *known* interstellar molecules, were found in Orion, as well as some 15 hydrogen ($\alpha, \beta, \gamma, \delta$) and helium ($\alpha$) recombination lines. Of the > 100 astronomically new detections in Orion, the majority is due to the asymmetric rotors CH_3COOH (in total 32 lines), $(CH_3)_2O$ (21), CH_3CH_2CN (19), CH_3OH (16), and SO_2 (17)- and the recombination lines. In IRC+10216 we observed $\sim$ 45 transitions from 12 already known interstellar species. Again, about half of these lines were new detections.

b. Line identifications

The *rare occurrence of unidentified lines* ($\sim$ 20 in Orion, $\sim$ 15 in IRC+10216), with only 5 definite ones in each object *(*)*, appears significant, as does our *assignment of most lines to already known interstellar species*. It should be pointed out,

---

(*) Our detection criteria have been quite hard, however, and a considerable number of possible lines have not been listed. For this reason all OSO scan spectra will be made publicly available.

---

however, that this identification work, not only included consultation of laboratory spectroscopy tabulations, but also required some new laboratory work. Among our Orion U-lines remained for a while three strong ones, at 84.424, 88.595, and 88.940 GHz, and also some weaker, but still conspicuous ones, which were finally assigned to high energy transitions of the abundant species CH_3OH, in close cooperation with R.M. Lees at University of New Brunswick, whose laboratory data now have been published (157).

Other important laboratory work in this connection was performed by R.C. Woods at University of Wisconsin. The strong U-line at 88.940 GHz fell close to the J=1 - 0 frequency of the ion HOC^+, 88.83 GHz, as *ab initio* calculated already in 1976 by Herbst et al. (158). Since its isomer HCO^+ was a widespread interstellar ion, and since HNC (the isomer of HCN, and isoelectronic with HOC$^+$) also was an abundant interstellar species, the isoformyl ion appeared to be an attractive candidate. Subsequently HOC$^+$ was successfully laboratory synthesized, but its frequency was measured to be 89.487 GHz, where no line appears in our Orion data. This ion was sought to low levels in many clouds at OSO and at FCRAO (Five College Radio Astronomy Observatory, University of Massachusetts) and a weak feature was finally detected at a low level in Sgr B2 by Woods et al. (159). Although this identification must be considered tentative, since it rests on only one transition, the low detection limits show that HOC$^+$ is at least two orders of magnitude less abundant than HCO$^+$, which may set useful constraints on chemical models (cf. De Frees et al. (160) and discussions in (161)). Although the HOC$^+$-case may be considered "a failure and a success", it certainly demonstrates, again, the importance of ab initio guidance for interstellar identifications as well as for laboratory work. Herbst et al. were accurate to within 0.7%. Recently Kraemer and Bunker (162) have performed an extended ab initio study of HOC$^+$ and its isotopes. Their best model fits with the laboratory data to within 0.05%.

If we exclude our two strongest (~ 0.3 K) Orion U-lines (which may tentatively be attributed to **torsionally excited** CH$_3$OH !), the remaining ones are all $\lesssim 0.2$ K, as observed with 1 MHz resolution. For comparison, we note that this is the intensity level of many of the detected CH$_3$OOCH and (CH$_3$)$_2$O transitions. At the present sensitivity level about 90% of the observed lines have been identified with known (interstellar) species, while ~ 10% are low intensity U-lines. A similar result has recently been reached in the 1.3 mm range at Owens Valley Radio Observatory by Sutton et al. (163), where also ~ 90% of their ~ 500 (!) lines detected towards Orion A could attributed to already known species. We will later on return to the question, whether these disappointing results mean that the "more important" molecular constituents now have been detected, or not. But first let us turn to some more positive aspects of the OSO spectral scan.

c. New species detected

We report (or confirm simultaneous results on) the existence in Orion of methyl formate (CH$_3$OOCH), ketene (H$_2$CCO), vinyl cyanide (CH$_2$CHCN), isocyanic acid (HNCO), cyanodiacetylene (HC$_5$N), and probably the formyl radical (HCO), as well as the first astronomical detection of the rare isotopic species ^{34}SO$_2$.

In IRC+10216 we have observed the first species containing a CH$_3$ group, methyl cyanide (CH$_3$CN) and report the first astronomical detections of ^{29}SiS and Si^{34}S. Harmonically related "U-lines", at 76.2 GHz (of peculiar line shape, probably an overlapping doublet, cf. Figure 38) and at 98.0 GHz (a clear doublet) are tentatively identified with the $^2\pi_{1/2}$, J=7/2 - 5/2 and 9/2 - 7/2 rotational transitions of C_3H - a new interstellar radical. (*)

(*) This assignment now seems more secure, after Thaddeus et al. (167) in a very recent communication have succeeded in identifying nine, comparatively strong (previously "irregularly scattered") U-lines in IRC+10216 as due to the asymmetric rotor SiC_2 - a "ring" species, and not a linear molecule as previously thought, as shown by a recent laser induced two-photon ionization experiment in a supersonic molecular beam by Michalopoulos et al. (168). Then a weaker, third C_3H-fitting doublet (J=13/2 - 11/2) remains at $\sim$ 141.7 GHz in the Bell Laboratories IRC+10216 data. In early June this year we detected at Onsala, with a maser preamplifier, in TMC-1 a doublet centered on 32.645 GHz, as predicted for C_3H (J=3/2 - 1/2).

The proposed identification was guided by the unusual carbon-chain chemistry in IRC+10216, the *ab initio* calculations of Green (164) and Cooper (165), the failure to detect harmonically related lines of other possible candidates such as HC_2NC, C_2NC and $HNSi$ (for which also ab initio information is available), and also by the non-detections towards Orion. Recent chemical model work by Nejad et al. (166) indicates that C_3H should be efficiently formed in the carbon rich stellar envelope.

d. Molecular excitation in different regions

The great number of lines detected for several species make possible comparatively precise determinations of *population distributions* (rotation temperatures). This is important not only for the accuracy of our molecular *abundance calculations*, but also provides useful estimates of *temperatures* and *densities*, in the various cloud features of the Orion Kleinmann-Low region, delineated by recent high spatial resolution single dish and aperture synthesis (VLA, Berkeley, and Caltech) data.

Even though the various sub-regions may not be resolved by a single radio telescope their emissions can be distinguished by their rather specific spectral signatures, as illustrated by Fig. 39. The ambient molecular cloud forms an elongated *ridge*, which across the Kleinmann-Low infrared cluster appears to be divided (*) into a "northern" and "southern" component (narrow lines of radial velocities $\sim$ 10 and $\sim$ 8 km s^{-1}, respectively). The latter region contains a localized, warmer, "clump" (the *compact ridge cloud*), where certain complex species seem to be concentrated. The *plateau* source is a joint notation for what is now interpreted to be an expanding *disk*, around a newly formed star, with a roughly orthogonal, bipolar-type, *high velocity outflow* (**). Semi-wide emission ($\Delta v \sim 10$ km s^{-1}) is seen from the *hot core* - a small, hot, dense region of yet uncertain physical

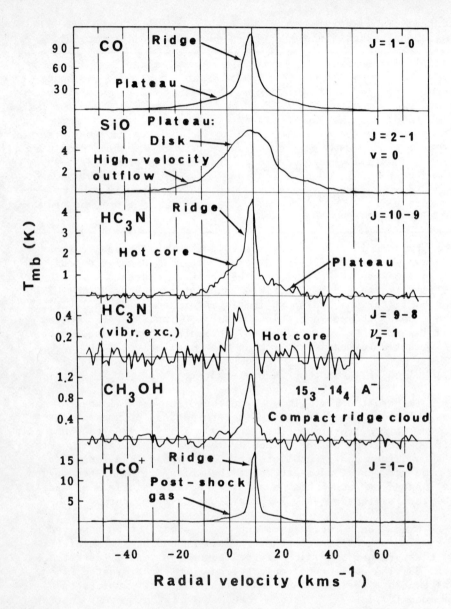

Figure 39: Spectra observed towards the Orion A region, illustrating the rather distinct spectral signatures of emissions from several separate sources near the Kleinmann-Low nebula (see § 16 d). The HCO+ spectrum, measured 80" north of K/L, displays broad emission ascribed to post-shock gas (cf. § 19 v).

origin. For further details and references we refer to the spectral scan report (156) and to recent papers by Friberg (171) and Olofsson (172).

(*) *An alternative interpretation, in terms of a slowly rotating, and contracting, accretion disk (surrounding the outflow region) very recently has been proposed (169), on the basis of CS (~ 6 mm) observations with the new Japanese 45 m telescope in Nobeyama. Although this interpretation certainly is interesting, further data seem to be needed to decide between the two alternatives. Here 3 mm work with the 45 m dish should provide the required angular resolution.*
(**) *This signpost of high mass-loss during the early evolution of massive, as well as small stars - deep inside their parental molecular clouds - is one of the important, but theoretically not yet well understood, discoveries of molecular radio astronomy, cf. Bally and Lada (170).*

Some examples of derived rotational temperatures ($T_{rot} \sim$ the average kinetic temperature over the emitting cloud in case of thermalization) may suffice to illustrate their importance for accurate abundance estimates. This knowledge also will be very important for our understanding of non-detections.

For the *extended ridge* we estimate (T_{rot} (CH_3C_2H) ~ 50 K, a value that according to the multi-level statistical equilibrium analysis by Askne et al. (173) should be close to the average cloud temperature (This is also close to the mean CO brightness temperature.). In contrast, the excitation temperature of the extended, optically thick HCN (J=1-0) emission is only ~ 20 K (cf. Rydbeck et al. 1981 (174)), so this species must be sub-thermally excited, if small scale clumping is not as important as in case of NH_3, (136, 175). On the other hand, our methanol rotation diagram in Fig. 40 yields T_{rot} (CH_3OH) ~ 140 K for the *compact ridge* cloud. Its lower energy transitions may, however, rather indicate ~ 90 K. In a similar way the *plateau* source temperature has been estimated from 14 SO_2 lines to be $\sim 90 - 150$ K, where the latter value relates to the higher energy transitions, cf. (176). In the *hot core* we find T_{rot} (CH_3CH_2CN) ~ 115 K. A recent statistical equilibrium analysis of the spectral scan CH_3CN data by Andersson et al. (177) - these transitions are not thermalized due to the large dipole moment (3.9 D, compared to 0.8 D for CH_3C_2H) - confirms the previous temperature estimates for the hot core and the compact ridge clouds.

For IRC+10216 the presumably radiatively excited high dipole moment species like HC_3N and HC_5N are estimated to have average rotational temperatures, $T_{rot} \sim 11-15$ K, from multi-line analysis, as well as from source mapping and modelling. The average excita-

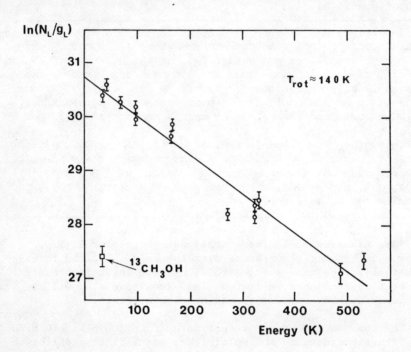

Figure 40: The population distribution of CH_3OH in Orion A, yielding $T_{rot} \sim 140$ K in the compact ridge cloud. The observed value for the $^{13}CH_3OH$ $5_{-1} - 4_0$ E transition indicates a $^{12}C/^{13}C$ ratio as low as 30-40 in this warm region.

tion temperatures for the optically thick J=1-0 transitions of HCN and CO, where the molecular envelope is partially resolved by the antenna beam, also fall in this range, cf. Olofsson et al. 1982 (178). Recently Rieu et al. (179) have modelled the excitation conditions for CO, ^{13}CO, SiO and SiS in this source.

e. Chemical composition

The information on molecular concentrations in various environments, together with the excitation information, perhaps must be considered the most fruitful harvest of the, consistently calibrated, OSO spectral scan data set. We will discuss the abundance results in some detail in a separate Section, 19, and will here only summarize "which species have been detected where".

In the extended Orion ridge emission was found from CO, C_2H, CH_3C_2H, HCO^+, HCS^+, HCN, HNC, HC_3N, HC_5N, SO, and SO_2. Spatially confined radiation - mainly originating in the "compact ridge cloud" at the "southern" edge of the Kleinmann-Low nebula, as recent mapping has verified, see (171) and (172) - was observed from HDO, CH_3OH, CH_3CN, $(CH_3)_2O$, CH_3OOCH, and OCS. Plateau emissions (broad lines) are present in CO, HDO, SO, SO_2, SiO, HCO^+, HCN, and HC_3N, while OCS and HNC exhibit semi-side features, which probably are mixtures of plateau and hot core radiation. Hot core emission lines are apparent in HDO, SO, SO_2, HCN, CH_3CN, HNCO, HC_3N, CH_2CHCN, and CH_3CH_2CN.

The following species were detected in the IRC+10216 envelope, viz., CO, CS, SiO, SiS, HCN, HNC, HC_3N, HC_5N, C_2H, C_4H, C_3N, $C_3H(?)$, and CH_3CN, where the two latter are new compounds.

f. Isotopic abundance ratios

Our most notable, and perhaps unexpected, results for the Orion ridge are $[CH_3OH]/[^{13}CH_3OH] \sim 20-40$ (*) (which should be quite reliable, since it is derived from the rotation diagram analysis, cf. Fig. 40), $[OCS]/[O^{13}CS] \sim 30$, and $[HC_3N]/[HC_3N(^{13}C)] \sim 40$ (using a composite average of the three ^{13}C species). These values agree with similarly low ratios for $[HCN]/[H^{13}CN]$ and $[HCO^+]/[H^{13}CO^+]$ in Orion, estimated by Rydbeck et al. (174) and Stark (180), see also the review paper by Wannier (181). Although some chemical fractionation might be possible, even at the high temperatures ($\gtrsim$ 100 K, cf. Smith and Adams (182)) estimated for the "compact ridge cloud" (to which the CH_3OH and OCS data are related), it is suggested that the remarkably similar ratios for very different species do instead indicate an *elemental* abundance ratio $[C]/[^{13}C] \sim 40$ in Orion.

(*) This low ratio has been confirmed by the very recent identification of the J=5 - 4 transition band of $^{13}CH_3OH$ as a result of a 1.3 mm spectral scan towards Orion A performed at Owens Valley and subsequent laboratory spectroscopy at Duke University (227).

To place these results in the "current trend" we note that the earlier galactic ring molecular cloud average $[C]/[^{13}C]\sim 60-70$ (Penzias (183), and Wannier (181)) could be an underestimate and fall closer to the terrestrial value of 89. Along these lines Henkel et al. (184), from an analysis of H_2CO data, suggest a positive radial gradient in the Galactic $C/^{13}C$ abundance ratio distribution, with an average value of ~ 70, increasing to ~ 90 in the outer parts of the galaxy (where, e.g., the solar system, Orion A, W3(OH), and NGC2264 are located). For nearby dark clouds Wilson et al. (185) report 75 ± 8. In a recent (1983) high precision analysis of CO, ^{13}CO, $C^{18}O$, and $^{13}C^{18}O$ in the giant molecular clouds NGC2264 and W3(OH) Penzias (186) estimates $[C]/[^{13}C] \sim 100 \pm 14$, "owing to new corrections implied by the data". In fact, a similar value (96 ± 5) has recently been deduced also in Orion A, from CO infrared measurements towards the Becklin-Neugebauer object by Scoville et al. (187). For the diffuse gas towards ζ Oph Vanden Bout and Snell (188) and Wannier et al. (189) estimate $[CH^+]/[^{13}CH^+]$ = 77 (+17, -12) and $[CO]/[^{13}CO] = 55 \pm 11$, respectively, from Copernicus and IUE observations. On the other hand, the Galactic center ratio (20-30) appears significantly smaller, presumably because of nuclear processing (181, 183). These examples may suffice to illustrate the present status and to point out the importance of continued efforts to study isotope ratios and their variation *among, and within,* interstellar clouds. Earlier we discussed recent HDO observations towards Orion A and W51M.

For the late type, carbon star IRC+10216 we estimate (cf. also Olofsson et al. (178)) from SiO and SiS lines $[Si]/[^{29}Si] \sim 16 \pm 3$, close to the terrestrial ratio (19.6), which is a sign that no substantial, explosive C or O burning has taken place. The lower limit (since the main isotopic lines are optically thick) $[HC_3N]/[HC_3N(^{13}C)] \gtrsim 20$ is compatible with the value 40 ± 8 derived from CS data by Wannier and Linke (190). Barnes el al. (191) find a ratio of 20, for the innermost part of the circumstellar shell, from their 4.6 μm CO observations.

As a curiosity we may mention that our non-detection of the radioactive isotope $H^{14}CN$ resulted in $[C]/[^{14}C] \gtrsim 3300$, while Barnes et al. from infrared CO data estimate $\gtrsim 10^4$. Similarly our limit for $HC^{15}N$ yielded $[N]/[^{15}N] \gtrsim 2000$ (terrestrial value ~ 270). Wannier et al. (192) in their dedicated, very sensitive study of $H^{13}CN$ and $HC^{15}N$ detected the latter species at a level corresponding to an abundance ratio of ~ 3000, a signpost of CNO burning. The latter clear detection by the 7 m Bell Laboratories telescope would have resulted in a $(20/7)^2 \sim 8$ times higher intensity (~ 0.06 K) at OSO 20 m telescope towards this spatially confined envelope. This sensitivity level for clear detections has been reached only in some limited regions of the OSO spectral scan, cf. Fig. 38. Our discovery of CH_3CN at this level may serve as an illustration that yet *we may only have "scraped the surface"*!

g. The helium abundance

An interesting digression from molecular lines is found in the many recombination lines automatically observed. Since the observed (velocity integrated) intensity ratios $H\alpha:H\beta:H\gamma:H\delta$ = 1:0.28:0.12:0.06 agree so well with those predicted for optically thin emission at LTE, viz., 1:0.28:0.12:0.07, we propose that our $He\alpha/H\alpha$ ratio ~ 0.084 should provide an accurate measure of the (cosmic) helium abundance. The He/H abundance by mass is estimated to be $28 \pm 2\%$, close to the predictions of big bang element synthesis. Such future observations of a number of HII regions may prove very valuable, although careful mapping is necessary to assure that the He ionization is "sufficient". (The ionization potentials of He and H are 24.6 and 13.6 eV, respectively). Such a survey at 22 GHz, with the Effelsberg 100 m telescope, by Thum et al. (193) has resulted in similar conclusions.

h. Non-detections, and why?

Upper limits were also listed for a number of "important" species. Although these limits, in terms of line intensity, may seem good, their conversion to abundance estimates is, in fact, *still encouraging for further search efforts*. The estimated concentration limits for non-detected species in Orion, e.g., formic acid, glycine, and urea, are not convincingly lower than the abundances of some known species (with strong emissions) like HNC and HC_3N. While the "methylated" version of HOOCH (formic acid), CH_3OOCH (methyl formate), is found to be an order of magnitude more abundant than the former species, our present limit for the isomer CH_3COOH (acetic acid) still only means that the two similar species may be equally concentrated. Another interesting example of *chemical selectivity* is provided by $(CH_3)_2O$, which is an order of magnitude more abundant than its non-detected isomer CH_3CH_2OH. This will be further illustrated in our Section 19 on molecular abundances.

Although the abundance ratio $[HCO^+]/[H_2] \lesssim 2 \times 10^{-9}$ in IRC+10216 (similar to the concentration of the detected species CH_3CN) is orders of magnitude lower than that of the dominant species in this source, this ratio is only marginally lower than the molecular cloud values. Other non-detection limits given for this (carbon rich) source, e.g., those for the oxygen species

SO, SO_2, and OCS, are just similar to the estimated concentrations $\sim 2 \times 10^{-7}$ of SiO and CS.

We have again stressed some negative aspects of the OSO spectral scan, but only to illustrate that *considerably increased sensitivity is required to reach really meaningful limits, and, hopefully, detections*. Apart from their possible non-existence or very low abundance, there are (at least) two obvious reasons why complex molecules may be so hard to detect: i) the *dilution* of the population *among many energy levels*, especially so for asymmetric rotors, and ii) the *excitation requirements* to populate higher energy states.

We may illustrate these conditions by two simple examples from the spectral scan:
i) Between HC_3N and its more hydrogen saturated asymmetric top analogue CH_2CHCN we find a "dilution" (ratio between their partition functions) of about 12 for a rotational temperature $T_{rot} \approx 100$ K. Assuming similar excitation requirements and optically thin lines we then may expect, for similarly abundant species, an order of magnitude weaker lines for CH_2CHCN. This fact is even more striking for the fully hydrogen saturated variant CH_3CH_2CN. In the spectral scan we easily detected CH_3CH_2CN and HC_3N, but just barely CH_2CHCN, in the hot core where $T_{rot} \gtrsim 100$ K, and estimate the former species to be nearly an order of magnitude more concentrated than the two latter ones.
ii) For observations at about the same frequency (90 GHz) of the cyanopolyynes HC_3N, HC_5N, and HC_7N we find that the main obstacle is due to high energy levels. For their J=10-9, 34-33, and 80-79 transitions, respectively, the level energies enter as $\exp(19/T_{rot}):\exp(72/T_{rot}):\exp(171/T_{rot})$. These dilution factors scale as 1:15:2000 for $T_{rot} \sim 20$ K, an estimate relevant to the Orion ridge as well as to IRC+10216. In the latter source we easily observe HC_3N and HC_5N, while HC_7N is invisible (but perhaps tentatively detected at a level of 0.015 K, as a composite average over 15 frequency bands!). However, HC_7N (J=21-20) has been detected at 24 GHz (at a level of about 0.4 K) with the Effelsberg 100 m telescope, and its abundance is estimated to be similar to those of HC_3N and HC_5N. In the Orion ridge HC_5N was estimated to be an order of magnitude less abundant than HC_3N, and was only detected (at 1 MHz resolution) as a composite average (over 7 frequency bands). HC_7N would be 100 times less intense at 90 GHz, but may well be detectable with a sensitive maser receiver at lower frequencies, where the lines would be stronger for a population distribution at about 20 K.

17. LINE STATISTICS - TOWARDS THE CONFUSION LIMIT

Fig. 41 roughly illustrates the density of molecular lines

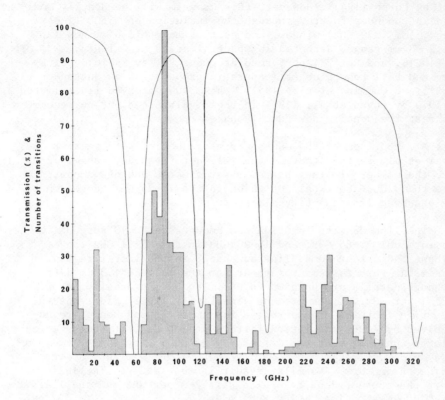

Figure 41: The number of observed interstellar molecular lines vs frequency in the atmospheric windows below 300 GHz. (Fine and hyperfine splitting transitions are omitted). The recent Texas 128-357 GHz line list (194) has been included here, while the results of Bell Laboratories (195) and Owens Valley (163) spectral scans are not yet available. The atmospheric transmission for an integrated water column of 4 mm is entered.

in the atmospheric windows below 300 GHz reported in the literature (including the OSO scan and a recent Texas 128-357 GHz line list by Loren (194)). However, this diagram will be heavily modified in the near future, through the inclusion of $\sim$ 500 lines in the 1.3 mm (230 GHz) window, and another manyfold of lines in the 2 - 3 mm range, detected in spectral scans of Orion A, IRC+10216, and Sgr B2, performed at Owens Valley Radio Observatory and Bell Laboratories (papers in preparation by Sutton et al. (163), and Cummins et al. (195)). The $\sim$ 400 observed lines listed in 1979 by Lovas et al. (196) in their valuable paper on recommended rest frequencies now roughly have *tripled*.

A comparison of the OSO spectral scan list with that of Lovas et al. reveals that in the relevant frequency range, 72-91 GHz, the number of lines has at least *doubled* and now amounts to more than 200; about half of which in the 72-85 GHz range, where the observed line quantity almost *tripled*.

The line density in Orion now amounts to 8 - 10 per GHz in the interval 72-85 GHz, and 15-20 GHz in the 85-91 GHz range, perhaps indicating some "line bunching" in the latter region. (A similar result seems to appear for the Caltech 1.3 mm data, although details are unknown to us). The U-line density in Orion is $\lesssim$ 1 per GHz, at the present sensitivity level. Since most of the Orion features are narrow (a few km s^{-1}), and hence are unresolved by 1 MHz resolution filters at 100 GHz, it seems reasonable to conclude that the *Orion spectra should be able to accomodate an almost tenfold number of lines before complete spectral confusion appears*. A similar result is obtained for IRC+10216, where the present density of lines is 2-3 per GHz and the lines are $\sim$ 30 km s^{-1} (i.e. $\sim$ 7-9 MHz) wide.

The *identification confusion limit would be encountered much earlier*, however, altough we tend to believe that, say, a *threefold increase* of the line density *could be handled by* means of sofisticated *pattern recognition* techniques. However, at such a stage, and probably even earlier, one may have to rely on statistical identification arguments. Such an approach recently has been discussed by Turner (197), in his matching of the calculated spectra of 613 molecular species to some 200 claimed 3 mm U-lines in Orion A and Sgr B2. A description of his U-line set is not yet published, however.

It may here be useful illustrate the already existing confusion by summarizing the *chance blends* and *re-identifications* found in the (Orion) spectral scan. These are:
i) CH_3OOCH lines at 76.702, 76.711 (are not the presumed CH_4 transitions), 85.927 (blend with NH_2D), and 86.224 GHz (blend with $(CH_3)_2O$)
ii) CH_3C_2H at 85.431 GHz (blend with CH_2CHCN)

iii) CH_3CH_2CN at 82.459 GHz (blend with $(CH_3)_2O$)
iv) CH_3OH at 84.744 GHz (blend with ^{30}SiO plateau)
v) H53β at 83.586 GHz (and not CH_3CHO)

In addition we have been informed that the Caltech 1.3 mm Orion scan has resulted in the *re-assignment* of lines previously attributed to CO^+ and O_3 to transitions from the ubiquitous species CH_3OH and $^{13}CH_3OH$ (Sutton et al. (163)). *At which sensitivity level would identification confusion appear?* Even if we assume that no new species exist in Orion, *isotopic variants* of the already known molecules are bound to *manyfold* the line density already before two orders of magnitude increased sensitivity is reached. Likewise, a multitude of higher order recombination lines would appear. While most parameters of these lines as well as of those due to the isotopic species are predictable, this is not yet the case for the frequencies of rarer isotopic variants of more complex species. What would be encountered at a sensititity level "only" 10 times lower than the present one in the OSO spectral scan may be hard to estimate, however, although is seems probably that our knowledge of the chemical composition of interstellar clouds would be considerably enriched. A useful example in this direction recently has been provided by Penzias (186), who during the previously discussed study of CO isotopes in NGC2264 and W3(OH), in his $^{13}C^{18}O$ spectra, at a level of 0.01 K, also "finds a feature due to a known transition of C_4H", as remarked in passing in his figure caption. The OSO limit for C_4H in Orion was given as $[C_2H]/[C_4H] > 50$, in contrast to a ratio close to one in the "carbon chain sources" IRC+10216 and TMC-1. Penzias' data required "unusually long integration times, 48 and 79 hours, respectively", with a system temperature similar to that achieved in the OSO spectral scan. The integration times are about an order of magnitude higher than those generally "afforded" in the scan, although it required in total about 6 months of antenna time.

With the telescopes and receivers now available, or under development, this order of magnitude sensitivity increase is indeed reachable even in spectral surveys. To investigate the possible outcome of such a deeper search it seems important to perform *test observations*, e.g., in frequency ranges where new lines are bound to appear (e.g., isotopic variants). Extensive laboratory spectroscopy support of such a project would be a necessity, however.

18. MOLECULES IDENTIFIED IN DIFFERENT REGIONS

We have in Fig. 42 tried to visualize which species have been observed in a number of prototype regions, viz., the massive, dense, and comparatively warm, giant molecular clouds in i) the *Galactic center* (Sgr B2), and ii) *Orion* (Orion A, Orion KL, OMC-1),

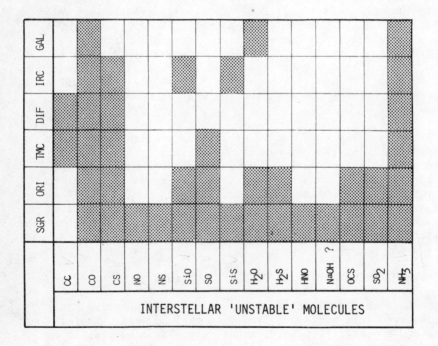

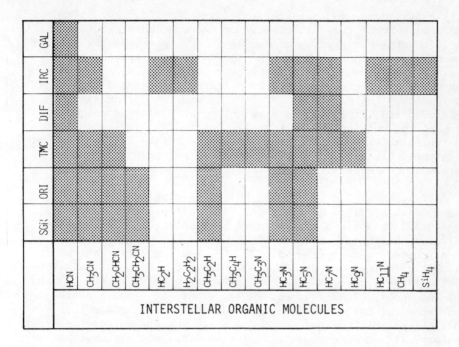

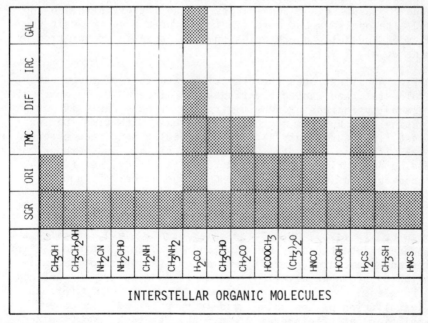

Figure 42: Molecular detections (dark areas) in a number of prototype regions.

iii) a less dense and massive, cold *dark cloud in Taurus* (TMC-1, the cyanopolyyne region), iv) *diffuse clouds* (more dilute regions, including the peripheries of denser clouds, where higher dipole moment species such as HCN, HCO$^+$ and CS are not appreciably excited above the 2.7 K cosmic background radiation and hence can only be detected in absorption against background continuum sources). v) the mass loss *envelope around* the carbon star *IRC+10216* and, finally, vi) external *galaxies* (as an ensemble). Many of the white areas of Fig. 42, hopefully, will be explored in the relatively near future with new sensitive instruments and receivers. Likewise, the presently known molecular collection is very much a sensitivity, and time, limited sample.

It appears that (excluding H_2) 48 species are known in Sgr B2, 33 in Orion A, 31 in dark clouds, 15 in diffuse regions, and 22 in IRC+10216, while 8 species have been detected in external galaxies. Of these compounds CO, HCN, and NH_3 have been observed in all source categories; CS, CN, C_2H, and HC_5N are reported in all but galaxies; and CH, OH, HCO$^+$, and H_2CO are found in all source types except IRC+10216. The latter source, together with the (unusual?) cloud fragment TMC-1, stand out because of the many abundant carbon chain species detected. We may conclude that the galactic center source is no longer the unique "molecular factory" it was some years ago. It is, e.g., unclear whether the detections of the 15 species found in Sgr B2, but not in Orion, are results of higher abundances in the former source, or just are due to insufficient sensitivity in the latter. Although the second alternative seems more likely, we here have to await the results of the Bell Laboratories spectral scan in preparation by Cummins et al. (195).

Fig. 43 illustrates the detection rate (per year) of new molecules (a), and isotopic variants (b) since 1970. In spite of the gradually increasing sensitivity the trend clearly is not an increasing one, but (if anything) a decrease with time. During the last five years "only" $\sim$ 15 new molecules have been observed and $\sim$ 13 isotopic variants. However, a positive, and perhaps more promising derivative may be found for dark clouds, see Fig. 44. In these objects (mainly due to TMC-1) the number of known species has doubled since 1979. Among the more recent discoveries of truly interstellar molecules those of C_3N and C_4H (Friberg et al. (137); Irvine et al. (138); Guélin et al. (139)), CH_3C_3N (Broten et al. (198), CH_3C_4H (Walmsley et al. (199), and also two independent detections by Canadian and US scientists), and C_3O (by Irvine and associates, immediately after its laboratory identification by Brown et al. (200)) were all made in TMC-1 - and all were *results of opening up new search frequency ranges* through the installation of low noise, widely tunable $\sim$ 20 GHz maser preamplifiers at Effelsberg, Green Bank, Haystack, and Onsala (here at about 20 and 30 GHz).

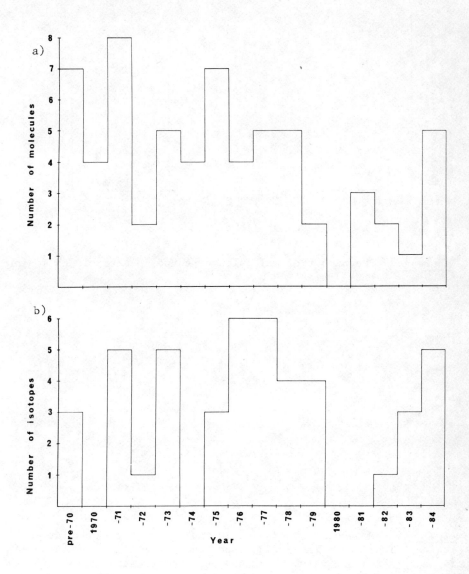

Figure 43: The number of new molecules (a) and isotopic variants (b) discovered since 1970.

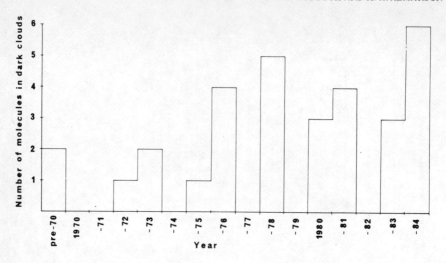

Figure 44: The number of molecules detected in dark clouds (mainly TMC-1) since 1970.

Figure 45: Molecular abundances in the Orion ridge and plateau regions, TMC-1, and IRC+10216. Upper and lower limits are denoted |← and |→, respectively. Although the concentrations often are only known to within an order of magnitude the ratios within a source may be rather accurate. Note that the upper limits are not far below the abundances of many known species.

RADIO OBSERVATIONS OF INTERSTELLAR MOLECULES 155

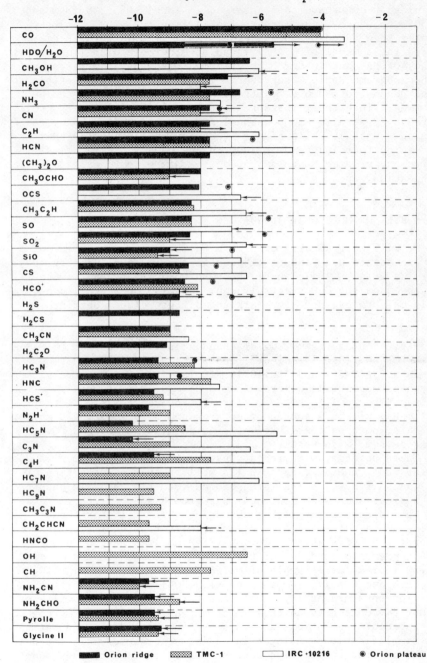

It should also be noted that perhaps the strongest (still remaining) U-lines have been observed in TMC-1, U85.338 (GHz) and U81.505, to which a new intense (4 K!) line, U45.379, has very recently been added by Suzuki et al (201). The latter was discovered (at a level of 0.4 K) in an ongoing molecular line survey of the 40 GHz band in Sgr B2, performed with the 45 m telescope of the Nobeyama Radio Observatory. These Japanese observations do indeed demonstrate how very efficient line searches now can be performed, since their instantaneous bandwidth of 2 GHz was fully covered by 8 parallel 2000 channel AOS receivers (frequency resolution 260 kHz), while higher frequency resolution (37 kHz) was achieved by four independent 2000 channel AOS's. *These recent results may provide important clues to future search strategy* (frequencies, sources).

19. SURVEY OF MOLECULAR ABUNDANCES

We will here heavily draw on a review paper, prepared for the "Protostars and Planets II" conference (Tucson, January 1984) by Irvine et al. (161), where the results presented very much were based on the OSO spectral scan, and also on recent, consistently calibrated, data for dark and diffuse clouds obtained at OSO and elsewhere.

Fig. 45 presents molecular abundances (with respect to H_2) for the Orion ridge ("ambient") and plateau clouds, compared with data for TMC-1 and IRC+10216. It is important to stress that these results in general should *only be relied upon to within an order of magnitude, although* abundance *ratios* (other than to H_2) within a source often are much *more accurately known*. With this, perhaps conservative, view in mind we proceed to summarize some main results:
i) Taken together with recent Onsala detections by Nyman (202, 203) of HCO^+, HCN, CS, and C_2H (as well as the non-detection of HN_2^+) in "diffuse", spiral arm clouds (cf. Fig. 32), in absorption against some weak continuum background sources, the data in Fig. 45 indicate a surprising abundance *uniformity* for these species among interstellar regions of different densities and temperatures, cf. Table 2. Some *earlier studies* have deduced a *strong inverse dependence* of the molecular concentration *on cloud density* (for a large sample of molecular clouds) for CO, H_2CO, HCO^+, HCN and C_2H (cf. Wootten et al. (204)), suggesting that depletion onto grains may cause this variation. Here we will not further advocate between these contrasting results, but merely note a recent theoretical analysis by Stenholm (205), indicating that simplifications employed in radiative transfer calculations, and the inverse correlation between errors in density and abundance, may *artificially* produce the latter results. These important but difficult matters certainly need extensive, future ob-

TABLE 2

Molecular abundances* vs H_2 in different regions

Species	Orion ridge	TMC-1/L134N	Diffuse
C_2H	2	$\gtrsim$ 1	5
HCN	2	2	0.5
HCO^+	0.3	0.8	0.5
HN_2^+	0.02	0.1	$\lesssim$ 0.05
CS	0.4	0.2	0.5
$n(H_2)[cm^{-3}]$	$10^4 - 10^6$	$10^3 - 10^5$	$10^2 - 10^3$
$T_{kin}[K]$	50 - 100	10 - 20	20 - 100

* unit: 10^{-8}

servational as well as theoretical attenuation.

ii) Relative to the Orion ridge and another dark cloud, L183 (L134N), the *carbon chains* HC_3N, C_3H, C_4H and HC_5N have considerably *increased abundances* in TMC-1. Differences in relative elemental abundances, UV radiation field, as well as cloud age, or history, have been suggested to explain the pronounced difference between TMC-1 and the other well studied dark cloud L183, which have similar temperature and density.

iii) The apparent *chemical selectivity* in the Orion ridge already has been mentioned in our earlier discussion of upper limits. Many of the species having more prominent emission in the "compact ridge cloud" - HDO, H_2CO, CH_3OH, $(CH_3)_2O$, CH_3OOCH, SO, SO_2, OCS, and CH_3CN - seem chemically related. The HDO data suggest that the H_2O concentration is only an order of magnitude lower than, or may even approach that of CO, cf. Olofsson (172).

iv) *Upper limits* to abundances of non-detected species *often do not have great significance*, since they are not considerably lower than the concentrations of similar, detected molecules. A good example is CH_3COOH (although its isomer, CH_3OOCH, has been detected), others are glycine, urea, as well as a number of cyclic compounds.

v) The already wellknown *abundance enhancement in the Orion plateau*, relative to the ridge, stands out even more apparent in the OSO spectral scan data: the SO, SO_2, and SiO concentrations increased by more than a factor 100, that of HCN by more than 10, while those of HDO, HC_3N, HCO^+, OCS, and HNC are greater by a factor of a few, or more. However, these conclusions are *mainly relevant to the disk outflow*, where, furthermore, our HDO data indicate that the amount of H_2O may approach that of CO, cf. Olofsson (172). The molecular *abundances in the bipolar* type, higher velocity *flow* seem at present essentially *unknown*, although such, weaker, wing emission clearly has been detected for CO, SiO, SO, SO_2, HCN, HCO^+, and H_2O. The only more quantitative result may appear from the HCO^+ mapping by Olofsson et al. (206), who find that, in the spatially extended region coinciding with the area of vibrationally excited H_2 emission, the ratio $[HCO^+]/[CO]$ increases by an *additional order of magnitude*, compared to the conditions in the disk, presumably reflecting *post-shock* conditions. The apparent plateau abundance enhancements have been suggested to be due to shock heating, but influences of "freeze out" during the expansion of an (oxygen rich) hot stellar atmosphere could be very important for the disk source, cf. (207, 208, 209).

vi) *Hot core* features have been detected in HDO, NH_3, HCN, HNC, HNCO, CH_3CN, HC_3N, CH_2CHCN, CH_3CH_2CN, SO_2, and SO (but in the last case the emission seems unusually extended, cf. Friberg (171)). The hot core abundances of species common with the plateau seem similarly *enhanced* with respect to the ridge. While the preponderance of *nitrogen-containing* species could perhaps be due to the comparatively high excitation requirements of many of these (high dipole moment) species, real chemical anomalies also seem to exist. The most *hydrogen saturated* carbon chain species, CH_3CH_2CN appears an order of magnitude more abundant than its less saturated relatives CH_2CHCN and HC_3N - completely different from the circumstances in TMC-1, IRC+10216 (and also the Orion ridge, although limits are less good here), where HC_3N dominates.

vii) As appears so clearly from Fig. 45, the *chemical composition of the IRC+10216 envelope is very different* from that in normal interstellar clouds, presumably reflecting a "freeze out" from the hot stellar atmosphere due to rapid expansion, cf. (210). The high abundances of HCN, SiO, SiS and a number of cyanopolyynes are especially conspicuous. The two species CO and HCN contain "almost all" the O and N, respectively. The high ratio $[HCN]/[HNC]$ $\sim$ 280 likewise is in agreement with expectations from hot stellar atmosphere chemistry. On the other hand, the observed abundances of SiO and SiS are an order of magnitude lower than expected, which may indicate that considerable amounts of Si are tied up in dust grains. A further argument in this direction has been given by Olofsson et al. (178), who observed that the SiO and SiS emission regions were considerably smaller than those of HCN, HNC, and HC_3N. Another important exception is the high abundance of CN observed by Wootten et al. (19), as mentioned already in Section 3. These authors suggest that CN is produced by *photodissociation* of HCN by the interstellar radiation field. Very recent Onsala maps of CN as well as C_2H (N=1 - 0) by Rieu et al. also seem to conform with this idea, where then C_2H_2 is the parent molecule for C_2H. Important theoretical contributions in this direction, where *photochemistry* produces *reactive radicals and ions* in the outer stellar envelope, have been given by Lafont et al. (211), and Huggins et al. (212). Although stressed earlier, we again point out that the *abundance limits achieved are not very satisfactory*. Our (comparatively good) limit for

HCO^+ in IRC+10216 is not significantly below that observed in interstellar clouds, and hence cannot be used as an argument against ion-molecule reactions in the former source. The concentration of the detected species CH_3CN is low in the stellar envelope and similar to that estimated in TMC-1 and the Orion A ridge.

In general it seems that the observed interstellar abundances of the simpler, few-atom molecules can be relatively well accomodated by today's rather elaborate chemical models, but the "boundary conditions" often are not observationally well constrained. The *elemental abundances* in a specific region, which may depend on cloud history as well as on depletion and desorption processes, still are almost "free parameters". Many reaction pathways and their efficiency also are uncertain, and the extrapolation of laboratory measured reaction rates to interstellar conditions may cause considerable uncertainties.

20. EXTERNAL GALAXIES

Although this topic is covered by a separate review, we would like to emphasize a few aspects, perhaps with an "OSO bias". During the last few years much attention, and telescope time, has been devoted to CO observaations of galaxies, resulting in a *doubling* of the number of detected galaxies and a *manyfolding* of observed positions, cf. Fig. 46 - to a considerable extent due to higher angular resolution studies performed at the new "sister-facilities" in Massachusetts (14 m diameter, FCRAO, Amherst) and Sweden (20 m, OSO). The dramatic increase of the number of map points since 1980 very much is a result of the development of cryogenic mixers at Bell Laboratories, FCRAO, NRAO and OSO, which has made this important, detailed mapping at all possible. It may be interesting to note that only five galaxies (M31, M51, M82, NGC6946, and IC342) represent about 80% of the, roughly 1000 positions observed, and that, as yet, only M82 is "fully sampled". In M51 the 150 positions hitherto observed at OSO will have to be more than doubled to cover the mapable part of that galaxy. Very recently the 10 m Leighton telescopes in California (OVRO, Caltech) have been used to observe galaxies, as single dishes at 230 GHz, and together as a successful synthesis instrument at 115 GHz.

While the Massachusetts emphasis very much has been on "larger scale phenomena", the Onsala view has been to study certain aspects in greater detail, which is only natural because of the narrower OSO antenna lobe (a factor of two smaller beam area). The *pointing requirement* (positioning accuracy) has turned out to be a major obstacle in such observations. Our own experience is that, at least in nuclei of galaxies with rapidly changing

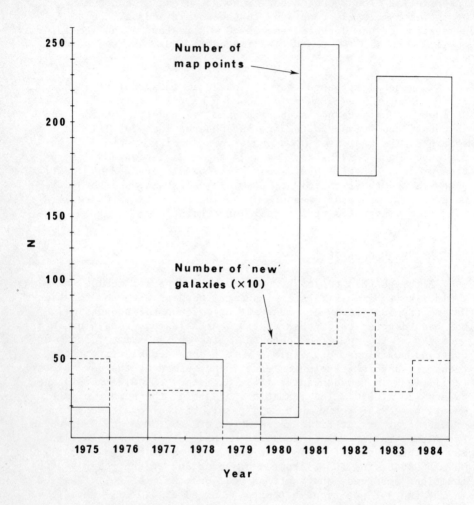

Figure 46: The number of galaxies detected in CO (x 10, dashed curve) and the number of CO map positions since 1975. Only five galaxies (M31, M51, M82, NGC6946, and IC342) represent ∼80% of the roughly 1000 map points.

velocity fields, offsets, or drifts, larger than 1/10 of the antenna beam (i.e., only 3") may indeed change the observed spectral shapes. Continuous pointing checks and updating therefore is a pressing "must" for such high resolution studies.

Recent CO observations at OSO, OVRO, and FCRAO of the irregular galaxy M82 have revealed new important features in its structure and kinematics, and even in the very nature of the molecular clouds, cf. Olofsson and Rydbeck (213), Sutton et al. (214), and Young and Scoville (215). The presumable close encounter with the more massive, neighbouring spiral galaxy M81 may be important in this context.

A rather detailed study of the spiral galaxy M51, performed at OSO by G. Rydbeck et al. (216), reveals, e.g., that non-circular motion seems to be necessary in its central regions to explain the spectral shapes observed. It also appears that the beam deconvolved radial distribution does *not* show a pronounced central peak, but instead better conforms with a central void (perhaps with a small "spike"), followed by a pronounced ridge and thereafter a rather flat, lower level, disk - very much similar to Boroson's (217) interpretation of his observed light distribution. This seems to contrast with the M51 CO data by Scoville and Young (218), but may mainly be a result of the smaller OSO antenna beam. In a very recent, fully sampled CO map ($\sim$ 100 positions) of the northeastern quadrant of the same galaxy, spiral arm features seem clearly discernible. However, since the CO observations essentially "count the number of clouds in the beam", all these studies have to be supported by measurements of its rarer isotopic variant, ^{13}CO, which supposedly responds more closely to the "amount of mass". Figure 47, depicting Onsala CO and ^{13}CO spectra towards the M51 center, may illustrate the data quality available today. To really understand how molecular clouds are formed at large scale (influences of density wave?), and to find out the differences (if any) between the "arm" and "interarm" cloud ensembles, we feel that accurate, time consuming, observations of *density sensing* species are required, e.g., ^{13}CO and HCN, as well as sub-mm transitions of CO. With the very low noise systems now available, or under development, such important studies are indeed feasible.

Finally, since we have here been discussing details rather than generalities, the importance of aperture synthesis observations should be stressed. The very recent 115 GHz CO maps of the center of IC342, produced by the Caltech interferometer by Lo et al. (219), have revealed a bar shaped distribution in this otherwise "normal" spiral. The importance of this discovery in connection with spiral arm theory is obvious.

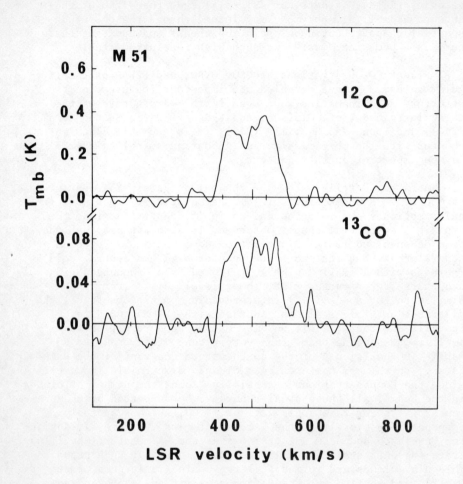

Figure 47: Onsala CO and ^{13}CO (J=1-0) spectra towards the M51 center, illustrating the data quality available today. The observed CO/^{13}CO integrated intensity ratio is ~ 5, a factor of two lower than previously observed with the larger beam of the Bell Laboratories telescope ($\sim 110"$, compared to $\sim 33"$ at OSO), and similar to the observed mean ratio in our Galaxy.

21. SUMMARY AND FUTURE PROSPECTS

Since the detection of OH only twenty years ago, molecular radio astrophysics has grown to a comprehensive area of research, with its tentacles in theoretical and experimental chemistry, and molecular physics, as well as in more conventional astronomy concerning cloud and stellar formation and evolution processes, galactic structure, dynamics and evolution, and also in cosmology. As in all new sciences the (carefully "planned" as well as "accidental") discoveries have been tied to the development of new, sensitive instruments and equipments, which we have tried to illustrate throughout this paper.

A very brief summary of the main development in this field may read as follows:

The radio molecules directly have demonstrated the existence of the dense clouds, in which star formation is ongoing, and make at all possible a study of the cloud characteristics - sizes, masses, structures, densities, temperatures, the ordered and stochastic internal motion, as well as the large scale motion and distribution of the denser gas in our own and external galaxies. The evolution of a molecular cloud is directly affected by the cooling and heating by the trace molecules (220, 221) as well as by the state of ionization (indirectly estimated via observations of HCO^+, and its isotopic variants, cf. (204, 222)), coupled to the magnetic field (223, 224). The latter has been estimated in denser regions by OH Zeeman splitting, as described earlier, and indicates that the field is "frozen-in" during cloud contraction, cf. (225).

Mass-loss phenomena have been studied not only during late stages of stellar evolution, but extensive such outflows ("high velocity wings", "plateaus", "bipolar flows") have been discovered from many newly formed stars deep inside the molecular clouds (170), a process not yet theoretically well understood.

Besides many important contributions to the understanding of OH, H_2O and SiO masers themselves and their physical environment, VLBI techniques has provided a new, model independent, distance measure as a result of the detection of proper (transverse) motion of H_2O masers in Orion A, W51M, and W51N. The latter series of VLBI experiments also gave a direct proof that the Orion plateau emission really was outflow, and not infall, or rotation (although the latter alternatives could be ruled out on a dynamical basis). The early, surprising, discoveries of excitation anomalies like the strong OH and H_2O masers, the refridgerated H_2CO, absorbing even the weak cosmic background radiation, and the everywhere population inverted CH, have led to extensive work towards the understanding of "pumping" processes. Such excitation analysis, including intricate theoretical and laboratory work on collision cross sections, also must be applied to "normal" radio molecules.

The chemical composition of interstellar and circumstellar clouds has grown from the three "optical" species, CH^+, CH and CN to include more than 60 molecules, the most heavy of which is $HC_{11}N$, twice as massive as glycine - which has not (yet) been seen in spite of several searches. The molecular abundances seem (at best) to be known to within an order of magnitude, although concentration ratios may be much more accurately determined.

Among the surprises of the interstellar medium appear abundant molecular ions like HCO^+ and HN_2^+ (the accidental detections of which "led" to the ion-molecule formation schemes that still are our common belief), the long carbon chains, including radicals such as C_3N, C_3H and C_4H, as well as the recently discovered methylated forms CH_3C_3N and CH_3C_4H, and, perhaps even more astonishingly, tricarbon monoxide, C_3O. The latter three species only have been detected in the cool dark cloud TMC-1. The present abundance limits for many species are *not* far below the detection levels of several known species.

Extensive observations of molecular isotopic variants have been used to study the elemental isotopes and their abundance variation in the Galaxy, with the ultimate goal to test theories of galaxy evolution. Notable signs of nuclear processing seem present in the Galactic center, and also in late type stellar envelopes, while the results on radial variations in the Galactic ring molecular clouds still are far from conclusive. Chemical fractionation (i.e., enrichment of the rarer species via ion-molecule reactions) presumably is a dominant problem in this context, and may be appreciable even in relatively warm clouds according to recent laboratory experiments.

Although the future of molecular radio astrophysics may not necessarily contain the many easily won victories of the past, it is clear that existing instruments and those under development or consideration - earth bound and space borne single dishes as well as interferometers - together with *very low noise receivers, multi-feed* (for mapping or frequency coverage) *systems and flexible, very broad band spectrometers* are bound to produce numerous, new, important building blocks towards a very much improved understanding of phenomena and processes in interstellar and circumstellar regions, as well as in external galaxies. Also, the unpredictable discoveries should not be forgotten here.

A more *precise* knowledge of molecular and isotopic *abundances* will require accurate multi-transition mapping in several isotopic species, but, on the other hand, such (time consuming in terms of analysis, even if observations can be performed in a very efficient way) fine tuning studies *simultaneously* will provide *upgraded information on physical parameters* of the clouds - *and vice versa*. The physical and chemical processes often are inseparable, e.g., a very good handle on cloud evolution towards star formation can only be obtained, if the abundances of importtant *coolants*, such as CO, H_2O, and C, are accurately known, which is, indeed, not the case. For many of these desires observations outside of the "radio regime" are also needed. For instance, to detect and study the very dense, really *"protostellar"* regions (rather than the "newly formed" stars discovered by near infrared techniques) submillimeter transitions, sensitive to higher densities and temperatures, probably will be very useful, as recently discussed e.g. by Harwit (226).

The number of *new interstellar molecules* certainly can be extended considerably by means of searches for specific molecules, as well as deep, unbiased spectral scans. While the "molecule hunting" via broad band searches is bound to be rather intricate and difficult, although still definitely possible, in already covered sources and frequency ranges (where the confusion limit may be approaching, e.g., due to isotopic variants and higher energy transitions of already known species, and where statistical identification criteria ultimately are needed), such discoveries might still be done in a *very profitable way in new types of sources* (e.g., the cold, dark clouds, cf. Fig. 44) *and/or in new frequency bands* (submillimeter, or below 50 GHz, or in bands not accessible to earthbound observations, perhaps looking for magnetic dipole lines of O_2). However, considering the small beam sizes available it should be remembered that even the *search positions* must be very carefully chosen. In Orion the necessary, initial guidance may have been provided by the Onsala spectral scan (156), supplemented by subsequent mapping (171, 172).

We may also "predict" that *laboratory spectroscopy in space* will continue to be as important in the future as it has been on numerous occasions in the past, for identification purposes as well as for molecular physics reasons. Short-lived species, such as C_5H, C_6H, C_5H and other possible carbon chains, immediately come into mind.

To better understand the structure, dynamics, and evolution of *galaxies* - linked to the formation of denser clouds, and, ultimately, of stars - continued, extensive single dish, as well as aperture synthesis, measurements obviously are necessary. Guided by published and forthcoming larger scale-size onservations with instruments already in operation, such studies also are bound to include *better density probes* than the CO (J=1-0) line, i.e., its higher rotational transitions, as well as ^{13}CO, and also HCN, HCO^+, CS,... Submillimeter transitions of H_2O - which according to the rare, existing H_2O and HDO data may be (almost) as abundant as CO in denser clouds - also seem very important in this context, as does mapping of the density regime between HI clouds and the molecular clouds by means of the sub-mm fine structure lines of neutral carbon, CI - another very abundant interstellar species. To increase the sample of galaxies, e.g., to include new morphological types, deep, (more or less) unbiased searches should be performed. As a successful project in this regard we may mention the ongoing Onsala search for CO in galaxies detected at 60 μm with the IRAS satellite.

Among the many desirable *VLBI-observations* of OH, H_2O, and SiO sources we especially want to emphasize an extreme goal, i.e., to extend the essentially model and correction independent distance scale provided by H_2O proper (transverse) motion experiments ultimately to nearby galaxies. If this were possible, perhaps when a longed-for VLBI-satellite has been launched, we could hopefully discriminate between the different "schools" of cosmological distance scales.

REFERENCES:

(1) Eddington, A.S.: 1926, Proc. R. Soc., 111, 424.
(2) Dunham, T., Jr. and Adams, W.S.: 1937, Publ. Am. Astr. Soc., 9, 5.
(3) Dunham, T., Jr. and Adams, W.S.: 1937, Publ. Astr. Soc. Pac., 49, 26.
(4) Swings, P., and Rosenfeld, L.: 1937, Astrophys. J., 86, 483.
(5) Rydbeck, O.E.H., Elldér, J., and Irvine, W.M.: 1973, Nature, 246, 466.
(6) Rydbeck, O.E.H.: 1974, Abh. Akad. d. Wiss. u. d. Literatur, Mainz, Nr. 1.
(7) McKellar, A.: 1940, Publ. Astr. Soc. Pac., 52, 187 & 312.
(8) Adams, W.S.: 1941, Astrophys. J., 93, 11.
(9) Douglas, A.E. & Herzberg, G.: 1941, Astrophys. J., 94, 381.
(10) Douglas, A.E. & Herzberg, G.: 1942, Can. J. Res. A. 20, 71.
(11) McKellar, A.: 1941, Publ. Dom. Astr. Obs., 7, 251.
(12) Penzias, A.A. & Wilson, R.W.: 1965, Astrophys. J., 142, 419.
(13) Field, G.B. & Hitchcock, J.L.: 1966, Phys. Rev. Lett., 16, 817.
(14) Shklovskii, I.S.: 1966, Astron. Tsirk, Nr. 371, 1.
(15) Thaddeus, P. & Clauser, J.F.: 1966, Phys. Rev. Lett., 16, 819.
(16) Penzias, A.A., Jefferts, K.B., and Wilson, R.W.: 1972, Phys. Rev. Lett., 28, 772.
(17 a) Meyer, D.M. & Jura, M.: 1984, Astrophys. J., 276, L1.

(17 b)   Hegyi, D., Traub, W., and Carlton, N.: 1972, Phys. Rev. Lett., 28, 1541.
(18 a)   Thaddeus, P.: 1972, Ann. Rev. Astr. Ap., 10, 305.
(18 b)   Weiss, R.: 1980, Ann. Rev. Astr. Ap., 18, 489.
(19)   Wootten, A., Lichten, S.M., Sahai, R., and Wannier, P.: 1982, Astrophys. J., 257, 151.
(20)   Dixon, T.A. & Woods, R.C.: 1977, J. Chem. Phys., 67, 3956.
(21)   Jack, D.: 1927, Proc. Roy. Soc. A., 115, 373.
(22)   Dieke, G.H. & Crosswhite, H.M.: 1962, J. Quant. Spectrosc. Radiat. Transfer, 2, 97.
(23)   Shklovskii, I.S.: Astron. Zhurnal XXVI, 10. 1949.
(24)   Shklovskii, I.S.: 1954, Memoires R. Sc. Liège, quatrième série, tome XV, fasc. unique.
(25)   Van Vleck, J.H.: 1929, Phys. Rev., 33, 467.
(26)   Mulliken, R. & Christy, A.: 1931, Phys. Rev., 38, 87.
(27)   Snow, T.P.: 1975, Astrophys. J., 201, L21.
(28)   Storey, J.W.V., Watson, D.M., and Townes, C.H.: 1981, Astrophys. J., 244, L27.
(29)   Washington Conference on Radio Astronomy - 1954: Journal of Geophysical Research, Vol. 59, No. 1, March 1954, p.198.
(30)   Sanders, T.M., Schawlow, A.L., Dousmanis, G.C., and Townes, C.H.: 1953, Phys. Rev., 89, L1158.
(31)   Winnberg, A., Walmsley, C.M., and Churchwell, E.: 1978, Astron. Astrophys., 66, 431.
(32)   Barrett, A.H. & Lilley, A.E.: 1956, NRL Report 4809, A Search for the Atmospheric and Cosmic Microwave Spectrum of OH; Naval Research Laboratory, Washington, D.C.
(33)   Hagen, J.P., Lilley, A.E., and McClain, E.F.: 1955, Astrophys, J., 122, 361.
(34)   Ehrenstein, G. and Townes, C.H., & Stevenson, M.J.: 1959, Phys. Rev. Lett., 3, 40.
(35)   Weinreb, S., Barrett, A.H., Meeks, M.L., and Henry, J.C.: 1963, Nature, 200, 829.
(36)   Van Vleck, J.H.: 1943, Report No. 51, Radio Research Laboratory, Harvard University.
(37)   Barrett, A.H. & Rogers, A.E.E.: 1964, Nature, 204. 62.
(38)   Ter Meulen, J.J. & Dymanus, A.: 1972, Astrophys. J., 172, L21.
(39)   Weaver, H., Williams, D.R.W., Dieter, N.H., & Lum, W.T.: 1965, Nature, 208, 29.
(40)   Weinreb, S., Meeks, M.L., Carter, J.C., Barrett, A.H., & Rogers, A.E.E.: 1965, Nature, 208, 440.
(41)   Rydbeck, O.E.H. and Kollberg, E.: 1968, IEEE Trans., MTT-16, No. 9, 799, and Research Report No. 89, March, 1968, *Rutile travelling wave masers for the frequency range 1300-3400*, Research Laboratory of Electronics, Chalmers University of Technology, Gothenburg.
(42)   Whiteoak, J.B., Gardner, F.F., and Höglund, B.: 1980, Mon. Not. R. astr. Soc., 190, 17P; and Gardner, F.F., Höglund, B., and Whiteoak, J.B.: 1980, ibid., 191, 19P.

(43) Rydbeck, O.E.H., Kollberg, E., and Elldér, J.: 1970, Astrophys. J., 161, L25.
(44) Rydbeck, O.E.H., Elldér, J., Kollberg, E., and Höglund, B.: 1972, Mém. Soc. Roy. des Sciences de Liège, 6e série, tome III.
(45) Elldér, J.: IAU Circular No. 2364, Oct. 19, 1971.
(46) Lo, K.Y., Walker, R.C., Burke, B.F., Moran, J.M., Johnston, K.J., and Ewing, M.S.: 1975, Astrophys. J., 202, 650.
(47 a) Moran, J., Burke, F., Barrett, A.H., Rydbeck, O.E.H., Hansson, B., Rogers, A.E.E., Ball, J.A., and Cudaback, D.D.: 1968, Astron. J. (Supplement) No. 1360, Number 5, Part II.
(47 b) Moran, J.M., Burke, B.F., Barrett, A.H., Rogers, A.E.E., Carter, J.C., Ball, J.A., and Cudaback, D.D.: 1968, Astrophys. J., 152, L97.
(48) Burke, B.F., Moran, J.M., Barrett, A.H., Rydbeck, O., Hansson, B., Rogers, A.E.E., Ball, J.A., and Cudaback, D.D.: 1968, Astron. J., 1968, December, (Supplement) No. 1365, Number 10, Part II (Papers presented at the 127th meeting of the American Astronomical Society, August 1968, at the Dominion Astrophysical Observatory, Victoria, British Columbia).
(49) Kellerman, K.I., Clark, B.G., Bare, C.C., Rydbeck, O., Elldér, J., Hansson, B., Kollberg, E., Höglund, B., Cohen, M.H., and Jauncey, D.L.: 1968, Astrophys. J. 153, L209.
(50) Cook, A.H.: 1968, M.N.R.A.S., 140, 299.
(51) Harris, S. and Scott, P.F.: 1976, M.N.R.A.S., 175, 371.
(52) Reid, M.J., Haschick, A.D., Burke, B.F., Moran, J.M., Johnston, K.J., and Swenson Jr., G.W.: 1980, Astrophys. J., 239, 89.
(53) Guilloteau, S.: 1982, Astron. Astrophys., 116, 101.
(54) Moran, J.M., Reid, M.J., Lada, C.J., Yen, J.L., Johnston, K.J., and Spencer, J.H.: 1978, Astrophys. J., 224, L67.
(55) Guilloteau, S., Stier, M.T., and Downes, D.: 1983, Astron. Astrophys., 126, 10.
(56) Winnberg, A., Walmsley, C.M., and Churchwell, E.: 1978, Astron. Astrophys., 66, 431.
(57) Haschick, A.D., Reid, M.J., Burke, B.F., Moran, J.M., and Miller, G.: 1981, Astrophys. J., 244, 76.
(58) Lo, K.Y., Burke, B.F., and Haschick, A.D.: 1975, Astrophys. J., 202, 81.
(59) Ho, P.T.P., and Barrett, A.H.: 1980, Astrophys. J., 237, 38.
(60) Haschick, A.D., Moran, J.M., Rodriguez, L.F., Burke, B.F., Greenfield, P.E., and Garcia-Barreto, J.A.: 1980, Astrophys, J., 237, 26.
(61) Ostriker, J.P. and Bodenheimer, P.: 1973, Astrophys. J., 180, 171.
(62) Baudry, A., Walmsley, C.M., Winnberg, A. & Wilson, T.L.: 1981, Astron. Astrophys., 102, 287.

(63)   Rydbeck, O.E.H., Elldér, J., & Yngvesson, K.S.: 1973, Report No. P15 (Detection of rotationally excited OH in absorption), Dept. of Electrical and Computer Engineering, Univ. of Mass., Amherst, Mass. (Communicated by Prof. William Irvine at the AAS June Meeting, Columbus, Ohio, 1973).
(64)   Kukolich, S.G.: 1969, J. Chem. Phys., 50, 3751.
(65)   Yngvesson, K.S., Cardiasmenos, A.G., Shanley, J.F., Rydbeck, O.E.H., and Elldér, J.: 1975, Astrophys. J., 195, 91.
(66)   Bertojo, M., Cheung, A.C., and Townes, C.H.: 1976, Astrophys. J., 208, 914.
(67)   Dewangan, D.P., and Flower, D.R.: 1982, M.N.R.A.S., 199, 457.
(68)   Varshalovich, D.A., and Khersonskii, V.K.: 1978, Sov. Astron., 22, 192.
(69)   Brown, J.M., Schubert, J.E. Evenson, K.M., and Radford, H.E.: 1982, Astrophys. J., 258, 899.
(70)   Litvak, M.M.: 1969, Astrophys. J., 156, 471. (See also Collisional and Radiative Processes by M.M. Litvak, Lecture Note No. 5, 1972, Onsala Space Observatory, Sweden).
(71)   Lucas, R.: 1980, Astron. Astrophys., 84, 36.
(72)   Guilloteau, S., Lucas, R., and Omont, A.: 1981, Astron. Astrophys., 97, 347.
(73)   Storey, J.W.V., Watson, D.M., and Townes, C.H.: 1981, Astrophys. J., 244, L27.
(74)   Watson, D.M., Storey, J.W.V., Townes, C.H., Haller, E.E., and Hansen, W.L.: 1980, Astrophys. J., 239, L129.
(75)   Kirchhoff, W.H., Lovas, F.J., and Johnson, D.R.: 1972, J. Phys. Chem. Ref. Data, $\underline{1}$ (4) (Formaldehyde), 1011.
(76)   Oka, Takeshi: 1960, J. Phys. Soc. Jap., 15, 2274.
(77)   Snyder, L.E., Buhl, D., Zuckerman, B., and Palmer, P.: 1969, Phys. Rev. Letters, 22, 679.
(78)   Townes, C.H. and Cheung, A.C.: 1969, Astrophys. J., 157, L103.
(79)   Evans II, N.J.: 1975, Astrophys. J., 201, 112.
(80)   Garrison, B.J., Lester Jr., W.A., Miller, W.H., and Green, S.: 1975, Astrophys. J., 200, L175.
(81)   Green, S., Garrison, B.J., Lester Jr., W.A., and Miller, W.H.: 1978, Astrophys. J. Suppl. S., 37, 321.
(82)   Kwan, J.: 1974, Astrophys. J., 191, 101.
(83)   Sume, A., Downes, D., and Wilson, T.L.: 1975, Astron. Astrophys., 39, 435.
(84)   Downes, D., Wilson, T.L., and Bieging, J.: 1976, Astron. Astrophys., 52, 321.
(85)   Evans, N.J., Zuckerman, B., Morris, G., and Sato, T.: 1975, Astrophys. J., 196, 433.

(86) Langer, W.D., Frerking, M.A., Linke, R.A., and Wilson, R.W.: 1979, Astrophys. J., 232, L169.
(87) Downes, D. & Wilson, T.L.: 1974, Astrophys. J., 191, L77.
(88) Forster, J.R., Goss, W.M., Wilson, T.L., Downes, D., and Dickel, H.R.: 1980, Astron. Astrophys., 84, L1.
(89) Rots, A.H., Dickel, H.R., Forster, J.R. & Goss, W.M.: 1981, Astrophys. J., 245, L15.
(90) Boland, W. & de Jong, T.: 1981, Astron. Astrophys., 98, 149.
(91) Snyder, L.E. & Buhl, D.: 1969, Astrophys. J., 155, L65.
(92) Cheung, A.C., Rank, D.M., Townes, C.H., Thornton, D.D. & Welch, W.J.: 1969, Nature, 221, 626.
(93) Van Vleck, J.H., & Weisskopf, V.F.: 1945, Theoretical Shape of Pressure Broadened Lines, Revs. Mod. Phys., 17, 227.
(94) Cheung, A.C., Rank, D.M., Townes, C.H., Thornton, D.D., and Welch, W.J.: 1968, Phys. Rev. Lett., 21, 1701.
(95) Matveenko, L.I., Kogan, L.R., & Kostenko, V.I.: 1980, Sov. Astron. Lett., 6(4), July-August.
(96) Abraham, Z., Cohen, N.L., Opher, R., Raffaeli, J.C., & Zisk, S.H.: 1981, Astron. Astrophys., 100, L10.
(97) Matveenko, L.I., Romanov, A.M., Kogan, L.R., Moiseev, I.G., Sorochenko, R.L. & Timofeev, V.V.: 1983, Sov. Astron. Lett., 9(4), July-August.
(98) Genzel, R., Downes, D., Schneps, M.H., Reid, M.J., Moran, J.M., Kogan, L.R., Kostenko, V.I., Matveenko, L.I. & Rönnäng, B.: 1981, Astrophys. J., 247, 1039.
(99) Prasad, S.S., Huntress, W.T. Jr.: 1980, Astrophys. J. Suppl. Ser., 43, 1.
(100) Scalo, J.M. & Slavsky, D.B.: 1980, Astrophys. J., 239, L73.
(101) Hall, R.T. & Dowling, J.M.: 1967, J. Chem. Phys., 47, 2454.
(102) Hagen, W., Tielens, A.G.G.M. & Greenberg, J.M.: 1983, Astron. Astrophys., 117, 132.
(103) Waters, J.W., Gustinic, J.J., Kakar, R.K., Kuiper, J.B.H., Roscoe, H.K., Swanson, P.N., Rodrigues Kuiper, E.N., Kerr, A.R. & Thaddeus, P.: 1980, Astrophys. J., 235, 57.
(104) Olofsson, H.: 1984, Astron. Astrophys., 134, 36.
(105) Phillips, T.H., Kwan, J. & Huggins, P.J.: 1980, IAU Symposium No. 87 (Interstellar Molecules), 21.
(106) Meeks, M.L., Carter, J.C., Barrett, A.H., Schwartz, P.R., Waters, J.W. & Brown, W.E.: 1969, Science, 165, 180.
(107) Turner, B.E., Zuckerman, B., Fourikis, N., Morris, M. & Palmer, P.: 1975, Astrophys. J., 198, L125.
(108) de Lucia, F.C., Helminger, P. & Kirchhoff, W.H.: 1974, J. of Physical and Chemical Reference Data, 3, No. 1, 211-219 (contains data also on H_2O & $H_2^{18}O$).
(109) Beckman, J.E., Watt, D.G., White, G.J., Phillips, J.P., Frost, R.L. & Davis, J.H.: 1982, Mon. Not. Roy. Astr. Soc., 201. 357.

(110 a) Genzel, R., Downes, D., Ho, P.T.P. & Bieging, J.: 1982, Astrophys. J., 259, L103.
(110 b) Watson, W.D. & Walmsley, C.M.: 1982, in "Regions of Recent Star Formation", Reidel, Dordrecht.
(111) York, D.C. & Rogerson, J.B.: 1976, Astrophys. J., 203,378.
(112) Thaddeus, P., Krishner, L.C. & Loubser, J.H.N.: 1964, J. Chem. Phys., 40, 257.
(113) Johansson, L.E.B., Andersson, C., Elldér, J., Friberg, P., Hjalmarson, Å., Höglund, B., Irvine, W.M., Olofsson, H. and Rydbeck, G.: 1984, Astron. Astrophys., 130, 227.
(114) Ho, P.T.P., Genzel, R. & Das, A.: 1983, Astrophys. J., 266, 596.
(115) Wilson, R.W., Penzias, A.A., Jefferts, K.B., Kutner, M., & Thaddeus, P.: 1971, Astrophys. J., 167, L97.
(116) Snyder, L.E. & Buhl, D.: 1974, Astrophys. J., 189, L31.
(117) Buhl, D., Snyder, L.E., Lovas, F.J. & Johnson, D.R.: 1974, Astrophys. J., 192, L97.
(118) Moran, J.M., Ball, J.A., Predmore, C.R., Lane, A.P., Huguenin, G.R., Reid, M.J. & Hansen, S.S.: 1979, Astrophys. J., 231, L67.
(119) Elitzur, M.: 1980, Astrophys. J., 240, 553.
(120) Bujarrabal, V. & Nguyen-Quang-Rieu: 1981, Astron. Astrophys., 102, 65.
(121) Olofsson, H., Hjalmarson, Å. & Rydbeck, O.E.H.: 1981, Astron. Astrophys., 100, L30.
(122) Genzel, R., Downes, D., Schwartz, P.R., Spencer, J.H., Pankonin, V. & Baars, J.W.M.: 1980, Astrophys. J., 239, 519.
(123) Genzel, R., Reid, M.J., Moran, J.M. & Downes, D.: 1981, Astrophys. J., 244, 884.
(124) Rydbeck, O.E.H., Hjalmarson, Å., Rydbeck, G., Elldér, J., Olofsson, H. & Sume, A.: 1981, Astrophys. J., 243, L41.
(125) Turner, B.E. & Zuckerman, B.: 1974, Astrophys. J., 187, L59.
(126) Robinson, B.J.: 1967, IAU Symposium No. 31, Radio Astronomy and the Galactic System, ed. H. van Woerden, p.49, (London, Academic Press).
(127) Rydbeck, O.E.H., Elldér, J., Irvine, W.M., Sume, A. & Hjalmarson, Å.: 1974, Astron. Astrophys., 34, 479.
(128) Rydbeck, O.E.H., Elldér, J., Irvine, W.M., Sume, A. & Hjalmarson, Å.: 1974, Astron. Astrophys., 33, 315.
(129) Rydbeck, O.E.H., Kollberg, E., Hjalmarson, Å., Sume, A., Elldér, J. & Irvine, W.M.: Radio observations of interstellar CH. I. Astrophys. J. Supplement Series, 31, No. 3, 1976 July; see also part II, of the same title, in Supplement Series, 35, 263, 1977.
(130) Bogey, M., Demuynck, C. & Destombes, J.L.: 1983, Chem. Phys. Letters, Vol. 100, No. 1, 105.
(131) Bouloy, D., Nguyen-Q-Rieu, and Field, D.: 1984, Astron. Astrophys., 130, 380.

(132) Johansson, L.E.B.: The galactic distribution of CH, 1979, Research Report No. 136, Res. Lab. of Electronics & Onsala Space Observatory, Chalmers Univ. of Technology; see also Johansson et al. 1979, IAU Symp. No. 84.
(133) Lovas, F.J., Suenram, R.D. & Evenson, K.M.: 1983, Astrophys. J., 267, L131.
(134) Kollberg, E.L. & Lewin, P.T.: 1976, IEEE Trans., MTT-24, 718.
(135) Rydbeck, O.E.H., Sume, A., Hjalmarson, Å., Elldér, J., Rönnäng, B.O. & Kollberg, E.: 1977, Astrophys. J., 215, L35.
(136) Ho, P.T.P. & Townes, C.H.: 1983, Ann. Rev. Astron, Astrophys., 21, 239.
(137) Friberg, P., Hjalmarson, Å., Irvine, W.M., and Guélin, M.: 1980, Astrophys. J., 241, L99.
(138) Irvine, W.M., Höglund, B., Friberg, P., Askne, J. and Elldér, J.: 1981, Astrophys. J., 248, L113.
(139) Guélin, M., Friberg, P., and Mezaoui, A.: 1982, Astron. Astrophys., 109, 23.
(140) Gottlieb, C.A., Gottlieb, E.W., Thaddeus, P. and Kawamura, H.: 1983, Astrophys. J., 275, 916.
(141) Buhl, D., and Snyder, L.E.: 1970, Nature, 228, 267.
(142) Klemperer, W.: 1970, Nature, 227, 1230.
(143) Woods, R.C., Dixon, T.A., Saykally, R.J., and Szanto, P.G.: 1975, Phys. Rev. (Letters), 35, 1269.
(144) Snyder, L.E., Hollis, J.M., Lovas, F.J., and Ulich, B.L.: 1976, Astrophys. J., 209, 67.
(145) Turner, B.E.: 1974, Astrophys. J. (Letters), 193, L83.
(146) Green, S., Montgomery, J.A., Jr., and Thaddeus, P.: 1974, Astrophys. J. (Letters), 193, L89.
(147) Saykally, R.J., Dixon, T.A., Anderson, T.G., Szanto, P.G., and Woods, R.C.: 1976, Astrophys. J. (Letters), 205, L101.
(148) Winnewisser, G., Churchwell, E., and Walmsley, C.M.: 1979, in Modern Aspects of Microwave Spectro-Scopy (ed. G.W. Chantry), Academic Press, p.313.
(149) Herbst, E., and Klemperer, W.: 1973, Astrophys. J., 185, 505.
(150) Watson, W.D.: 1974, Astrophys. J., 188, 35.
(151) Snyder, L.E., and Buhl, D.: 1974, Astrophys, J. (Letters), 189, L31.
(152) Lovas, F.J., Johnson, D.R., Buhl, D., Snyder, L.E.: 1976, Astrophys. J., 209, 770.
(153) Elldér, J., Friberg, P., Hjalmarson, Å., Höglund, B., Irvine, W.M., Johansson, L.E.B., Olofsson, H., Rydbeck, G., Rydbeck, O.E.H., and Guélin, M.: 1980, Astrophys. J. (Letters), 242, L93.
(154) Clark, F.O., Lovas, F.J., and Johnson, D.R.: 1979, Astrophys. J., 229, 553.
(155) Bauder, A.: 1979, J. Phys. Chem. Ref. Data, 8, 583.

(156) Johansson, L.E.B., Andersson, C., Elldér, J., Friberg, P., Hjalmarson, Å., Höglund, B., Irvine, W.M., Olofsson, H., and Rydbeck, G.: 1984, Astron. Astrophys., 130, 227.
(157) Sastry, K.V.L.N., Lees, R.M., and De Lucia, F.C.: 1984, J. Mol. Spectrosc., 103, 486.
(158) Herbst, E., Norbeck, J.M., Certain, P.R., and Klemperer, W.: 1976, Astrophys. J., 207, 110.
(159) Woods, R.C., Gudeman, C.S., Dickman, R.L., Goldsmith, P.F., Huguenin, G.R., Irvine, W.M., Hjalmarson, Å., Nyman, L.-Å., and Olofsson, H.: 1983, Astrophys. J., 270, 583.
(160) De Frees, D.J., McLean, A.D., and Herbst, E.: 1984, Astrophys. J., 279, 322.
(161) Irvine, W.M., Schloerb, F.P., Hjalmarson, Å., and Herbst, E.: 1984, in Protostars and Planets II (eds. D.C. Black and M.S. Matthews), Univ. Arizona Press.
(162) Kraemer, W.P., and Bunker, P.R.: 1983, J. Mol. Spectrosc., 101, 379.
(163) Sutton, E.C., Blake, G., Masson, C., and Phillips, T.G.: 1984, in preparation.
(164) Green, S.: 1980, Astrophys. J., 240, 962.
(165) Cooper, D.L.: 1983, Astrophys. J., 265, 808.
(166) Nejad, L.A.M., Millar, T.J., and Freeman, A.: 1984, Astron. Astrophys., 134, 129.
(167) Thaddeus, P., Cummins, S.E., and Linke, R.A.: 1984, preprint.
(168) Michalopoulos, D.L., Geusic, M.E., Langridge-Smith, P.R.R., and Smalley, R.E.: 1983, J. Chem. Phys., submitted.
(169) Hasegawa, T., Kaifu, N., Inatani, J., Morimoto, M., Chikada, Y., Hirabayashi, H., Iwashita, H., Morita, K., Tojo, A., and Akabane, K.: 1984, Astrophys. J. (Letters), July 15.
(170) Bally, J., and Lada, C.F.: 1983, Astrophys. J., 265, 824.
(171) Friberg, P.: 1984, Astron. Astrophys., 132, 265.
(172) Olofsson, H.: 1984, Astron. Astrophys., 134, 36.
(173) Askne, J., Höglund, B., Hjalmarson, Å., and Irvine, W.M.: 1984, Astron, Astrophys., 130, 311.
(174) Rydbeck, O.E.H., Hjalmarson, Å., Rydbeck, G., Elldér, J., Olofsson, H., and Sume, A.: 1981, Astrophys. J. (Letters), 243, L41.
(175) Batrla, W., Wilson, T.L., Bastien, P., and Ruf, K.: 1983, Astron. Astrophys., 128, 279.
(176) Schloerb, F.P., Friberg, P., Hjalmarson, Å., Höglund, B., and Irvine, W.M.: 1983, Astrophys. J., 264, 161.
(177) Andersson, M., Askne, J., and Hjalmarson, Å.: 1984, Astron. Astrophys., in press.
(178) Olofsson, H., Johansson, L.E.B., Hjalmarson, Å., and Nguyen-Quang-Rieu: 1982, Astron. Astrophys., 107, 128.
(179) Nguyen-Quang-Rieu, Bujarrabal, V., Olofsson, H., Johansson, L.E.B., and Turner, B.: 1984, Astrophys. J., in press.

(180) Stark, A.A.: 1981, Astrophys. J., 245, 99.
(181) Wannier, P.G.: 1980, Ann. Rev. Astron. Astrophys., 18, 399.
(182) Smith, D., and Adams, N.G.: 1980, Astrophys. J., 242, 424.
(183) Penzias, A.: 1980, Science, 208, 663.
(184) Henkel, C., Wilson, T.L., and Bieging, J.: 1982, Astron. Astrophys., 109, 344.
(185) Wilson, R.W., Langer, W.D., and Goldsmith, P.F.: 1981, Astrophys. J. (Letters), 243, L47.
(186) Penzias, A.A.: 1983, Astrophys. J., 273, 195.
(187) Scoville, N.Z., Kleinman, S.G., Hall, D.N.B., and Ridgway, S.T.: 1983, Astrophys. J., 275, 201.
(188) Vanden Bout, P.A., and Snell, R.L.: 1980, Astrophys. J., 236, 460.
(189) Wannier, P.G., Penzias, A.A., and Jenkins, E.B.: 1982, Astrophys. J., 254, 100.
(190) Wannier, P.G., and Linke, R.A.: 1978, Astrophys. J., 225, 130.
(191) Barnes, T.G., Beer, R., Hinkle, K.H., and Lambert, D.L.: 1977, Astrophys. J., 213, 71.
(192) Wannier, P.G., Linke, R.A., and Penzias, A.A.: 1981, Astrophys. J., 247, 522.
(193) Thum, C., Mezger, P.G., and Pankonin, V.: 1980, Astron. Astrophys., 87, 269.
(194) Loren, R.B.: 1984, MWO Spectral Line Detections from 128 to 357 GHz, 1979-1984, Technical Report No. AST8116403-1, Electrical Engineering Research Laboratory, University of Texas.
(195) Cummins, S.E., Linke, R.A., and Thaddeus, P.: 1984, in preparation.
(196) Lovas, F.J., Snyder, L.E., and Johnson, D.R.: 1979, Astrophys. J. Suppl. Ser., 41, 451.
(197) Turner, B.E.: 1983, Astrophys. Letters, 23, 217.
(198) Broten, N.W., McLeod, J.M., Avery, L.W., Irvine, W.M., Höglund, B., Friberg, P., and Hjalmarson, Å.: 1984, Astrophys. J. (Letters), 276, L25.
(199) Walmsley, C.M., Jewell, P.R., Snyder, L.E., and Winnewisser, G.: 1984, Astron. Astrophys. (Letters), 134, L11.
(200) Brown, R.D., Eastwood, F.W., Elmes, P.S., and Godfrey, P.: 1983, J. Am. Chem. Soc., 105, 6496.
(201) Suzuki, H., Kaifu, N., Miyaji, T., Morimoto, M., Ohishi, M., and Saito, S.: 1984, Astrophys. J. (Letters), submitted (Nobeyama Radio Observatory Report No. 21, 1983).
(202) Nyman, L.-Å.: 1983, Astron. Astrophys., 120, 307.
(203) Nyman, L.-Å.: 1984, Astron. Astrophys., in press.
(204) Wootten, A., Loren, R.B., and Snell, R.L.: 1982, Astrophys. J., 255, 160.
(205) Stenholm, L.G.: 1983, Astron. Astrophys., 117, 41.
(206) Olofsson, H., Elldér, J., Hjalmarson, Å., and Rydbeck, G.: 1982, Astron. Astrophys., 113, L18.

(207) Scalo, J.M., and Slavsky, D.B.: 1980, Astrophys. J. (Letters), 239, L73.
(208) Dalgarno, A.: 1981, Phil. Trans. R. Soc. London, A303, 513.
(209) Watson, W.D., and Walmsley, C.M.: 1982, in Regions of Recent Star Dormation (eds. Roger, R.S. & Dewdney, P.E.), Reidel, Dordrecht, p.357.
(210) McCabe, E.M., Smith, R.C., and Clegg, R.E.S.: 1979, Nature, 281, 263.
(211) Lafont, S., Lucas, R., and Omont, A.: 1982, Astron. Astrophys., 106, 201.
(212) Huggins, P.J., Glassgold, A.E., and Morris, M.: 1984, Astrophys. J., 279, 284.
(213) Olofsson, H., and Rydbeck, G.: 1984, Astron. Astrophys., in press.
(214) Sutton, E.C., Masson, C.R., and Phillips, T.G.: 1983, Astrophys. J. (Letters), 275, L49.
(215) Young, J.S., and Scoville, N.Z.: 1984, Astrophys. J., submitted.
(216) Rydbeck, G., Hjalmarson, Å., and Rydbeck, O.E.H.: 1984, Astron. Astrophys., submitted.
(217) Boroson, T.: 1981, Astrophys. J. Suppl. Ser., 46, 177.
(218) Scoville, N.Z., and Young, J.S.: 1983, Astrophys. J., 265, 148.
(219) Lo, K.Y., Berge, G.L., Claussen, M.J., Heiligman, G.M., Leighton, R.B., Masson, C.R., Moffet, A.T., Phillips, T.G., Sargent, A.I., Scott, S.L., Wannier, P.G., and Woody, D.P.: 1984, Astrophys. J. (Letters), submitted.
(220) Goldsmith, P.F., and Langer, W.D.: 1978, Astrophys. J., 222, 881.
(221) Takashi, T., Hollenbach, D.J., and Silk, J.: 1983, Astrophys. J., 275, 145.
(222) Guélin, M., Langer, W.D., and Wilson, R.W.: 1982, Astron. Astrophys., 107, 107.
(223) Langer, W.D.: 1978, Astrophys. J., 225, 95.
(224) Paleologou, E.V., and Mouschovias, T.Ch.: 1983, Astrophys. J., 275, 838.
(225) Davies, R.D.: 1981, Phil. Trans. R. Soc. London, A303, 581.
(226) Harwit, M.: 1984, Comments Astrophys., 10, No. 2, p.65.
(227) Blake, G.A., Sutton, E.C., Masson, C.R., Phillips, T.G., Herbst, E., Plummer, G.M., and De Lucia, F.C.: 1984, preprint.

ACKNOWLEDGEMENT

We want to thank T. Wiklind for careful help with preparation of figures and tables and with final editing, and K. Lundquist for rapid and excellent typing of this review paper.

SUB-MILLIMETER ASTRONOMY AND ASTROPHYSICS

T. L. Wilson, C. M. Walmsley

Max-Planck-Institut für Radioastronomie, Bonn, F.R.G.

As of 1984, the submm band ($\nu > 300$ GHz) is relatively unexplored. The word "relatively" should be stressed since, evidently, a great deal of work has been done. However, much of this is of the "discovery" or "first detection" category with emphasis on the priority of a given result, and less thought as to astrophysical consequences. It therefore seems to be a useful exercise to examine what astrophysics one should expect from submm measurements. A great deal of these measurements will be the extension of work going on at longer wavelengths, and we will base much of our discussion on our own recent research and comment on its consequences for observations at frequencies above 300 GHz. (Summaries of frequency scans in the 200-260 GHz region are Erickson et al. 1980 and G. Blake, this volume.) We will emphasize the way in which observations in the various wavelength bands complement one another, and outline the connections by means of a few examples. Finally, we will present a short description of the 10-m submm dish to be built by the MPIfR and to be operated by the University of Arizona and the MPIfR on Mt. Graham, near Tucson, Arizona.

As in studies done at longer wavelengths, the goals of submm measurements are to: (1) determine the physical conditions (kinetic temperature, T_K, H_2 density, $n(H_2)$, radiation field, magnetic field strength, B, velocity field, V, etc.) in molecular clouds and circumstellar envelopes; (2) measure the molecular, elemental and isotopic ratios in such regions, and (3) determine their spatial structure.

In this review we will concentrate on points (1) and (3), since an understanding of cloud parameters is necessary in order to properly evaluate the isotopic results. Using this information, one can attempt to understand molecular formation schemes, determine the actual isotopic and chemical constitution of molecular clouds, describe the history of these clouds (including the ever-elusive protostar phase), and attempt to pick out the best sites in which to search for new molecular (perhaps biologically important) species.

CONTINUUM EMISSION FROM DUST GRAINS

One of the most obvious reasons for doing submm astronomy is the study of dust radiation from molecular clouds in our galaxy and external galaxies. As pointed out by many authors (see e.g. Whitcomb et al. 1981, Hildebrand 1983), the emission at 400 µm provides an estimate of the column density of dust and, assuming a dust-to-gas ratio, the column density of H_2. This last number cannot yet be reliably estimated from molecular line data. The rapid improvements in FIR detectors and observing techniques (see e.g. Chini et al. 1984) allow measurements of a wide variety of sources. In particular, the IRAS results (see March 1984, Ap.J. Letters L1 ff.) show that most spiral galaxies radiate a large fraction of their total luminosity at wavelengths longwards of 100 µm; in this range, the radiation is optically thin and estimates of the total emitting dust mass can be made. The largest uncertainties are in the composition and size distribution of the dust grains. First guesses for these parameters can be made using measurements of dust grains in the solar neighborhood where we know something about grain sizes. These results are extrapolated to more distant objects and used to estimate cloud masses. One can then compare with mass estimates based on the column density and extent of CO (which is usually optically thick) using observations of the rare isotopes of CO (which are optically thin, but for which the line emission is weak) and the angular extent of the line emission. As will be discussed later, one must determine the total population of CO from two or three line measurements and a quasi-LTE partition function. To obtain the column density of H_2 from CO data, one must also assume the ratio of CO to H_2. Estimates by R.W. Wilson et al. (1982), based on star counts and measurements of the J=1-0 lines of CO, ^{13}CO and $C^{18}O$ for a dust cloud in Taurus (160 pc from the sun), show some variation in this ratio with increasing visual extinction. A larger sample would be useful: extensions to darker ($A_v > 8^m$) regions and to higher angular resolution are also needed (the previous results are averages over $\geq 2'$).

SPECTRAL LINE DATA
BACKGROUND

The interpretation of spectral line results in the submm range will be at once more challenging but less straightforward than the study of dust emission. The shorter wavelength implies that the lines are emitted from hot and dense regions. In the simplest case of a single pair of energy levels (Spitzer 1978, p. 82) the critical density n_{CR}, needed for the collisional excitation of optically thin lines is

$$n_{CR} = \frac{A}{\gamma} \qquad (1)$$

where A is the spontaneous decay rate out of the upper level and $\gamma = <\sigma v>$ is collision rate; typically this is 10^{-11} cm^3 sec^{-1}. For a dipole moment of 1 Debye, the value of A for a 3 mm line is 6×10^{-6} sec^{-1}, while for a 0.3 mm line it is 6×10^{-3} sec^{-1}. Thus n_{CR} at 300 µm is 10^3 times larger than at 3 mm. Also $h\nu/k$ for the 300 µm line is ~ 20 K, i.e. the upper energy level is at least 20 K above the ground state. This requires a moderately warm gas in order to obtain a reasonable population in the upper level.

In astronomical terms, this implies a change of theme, from cold (~ 10 K) molecular clouds to warm dense regions where young stars have formed. One also expects to observe at the shorter wavelengths circumstellar envelopes, regions shocked by supernovae, and regions within cold clouds where cloud collisions and thus shock waves are present. Aside from molecular radiation in shocked regions there should be a large number of atomic fine structure lines. An example is the 610 µm line of CI (Phillips et al. 1980). This seems to be optically thick (Phillips and Huggins 1981), indicating a large abundance of atomic carbon, even in the interiors of molecular clouds. In general, most violent events in the interstellar medium lead to shock waves which cause radiation in the atomic fine structure lines and rotational transitions of molecules. Observations of these lines will give a more complete understanding of the phenomena mentioned above.

The actual world is more complicated (and more interesting) than the simplified viewpoint presented above. Ignoring the complications can lead to qualitatively wrong answers. Molecular clouds are inhomogeneous. Thus the space density cannot be obtained from the column density divided by the cloud dimensions. Moreover, different transitions are weighted towards different density and temperature regimes in the cloud. Each transition tends to see gas with density close to its particular n_{CR} and comparison of different transitions does not yield a unique value for the gas density. A curious characteristic of molecular clouds is that the column density characteristic of different density regimes is roughly density independent (see e.g. Larson (1981)). This tends to accentuate the difficulties mentioned above although it is not clear whether the law holds for densities much above 10^5 cm^{-3}. One can, in any case, protect oneself to some extent by comparing line profiles in order to ensure that different transitions "see" essentially the same gas volume. This approach has been applied, for example, by Henkel et al. (1980) and by Linke and Goldsmith (1980) in studies of H_2CO and CS respectively. The results are relatively convincing and, in fact, a rather good agreement is obtained between the profiles in various transitions. Nevertheless, VLA formaldehyde results (see e.g. Martin-Pintado et al. (1984)) show that single dish absorption profiles are a composite of contributions from different size scales. Hence, the densities derived in single dish studies should be looked on as crude averages.

This sort of situation is not, of course, unique to molecular cloud studies. Analogous problems are encountered when deriving electron densities in emission line studies of HII regions and quasars. However, molecular line studies do have the extra complication that many of the important transitions are optically thick and hence crude models of the radiative transport in the lines have to be used in order to interpret the observations. In the simplest situation, the effective A value in equation (1) decreases roughly proportionally to the optical depth of the transition in question. n_{CR} can drastically be reduced and, in this limit, it becomes more or less transition independent. That is to say, the difference between n_{CR} for submm and longer wavelength transitions referred to earlier, more or less disappears. One practical consequence of this is that lines with relatively high excitation requirements (i.e. high n_{CR} according to eq. (1)) can be observed as extended sources in molecular clouds. Examples are HCO^+ and HCN for which equation (1) would predict values close to 10^6 cm^{-3} (for J=1→0) but which can be mapped extensively in local dust clouds. These "trapping" effects are a considerable complication and their magnitude is dependent both on geometry and velocity field.

Faced with these problems, a number of attitudes are possible. One is to attempt rather detailed radiative transport models with some assumptions (spherical symmetry) about the geometry and velocity field in the cloud. The output of such models can be fit to observations using as large a selection of data as possible (see e.g. Stenholm (1983) and references therein). The difficulties with this approach are on the one hand the justification of the assumed geometry and on the other hand the complexity of the modelling procedure. Less complex but perhaps oversimplified is the LVG (large velocity gradient) procedure adopted by several groups which reduces the radiative transport to a "local problem". Goldsmith (1982) has given a good account of the perils of LVG calculations. Such an "ansatz" works best when either the velocity gradient really is high or alternatively when optical depths are almost negligible. The last possibility - an essentially optically thin situation - of course only occurs for the weakest lines and is one reason why one would like "quantum-limited" systems right up to submm wavelengths. While waiting for this, it is

clear in any case that testing the models requires measurements
of many lines with a large range of excitation requirements. The
submm lines have different excitation requirements than lines at
lower frequency and are in this sense complementary to existing
measurements. In the following discussion, we will try to illustrate this complementarity with some examples of regions which
have been recently under study. In most cases but not in all, one
can expect that higher frequency measurements will provide new
insight into the structure of the molecular cloud.

LOW DENSITY REGIONS AND DARK CLOUDS

Even low density molecular material far from regions of rapid
star formation show abundant evidence for structure. For instance,
the molecular clouds seen towards the supernova remnant Cas A
(Batrla et al. 1983, 1984) are found to have densities in the
range 10^3–10^4 cm^{-3}. From this H_2 density and estimates of the
column density, the line-of-sight dimensions are 10% of the diameter. The most straightforward conclusion is that the clouds
seen in molecules consist in reality of a large number of smaller
structures, or clumps. It appears that the absorption lines arise
in regions near the edges of two large molecular clouds which
happen by chance to be located along the line-of-sight to Cas A.
These clouds are seen in CO emission (Scoville et al. 1977). The
only function of Cas A (in absorption line observations) is to
serve as a convenient background source.

Other observations show evidence for gradients in the H_2 density
in molecular clouds. These gradients could lead to self-absorption
in optically thick lines. For example, the analysis of centimeter

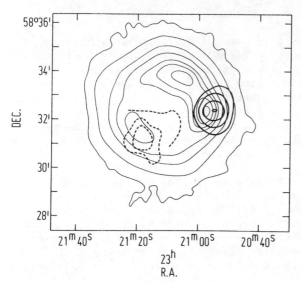

Fig. 1. The locations and sizes of the two molecular clouds seen in the 2_{11}-2_{12} absorption line of H_2CO toward the supernova remnant Cas A (Batrla et al. 1983). The coordinates are epoch 1950.0. The dashed curves refer to the cloud at a radial velocity -47 km s^{-1}, the solid curves to the cloud at -39 km s^{-1}.

and millimeter lines of H_2CO for the high-density (n =10^5 to 10^6 cm^{-3}, T_K =15 to 20 K) region in the ρ Oph cloud, called ρ Oph B (Loren et al. 1980, Martin-Pintado et al. 1983, Zeng et al. 1984) shows that this region is behind a layer of lower density gas. Such a conclusion is based on detailed models of the excitation of K-doublet lines of H_2CO. The presence of such density gradients greatly affects any analysis based on optically thick lines connected to the ground state. Measurements of a series of rotational lines of H_2CO or of other molecules may allow us to check this argument and better specify the physical parameters of the regions. The situation for TMC-1, another fragment in a dust cloud (also 160 pc from the sun) is not so clear (see Toelle et al. 1981, Henkel et al. 1981). TMC-1 is a highly interesting source because of the presence of long chain carbon molecules. The H_2 density in TMC-1 appears to be lower than that in ρ Oph B; however the abundance of long chain carbon bearing molecules is larger in TMC-1 (see also the discussion by H. Kroto in this volume). The ground state K-doublet line from ortho-H_2CO, at 6 cm, has a larger linewidth than that from the higher 2 cm line. The lower density material (sampled by the 6 cm line) fills a larger volume, but could be mostly behind TMC-1. If this is so, the J=2-1 rotational line of H_2CO should be seen at the strength predicted by simple homogeneous models. If the lower density material is in front, the optically thick J=2-1 line of H_2CO will be absorbed and scattered by the foreground gas. Inside TMC-1 itself, the gas seems to be cold (T_K ∼10 K). Thus, it is unlikely that levels far above the ground state will be populated significantly and searches for molecules in the mm and submm region will not be useful. On the other hand, a map of the 400 μm continuum would be very useful to determine the column density of dust.

HIGH DENSITY CLUMPS SEEN IN AMMONIA AND FORMALDEHYDE

The non-metastable levels of ammonia have lifetimes of order 10-100 sec. and correspondingly, from equ. (1), n_{CR} is more than 10^9 cm^{-3}. This is far higher than normal molecular cloud densities and hence one might expect such regions to be tracers for hot (the levels of interest are all more than 50 K above ground) dense condensations. However, the trapping effects mentioned earlier (see Sweitzer 1978) change the picture and can cause the <u>effective</u> n_{CR} to be as much as two orders of magnitude lower. Nevertheless, relatively high densities are required even for the lowest non-metastable inversion line, the (2,1). It is therefore not surprising that prior to 1982, the (2,1) line had been found in only 3 sources; recently however, with the new K-band maser receiver and the 100-m telescope, (2,1) line radiation has been detected in two clouds in the Orion region outside of Orion-KL (Batrla et al 1983) and the 8 additional sources shown in Fig. 2 (Mauersberger et al. 1983). Detailed calculations based on the (2,1) line line data give H_2 densities which agree roughly with values ob-

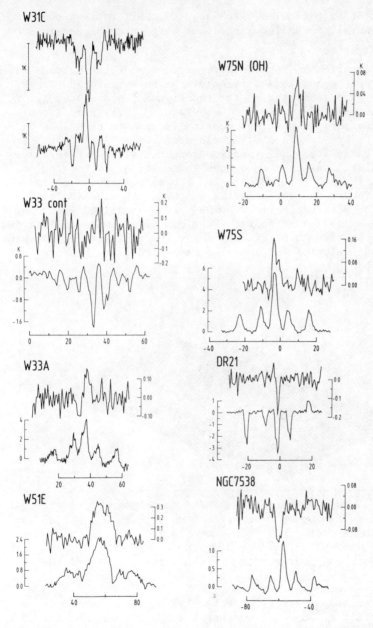

Fig. 2. The $(J,K) = (2,1)$ and $(1,1)$ line profiles (above and below) for the sources shown. The temperatures are on a main-beam brightness scale. The $(2,1)$ absorption line toward W31 exhibits hyperfine splitting; the $(2,1)$ line toward W33 continuum should be regarded as marginal.

tained from other molecules, such as H_2CO, CS, etc. and show that, in general, hot molecular clouds have large H_2 densities.

The non-metastable NH_3 results give estimates of the H_2 density, and ratios of the column densities of the metastable NH_3 lines (where J=K) give estimates of the rotational temperature, T_R, which can be related to kinetic temperature, T_K. The metastable lines are thermalized when the density is greater than $10^{4.5}$ cm^{-3}. These results are combined with values of the optical depth (from least-squares fits to the hyperfine components) to get estimates of the angular sizes of the sources in question. In the region between OMC-1 and 2 (Fig. 3), the NH_3 appears to have a clumpy structure. Toward the three maxima in the south, the (2,1) line is found, the metastable lines of NH_3 are thermalized (see e.g. Ho and Townes 1983), and source sizes can be estimated, since the optical depths (from hyperfine ratios), the excitation temperature (equal to the value of T_R) and line temperature are known.

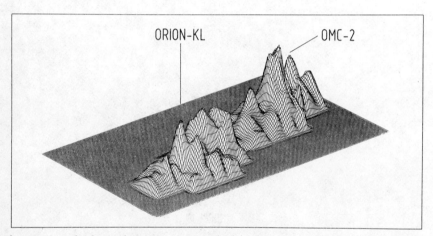

Fig. 3. A shadow plot of the peak line temperature of the (1,1) line of NH_3 in the Orion region, produced by Dr. W. Batrla. The temperatures were obtained from gaussian fits to the data, restricting the results to lines narrower than 5 km s^{-1}. Thus the broad NH_3 velocity features in Orion-KL are not present in this plot.

Three open questions are (1) whether the H_2 distribution is as clumped as that of NH_3, (2) whether the NH_3 regions consist of many clumps (all smaller than the telescope beam) which fill the telescope beam at each position, and finally, (3) how much of the mass of a molecular cloud is associated with this high density material. Question (2) can be answered for one source by a comparison with the high resolution (20") VLA map of Harris et al. (1983) shown in Fig. 4. As with the 100-m map, there is direct

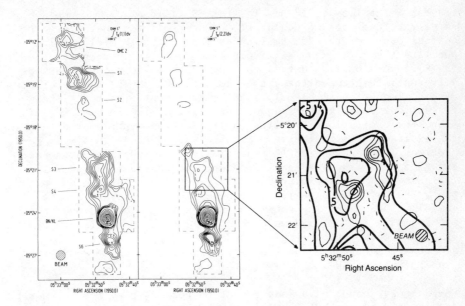

Fig. 4. On the left, maps of the (1,1) and (2,2) lines of NH₃, made using the 100-m telescope (resolution 43", Batrla et al. 1983b). On the right, a VLA map in the (2,2) line, shown as thin contours (Harris et al. 1983), superimposed on the 100-m data, shown as thick contours.

evidence for clumping in the VLA data. Surprisingly, however, the maxima seen with the VLA are offset from the maxima seen with the 100-m. The maps are consistent, since the peak temperatures seen with the VLA (when diluted in a 43" beam) are less than those seen with the 100-m, and the VLA map has ∼25% of the line flux density seen with a single dish. Harris et al. (1983) report that the condensations are the clumps rotating. These may be collapsing molecular cloud fragments.

A comparison of the H_2 mass obtained from the 400 μm map of Smith et al. (1979) with virial mass estimates shows that the NH_3 regions near Orion-KL contain nearly all of the molecular mass. This seems not to be the case in some of the northern regions. However, the NH_3 is probably still thermalized, and source sizes can be obtained from the optical depth, T_R, and line temperatures. All of these conclusions should be checked by 20"-40" resolution observations of dust emission at 400 μm and maps of the $C^{18}O$ emission to determine whether the NH_3 peaks are (as we believe) density and temperature enhancements.

Another possible mechanism which could populate the (2,1) levels of NH_3 is the far IR continuum radiated by dust grains. In the case of NH_3, the rotational transition between the J=2 and 1 levels for K=1 falls at 252 μm. The far IR field at 252 μm could be large, at least for molecular clouds near HII regions. Since the optical depth of the 252 μm line is large, the IR field could populate the (2,1) levels. A comparison of other non-metastable (J >K) inversion line data with model calculations could perhaps differentiate between these two excitation schemes. A more direct method (Townes et al. 1983) is the measurement of a rotational transition. These authors argue that if the line is seen in emission, the excitation is collisional; if in absorption or absent, the excitation is radiative. This analysis depends on simple geometries, i.e. no beamed IR radiation, etc., and also on the assumption that other excitation paths, such as near-IR radiation (which could populate vibrationally excited states) are unimportant. The (4,3)→(3,3) emission line was detected by Townes et al. (1983) in Orion KL. The estimated H_2 density, averaged over the $\sim 1'$ beam, is 10^7 cm^{-3}. However, as will be discussed later, Orion-KL is an inhomogeneous region, and this complex structure may affect arguments based on the very optically thick ($\tau \geq 10^3$) rotational lines of NH_3.

Fortunately, we do have independent methods of sampling these high density clumps. One of these is using the formaldehyde molecule which, due to its slight asymmetry, has observable transitions at cm, mm, and submm wavelengths. The cm transitions have the useful property that at moderate densities (below 10^5 cm^{-3}), they are "cooled" below the microwave background temperature (3 K) and hence appear in absorption even in the absence of background radio continuum. Hence, the mere presence of 6 cm or 2 cm formaldehyde in _emission_ has been interpreted to imply the presence of high density clumps ($\sim 10^6$ cm^{-3}). The situation has been complicated slightly by the discovery that 6 cm formaldehyde can exhibit maser emission (Forster et al. (1980)). However, this can be checked for by interferometer measurements and, in general, it does seem that 2 cm emission is a good tracer for high density clumps. It can be excited at low temperatures and hence provides complementary information to the $NH_3(2,1)$ line.

Near W75S (see Fig. 5), Wilson et al. (1982) detected such an emission line from the $2_{11}-2_{12}$ transition of H_2CO. A map of the source made with the VLA (Johnston et al. 1984) shows the small scale structure of the H_2CO emission and the association of this quasi-thermally excited gas with OH and H_2O masers. From NH_3 observations, the kinetic temperature of the gas is ~ 60 K. Altogether, these measurements reinforce the picture of fragmentation of the column of gas into smaller high density clumps. The density and kinetic temperature of W75S are similar to those found in Orion-KL. Since Orion-KL is much closer to the sun, more spatial details can be obtained for a given angular resolution. We show a superposition of high (<25") resolution molecular line results for Orion-KL in Fig. 6. From this, it would appear that dif-

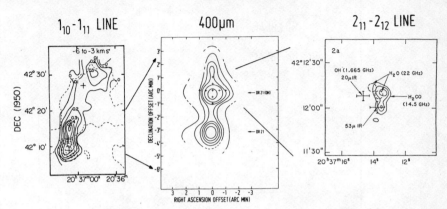

Fig. 5. On the left, a map of the $1_{10}-1_{11}$ absorption line of H_2CO at 6 cm (resolution 2.6'; Bieging et al. 1982). In the middle, the continuum dust radiation measured at 400 μm (Werner et al. 1977), and on the right, a VLA map (resolution 10"; Johnston et al. 1984) of 2 cm emission from the DR21(OH)/W75S source. For comparison, in the VLA map, the positions of OH and H_2O masers and IR sources are shown.

ferent molecules are formed in different locations. However, excitation effects could have an unexpectedly large influence. A region which is prominent in K-doublet emission lines of H_2CO is centered on IRc5 (Johnston et al. 1983). This source is also prominent in di-methyl ether, $(CH_3)_2O$ (Friberg 1984). About 90" south of Orion-KL is a cloud found by Johnston et al. (1983) in H_2CO absorption at 6 cm; it is prominent in CH_3OH (see our Fig. 7 and Boland et al. 1982), HCN and N_2H^+ (Turner and Thaddeus 1977) and in the submm continuum (Keene et al. 1982). It may be inside the HII region and exist as a PIG (Partially Ionized Globule) of dense gas.

Evidently, BN/KL has the advantage of proximity and the disadvantage of complexity. Our hope is that submm observations will allow us to unravel the tangle. We note that in a long term future, this will imply angular resolutions of a few arc sec and hence probably interferometry at millimeter and submm wavelengths. In the short term, however, single dish measurements with, say, 20" resolution will be immensely valuable.

INTERSTELLAR MASERS

Interstellar masers are, at best, partially understood but it is clear that they are found in regions where massive stars are presently forming. Because they are so strong, their study has proved

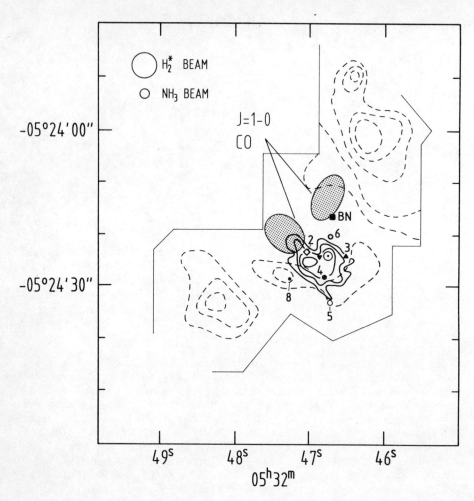

Fig. 6. The distribution of vibrationally excited H_2 (labelled H_2^* and shown as dashed lines) from Beckwith et al. 1979, of high velocity CO in the J=1-0 line (33" resolution; Olofsson et al. 1983) shown shaded, and of hot NH_3 (Pauls et al. 1983) shown as solid contours. A ridge of moderately broad HCO^+ emission is north of the western H_2^* maximum (Olofsson et al. 1983). Measurements of the high velocity J=3-2 lines of CO (Erickson et al. 1982) show reasonable agreement with the results of Olofsson et al. (1983).

useful even although we do not in general understand the pump mechanism. It is consequently still quite unclear what, for example, the basic parameters of OH masing regions are. However, some progress in this field is being made. In particular, it is of importance that it has become possible to study the OH maser

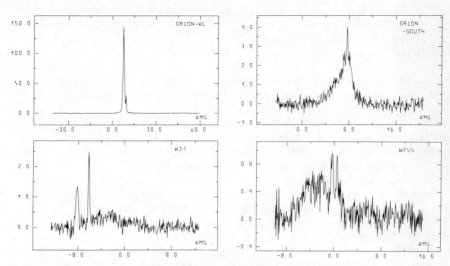

Fig. 7. Some of the $6_2 \rightarrow 6_1$ lines of E-type methanol detected with the 100-m telescope. The line temperatures are on a main-beam brightness scale. The emission from Orion-KL is caused by masering (Barrett et al. 1971, Matsakis et al. 1980). Emission from the other 3 sources is probably also caused by maser emission. These are the first masers in E-type methanol found outside of Orion-KL.

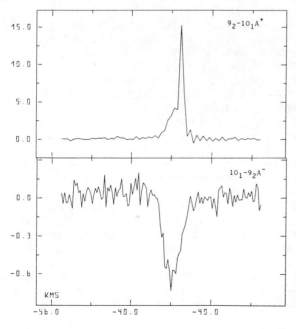

Fig. 8. The $9_2-10_1 A^+$ and $10_1-9_2 A^-$ lines of methanol, measured toward the compact HII region W3(OH). The velocity width of the absorption matches that of the emission. From mapping results, the A^- line absorbs only the continuum source. Thus, the A^+ line is a weak maser which amplifies the background continuum source.

regions in non-masering transitions of both NH_3 and of OH itself
(see Guilloteau et al. (1983), (1984)). Also of importance is
the detection of the rotational transitions of OH in the far
infrared (Harwit 1984).

However, we wish now to underline the importance of methanol
(CH_3OH), which is another molecule with lines in the centimeter
and submm range. In a sense, methanol, being an asymmetric top
molecule, is more typical than either NH_3 or OH. From the spin
statistics of the 3 identical hydrogen nuclei attached to the C
atom, there are A and E species. In ordinary molecular collisions,
or in radiative transitions, these species are not mixed. In
Orion-KL, maser emission has been found in the $\Delta J=0$, $\Delta K=1$ transitions from the E species (Barrett et al. 1971). An intensive
search led to no detections of other sources (Buxton et al. 1977).
With the 100-m telescope (see Fig. 7), we have detected several
new sources of emission in the E species. Two of these, in W31
and W75S, seem to be masers. Thus CH_3OH masers are not unique to
Orion-KL; in other sources, these masers have comparable luminosities , but the measured emission is weaker because of the
greater distances. Searches for CH_3OH in late-type stars were
unsuccessful. Thus, unlike OH or H_2O masers, CH_3OH masers are
found only near regions of active star formation. The cause of
the inversion of the methanol lines is unknown. From time variations of the features in Orion-KL, Matsakis et al. (1980) argued
for a radiative pumping scheme. Direct measurements of the $\Delta J=1$
transitions, in the submm range, may help to decide whether collisional or radiative processes are chiefly responsible.

For centimeter observers, methanol is interesting because it is
possible to observe different lines (which are close in frequency)
and which are between related levels. Deviations from LTE populations may lead to masering if the upper level is overpopulated,
and to absorption if the lower level is overpopulated. In Figure
8, the $9_2-10_1A^+$ emission toward the compact HII region W3(OH) is
compared with the $10_1-9_2A^-$ absorption profile. The velocity range
of emission (in A^+) and absorption (in A^-) are the same, leading
us to conclude that the A^+ emission is caused by an amplification
of the continuum background. The emission and absorption line-
shapes are not the same. The sharp spike at -43.9 km s^{-1} is probably caused by a long coherence length in the molecular cloud,
which leads to a build-up of intensity. Such velocity coherence
in an absorption line has no such effect and since the $9_2-10_1A^+$
line amplifies the background, one can use the line-to-continuum
ratio to estimate the optical depth, τ. If the whole of W3(OH) is
covered, τ is between 1 and 2. It is likely that the CH_3OH covers
only a part of W3(OH). If it covers that part of the source where
the OH masers are found (Reid et al. 1980), the maximum optical
depth is ~ 3. The measurements of the centimeter lines could be
explained if the K=2 ladder is overpopulated relative to the K=1

ladder. It might be possible to check this idea by measurements of the $\Delta J = 1$ rotational transitions, connecting the 9_2 and 10_1 levels. These lines fall at ~ 600 μm; a high angular resolution map might help to understand the population distribution. It is probably easier to understand the maser pumping process for methanol than for OH or H_2O because the optical depths of the centimeter methanol lines are smaller. For example, from the brightness temperatures of the 18 cm line of OH observed toward W3(OH), the optical depth must be ~ 20; the optical depths of H_2O maser lines must be even larger. Furthermore the pumping schemes for H_2O masers are particularly difficult to test, since there is only one line observed with a high angular resolution, in the centimeter range, at 1.35 cm and two others in the millimeter range. Observations of rotational lines of H_2O are difficult from the earth's surface, since these are attenuated by H_2O in the earth's atmosphere.

HOT COMPACT REGIONS IN MOLECULAR CLOUDS

In this section, we highlight some of the facts known about hot compact sources of molecular line emission in the Galaxy. Such regions are thought to be tracers for the birthplace of massive stars and are often associated with compact HII regions. They typically also have luminosities $>10^4$ $L_\odot$ in the far IR and are associated with H_2O and OH masers. The prime example of such a region is the BN/KL nebula in Orion and we will briefly review some of the characteristics of this source.

Orion is firstly notable for the cluster of infra-red sources nestling behind the Orion HII region itself. The Becklin-Neugebauer object (see e.g. Scoville this volume) appears to be an early B type star which is losing mass at a rate of $\sim 10^{-6} M_\odot$ yr^{-1}. The infra-red source IRc2 is coincident with one of the most powerful SiO masers in the galaxy and may be the source which is providing the energy to power the motion of the cluster of H_2O masers which appear to be expanding away from the K-L nebula (Genzel et al. 1981). Downes et al. (1981) argue that IRc2 is responsible for the high velocity wings seen in CO and H_2^* toward BN-KL (see our Fig. 6). These are generally referred to as the plateau feature; whether IRc2 is really the sole source of this rapidly (100 km s^{-1}) flowing gas is not entirely clear. It seems certain that the interaction of the outflow with surrounding material gives rise to the hot shocked gas which may be observed in the vibration-rotational transitions of molecular hydrogen. It appears paradoxically to be the case that young stars make themselves most readily apparent due to the mass-loss they undergo (rather than by collapse). With one exception, all the above features (broad CO lines, excited H_2 emission, water masers) are seen in a variety of sources. The exception is the SiO maser in Orion which continues to be unique (see e.g. Jewell et al. 1984). HII regions, broad

line CO sources, etc. have been searched in v=1 SiO down to limits
(in luminosity!) an order of magnitude below that of Orion and
there is still no sign of any comparable SiO maser source, although v=0 emission from SiO is seen toward many HII regions. On
the other hand, many Mira variables, supergiants and OH-IR stars
show maser emission in the v=1 transition of SiO at some level.
This has led to the speculation that IRc2 may be a supergiant in
disguise. The main argument against this is simply IRc2's position
immersed in the BN/KL nebula.

The rapid gas flows which take place in BN/KL presumably compress
and heat the surrounding gas condensations. This at least is the
simplest explanation of the hot dense clumps which are observed
in interferometer maps of Orion K-L. The VLA maps of hot NH_3
(Fig. 6) suggest that there are structures on a scale of $\leq 5''$
(= 1.2×10^{-2} pc at 500 pc). From measurements of the (J,K) =
(1,1), (2,2), (3,3), (4,4), (5,5), and (6,6) inversion lines of
$^{15}NH_3$ (which are all optically thin, Hermsen et al. (1984) find
that the rotational temperature T_R of the hot NH_3 is 110 ± 10 K in
both the ortho- and para-species. At the H_2 densities of 10^6 cm^{-3}
thought to be present in Orion K-L, this is the kinetic temperature (Walmsley and Ungerechts 1983) . Since the NH_3 lines are
thermalised and optically thick, the beam filling factor is 0.5-
0.7 even for a 2" beam! This is a strong indication that structure
is present in the molecular gas on a scale of 2" (= $5 \ 10^{-3}$ pc =
1000 A.U.). Kinetic temperature estimates can also be obtained
from rotational transitions of other symmetric top molecules (see
Churchwell and Hollis 1983) from high lying levels of CO in the
far IR (Storey et al. 1980, 1981 , Stacey 1982) , or from rotationally and/or vibrationally excited H_2 (see Beckwith et al.
1979, 1983). Other possibilities are transitions of methanol
(see Wilson et al. 1984) for a collection of references), SO
(see Olofsson 1984, for a summary) or SO_2 (see Schloerb et al.
(1983). Measurements have been made of the transitions from the
vibrationally excited states of various species (see Goldsmith
et al. 1983) and of torsionally excited methanol (see Hollis
et al. 1983).

An analysis of the inversion lines of $^{15}NH_3$ cannot give a completely
unambiguous value for the ortho-to-para ratio in ammonia, since
the (J,K) = (1,0) or (2,0) lines (which cannot undergo inversion
transitions) were not measured. (The J = 1→0, K = 0 line has been
measured by Keene et al. (1983) and a value for this transition
for $^{15}NH_3$ would probably allow good limits to be set on the populations in the K = 0 ladder.) From the available inversion line
data for $^{15}NH_3$, Hermsen et al. (1984) find a ratio which is only
slightly below the LTE value.

The optical depths in the ammonia lines in Orion-KL are very high
and suggest ammonia abundances which are locally much larger than

the typical value of $NH_3/H_2 \approx 10^{-8}$ in nearby molecular clouds (see e.g. Watson 1976). This could be due either to processing through shocks or evaporation of ammonia ice in grain mantles. Two other sources with large (NH_3/H_2) ratios are W31 and NGC7538 (Wilson et al. 1983). These molecular clouds are hot (the (6,6) inversion line is easily seen), dense (the non-metastable (2,1) and (3,2) lines are present), and have large column densities of NH_3. Thus, for these sources and Orion-KL, temperature maxima are NH_3 abundance maxima.

A contrast to the above mentioned regions are the molecular cloud regions close to the galactic center where temperatures are certainly very high but it is not clear whether H_2 densities are very high also. In fact, a recent study shows that temperatures are close to 100 K not only in the Sgr A cloud but also in regions as far as 300 pc from the galactic center (Güsten et al. 1984), since the $NH_3(5,5)$ transition, which is 293 K above the ground state, is clearly seen in many galactic center clouds.

In the Sgr B2 cloud, the $NH_3(9,9)$ inversion line has been seen in absorption toward the most intense HII region, Sgr B2(S) (Wilson et al. 1982). This line is 850 K above the ground state. This detection implies that the molecular cloud, at least toward Sgr B2(S), is very hot. Wilson et al. (1982) deduced an NH_3 rotational temperature, T_R, of 175 K (±20%). Because of the low H_2 density, the value of T_K may be higher. A map of NH_3 inversion lines from the (6,6) level at 405 K above ground (Fig. 9), show that this hot gas is extended over >3' (= 8.8 pc at 10 kpc). In contrast, while the NH_3 in Orion-KL has T_R of 110 K, the region has a size of <0.1 pc.

Although the ammonia excitation in the galactic center regions is far from clear, there is no doubt that temperatures are much higher than in comparably sized regions in the solar neighbourhood molecular clouds. It also seems that the gas temperature in these regions is higher than the dust temperature (see Güsten et al 1984). Hence a direct heating mechanism for the gas is required which has not however been identified. Proposals for this range from ionization by low energy cosmic rays (Güsten et al. 1982) to dissipation of turbulence driven by differential rotation in the center region (Wilson et al. 1982). The latter idea amounts to supposing the widescale presence of many weak shocks. Such shocks would presumably cool mainly in submillimeter transitions of CO and CI and hence one can presume that measurements of these lines will be an important contribution to our understanding of molecular cloud thermodynamics.

The results in the previous sections show that there are a large number of hot ($T_K \geq 100$ K) molecular regions in the galaxy, and that there are $\sim 2"$ structures in these regions. At least in the case of Orion-KL, the region outlined in NH$_3$ (shown in Fig. 6) can be identified with sources of complex molecules, such as Ethyl Cyanide (CH$_3$CH$_2$CN), (see Johansson et al. 1984). A meaningful investigation of line and continuum radiation from these regions will require high angular resolution, and thus either large single dishes or interferometers.

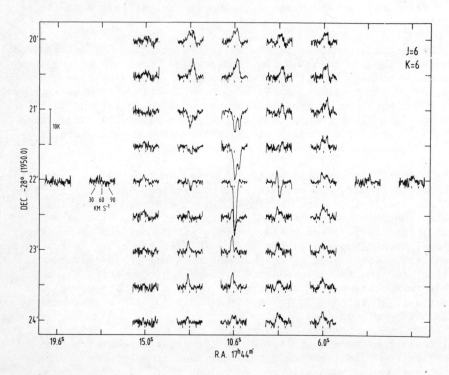

Fig. 9. Line profiles of the (J,K) = (6,6) inversion line of NH$_3$ toward Sgr B2. The angular resolution is 43". The temperatures are on a main beam brightness scale. The NH$_3$ absorbs the continuum sources Sgr B2(S), at R.A. = $17^h44^m10.6^s$, Dec. = $-28°22'$ (1950.0), and Sgr B2(N) which is 1' to the north.

DESIGN DETAILS OF THE MPIfR-UA 10-m SUBMILLIMETER TELESCOPE

The MPIfR and the University of Arizona will collaborate on the construction of a 10-m diameter submillimeter telescope (SMT). It will be used at wavelengths as short as 350 μm. The telescope will be an on-axis paraboloid, with an f/D ratio of 0.35. The RMS surface accuracy is planned to be ≤17 μm; an artists conception is shown in Fig. 10. The axes are in azimuth and elevation. The receivers will be located at the Nasmyth focus. From this focus, the f/D ratio is 9.7. To bring the radiation to this focus requires 4 reflections; however the receiver can be left in a horizontal position; there is much more space than at the prime or cassegrain focus, and the change from one receiver to another should require only a few minutes.

At the shortest wavelength, the diffraction limited beamsize will be about 8"-10". Thus the telescope pointing should be accurate to 0.5"-1". This requirement must be met even in the presence of wind gusts of up to 27 mph. This is a great challenge, but this problem is also present in the case of the 30-m millimeter telescope. The demands on the SMT servo control for pointing should be somewhat less, since the telescope is to be housed in a shelter which co-rotates in azimuth. An artists conception of this shelter, named a "cow barn" by the Univ. of Arizona group, is shown in Fig. 10.

The material to be used in construction of the SMT will be revolutionary: instead of the usual steel and aluminium, the surface and backup structure will be made from carbon fiber. The advantages, of carbon fiber material are twofold: first, it is as strong as

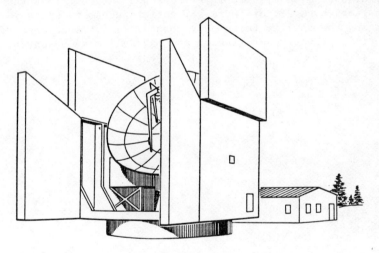

Fig. 10. An artist's sketch of the MPI/UofA 10-m submillimeter telescope in the "cow barn" shelter.

steel and as lightweight as aluminium, and second, it has a very low coefficient of expansion. The second property is the more important, since the design is based on homologous deformations of the structure. Thus, if wind forces and differential temperature differences are unimportant, the accuracy depends only on the quality of the panels and their setting. First estimates indicate that if the accuracy of the pyrex form (from which the panels are molded) is 2 μm, the accuracy of the carbon fiber panels will be

8 μm – 10 μm. This shape will be maintained in an environment with temperatures up to 50° C. The backup structure itself will be more of a problem because carbon fiber rods must be joined. One possibility is the use of stainless steel ball-joints. Although the coefficient of expansion of steel is different from that of carbon fiber, compensation schemes are possible.

The site of the telescope is important, since the atmospheric transmission in the wavelength range between 1 mm and 350 μm is not large, even at the best sites. The SMT will be placed on Mt. Graham, at 3300 m altitude. This site compares favorably with the peak of Mauna Kea, 4200 m altitude, where the CalTech group are erecting a 10-m telescope and the British-Dutch groups will operate a 15-m telescope. The CalTech dish will be a submm telescope; the British-Dutch dish is designed to function down to 0.8 mm, but may be used in the "light bucket" mode at 350 μm. It is rather difficult to estimate the number of nights which are useful for observing at 350 μm either at Mauna Kea or Mt. Graham: it is probable that the fraction is small anywhere on earth. It has been claimed that the number of nights where useful measurements can be made at 350 μm wavelength in Arizona are somewhat less than at Mauna Kea. Even if this were so, easier access to sites in Arizona and the ability to quickly change receivers on the SMT will optimize the time available, and may well make up any differences.

REFERENCES

Barrett, A. H., Schwartz, P. R., Waters, J. W. 1971, Astrophys. J. 168, L101
Batrla, W., Wilson, T. L., Martin-Pintado, J. 1983a, Astron. Astrophys. 119, 139
Batrla, W., Wilson, T. L., Bastien, P., Ruf, K. 1983b, Astron. Astrophys. 128, 279
Batrla, W., Walmsley, C. M., Wilson, T. L. 1984, Astron. Astrophys. 136, 127
Beckwith, S., Persson, S. E., Neugebauer, G., Becklin, E. E. 1979, Astrophys. J. 223, 464
Beckwith, S., Evans, N. J., Gatley, I., Gull, G., Russell, R. W. 1983, Astrophys. J. 264, 152

Boland, W., de Graauw, T., Lidholm, S., Lee, T. J. 1983, Astrophys. J. 271, 183
Buxton, R. B., Barrett, A. H., Ho, P. T. P., Schneps, M. H. 1977, Astron. J. 82, 985
Chini, R., Mezger, P. G., Kreysa, E., Gemünd, H.-P. 1984, Astron. Astrophys. 135, L14
Churchwell, E., Hollis, J. M. 1983, Astrophys. J. 272, 591
Downes, D., Genzel, R., Becklin, E. E., Wynn-Williams, C. G. 1981, Astrophys. J. 244, 869
Erickson, N. R., Davis, J. H., Evans, N. J. II, Loren, R. B., Mundy, L., Peters, W. L. III, Scholtes, M., Vanden Bout, P. 1980, in Interstellar Molecules, Proc. of IAU Symp. 87, ed. B. Andrew, Reidel, Dordrecht
Erickson, N. R., Goldsmith, P. F., Snell, R. L., Berson, R. L., Huguenin, G. R., Ulich, B. L., Lada, C. J. 1982, Astrophys. J. 261, L103
Forster, J. R., Goss, W. M., Wilson, T. L., Downes, D., Dickel, H. R. 1980, Astron. Astrophys. 84, L1
Frerking, M. A., Langer, W. D., Wilson, R. W. 1982, Astrophys. J. 262, 590
Friberg, P. 1984, Astron. Astrophys. 132, 265
Genzel, R., Reid, M., Moran, J. M., Downes, D. 1981, Astrophys. J. 244, 884
Goldsmith, P. F. (1982), in "Galactic and extragalactic infrared spectroscopy" (ed. Kessler, M. F., Phillips, J. P.; publ. D. Reidel, Dordrecht, pp233-250)
Güsten, R. Walmsley, C. M., Pauls, T. 1981, Astron. Astrophys. 103, 197
Güsten, R., Walmsley, C. M., Ungerechts, H., Churchwell, E. 1984, Astron. Astrophys. (submitted)
Guilloteau, S., Stier, M. T., Downes, D. 1983, Astron. Astrophys. 126, 10
Guilloteau, S., Baudry, A., Walmsley, C. M., Wilson, T. L., Winnberg, A. 1984, Astron. Astrophys. 131, 45

Harris, A., Townes, C. H., Matsakis, D. N., Palmer, P. 1983,
 Astrophys. J. 265, L63
Harwit, M. 1984, unpublished
Henkel, C., Walmsley, C. M., Wilson, T. L. 1980, Astron. Astrophys. 82, 41
Henkel, C., Wilson, T. L., Pankonin, V. 1981, Astron. Astrophys. 99, 270
Henkel, C. 1984, Astrophys. J. (submitted)
Hermsen, W., Wilson, T. L., Walmsley, C. M., Batrla, W. 1984,
 Astron. Astrophys. (submitted)
Hildebrand, R. H. 1983, Q.J.R.A.S. 24, 267
Ho, P. T. P., Townes, C. H. 1983, Ann. Rev. Astron. Astrophys.
 (ed. Burbidge, Layzer, Phillips, 21, p. 239
Hollis, J. M., Lovas, F. J., Suenram, R. D., Jewell, P. R.,
 Snyder, L. E. 1983, Astrophys. J. 264, 543
Jewell, P., Batrla, W., Walmsley, C. M., Wilson, T. L. 1984,
 Astron. Astrophys. 130, L1
Johansson, L. E. B., Andersson, C., Ellder, J., Friberg, P.,
 Hjalmarsson, Å., Höglund, B., Irvine, W. M., Olofsson, H.,
 Rydbeck, G. 1984, Astron. Astrophys. (in press)
Johnston, K. J., Palmer, P., Wilson, T. L., Bieging, J. 1983,
 Astrophys. J. 271, L89
Johnston, K. J., Henkel, C., Wilson, T. L. 1984, Astrophys. J.
 (in press)
Keene, J., Hildebrand, R., Whitcomb, S. E. 1982, Astrophys. J.
 252, L11
Keene, J., Blake, G., Phillips, T. G. 1983, Astrophys. J. 271,
 L27
Larson, R. B. 1981, Mon. Not. Roy. Astr. Soc. 194, 809
Linke, R. A., Goldsmith, P. F. 1980, Astrophys. J. 235, 437
Loren, R. B., Wootten, A., Sandquist, A., Bernes, C. 1980,
 Astrophys. J. 240, L165
Martin-Pintado, J., Wilson, T. L., Gardner, F. F., Henkel, C.
 1983, Astron. Astrophys. 117, 145
Martin-Pintado, J., Wilson, T. L., Johnston, K. J., Henkel, C.
 1984, Astrophys. J. (submitted)
Matsakis, D. N., Cheung, A. C., Wright, M. C. H., Askne, J. E. H.,
 Townes, C. H., Welch, W. J. 1980, Astrophys. J. 236, 481
Mauersberger, R., Wilson, T. L., Walmsley, C. M., Batrla, W.
 1983, Mitt. Astron. Ges. 57,
Oka, T. 1980, in Interstellar Molecules, Proc. of IAU Symp. 87,
 ed. B. Andrew, Reidel, Dordrecht, p. 221
Olofsson, H., Elldér, J., Hjalmarsson, Å., Rydbeck, G. 1982,
 Astron. Astrophys. 113, L18
Olofsson, H. 1984, Astron. Astrophys. 134, 36
Phillips, T. G., Huggins, P. J., Kuiper, T. B. H., Miller, R. E.
 1980, Astrophys. J. 238, L103
Phillips, T. G., Huggins, P. J. 1981, Astrophys. J. 251, 533
Reid, M. J., Haschick, A. D., Burke, B. F., Moran, J. M.,
 Johnston, K. J., Swenson, G. 1980, Astrophys. J. 239, 89

Schloerb, F. P., Friberg, P., Hjalmarsson, Å., Höglund, B.,
  Irvine, W. M. 1983, Astrophys. J. 264, 161
Scoville, N. Z., Irvine, W. M., Wannier, P. G., Predmore, C. R.
  Astrophys. J. 216, 320
Scoville, N. Z. 1983, Q.J.R.A.S.
Smith, J., Lynch, D. K., Cudaback, D., Werner, M. W. 1979,
  Astrophys. J. 234, 902
Spitzer, L. 1978, Physical Processes in the Interstellar Medium,
  Wiley, New York, p. 77ff
Stacey, G. J., Kurtz, N. T., Smyers, S. D., Harwit, M., Russel, R.
  W., Melnick, G. 1982, Astrophys. J. 257, L37
Stenholm, L. G. 1983, Astron. Astrophys. 117, 41
Storey, J. W. V., Townes, C. H., Haller, E. E. 1980, Astrophys.
  J. 241, L43
Storey, J. W. V., Watson, D. M., Townes, C. H., Haller, E. E.,
  Hansen, W. L. 1981, Astrophys. J. 247, 136
Sweitzer, J. 1978, Astrophys. J. 225, 116
Toelle, F., Ungerechts, H., Walmsley, C. M., Winnewisser, G.,
  Churchwell, E. 1981, Astron. Astrophys. 95, 143
Townes, C. H., Genzel, R., Watson, D. M., Storey, J. W. V. 1983,
  Astrophys. J. 269, L11
Turner, B., Thaddeus, P. 1977, Astrophys. J. 211, 755
Walmsley, C. M., Ungerechts, H. 1983, Astron. Astrophys. 122, 164
Watson, W. D. 1976, Rev. Mod. Phys. 48, 513
Whitcomb, S. E., Gatley, I., Hildebrand, R. H., Keene, J.,
  Sellgreen, K., Werner, M. W. 1981, Astrophys. J. 246, 416
Wilson, T. L., Martin-Pintado, J., Gardner, F. F., Henkel, C.
  1982, Astron. Astrophys. 107, L10
Wilson, T. L., Ruf, K., Walmsley, C. M., Martin, R. N., Pauls, T.,
  Batrla, W. 1982, Astron. Astrophys. 115, 185
Wilson, T. L., Mauersberger, R., Walmsley, C. M., Batrla, W.
  1983, Astron. Astrophys. 127, L19
Zeng, Q., Batrla, W., Wilson, T. L. 1984, Astron. Astrophys.,
  in press
Zuckerman, B., Morris, M., Turner, B. E., Palmer, P. 1981,
  Astrophys. J. 169, L105

INFRARED SPECTROSCOPY OF INTERSTELLAR MOLECULES

N. Z. Scoville

Astronomy Department
California Institute of Technology
Pasadena, CA 91125, U.S.A.

Abstract

Infrared spectroscopy of molecules in regions of star formation holds the greatest potential for probing the high excitation gas adjacent to protostars and the shock fronts generated as a result of star formation activity. The current observational data at $\lambda = 2 - 300$ μm are reviewed and methods for analysis are developed. Future studies which will become possible with spectrometers in space are briefly discussed.

1. Introduction

Infrared spectroscopy offers a unique opportunity for observations in areas of recent star formation and incipient star formation. Spectroscopy in the wavelength band $\lambda = 2$-20 μm can be used to probe the environs of compact, luminous embedded infrared sources; high velocity outflows from these objects; and the shock fronts occurring where the high velocity flows hit the ambient molecular cloud.

The principal probes available in the near infrared band include recombination lines of ionized Hydrogen (HII), the vibrational transitions of CO and H_2 and the pure rotation, quadruple transitions of H_2. At longer wavelengths ($\lambda \sim 70$-200 μm) there exist the pure rotational transitions in CO, OH, and NH_3 (cf. Watson 1982). Low resolution spectroscopy in the near infrared has also been useful in analyzing the composition of the dust grains associated with the dense

G. H. F. Diercksen et al. (eds.), Molecular Astrophysics, 201–213.

gas. A resonance associated with silicates has been clearly identified at 9.8 μm. Narrower features at λ = 2.98, 3.1, 3.4, 4.6, and 6.8 μm have been attributed to the stretching modes of NH, OH, CH, CO, CH_2, and CH_3 (cf. Allamandola 1982).

The recent observational studies in the core of the Orion molecular cloud provide maps of the distribution, kinematics and temperature for the high excitation molecular gas associated with an extended system of shock fronts. Near infrared spectroscopy of the Becklin-Neugebauer continuum source have indicated its spectral type and dynamics, in addition to revealing a dense, high temperature circumstellar nebula.

An additional hope of infrared spectroscopy has been that it would provide a probe of protostellar objects. To date no such object has been unequivocably identified from the observational data, either spectroscopic or photometric. The lack of success in this area must probably be attributed to the enormous extinction expected for such objects and their low temperatures (prior to nuclear ignition). For a reasonable mass protostellar nebula (1 - 10 $M_\odot$) with size comparable to the solar system, the dust opacity will probably exceed unity at all wavelengths λ < 1 mm.

In this contribution a brief review is presented for the infrared spectroscopy of recently formed stars still embedded in molecular clouds and the shock fronts associated with star formation activity. After an introductory discussion of the observed transitions and their excitation requirements, the techniques and analysis are illustrated with data pertaining to the core of the Orion molecular cloud (OMC-1) and the Becklin-Neugebauer object (BN). A more general treatment of infrared astronomical spectroscopy may be found in the recent volume Galactic and Extragalactic Infrared Spectroscopy (Kessler and Phillips 1982).

2. Molecular Spectroscopy.

To illustrate the considerations involved in analyzing the molecular observations, the energy levels and spontaneous decay rates for the vibrational transitions of CO are presented in Figure 1. Within each vibrational state (e.g. v = 0) there exists a ladder of rotation levels specified by quantum number J. The commonly observed 2.6 mm CO transition arises between the states J' = 1 and J = 0. The highest pure rotational transition observed in space is the J = 34-33 line at 77 μm (Watson et al. 1984) located at an energy of approximately E/k = 3300 K. The v = 0-1 and v = 0-2 bands at 4.6 and 2.3 μm have been de-

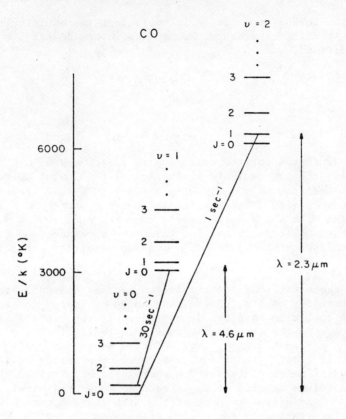

Figure 1: Energy levels and radiative decay rates are shown for CO in the first three vibrational states. (Rotational energies are not drawn to scale).

tected in absorption in front of embedded infrared sources (Scoville et al. 1983).

The physical conditions (n_{H_2} and T_K) required to produce appreciable emission in the various CO transitions are radically different (cf. Scoville 1982 and Scoville, Krotkov, and Wang 1980). To establish a significant population in the upper state of a transition there must be a collisional or radiative pump occurring at a rate comparable with the radiative decay rate in the same transition. In the case of an <u>optically thin transition</u>, the latter rate is specified by the Einstein spontaneous decay coefficient, A_{ul}. If the pump is collisional then the rate of collisions, presumably by impact with H or H_2, must then be comparable with the spontaneous decay rate. It is easily shown then that the condition for appreciable excitation in the upper state is equivalent to requiring that the gas kinetic temperature, T_K and gas volume density, n satisfy the conditions.

$$T_K > (E_u - E_l)/k$$

and

$$n_H, n_{H_2} > A_{ul}/(\sigma_v)_{ul}.$$

Typical estimates for the collision rate, (σ_v), are $\sim 10^{-11}$ cm^{3}sec^{-1}. Thus appreciable excitation of the lowest CO rotational transition (J = 1-0) with A = 6 x 10^{-8} sec^{-1} requires a hydrogen density of approximately 10^3 cm^{-3} and the J = 34-33 transition with A = 4 x 10^{-3} sec^{-1} requires $n_H \gtrsim 10^6$ cm^{-3}. The vibrational transitions for which the spontaneous decay rates are approximately 30 sec^{-1} require densities exceeding 10^{11} cm^{-3}.

In the event that the transition becomes optically thick to the transfer of line photons the above estimates for the required collision rate become considerably reduced. This arises because the important criteria is really that the upward pump rate exceed the <u>effective</u> radiative decay rate which for an optically thick transition is reduced by a factor of $1/\tau$. For the millimeter CO transition where typically $\tau \geq 10$ the required volume density of H_2 is reduced to approximately 300 cm^{-3}. Radiative trapping of the line photons probably plays a important role for molecules such as CO and OH but is clearly unimportant in the weak quadrupole transitions of H_2.

Table 1 lists a sample of molecular transitions observed in the infrared together with a qualitative indication of their excitation requirements and optical depths in Orion. In terms of temperature the transitions are separated into three regimes; a low energy regime (< 200 K), a mid-energy regime (200-1,000 K), and a high energy regime (> 1,000 K). Within each regime it can be seen from Table 1 that there exists a number of molecular transitions requiring quite different characteristic densities for excitation of emission. These critical densities range from 300 cm^{-3} up to > 10^{10} cm^{-3}.

3. The Orion Molecular Cloud

In this section recent infrared spectroscopy relating to the shock fronts in the core of the Orion molecular cloud (OMC-1) and the Becklin-Neugebauer object are summarized. Although similar spectroscopy exists for other cloud core

TABLE 1: Selected Spectroscopic Probes

	λ	τ	Critical n_{H_2} (cm^{-3})	Obs. Ref.
LOW ENERGY (T < 200K)				
Low J CO pure rotational emission	$\sim$1 mm	> 10 (CO)	$\sim$300	a
		<1 (^{13}CO, C^{18}O)	$\sim$3000	a
Low J CO rotation-vibration absorption	4.6μm	$\sim$100(CO)	$\sim$300	b
		$\sim$1(^{13}CO)	$\sim$3000	b
	2.3μm	$\sim$1(CO)	$\sim$300	b
MID ENERGY (T $\simeq$ 200-1000 K)				
H$_2$ S(2)	12μm	<<1	$\sim$10^3	c
High J CO pure-rotational emission	75-157μm	<1	$\sim$10^6	d
OH rotational emission	120μm	>1	$\sim$10^8(R)	d
Mid J CO rotation-vibration emission	4.6μm	$\sim$1	$\sim$10^{10}(R)	b
HIGH ENERGY (T > 1000 K)				
H$_2$ rotation-vibration emission	2-4μm	$\sim$10^{-6}	10^5	e
CO overtone bandhead emission	2.3μm	$\sim$1	10^{10}(R)	b

NOTES:

"R" indicates transitions <u>possibly</u> pumped by radiation or alternatively the observed line photons are in fact continuum photons from a central source scattered in an assymetric gas envelope.

a - Goldsmith (1982).

b - Scoville et al. (1983).

c - Beck et al. (1979).

d - Storey, Watson, and Townes (1981); Watson (1982).

e - Beckwith, Persson, and Neugebauer (1979), Scoville et al. (1982).

regions, the observations in Orion are by far the most complete.

3.1 OMC-1 Shock Fronts.

Though evidence of high velocity gas in the core of the Orion molecular cloud was first derived from radio studies of H_2O masers and CO thermal emission, the data most specifically defining the conditions in these shock fronts is derived in the near infrared via the vibration-rotation transitions of H_2 and in the far infrared via the high J, pure rotational transitions of CO. The near infrared H_2 transitions have been studied with grating and Fourier transform spectrometers on large ground based optical telescopes. The far infrared transitions have been studied using a tandem Fabry-Perot spectrometer carried aboard the NASA C-141 airborne observatory.

The original detection by Gautier et al. (1976) of the 2 μm H_2 quadrupole lines from OMC-1 occurred at about the same time as millimeter observers discovered the presence of high velocity CO emission. The initial interpretation -- that the H_2 emission (v = 1 $\longrightarrow$ 0 and v = 2 $\longrightarrow$ 1) arises in shock fronts where the high velocity gas detected in the millimeter CO line impinges on the ambient cloud -- has been generally supported by all subsequent observations. More recent H_2 data has provided better angular resolution (5"; Beckwith et al. 1978) and better kinematic resolution (down to 20 km sec^{-1}; Nadeau and Geballe 1979 and Scoville et al. 1982). The former study showed marked clumping on the scale of 5" for the emission distributed over a region of approximately 40" in extent. The kinematic measurements clearly demonstrated an enormous velocity extent of the H_2 emission ($\geq$ 100 km sec^{-1}) in addition to showing large-scale kinematic structure suggestive of an expanding shock front system (Scoville et al. 1982).

Estimates of the extinction in front of the H_2 emitting region have been derived by several investigators based upon the ratio of lines arising from the same upper state or from states of similar excitation energy. The current estimates place the extinction at 1.2-2.9 magnitudes at 2 μm (Scoville et al. 1982, Davis, Larson, and Smith 1982). Variations in the extinction are seen to correlate inversely with the observed intensities in a manner suggestive that much of the observed intensity structure could be due to variable foreground extinction in the front of the cloud (Scoville et al. 1982). Using the measured excitation temperature of approximately 2,000 K for the H_2, the observed intensities in the S(1) line may be corrected by the deduced extinction to infer a total H_2 luminosity in all lines of approximately 100 L_o. If the source is assumed to be spherically symmetric with a diameter of 40 arcsec then the total H_2 luminosity could approach 10^3 L_o.

Lower temperature shock-heated molecular gas has been detected in the pure rotation S(2) line by Beck, Lacy and Geballe 1979 for which the excitation temperature is estimated to be 1,000 K. The high J, pure rotational lines of CO (Watson et al. 1983) may originate from still colder (T $\sim$ 750 K) gas further downstream from the shock fronts. The total mass of gas in the two temperature regimes (2,000 K and 700-1,000 K) are $\sim 10^{-2}$ $M_\odot$ and $\sim$ 1.5 $M_\odot$ respectively. The relative amount of gas at the two temperatures is consistent with the shock models in which the cooling rate is considerably reduced at lower temperatures (Kwan 1977, Hollenbach and Schull 1977, Draine, Roberge, and Dalgarno 1982). The total energy estimated for the 2 μm H_2 emission, after correction to a spherically symmetric source, is in fact consistent with the energy expected to be liberated when the high velocity cold molecular gas seen in the millimeter transitions is thermalized in the ambient cloud. The latter may be estimated taking the total mass of gas seen in the 2.6 mm CO emitting region ($\sim$ 5 $M_\odot$) moving at $\sim$ 30 km sec^{-1} with an expansion time scale R/V $\sim$ 2,000 years (Kwan and Scoville 1976, Snell et al. 1984).

4. Spectroscopy of the Becklin-Neugebauer Object

In view of the highly energetic activity seen in the Orion cloud core it is of great interest to define the nature of the luminosity sources in this region. It is commonly assumed that one of the two principal luminosity sources (BN or IRC-2) must be responsible for generating the high velocity gas and shock fronts. At present extensive near infrared spectroscopy has been performed for the BN object; the present instrumental sensitivity is not quite sufficient to reach IRC-2.

The richness of the CO spectrum in BN is illustrated by the profiles shown in Figure 2. In this figure the vibration rotation transitions in each of three bands [the fundamental bands (v = 0-1) of both CO and ^{13}CO, and the overtone band (v = 0-2) of CO] have been summed coherently in velocity space to show the separate kinematic features present. Strong CO absorption is seen at +9 and -17 km sec^{-1} due to low excitation gas in the line of sight to BN. The rather modest temperature (T_R = 150 K) derived from the line strengths as a function of J implies that this gas must be at > 1000 AU from the central source. The remaining two features (the absorption at + 30 km sec^{-1} and the emission at +20 km sec^{-1}) have a high rotational temperture (T_R = 600 K) and are probably associated with the immediate circumstellar nebula surrounding the BN object. The fact that the absorption here appears at more positive velocities than the emission is suggestive of a accre-

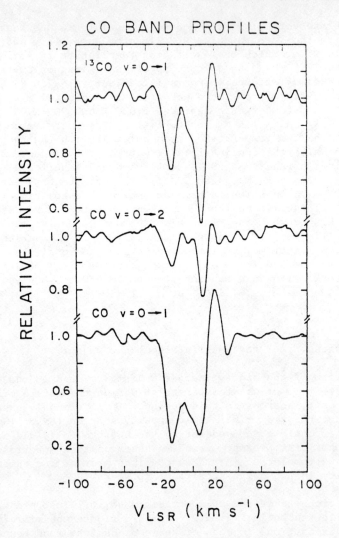

Figure 2: Kinematic profiles are shown for the ^{13}CO v = 0-1, CO v = 0-2 and 0-1 bands. Each profile was obtained by averaging all detected rotational transitions in each of the separate bands. The velocity resolution was 7 km sec^{-1}.

ting shell collapsing towards the central star (Hall et al. 1978, Scoville et al. 1983).

An additional CO emission feature was observed in the overtone bandheads (Scoville et al. 1983) and is thought to arise in a compact region < 2/3 of an AU in size with density exceeding 10^{10} cm^{-3}. The interpretation of this feature is very uncertain but two possibilities are that it is produced in a dense circumstellar disk within a few AU of the central star or in a shock front where the star moves through the molecular cloud.

The properties of the star itself are best defined by the Hydrogen recombination lines. The Brackett lines show high velocity wings extending to ±100 km sec^{-1} with a total emission measure, corrected for dust extinction, suggestive of a B0.5 main sequence star.

The richness of the phenomena observed in the molecular lines associated with BN underscore the need for more extensive infrared spectroscoopy of other embedded sources. At present, the only instrument capable of operation in the 4.6 µm CO band with the resolution required to separate the telluric and interstellar CO lines is the FTS on the 4m telescope at Kitt Peak National Observatory. Several of the other sources observed with this instrument include UOA 27 (GL 2591), LkHα101, and MWC 297 (Kleinmann, Hall and Scoville 1984). In the former source the CO rotational temperature is found to be comparable with that observed in OMC-1, but in the latter two sources the CO appears to be at much lower temperatures $\sim$10 K. An interesting correlation is found between the column density of CO as indicated by the absorption lines and the dust extinction as indicated by the hydrogen line ratios in these sources. In the hotter sources, (UOA 27 and BN) approximately 100% of the available carbon must be in carbon monoxide while in the two colder sources only 10% is required. These measurements suggest that the CO to hydrogen ratio may be a variable from cloud to cloud especially between cloud cores and the cloud envelopes. It is possible that in the colder clouds a large fraction of the CO is depleted onto dust grains but in the warmer regions where the dust is above 50 K the CO sublimates back into the gas.

5. Future Prospects for Infrared Spectroscopy

Within the next ten years major advances will be realized in infrared spectroscopy as a result of the cooled telescopes expected to be launched into space (ISO and SIRTF). These telescopes will open up new windows for astronomical observations (λ = 10-600 µm) which include the pure rotational transitions of H_2 and the pure rotational transitions of H_2O, OH, and NH_3. At shorter wavelengths it will become possible to detect important hydrocarbons such as methane which have **no**

TABLE 2

Sample Trace Molecule Transitions

Transition		λ	Comments
CO	$\Delta v = 1$	4.6	Emission from shocks
H_2O	ν_1, ν_2, ν_3	2.7, 6.3, 2.7	Emission from shocks
O_2	$\Delta v = 1$	6.3	
OH	$J = 5/2 \to 3/2$	35	Absorption and emission
	$\Delta v = 1$	2.7	Emission from shocks
NH_3	$\nu_4, \nu_2, 2\nu_2$	7, 10, 16	Emission in cloud core sources
CH_4	ν_1, ν_2, ν_3	3.4, 6.5, 3.3	Probe of C/O abundance and shock chemistry

radio frequency transitions (see Table 2).

Shown in Figure 3 are the emission rates expected in H_2 pure rotation lines estimated as a function of temperature. Here it may be seen that the choice transition is probably the S(1) line at 17 μm. This line becomes readily observable as the gas temperature rises above 100 K. The sensitivities expected for the cooled space telescopes (ISO and SIRTF) together with the detectors anticipated for them are such that one solar mass of molecular gas at a temperature of 200 K could be detected in the S(1) line from any cloud in our galaxy with a few minutes of integration. (The one complicating factor is the requirement that the H_2 and dust not be in thermal equilibrium, otherwise the dust continuum emission will swamp the line emission).

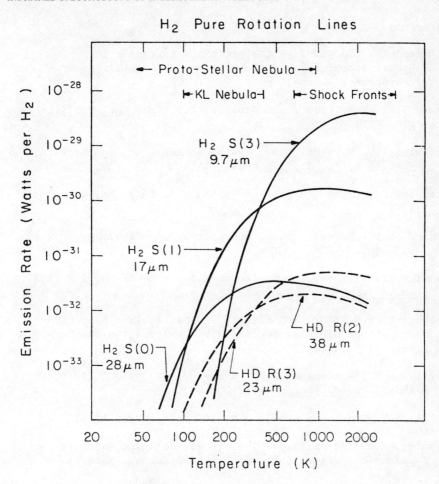

Figure 3: The emission rates are shown for H_2 molecules as a function of temperature. The H_2 rotational levels were assumed to be in equilibrium and the emission rates were calculated using the spontaneous decay rates tabulated by Turner, Kirby-Docken and Dalgarno (1977).

References

Allamandola, L.J. 1982 in "Galactic and Extragalactic Spectroscopy" ed. M.F. Kessler and J.P. Phillips (Dordrecht: Reidel) p. 5.

Beck, S.C., Lacy, J.H., and Geballe, T.R. 1979, Ap. J. (Letters), 234, L213.

Beckwith, S., Persson, S.E., and Neugebauer, G., and Becklin, E.E. 1978, Ap. J., 223, 464.

Davis, D.S., Larson, H.P., and Smith, H.A. 1982, Ap. J., 259, 166.

Draine, B.T., Roberge, W.G., and Dalgarno, A. 1982, Ap. J., 264, 485.

Gautier, T.N., Fink, U., Treffers, R.R., and Larson, H.P. 1976, Ap. J. (Letters), 207, L29.

Goldsmith, P.G. 1982 in "Galactic and Extragalactic Spectroscopy" ed. M.F. Kessler and J.P. Phillips (Dordrecht: Reidel) p. 231.

Hall, D.N.B., Kleinmann, S.G., Ridgway, S.T., and Gillett, F. 1978. Ap. J. (Letters), 223, L47.

Hollenbach, D.J., and Shull, J.M. 1977, Ap. J., 216, 419.

Kessler, M.F. and Phillips, J.P. 1982, "Galactic and Extragalactic Infrared Spectroscopy" (Dordrecht: Reidel).

Kleinmann, S.G., Hall, D.N.B., and Scoville, N.Z. 1984 (in preparation).

Kwan, J. 1977, Ap. J., 216, 713.

Kwan, J. and Scoville, N.Z. 1976, Ap. J. (Letters), 210, L39.

Nadeau, D. and Geballe, T.R. 1979, Ap. J. (Letters), 230, L169.

Scoville, N.Z. 1982 in "Galactic and Extragalactic Spectroscopy" ed. M.F. Kessler and J.P. Phillips (Dordrecht: Reidel) p. 165.

Scoville, N.Z., Hall, D.N.B., Kleinmann, S.G., and Ridgway, S.T. 1982, Ap. J., 253, 136.

Scoville, N.Z., Kleinmann, S.G., Hall, D.N.B., and Ridgway, S.T. 1983, Ap. J., 275, 201.

Scoville, N.Z., Krotkov, R., and Wang, D. 1980, Ap. J., 240, 929.

Snell, R.L., Scoville, N.Z., Sanders, D.B., and Erickson, N.R. 1984, Ap. J. (in press).

Storey, J.W.V., Watson, D.M., and Townes, C.H. 1981, Ap. J. (Letters), 224, L27.

Watson, D.M. 1982 in "Galactic and Extragalactic Spectroscopy" ed. M.F. Kessler and J.P. Phillips (Dordrecht: Reidel) p. 193.

Watson, D.M., Genzel, R., Townes, C.H., and Storey, J.W.V. 1984, Ap. J. (in press).

VISIBLE AND ULTRAVIOLET STUDIES OF INTERSTELLAR MOLECULES

John H. Black

Steward Observatory
University of Arizona
Tucson, AZ 85721 USA

The technique of optical absorption line spectroscopy, with improvements in sensitivity and extensions into the ultraviolet and infrared, has continued to yield valuable information about interstellar molecules and the clouds that contain them. Abundances in diffuse clouds provide useful tests of theories of molecule formation. Rotational population distributions can be measured directly and can be used as diagnostic probes of densities, temperatures, and radiation fields within interstellar clouds. Accurate molecular data as well as elaborate theoretical cloud models are needed for the interpretation of the observations.

1. INTRODUCTION

By 1941, three interstellar molecules, CH, CN, and CH^+, had been identified by means of the narrow absorption lines that they superimpose upon the spectra of distant stars (1,2). Since that time, H_2 (3), CO (4), OH (5,6), C_2 (7), and tentatively CS^+ (8) have been observed by optical methods in the visible and ultraviolet regions of the spectrum. Upper limits have been placed upon the abundances of a number of other species. Although the vast majority of interstellar molecules have been identified by means of millimeter and microwave lines, optical observations continue to provide interesting problems in spectroscopy and molecular physics as well as valuable information about interstellar matter. This paper will survey developments of the last five years and discuss some of the more important scientific problems: the reviews of Snow (9) and of Spitzer and Jenkins (10) are useful guides to the literature prior to 1980.

G. H. F. Diercksen et al. (eds.), Molecular Astrophysics, 215–236.

2. BASIC OBSERVATIONAL CONSIDERATIONS

Absorption lines of interstellar molecules can be observed where a molecule-containing volume of interstellar matter lies along the line of sight to a suitable background star. By suitable star we mean one which is bright enough for sensitive, high-resolution spectroscopy and which offers little chance for confusion between stellar and interstellar absorption features. Ordinarily the most suitable stars are those of spectral types O and B. These are stars of high surface temperature ($T \gtrsim 10000$ K) and high luminosity, which can be seen to fairly great distances. They have approximately the same scale height with respect to the galactic plane as the interstellar matter. Often, but not always, the absorption lines arising in their atmospheres are significantly broader than typical interstellar absorption lines. The OB stars are also the rarest stars, and relatively few of them are both bright and positioned behind molecule-rich regions. It is customary to speak of interstellar "clouds", even though this concept is somewhat poorly defined. In the context of optical absorption line studies, it is appropriate to identify a "cloud" merely as a distinct, narrow Doppler-shifted velocity component in the interstellar line spectrum. It is also customary to distinguish between diffuse interstellar clouds, for which the visual extinction $A_V < 3$ magnitudes, and the thicker, dark molecular clouds. In the past this distinction corresponded to an operational definition of diffuse clouds as those that could be studied by optical absorption line methods as opposed to dark clouds which could not. Recently it has become possible to observe regions of high extinction, including **bona fide** molecular clouds, by optical techniques (7,11-15).

The ideal circumstance for optical observations of interstellar molecules is a thick foreground molecular cloud and a fortuitously located, <u>bright</u> background star. Nature conspires to make this rare: the light from stars that lie behind clouds is severely extinguished by the interstellar dust and the amount of this extinction (that is, the opacity) is strongly correlated with the total column density of interstellar gas (16). The visual extinction, A_V, in magnitudes at a wavelength of 5500 Å is related to the total gas column density, N_H, by

$$N_H = N(H) + 2N(H_2) = 1.59 \times 10^{21} \, A_V \quad \text{cm}^{-2} \tag{1}$$

based upon an average of data for diffuse clouds (17). The optical depth for extinction, $\tau_\lambda = A_\lambda / 1.086$, has a characteristic wavelength dependence (extinction is greatest at the shortest wavelengths), but shows localized anomalies.

The integrated intensity of an absorption line, called the equivalent width W_ν, depends upon the column density of the absorbing molecule in its initial (lower) state N_1, the absorption oscillator strength f_{21} of the transition $2 \leftarrow 1$, and the line width Δv, which is usually given in velocity units. For very weak lines, $W_\nu \propto N_1 f_{21}$, but for stronger lines the effects of saturation (curve-of-growth effects) must be taken into account. These effects will exceed 10 per cent in the derived column density when $N_1 f_{21} \lambda/b \gtrsim 1.87 \times 10^{14}$ s^{-1}, where λ is the wavelength of the transition in Å and b, the Doppler broadening parameter in km s^{-1}, is related to the full-width at half-peak of the line profile by $b = \Delta v/1.665$. In most interstellar clouds, the line width is significantly larger than that expected from thermal Doppler broadening at cloud temperatures, $T \lesssim 100$ K. This broadening can be ascribed to macroscopic gas motions, with typical values of $\Delta v = 1-3$ km s^{-1}. In the limit of very weak, unsaturated absorption,

$$W_\nu \equiv \int (I_0 - I)/I_0 \, d\nu \equiv \frac{c}{\lambda^2} W_\lambda \approx \frac{\pi e^2}{mc} N_1 f_{21} \quad \text{Hz} \quad (2)$$

where I_ν and I_0 are the intensity in the line and in the adjacent continuum, respectively, λ is the wavelength is cm, N_1 is in cm^{-2}, and the other symbols have their usual meanings. The observability of a line can be expressed in terms of a minimum detectable equivalent width

$$W_\nu(\min) \approx \Delta \nu \, (S/N)^{-1} \quad (3)$$

which depends upon the resolution, $\Delta \nu$, and signal-to-noise ratio (S/N) achieved in the observation. It is instructive to consider the abundances of various molecules that can be measured by optical techniques from the far ultraviolet to the mid-infrared. Table 1 summarizes information for some observed or potential interstellar absorption lines. The first column contains the wavelength in Å; the second column, the molecule and transition; the third column, the oscillator strength; and the fourth column, the ratio of extinction optical depth to visual extinction for a standard extinction curve. The sources of the oscillator strengths are indicated in the final column. With the exception of the vibrational transition of H_2 at 22200 Å, for which an individual line is specified, these are band oscillator strengths. Experience has shown that it is surprisingly easy to come to grief (usually by factors of precisely 2) in relating band oscillator strengths to those of individual lines and to transition moments unless internally consistent conventions are adhered to (18). The adopted extinction curve is derived from (19) corrected for a ratio of total-to-selective-extinction $A_V = 3.1 E(B-V)$ and (20). Table 1 also contains calculated minimum detectable abundances expressed as a logarithmic ratio $\log(3N_1/N_H)$, where N_1 is evaluated by combining eqs. (2) and (3), N_H is defined in eq. (1), and the factor of 3 comes from the arbitrary requirement that a line appear at the level of 3 times the rms noise. Results are presented for three illustrative cases of the same star located

Table 1. Detectability of Optical Transitions

λ Å	Transition	f_{21}	τ_λ/A_V	$\log(3N_1/N_H)$ $A_V=1$	$A_V=3$	$A_V=10$	Ref
1092	H_2 B-X (1,0)	5.79-3	3.40	- 9.36	- 8.36	*	22
1340	C_2 F-X (0,0)	2.0 -2	2.56	-10.14	- 9.50	*	23
1510	CO A-X (1,0)	2.32-2	2.32	-10.28	- 9.75	*	24
2312	C_2 D-X (0,0)	5.4 -2	2.56	-10.72	-10.09	*	25
3078	OH A-X (0,0)	1.10-3	1.58	- 9.28	- 9.07	*	26
3876	CN B-X (0,0)	3.38-2	1.29	-10.87	-10.79	*	27
4232	CH^+ A-X (0,0)	5.5 -3	1.20	-10.17	-10.13	- 8.82	**
4300	CH A-X (0,0)	5.3 -3	1.17	-10.12	-10.09	- 9.84	28
8760	C_2 A-X (2,0)	1.67-3	0.431	- 9.83	-10.12	- 9.99	29
10970	CN A-X (0,0)	3.25-3	0.310	- 9.88	-10.25	-10.37	27
12090	C_2 A-X (0,0)	2.70-3	0.265	-10.09	-10.46	-10.58	29
22200	H_2 v=1-0 S(0)	9.37-14	0.103	-	-	- 0.48	30
23450	CO v=2-0	7.5 -8	0.093	- 5.59	- 6.03	- 6.40	31
46570	CO v=1-0	1.0 -5	0.023	- 7.73	- 8.20	- 8.67	31
	Apparent Magnitude V =			2.5	4.5	11.5	

*(S/N) < 5 in the continuum; i.e. lines are not detectable.
** see section 3.1

behind different clouds having values of extinction $A_V=1$, 3, and 10. We have assumed the same telescope aperture, D=240 cm, the same resolving power $R=\lambda/\Delta\lambda = \nu/\Delta\nu = 10^5$, and the same overall photon-counting efficiency, $\varepsilon=0.005$, at all wavelengths, in order to evaluate the signal (i.e. the number of photons detected in an integration time t=1000 s):

$$S = \pi(D/2)^2 \varepsilon F_\lambda (\lambda^2/hcR) t \exp(-\tau_\lambda) \qquad (4)$$

where F_λ in ergs cm^{-2} s^{-1} cm^{-1} is the received flux from the star in the absence of extinction. The star is taken to have an unreddened apparent magnitude $V-A_V = 1.5$ with an energy distribution given by the T=30000 K, log g=4 model atmosphere of Kurucz (21). This is appropriate for a star of approximate spectral type B0. It is assumed that the noise is due entirely to photon-counting statistics, in which case $S/N \simeq S^{1/2}$. It is more than coincidental that the aperture of Space Telescope is 2.4 meters and that its High Resolution Spectrograph will have a maximum resolving power of $R=10^5$ (32). This resolving power also corresponds to a resolution in velocity units of 3 km s^{-1}, which is well-matched to the line width in many interstellar clouds. The results in Table 1 for V=2.5 and $A_V=1$ are appropriate for a cloud like that toward ζ Oph and show that molecules with fractional abundances of the order of 10^{-10} can be detected throughout the ultraviolet and visible regions. The intermediate case, V=4.5 and A_V=3 is overoptimistic since known stars that show this much foreground extinction have apparent magnitudes in the range V=6 to 8 at best. The final column of results, for V=11.5 and A_V=10, illustrates conditions very similar to those for the highly-reddened supergiant star Cyg OB2 No.12. These results show that the ultraviolet becomes relatively less favorable and eventually useless for the study of minor interstellar species as the cloud thickness and extinction increase. Danks and Lambert (36) have claimed that "as a probe of interstellar C_2 molecules, the Mulliken system [D-X in the ultraviolet] is to be preferred to the Phillips system [A-X in the red] because of the factor of 50 to 60 increase in the f-value". It is clear from Table 1 that this assertion is false: it can apply only for relatively thin clouds with $A_V \lesssim 3$. Molecules like C_2 and CN can be studied better in the thicker clouds by means of their transitions in the near infrared.

3. ABUNDANCES OF MOLECULES

All of the interstellar molecules detected by optical means have been seen in the principal cloud toward the star ζ Oph. The measured column densities in this cloud are listed in Table 2 along with a few upper limits. Snow (9) has summarized results of searches for various other species. More recently, sensitive

Table 2. Molecular Column Densities Toward ζ Oph

Molecule	Column Density (cm^{-2})	Reference
H	5.2 (20)	34
H_2	4.2 (20)	34
HD	1.6 (14)	34
CO	2.4 (15)	35
CH	2.3 (13)	33,37
CH^+	3.0 (13)	38,39
CN	2.7 (12)	40,41
C_2	1.5 (13)	29,36,71
OH	5.1 (13)	5,6,42
C_3	<1.5 (12)	43
H_2O	<4.0 (13)	44,45
HCl	<4.6 (11)	46

searches have been reported for H_2O (44), H_2O^+ (47), and C_3 (43). Results of surveys in a number of clouds have also been published for CO (48), CH^+ (49), CH (33,49), CN (40), and for various **carbon-bear**ing molecules (50). Absorption line studies of molecule abundances are particularly important as tests of theories of interstellar molecule formation. Indeed, the optical absorption technique offers several distinct advantages over radio emission line measurements in this connection. The background star isolates an absorbing column that has the same extent in the line-of-sight direction for all observations and that has a very small angular extent on the sky. In contrast, the relatively large antenna beam patterns involved in radio measurements often average over a considerable amount of structure in abundances and excitation conditions. In cases where excited rotational levels of molecules are substantially populated, different measurements at quite different frequencies are usually required in order to infer the population distribution and total abundance of a molecule from radio emission lines. Radio emission lines are sometimes highly saturated, so that the derivation of column densities from measured line intensities is complicated at best and often severely model-dependent. In optical transitions, the rotational structure of a band can usually be seen in a single observation. Provided that the rotational structure can be resolved by the spectrometer and that the line widths are smaller than the line separations, the rotational population distribution can be extracted directly. Although ultraviolet lines of H_2 can be highly saturated and those of CO can be severely blended, the absorption lines of the rare species are usually rather weak so that corrections for curve-of-growth effects are small. The principal molecular constituent, H_2, has not yet been observed directly in a quiescent, dense molecular cloud, and direct measurements of its abundance by means of forbidden infrared absorption lines have only recently become truly feasible (51). In contrast, ultraviolet lines of H_2 and H have been observed in a number of diffuse clouds. In the best studied diffuse clouds, quite a lot of additional information on atomic abundances is also available, with the result that the parameter space available to theoretical models can be considered more restricted than is usually true for thick, dense clouds.

3.1 CH^+

The large abundance of the methylidyne ion, CH^+, remains one of the outstanding puzzles of interstellar chemistry. There is a long history of attempts to explain the interstellar CH+ in terms of a steady state among microscopic processes of formation and destruction (52,53). The abundances of CH and CH^+ are comparable in many diffuse clouds, but CH^+ is expected to be destroyed so much more rapidly in reactions with electrons and with H_2 that normal schemes of molecule formation have great difficulty in explaining the observations. More recently, Elitzur and Watson (54) have proposed that CH^+ is formed efficiently by the reaction

$$C^+ + H_2 \rightarrow CH^+ + H \tag{5}$$

at the elevated temperatures to be found in shock-heated gas. With favorable choices of shock speed, pre-shock density and fractional abundance of H_2, Elitzur and Watson have succeeded in reproducing CH^+ column densities similar to those observed in diffuse clouds. Moreover, Federman (49,55-56) has suggested that small (i.e. comparable to or slightly larger than the resolution) apparent differences in radial velocity between CH and CH^+ absorption features toward the same stars provide further evidence for shock chemistry. Frisch and Jura (57) have also discussed the effects of shocks on the abundance of CH^+ and the excitation of H_2 in the gas near the Pleiades. Very recently, White (58) has presented additional observations of interstellar lines toward Pleiades stars and has proposed a model to explain the CH^+ abundance in which the reaction above (eq. 5) occurs rapidly in gas heated photoelectrically by starlight rather than by shocks. Further information concerning a small molecular cloud near the Pleiades has been assembled by Federman and Willson (59). It should be kept in mind that the successes of the shock chemistry model for CH^+ refer only to diffuse clouds with $A_V \lesssim 1$. Souza (14) observed CH^+ toward several stars of the Cygnus OB2 association and found that the column density of the ion reaches values as large as 10^{14} cm^{-2} and continues to increase linearly with A_V up to 10 magnitudes. This result is difficult to reconcile with the shock chemistry models unless the <u>number</u> of shocked regions is strongly correlated with total visual extinction, because each shocked region is able to account for little more than 10^{13} cm^{-2} of CH^+. It is evident that more work needs to be done: observations of CH and CH^+ line profiles with resolution better than 1 km s^{-1} would be valuable; further studies of CH^+ toward very highly-reddened stars are needed; and other observable consequences of shock chemistry must be tested further.

Microscopic processes involving CH^+ continue to attract attention. The recent investigations of the $A^1\Pi-X^1\Sigma^+$ oscillator strengths all seem to be converging on a consistent result for the (0,0) band: $f=6.82\times10^{-3}$ (60), 6.45×10^{-3} (61), 5.66×10^{-3} (39, corrected for variation of the transition moment with internuclear distance), and 5.45×10^{-3} (62). The rate coefficient for formation of CH^+ by radiative association of C^+ and H has increased at low temperatures with the inclusion of resonant processes (63). Photodissociation processes have been studied (64). Non-equilibrium chemical effects relevant to shock chemistry have been considered (65). An interstellar carbon isotope abundance ratio has been determined from observations of $^{12}CH^+$ and $^{13}CH^+$ lines in diffuse clouds (38).

3.2 OH, H_2O, H_2O^+, and O_2

Although only a few optical observations of OH in diffuse clouds have been made, they are important because the OH abundance, together with that of HD, can be used as a measure of the cosmic ray ionization rate (66,67). OH has been observed both in the $A^2\Sigma^+-X^2\Pi$ (0,0) band near 3080 Å (5) and in the $D^2\Sigma^--X^2\Pi(0,0)$ band near 1220 Å (6). The astronomically determined oscillator strength (42), f=0.0108, of the latter system has been confirmed by **ab initio** theoretical calculations (68), f=0.012. A careful quantum mechanical treatment of the H^++O charge transfer process (69) and a thorough investigation of photodissociation processes for OH (70, and references therein) have made it possible to model the interstellar OH abundance with some confidence. Existing observations of OH in diffuse clouds have been discussed with reference to the new molecular data and detailed cloud models, and cosmic ray ionization rates have been derived (71). Possible contributions of shock-heated gas to the observed OH abundances still need to be evaluated more thoroughly. It can be hoped that the High Resolution Spectrograph on Space Telescope will extend optical observations of OH to more and thicker clouds.

Interstellar H_2O is expected to be closely related chemically to OH: both can result from dissociative recombination of H_3O^+, and OH is the principal photodissociation product of H_2O. The ultraviolet lines near 1239 Å and 1115 Å have been searched for in interstellar absorption, and the oscillator strengths are well determined (45,72). Snow and Smith (44) placed a very sensitive upper limit on a line arising from the 0_{00} ground state of the para species of H_2O toward ζ Oph. Because it is not known whether there is an effective ortho/para interchange process for interstellar H_2O, the resulting upper limit on the total column density is still uncertain (45); even so, the current upper limit is near that needed to make a significant test of the suggestion that much of the CH^+ and OH in the ζ Oph cloud result from high-temperature chemistry in shock-heated gas (54). The molecular ion H_2O^+ is formed during the ion-molecule reaction sequence that produces OH and H_2O. Results of a search for interstellar H_2O^+ lines near 6150 Å have recently been published (47). Discrepancies between existing radiative lifetime measurements for the $\tilde{A}$ state of H_2O^+ need to be resolved before full use can be made of the limits on interstellar line strengths. These lifetimes are also of interest for the interpretation of cometary emission spectra.

One major uncertainty in the chemistry of interstellar oxygen concerns the abundance of O_2 (73). Smith **et al.** (74) have measured line oscillator strengths in the B-X Schumann-Runge system to facilitate the future study of interstellar O_2 with Space Telescope. Much remains to be learned about some basic processes that affect the abundances of some oxygen-bearing molecules. The low-temperature ($T \lesssim 100$ K) behaviors of the reactions

and
$$O + OH \rightarrow O_2 + H \tag{6}$$
$$CO + OH \rightarrow CO_2 + H \tag{7}$$

need to be measured before the abundances of O_2 and CO_2 can be predicted confidently (73,75). Uncertainty about the product branching ratios in dissociative recombination of H_3O^+ with electrons is a clear reminder that no such branching ratio has been measured yet and that statistical reaction models (76) provide plausible but not precise predictions.

3.3 C_2, C_3 and Other Carbon Chain Molecules

A thorough discussion of the remarkably abundant interstellar cyanopolyyne molecules is beyond the scope of this paper, but studies of the chemistry of the smallest carbon-chain species in diffuse clouds may help account for the larger molecules (77). Indeed, the observation of microwave absorption lines of HC_5N and HC_7N in diffuse cloud material (78) emphasizes this point. The C_2 molecule is becoming widely observed in interstellar clouds (see section 4.2 below) and there is evidence that its column density is strongly correlated with visual extinction (14,15). The chemically related molecule C_3 has been sought but not found in several diffuse clouds (43). Further searches for C_3 in the thickest clouds where C_2 has been found will help constrain uncertain aspects of the carbon chemistry. It is worthwhile to draw attention to the comment of Clegg and Lambert (43) that the $^1\Sigma_u^+ - X^1\Sigma_g^+$ transition of C_3 near 1580 Å is a potentially valuable interstellar feature because of its large oscillator strength (79). There is an urgent need for a vibrational and rotational analysis of this system.

3.4 CO

Carbon monoxide, CO, is generally the second most abundant interstellar molecule after H_2 (although this is not true in some diffuse clouds). Because of its high abundance and the relative ease with which its radio emission lines are excited, CO emission is used as a tracer of molecular gas throughout our galaxy and in other galaxies. In order to use the integrated brightness of CO lines as a measure of the total mass of a molecular cloud, it is necessary to understand the distribution of CO within the cloud, the radiative transfer affecting the line formation, and the abundance of CO relative to that of the principal constituent, H_2. It is thus disturbing that there are fundamental gaps in our theoretical understanding of the CO abundance and that adopted CO/H_2 abundance ratios are based upon indirect determinations. In the first case, studies of the chemistry of CO in diffuse clouds (e.g. 80) can provide good tests of methods for describing the CO abundance through the photochemical region of a molecular cloud. In the second case, the extension of absorption line techniques into the infrared is able to provide stringent, <u>direct</u>, limits on the CO/H_2 ratio in thick molecular clouds (51).

The most serious voids in our knowledge of the CO molecule concern photodissociation, which is the principal sink of CO throughout diffuse clouds and in the photochemical layers of dense molecular clouds. The dissociation energy of CO is extremely large, $D_0^0 = 11.1$ eV corresponding to a threshold wavelength of 1118 Å. Between this threshold and 1060 Å, no continuous absorption is measured experimentally (81), nor has any repulsive state been identified that will provide a significant cross section for continuous absorption out of the ground state at wavelengths longer than 912 Å where the interstellar radiation field cuts off. Two fairly strong discrete bands, $E^1\Pi - X^1\Sigma^+$ (0,0) and $C^1\Pi - X^1\Sigma^+$ (1,0), are known to predissociate, and it is likely that these will dominate the interstellar photodissociation of CO. Unfortunately, there is serious disagreement in the literature about the oscillator strengths of these bands: recent absorption measurements (81), lifetime measurements (24,82), and theoretical calculations (83) give results that range over an order of magnitude for both bands.

A rigorous treatment of the depth dependence of the CO abundance is of some interest. In the outer layers of a molecular cloud, the photodissociation rate of CO, the photoionization rate of C, and the OH concentration will all influence the CO abundance, and all three are expected to vary with depth. Despite the attention devoted to the possibility of an isotope-selective photodissociation process in CO (84), the depth-dependence of CO photodissociation has not been treated adequately. New theoretical models of the photochemical regions of molecular clouds have been developed in which self-shielding in the pre-dissociating lines of CO is calculated carefully for a realistic population distribution among rotational levels (85). By trying to model the observed column densities of C, C^+, H_2, CO and other species in diffuse clouds, it is possible to test the theory that would describe the CO abundance in the photochemical regions of thick molecular clouds as well. Previously, Federman, Glassgold, Jenkins and Shaya (48) analyzed their ultraviolet CO survey in this way. Their analysis is, however, internally inconsistent: they adopted oscillator strengths of f=0.089 and 0.12 for the C-X (0,0) and E-X (0,0) bands, respectively, to determine the CO column densities, but they used an unshielded photodissociation rate of only 2×10^{-11} s^{-1} to compute the CO abundance. Their adopted oscillator strength and radiation field require a rate 1.7×10^{-10} s^{-1} due to predissociation in the E-X (0,0) band alone.

As indicated in Table 1, the ultraviolet lines of CO are sensitive to very small abundances of the molecule. However, the determination of column densities of CO from existing ultraviolet absorption line data can be complicated: in many measurements of interstellar CO lines with the Copernicus ultraviolet spectrometer and with IUE the rotational structure of the bands is not fully resolved. The total column density is not directly related to the equivalent width of an unresolved band if any of the lines is saturated: the rotational population distribution determines how many levels and lines contribute to the absorption in the band, and the velocity dispersion determines the column densities **at**

which the various lines begin to be saturated. The structure of
part of the curve of growth of a band can be dominated by absorption in the lines arising in the less populous levels, which
remain optically thin. These complications should be dealt with
by fitting theoretical band profiles to data or by the use of properly computed "effective curves of growth" (35,86-89). Some recent
discussions of IUE observations of CO evidently fail to take these
precautions and treat equivalent widths of unresolved but saturated
bands as though they were formed by single lines (90-95).

As mentioned above, there is considerable disagreement in the
literature about ultraviolet oscillator strengths of CO. This
situation also applies to the fourth-positive system, $A^1\Pi-X^1\Sigma^+$:
the oscillator strengths derived from electron collisional energy
loss spectra (96) are still often adopted in the astrophysical
literature, but these are almost a factor of 2 larger than the more
recent values derived from lifetime measurements (24), from absorption measurements (81), and from theory (83). There is a very
great need for definitive laboratory measurements and further
theoretical studies of the ultraviolet oscillator strengths of CO,
especially for the C-X and E-X systems. In the meantime, it is
feasible to make an astronomical determination of them. In some
diffuse clouds where the CO column density is at least 10^{15} cm^{-2},
the infrared lines of the v=1-0 vibrational band near 4.6 microns
wavelength should be detectable with existing instruments. By
comparing the intensities of ultraviolet bands with those of the
infrared bands, the ultraviolet oscillator strengths can, in
principle, be determined in relation to the vibration-rotation
oscillator strengths, which are very accurately known (31,97).
Such infrared absorption line observations might also be useful
for confirming the tentative identification of circumstellar CO
around B stars (91,93,98).

4. MOLECULES AS DIAGNOSTIC PROBES

The relative populations of rotational levels are determined
by the rates of collisional excitation and de-excitation, the rates
of spontaneous emission, and the rates of absorption of and stimulated emission by the interstellar radiation field. Measurements
of population distributions in interstellar molecules can therefore be interpreted to determine densities, kinetic temperatures
and intensities of radiation fields. Optical absorption line
observations are often very useful in this connection because the
population distribution can be extracted from a <u>single</u> observation
of a resolved band. The simple interstellar diatomic hydrides
(other than H_2), which have large rotational level splittings and
large permanent dipole moments, tend to exist predominantly in
their lowest rotational states at the low temperatures and densities
of typical interstellar clouds. In contrast, homonuclear species
like H_2, C_2, and O_2 can have significant populations in **excited**

states because of their long lifetimes for radiative decay by forbidden rotational transitions. The interpretation of population distributions in interstellar molecules requires accurate molecular data as well as complex models of excitation. A few examples are discussed below.

4.1 Molecular Hydrogen

Ultraviolet absorption line observations of H_2 using the Copernicus spectrometer showed rotational population distributions in many diffuse clouds characterized by a low excitation temperature, 50-100 K, for the lowest levels J=0,1 and 2 and by a much higher excitation temperature, $\gtrsim$ 200 K, for levels J=3-6 (10). This has been interpreted as a combination of collisional processes that thermalize the populations of the lowest levels at approximately the kinetic temperature and in terms of radiative excitation (absorption and fluorescence) that controls the higher excited state populations (66,99,100,101). The populations of the J=0 and J=1 ground states of the para and ortho nuclear spin species can be thermalized efficiently by proton-interchange reactions even though free protons are a minor constituent of interstellar clouds (102). The ratio of observable column densities in those levels will, however, also reflect the effect of temperature gradients along the line of sight (99) and cannot be used blindly as measures of kinetic temperature. Absorption in the Lyman and Werner systems of H_2 is followed with high probability by fluorescence into excited vibration-rotation levels of the ground electronic state, which then decay by quadrupole transitions to levels of the lowest vibrational state. The calculated probabilities of redistributing rotational populations by this mechanism of absorption and fluorescence (103) are able to account for the observed populations of the higher J levels in diffuse clouds. Because some fraction of the fluorescence goes into the vibrational continuum of the $X^1\Sigma_g^+$ state and dissociates the molecule, the population distribution and the abundance of the molecules are closely coupled in steady state. The ultraviolet absorption lines readily become optically thick; therefore the rates of absorption, excitation and dissociation are complicated functions of depth through a cloud. In other words, H_2 molecules nearest the cloud boundary effectively shield the molecules in the interior. Because the observable properties (i.e. column densities) are integrals over line-of-sight distance, the observations must be interpreted by comparison with theoretical models that accurately represent this internal structure of clouds (99,104). It is worth emphasizing that the accuracy of derived densities is still limited in part by persistent uncertainties about cross sections for low-energy rotationally inelastic scattering of H and H_2 by H_2.

4.2 Diatomic Carbon

The rotational populations in C_2, like those in H_2, are evidently determined by a competition between collisional and radiative excitation and thus provide similar diagnostic information. C_2 also has two distinct advantages over H_2 for interstellar absorption line studies: a) its Phillips system $A^1\Pi_u - X^1\Sigma_g^+$ has favorable transitions in the 8750-12000 Å region and it is observable from Earth, whereas the ultraviolet transitions of H_2 require a spectrometer aboard a spacecraft, and b) the interstellar C_2 lines are weak enough that derived column densities are at worst only slightly affected by saturation. Interstellar C_2 has been observed both in diffuse clouds (7,36,37,104-110) and in a **bona fide** molecular cloud (13,15). Following the suggestion by Lutz and Souza (104) that the rotational populations in C_2 might provide a good thermometer, and the suggestion by Chaffee **et al.** (108) that radiative excitation might be important, van Dishoeck and Black (111) devised a theory of the excitation of interstellar C_2. Detailed theoretical models of interstellar clouds have been developed to interpret the observations of clouds in which both H_2 and C_2 are seen (71,85,110).

The interpretation of C_2 observations requires accurate oscillator strengths, photodissociation cross sections and collisional excitation cross sections. Recent determinations of the Phillips system oscillator strengths yield conflicting results. Three independent **ab initio** theoretical calculations are in excellent mutual agreement, with $f=2.7 \times 10^{-3}$ for the (0,0) band (23,25,29), and are in harmony with older experimental results (112). On the other hand, recent lifetime measurements (113) and astronomical determinations based upon C_2 lines in the solar spectrum (114), suggest oscillator strengths approximately a factor of 2 times smaller. Lien (37) has recently reported detection of the $F^1\Pi_u - X^1\Sigma_g^+$ (0,0) band near 1340 Å in interstellar absorption toward the star X Per. The strength of this feature and an upper limit on the A-X (2,0) lines suggest an F-X oscillator strength rather larger than the value computed by Pouilly **et al.** (23). The oscillator strengths of C_2 affect both the determination of its column density and the calculation of the effects of radiative excitation: needless to say, more work is urgently needed to resolve existing discrepancies. It is also very important to know the cross sections for collisional excitation of C_2 by H_2 at very low energies in order that densities can be derived accurately from the excitation analysis.

4.3 CN and CH

In diffuse clouds of low density, the rotational populations of CN are determined entirely by the rates of spontaneous emission and of stimulated emission and absorption due to the ambient radiation field. Because the radiation field at the wavelengths of the N=1-0 (2.64 mm) and N=2-1 (1.32 mm) transitions is the cosmic microwave background, the relative populations in N=0,1, and 2 provide a good measure of the brightness temperature (i.e. intensity) of this radiation. Thus interstellar absorption line measurements of CN can establish whether the temperature of the cosmic background near its peak deviates from its long-wavelength value of 2.74±0.087 K (115). This has important cosmological implications. Recent observations by Meyer and Jura (41) yield excitation temperatures of 2.73±0.04 and 2.8±0.3 K at 2.64 and 1.32 mm, respectively, consistent with a 2.7 K blackbody spectrum. Federman, Danks and Lambert (40) have recently performed a survey of CN in diffuse clouds.

CH is not expected to be significantly rotationally excited in normal interstellar clouds. It is possible, however, to measure directly the populations of its ground-state lambda-doubling levels by means of optical absorption line observations as pointed out by Lien (37). In two cases, Lien has measured lambda-doubling excitation temperatures less than zero, indicating that the populations are inverted. This provides an independent measure of the excitation temperature of the weak maser emission observed in the radio lines of CH at 9 cm wavelength (116).

5. RELATED SUBJECTS

5.1 Nebular Molecules

Unidentified emission bands that appear to be molecular in origin have been observed in the visible spectrum of a nebula called the "Red Rectangle"=HD44179 (117,118). More recently, additional molecular features have been reported in the ultraviolet spectrum of this unusual nebula (119,120). Although no specific identification of the emitting species has been made, various constituents of volatile grain mantles have been suggested (118, 120,121). There is also a possible molecular feature in the visible spectrum of V645 Cyg (122).

A noteworthy case of molecular emission lines is the identification of the Lyman system of H_2 in the ultraviolet spectrum of the young star T Tau and its associated nebula (123). The observed lines all arise in $B^1\Sigma_u^+$ (v=1,J=4) which is excited by the absorption of H Lα photons (1215.670 Å rest wavelength) by molecules in (v=2,J=5) of the ground electronic state in the B-X (1,2) P(5) line at 1216.0753 Å, which is within 100 km s^{-1} of exact resonance with H Lα. This fluorescence mechanism requires a broad intense H Lα line and a substantial population of excited H_2 in the v=2 levels, both of which are supplied in T Tau (124). The significance of this fluorescence mechanism had previously been discussed by Jordan **et al.** (125) and by Shull (126). Another fluorescence mechanism is possible in which the ground state of ortho-hydrogen, $X^1\Sigma_g^+$ (v=0,J=1) absorbs H Lβ photons (1025.722 Å)

in the near-resonant B-X (6,0) P(1) line at 1025.935 Å (127).
This mechanism has been discussed in the context of possible H_2
emission from supernova remnants (128,129). In general, such
fluorescent emission of H_2 can provide excellent diagnostics
both of the H Lyman line profiles and intensities and of the
excitation conditions of H_2 in the emitting region.

The subject of molecules in planetary nebulae has been reviewed recently (130). A number of planetary nebulae and related
objects exhibit H_2 emission lines in the infrared (131-133). A
particularly interesting example is the pre-planetary nebula, AFGL
2688, also known as the "Egg Nebula" (134,135), which also shows
features due to C_2, C_3, and SiC_2 in its visible spectrum (136,137).
Further studies of these molecular features at higher resolution
will be valuable: resolved rotational structure in the C_2 and C_3
lines can be analyzed to yield information about physical conditions
in the nebula. Possible absorption features in the ultraviolet
spectra of central stars of planetary nebulae have been attributed
to nebular H_2^+ (138). Although the reality of these featues has
been challenged (remarks by R.E.S. Clegg: see ref. 130, page 101),
the existence of detectable amounts of such molecules in young
planetary nebulae is not implausible (130). Tentative evidence
has been presented for the identification of the 5600 Å band of H_3 in
emission in the planetary nebula NGC 7027 (139). If verified
by observation of the stronger H_3 feature near 7100 Å, this would
confirm Herzberg's (140) suggestion that H_3 might appear in nebular spectra as a byproduct of the recombination of H_3^+ (see also
ref. 14**1**). It is worth emphasizing that studies of molecules in
young planetary nebulae are especially important because isotope
abundances are more easily determined from molecular spectra than
from atomic spectra. Isotope abundances can be useful clues to
the evolutionary state of the star that forms a planetary nebula
because they identify the processes of nucleosynthesis and indicate
the extent to which the interior and surface layers have been
mixed.

5.2 Molecular Lines in QSO Spectra

It is possible to look for absorption lines of interstellar
or intergalactic molecules in the spectra of quasi-stellar objects
(QSOs). In particular, some QSOs show absorption line redshifts
due to cosmic expansion that are large enough to shift the strong
ultraviolet systems of H_2 and CO to wavelengths greater than
3300 Å where observations can be made from the Earth's surface.
The problem of line identification in the spectra of high-redshift
QSOs is difficult because they often show hundreds of narrow
absorption lines whose redshifts are not known **a priori**, and
because the multiplicity of possible H_2 and CO lines encourages
chance coincidences. Carlson (142) noticed a number of coincidences
with red-shifted H_2 bands in the spectrum of 4C05.34. Aaronson,
Black and McKee (143) found perhaps a more striking case in the

spectrum of PHL 957 where coincidences with H_2 lines occur at the same redshift as a well known low-excitation atomic absorption line system of large H column density. Several additional cases of possible H_2 and CO lines in QSO spectra have been reported more recently by Varshalovich and Levshakov (144). Now that it is possible to obtain spectra of faint QSOs with high S/N at resolutions of 10-30 km s^{-1}, it should be possible to make the definitive tests for the presence of H_2 absorption by trying to resolve the rotational structure of several adjacent bands in the spectrum of a QSO like PHL 957. If the abundance and excitation of H_2 can be measured, a valuable diagnostic probe will be available for studying conditions in diffuse matter at a much earlier epoch in the history of the universe.

6. SUMMARY

Optical observations of interstellar molecules can provide valuable information about interstellar clouds. Much important work remains to be done. It is likely that a few stars lying behind genuine molecular clouds (e.g. HD 29647) will be accessible in the ultraviolet to Space Telescope. We can also anticipate interesting new results as absorption line methods are extended to more and more highly obscured stars and are extended with greater sensitivity into the infrared. Attempts to confirm the identification of interstellar CS$^+$ and searches in dark clouds for such species as NH, NaH, MgH, MgO, SH, SH$^+$, SiH$^+$, and C$_3$ are especially to be encouraged. Optical observations of such molecules as C_2, CH, OH, and CN in dark clouds can be valuable complements to radiofrequency emission line observations. Further study of the abundance of interstellar CO, both by high resolution infrared spectroscopy and by theoretical means is urgently required.

In the course of this discussion, many gaps in our knowledge of basic molecular processes have been mentioned. The number of opportunities for related studies in spectroscopy and molecular physics can only be expected to increase; it is likely that many of these problems will be of intrinsic interest aside from the astronomical applications. The need for a better understanding of photodissociation processes in CO cannot be overemphasized. Discrepancies in the literature concerning oscillator strengths of CO, C_2, and H_2O^+ need to be resolved. More work on collisional excitation of H_2 and C_2 is clearly needed. There is, of course, a continuing need for data on low-temperature behavior of many chemical reactions. There are many astrophysically interesting molecules for which there exist no rotational analyses of strong ultraviolet band systems.

REFERENCES

1. Adams, W.S. 1941, Astrophys. J. 93, pp. 11-23.
2. Douglas, A.E. and Herzberg, G. 1941, Astrophys. J. 94, p. 381.
3. Carruthers, G.R. 1970, Astrophys. J. 161, pp. L81-L85.
4. Smith, A.M. and Stecher, T.P. 1971, Astrophys. J. 164, pp. L43-L47.
5. Crutcher, R.M. and Watson, W.D. 1976, Astrophys. J. 203, pp. L123-L126.
6. Snow, T.P. 1976, Astrophys. J. 204, pp. L127-L130.
7. Souza, S.P. and Lutz, B.L. 1977, Astrophys. J. 216, pp. L49-L51; erratum, 218, p. L31.
8. Ferlet, R., Roueff, E., Horani, M., and Rostas, J. 1983, Astron. Astrophys. 125, pp. L5-L8.
9. Snow, T.P. 1980, in "Interstellar Molecules", IAU Symposium No. 87, B.J. Andrew, ed., (Dordrecht: D. Reidel), pp. 247-254.
10. Spitzer, L. and Jenkins, E.B. 1975, Ann. Rev. Astron. Astrophys. 13, pp. 133-164.
11. Cohen J.G. 1973, Astrophys. J. 186, pp. 149-163.
12. Crutcher, R.M. 1980, Astrophys. J. 239, pp. 549-554.
    Crutcher, R.M. 1983, preprint.
13. Hobbs, L.M., Black, J.H., and van Dishoeck, E.F. 1983, Astrophys. J. 271, pp. L95-L99.
14. Souza, S.P. 1979, Ph.D. Thesis, State University of New York at Stony Brook.
15. Lutz, B.L. and Crutcher, R.M. 1983, Astrophys. J. 271, pp. L101-L105.
16. Savage, B.D. and Mathis, J.S. 1979, Ann. Rev. Astron. Astrophys. 17, pp. 73-111.
17. Savage, B.D., Bohlin, R.C., Drake, J.F. and Budich, W. 1977, Astrophys. J. 216, pp. 291-307.
18. Larsson, M. 1983, Astron. Astrophys. 128, pp. 291-298.
    Whiting, E.E., Schadee, A., Tatum, J.B., Hougen, J.T. and Nicholls, R.W. 1980, J. Molec. Spectr. 80, pp. 249-256.
19. Seaton, M.J. 1979, Mon. Not. R. Astron. Soc. 187, pp. 73P-76P.
20. Rieke, G.H. and Lebofsky, M.J. 1984, preprint.
21. Kurucz, R.L. 1979, Astrophys. J. Suppl. 40, pp. 1-340.
22. Allison, A.C. and Dalgarno, A. 1970, Atomic Data 1, pp. 289-304.
23. Pouilly, B., Robbe, J.M., Schamps, J. and Roueff, E. 1983, J. Phys. B 16, pp. 437-448.
24. Carlson, T.A., Duric, N., Erman, P. and Larsson, M. 1978, Zeits. f. Phys. A 287, pp. 123-136.
25. Chabalowski, C.F., Peyerimhoff, S.D. and Buenker, R.J. 1983, Chem. Phys. 81, pp. 57-72.
26. Langhoff, S.R., van Dishoeck, E.F., Wetmore, R. and Dalgarno, A. 1982, J. Chem. Phys. 77, pp. 1379-1390.
27. Duric, N., Erman, P. and Larsson, M. 1978, Physica Scripta 18, pp. 39-46.
    Danylewich, L.L. and Nicholls, R.W. 1978, Proc. Roy. Soc. A 360, pp. 557-580.

Cartwright, D.C. and Hay, P.J. 1982, Astrophys. J. 257, pp. 383-387.
Larsson, M., Siegbahn, P.E.M. and Ågren, H. 1983, Astrophys. J. 272, pp. 369-376.
28. Larsson, M. and Siegbahn, P.E.M. 1983, J. Chem. Phys. 79, pp. 2270-2277.
Brzozowski, J., Bunker,P., Elander, N. and Erman, P. 1976, Astrophys. J. 207, pp. 414-424.
29. van Dishoeck, E.F. 1983, Chem. Phys. 77, pp. 277-286.
30. Turner, J., Kirby-Docken, K. and Dalgarno, A. 1977, Astrophys. J. Suppl. 35, pp. 281-292.
31. Kirby-Docken, K. and Liu, B. 1978, Astrophys. J. Suppl. 36, pp. 359-387.
32. Brandt, J.C. **et al.** 1979, Proc. Soc. Photo-Opt. Engr. 172, pp. 254-263.
33. Danks, A.C., Federman, S.R. and Lambert, D.L. 1984, Astron. Astrophys. 130, pp. 62-66.
34. Morton, D.C. 1975, Astrophys. J. 197, pp. 85-115.
35. Wannier, P.G., Penzias, A.A. and Jenkins, E.B. 1982, Astrophys. J. 254, pp. 100-107.
36. Danks, A.C. and Lambert, D.L. 1983, Astron. Astrophys. 124, pp. 188-196.
37. Lien, D.J. 1984, preprints.
38. Vanden Bout, P.A. and Snell, R.L. 1980, Astrophys. J. 236, pp. 460-464; erratum 246, p. 1045.
39. Mahan, B.H. and O'Keefe, A. 1981, Astrophys. J. 248, pp. 1209-1216.
40. Federman, S.R., Danks, A.C. and Lambert, D.L. 1984, preprint.
41. Meyer, D.M. and Jura, M. 1984, Astrophys. J. 276, pp. L1-L3.
42. Chaffee, F.H. and Lutz, B.L. 1977, Astrophys. J. 213, pp. 394-404.
43. Clegg, R.E.S. and Lambert, D.L. 1982, Mon. Not. R. Astron. Soc. 201, pp. 723-733.
44. Snow, T.P. and Smith, W.H. 1981, Astrophys. J. 250, pp. 163-165.
45. Smith, P.L., Yoshino, K., Griesinger, H.E. and Black, J.H. 1981, Astrophys. J. 250, pp. 166-174; erratum 256, p. 798.
46. Jura, M. 1984, in preparation.
47. Smith, W.H., Schempp, W.V. and Federman, S.R. 1984, Astrophys. J. 277, pp. 196-199.
48. Federman, S.R., Glassgold, A.E., Jenkins, E.B. and Shaya, E.J. 1980, Astrophys. J. 242, pp. 545-559.
49. Federman, S.R. 1982, Astrophys. J. 257, pp. 125-134.
50. Dickman, R.L., Somerville, W.B., Whittet, D.C.B., McNally, D. and Blades, J.C. 1983, Astrophys. J. Suppl. 53, pp. 55-72.
51. Black, J.H. and Willner, S.P. 1984, Astrophys. J. 279, pp. 673-678.
52. Kramers, H.A. and ter Haar, D. 1946, B. A. N. 10, pp. 137-146.
Bates, D.R. and Spitzer, L. 1951, Astrophys. J. 113, pp. 441-463.
Solomon, P.M. and Klemperer, W. 1972, Astrophys. J. 178, pp. 389-421.

53. Dalgarno, A. 1976, in "Atomic Processes and Applications", P.G. Burke, ed., (Amsterdam: North-Holland), pp. 109-132.
54. Elitzur, M. and Watson, W.D. 1978, Astrophys. J. 222, pp. L141-L144; erratum 226, p. L157.
    Elitzur, M. and Watson, W.D. 1980, Astrophys. J. 236, pp. 172-181.
    Elitzur, M. 1980, Astron. Astrophys. 81, pp. 351-353.
55. Federman, S.R. 1980, Astrophys. J. 241, pp. L109-L112.
56. Federman, S.R. 1982, Astrophys. J. 253, pp.601-605.
57. Frisch, P.C. and Jura, M. 1980, Astrophys. J. 242, 560-567.
58. White, R.E. 1984a,b, Astrophys. J., in press.
59. Federman, S.R. and Willson, R.F. 1984, preprint.
60. Uzer, T. and Dalgarno, A. 1978, Chem. Phys. 32, pp. 301-303.
61. Yoshimine, M., Green, S. and Thaddeus, P. 1973, Astrophys. J. 183, pp. 899-902.
62. Larsson, M. and Siegbahn, P.E.M. 1983, Chem. Phys. 76, pp. 175-184.
63. Graff, M.M., Moseley, J.T. and Roueff, E. 1983, Astrophys. J. 269, pp.726-802.
64. Kirby, K., Roberge, W.G., Saxon, R.P. and Liu, B. 1980, Astrophys. J. 239, pp. 855-858.
65. Herbst, E. and Knudson, S. 1981, Astrophys. J. 245, pp. 529-533.
66. Black, J.H. and Dalgarno, A. 1973, Astrophys. J. 184, pp. L101-L104.
67. O'Donnell, E.J. and Watson, W.D. 1974, Astrophys. J. 191, pp. 89-92.
68. van Dishoeck, E.F., Langhoff, S.R. and Dalgarno, A. 1983, J. Chem. Phys. 78, pp. 4552-4561.
69. Chambaud, G., Launay, J.M., Levy, B., Millie, P., Roueff, E. and Tran Minh, F. 1980, J. Phys. B 13, pp. 4205-4216.
70. van Dishoeck, E.F. and Dalgarno, A. 1983, J. Chem. Phys. 79, pp. 873-888.
    van Dishoeck, E.F. and Dalgarno, A. 1984, Astrophys. J. 277, pp. 576-580.
71. van Dishoeck, E.F. 1984, Ph.D. Thesis, Leiden.
72. Smith, P.L. and Parkinson, W.H. 1978, Astrophys. J. 223, pp. L127-L130.
73. Black, J.H. and Smith, P.L. 1984, Astrophys. J. 277, pp. 562-568.
74. Smith, Griesinger, H.E., Black, J.H., Yoshino, K. and Freeman, D.E. 1984, Astrophys. J. 277, pp. 569-575.
75. Prasad, S.S. and Huntress, W.T. 1979, Astrophys. J. 228, pp. 123-126.
76. Herbst, E. 1978, Astrophys. J. 222, pp. 508-516.
77. Black, J.H. and van Dishoeck, E.F. 1982, in "The Scientific Importance of Submillimetre Observations", ESA SP-189, T. de Graauw and T.D. Guyenne, eds., pp. 13-20.
78. Bell, M.B., Feldman, P.A. and Matthews, H.E. 1981, Astron. Astrophys. 101, pp. L13-L16.
79. Shinn, J.L. 1982, in "Thermophysics of Atmospheric Entry",

T.E. Horton, ed., Prog. Astronaut. Aeronaut. 82, pp. 68-80.
80. Glassgold, A.E. and Langer, W.D. 1976, Astrophys. J. 206, pp. 85-99.
81. Lee, L.C. and Guest, J.A. 1981, J. Phys. B 14, pp. 3415-3421.
82. Smith, W.H. 1978, Physica Scripta 17, pp. 513-515.
83. Cooper, D.M. and Langhoff, S.R. 1981, J. Chem. Phys. 74, pp. 1200-1210.
84. Bally, J. and Langer, W.D. 1982, Astrophys. J. 255, pp. 143-148.
Chu, Y.-H. and Watson, W.D. 1982, Astrophys. J. 267, pp. 151-155.
85. van Dishoeck, E.F. and Black, J.H. 1984, in preparation.
86. Smith, A.M., Krishna Swamy, K.S. and Stecher, T.P. 1978, Astrophys. J. 220, pp. 138-148.
87. Black, J.H. 1980, in "Interstellar Molecules", IAU Symposium No. 87, B.H. Andrew, ed., (Dordrecht: D. Reidel), pp. 257-260.
88. Snow, T.P. and Jenkins, E.B. 1980, Astrophys. J. 241, pp. 161-172.
89. Black, J.H. and Raymond, J.C. 1984, Astron. J. 89, pp. 411-416.
90. Gondhalekar, P.M. and Phillips, A.P. 1980, Mon. Not. R. Astron. Soc. 191, pp. 13P-17P.
91. Tarafdar, S.P. and Krishna Swamy, K.S. 1981, Mon. Not. R. Astron. Soc. 196, pp. 67-71.
92. Tarafdar, S.P. and Krishna Swamy, K.S. 1982, Mon. Not. R. Astron. Soc. 200, pp. 431-444.
93. Tarafdar, S.P. 1983, Mon. Not. R. Astron. Soc. 204, pp. 1081-1089.
94. Welsh, B.Y. 1984, Mon. Not. R. Astron. Soc. 207, pp. 167-172.
95. McLachlan, A. and Nandy, K. 1984, Mon. Not. R. Astron. Soc. 207, pp. 355-360.
96. Lassettre, E.N. and Skerbele, A. 1971, J. Chem. Phys. 54, pp. 1597-1607.
97. Kirschner, S.M., LeRoy, R.J., Ogilvie, J.F. and Tipping, R.M. 1977, J. Molec. Spectr. 65, pp. 306-312.
Werner, H.-J. 1981, Mol. Phys. 44, pp. 111-123.
Chackerian, C. and Tipping, R.H. 1984, preprint.
98. Tarafdar, S.P., Krishna Swamy, K.S. and Vardya, M.S. 1980, Mon. Not. R. Astron. Soc. 192, pp. 417-426.
99. Black, J.H. and Dalgarno, A. 1977, Astrophys. J. Suppl. 34, pp. 405-423.
100. Jura, M. 1975, Astrophys. J. 197, pp. 575-580, 581-586.
101. Spitzer, L. and Zweibel, E.G. 1974, Astrophys. J. 191, pp. L127-L130.
102. Dalgarno, A., Black, J.H. and Weisheit, J.C. 1973, Astrophys. Lett. 14, pp. 77-79.
103. Black, J.H. and Dalgarno, A. 1976, Astrophys. J. 203, pp. 132-142.
104. Federman, S.R., Glassgold, A.E. and Kwan, J. 1979, Astrophys. J. 227, pp. 466-473.
105. Chaffee, F.H. and Lutz, B.L. 1978, Astrophys. J. 221, pp. L91-L93.
106. Snow, T.P. 1978, Astrophys. J. 220, pp. L93-L96.

107. Hobbs, L.M. 1979, Astrophys. J. 232, pp. L175-L177.
     Hobbs, L.M. 1981, Astrophys. J. 243, pp. 485-488.
108. Chaffee, F.H., Lutz, B.L., Black, J.H., Vanden Bout, P.A. and Snell, R.L. 1980, Astrophys. J. 236, pp. 474-480.
109. Hobbs, L.M. and Campbell, B. 1982, Astrophys. J. 254, pp. 108-110.
110. van Dishoeck, E.F. and de Zeeuw, T. 1984, Mon. Not. R. Astron. Soc. 206, pp. 383-406.
111. van Dishoeck, E.F. and Black, J.H. 1982, Astrophys. J. 258, pp. 533-547.
112. Cooper, D.M. and Nicholls, R.W. 1975, J. Q. S. R. T. 15, pp. 139-150.
     Cooper, D.M. and Nicholls, R.W. 1976, Spectr. Lett. 9, pp. 139-155.
     Roux, F., Cerny, D. and d'Incan, J. 1976, Astrophys. J. 204, pp. 940-943.
113. Erman, P. Lambert, D.L., Larsson, M. and Mannfors, B. 1982, Astrophys. J. 253, pp. 983-988.
114. Brault, J.W., Delbouille, L., Grevesse, N., Roland, G., Sauval, A.J. and Testerman, L. 1982, Astron. Astrophys. 108, pp. 201-205.
115. Weiss, R. 1980, Ann. Rev. Astron. Astrophys. 18, pp. 489-535.
116. Rydbeck, O.E.H., Kollberg, E., Hjalmarson, Å., Sume, A., Ellder, J. and Irvine, W.M. 1976, Astrophys. J. Suppl. 31, pp. 333-415.
117. Schmidt, G.D., Cohen, M. and Margon, B. 1980, Astrophys. J. 239, pp. L133-L135.
118. Warren-Smith, R.F., Scarrott, S.M. and Murdin, P. 1981, Nature 292, pp. 317-319.
119. Sitko, M.L., Savage, B.D. and Meade, M.R. 1981, Astrophys. J. 246, pp. 161-183.
120. Sitko, M.L. 1983, Astrophys. J. 265, pp. 848-854.
121. Wdowiak, T.J. 1981, Nature 293, pp. 724-725.
122. Humphreys, R.M., Merrill, K.M. and Black, J.H. 1980, Astrophys. J. 237, pp. L17-L20.
123. Brown, A., Jordan, C., Millar, T.J., Gondhalekar, P. and Wilson, R. 1981, Nature 290, pp. 34-36.
124. Beckwith, S., Gatley, I., Matthews, K. and Neugebauer, G. 1978, Astrophys. J. 223, pp. L41-L43.
125. Jordan, C., Brueckner, G.E., Bartoe, J.-D. F., Sandlin, G.D. and Van Hoosier, M.E. 1978, Astrophys. J. 226, pp. 687-697.
126. Shull, J.M. 1978, Astrophys. J. 224, pp. 841-847.
127. Feldman, P.D. and Fastie, W.G. 1973, Astrophys. J. 185, pp. L101-L104.
128. Raymond, J.C., Black, J.H., Dupree, A.K., Hartmann, L. and Wolff, R.S. 1981, Astrophys. J. 246, pp. 100-109.
129. Benvenuti, P., Dopita, M. and D'Odorico, S. 1980, Astrophys. J. 238, pp. 601-603.
130. Black, J.H. 1983, in "Planetary Nebulae", IAU Symposium No. 103, D.R. Flower, ed., (Dordrecht: D. Reidel), pp. 91-102.

131. Shull, J.M. and Beckwith, S. 1982, Ann. Rev. Astron. Astrophys. 20, pp. 163-190.
132. Isaacman, R. 1984, Astron. Astrophys. 130, pp. 151-156.
133. Storey, J.W.V. 1984, Mon. Not. R. Astron. Soc. 206, pp. 521-527.
134. Beckwith, S., Beck, S.C. and Gatley, I. 1984, Astrophys. J. 280, pp. 648-652.
135. Thronson, H.A. 1982, Astron. J. 87, pp. 1207-1212.
136. Cohen, M. and Kuhi, L.V. 1977, Astrophys. J. 213, pp. 79-92.
    Cohen, M. and Kuhi, L.V. 1980, Publ. Astron. Soc. Pacific 92, pp. 736-745.
137. Humphreys, R.M., Warner, J.W. and Gallagher, J.S. 1976, Publ. Astron. Soc. Pacific 88, pp. 380-387.
138. Heap, S.R. and Stecher, T.P. 1981, in "The Universe at Ultraviolet Wavelengths", R.D. Chapman, ed., NASA Conference Publ. 2171, pp. 657-661.
    Feibelman, W.A., Boggess, A., McCracken, C.W. and Hobbs, R.W. 1981, Astron. J. 86, pp. 881-884.
139. Pritchet, C.J. and Grillmair, C.J. 1984, Publ. Astron. Soc. Pacific 96, 349-353.
140. Herzberg, G. 1979, J. Chem. Phys. 70, pp. 4806-4807.
141. Dabrowski, I. and Herzberg, G. 1980, Can J. Phys. 58, pp. 1238-1249.
    Herzberg, G. and Watson, J.K.G. 1980, Can. J. Phys. 58, pp. 1250-1258.
142. Carlson, R.W. 1974, Astrophys. J. 190, pp. L99-L100.
143. Aaronson, M., Black, J.H. and McKee, C.F. 1974, Astrophys. J. 191, pp. L53-L56.
144. Varshalovich, D.A. and Levshakov, S.A. 1982, Comments Astrophys. 9, pp. 199-209, and references therein.

THE PRODUCTION OF COMPLEX MOLECULES IN DENSE INTERSTELLAR CLOUDS

Eric Herbst

Duke University, Durham, N.C., U.S.A.

The synthesis of gas phase molecules in dense interstellar clouds is discussed. Detailed models that utilize gas phase reactions to both produce and deplete smaller molecules are explained. Problems in extending gas phase, ion-molecule synthetic pathways to complex molecules are clarified and the role of radiative association processes is emphasized. Reaction pathways for gas phase complex molecule production that include radiative association reactions are depicted, especially for hydrocarbons and cyanoacetylenes. Recent models of the gas phase chemistry of dense clouds that contain complex molecular species are examined. Finally, avenues for future investigation are explored.

INTRODUCTION

To understand the formation processes of the complex molecules found in dense interstellar clouds, it is first necessary to determine where the molecules are formed. Either the molecules are formed in the low density, low temperature, non-equilibrium regions where they are observed or they are formed in higher density, higher temperature, equilibrium sources where collision time scales are shorter and the conditions would seem in general to be more favorable for complex molecule development. There are strong indicators that the observed interstellar molecules are formed where we observe them. These indicators include strong isotope fractionation effects, high abundances of the metastable isomer HNC, and significant abundances of molecular ions. In contrast, molecular abundances in high temperature sources such as the stellar evelope IRC + 10216 reflect to some extent a high temperature equilibrium chemistry (1). In addition, the large

G. H. F. Diercksen et al. (eds.), Molecular Astrophysics, 237–254.

scales over which interstellar molecules are observed and the difficulty in protecting most species from photodissociation via the general stellar background radiation field argue for molecule formation in the protected environment of dense clouds. Finally, depletion processes for molecules exist in dense clouds requiring the existence of "rapid" formation processes. It is therefore likely that most observed interstellar molecules of some complexity have been synthesized from precursor atomic material in dense clouds. One cannot ignore, however, the detection of complex species via radio absorption studies in regions with relatively little visual extinction (2). Perhaps some species are indeed quite resistant to photodissociation.

There are two possible mechanisms by which molecules can be produced under dense cloud conditions. They can be formed via synthetic gas phase reactions or via reactions occurring either on the surface of or inside dust particles. In general, both of these processes will occur and it is difficult to sort out their respective contributions. The simplest view of processes on dust grains envisages the surface as a rather passive site on which atoms and smaller reactive species will adsorb at low temperatures, and subsequently migrate from site to site until they form chemical bonds with each other. The newly-formed molecules must then leave the grain to be detected by normal radioastronomical methods. Suggested mechanisms for leaving include thermal evaporation, photo-ejection, interstellar shock waves, grain-grain collisions, and grain explosions. For a detailed model of grain formation of gas phase species, the reader is advised to see the work of Allen and Robinson (3) in which the mechanism for grain ejection is the disruption of very small grains caused by the heat generated in the exothermic reaction process of forming a more complex molecule. On the other hand, Tielens and Hagen (4) have recently published a model of grain surface chemistry in which the newly formed molecules (with the exception of H_2) remain on the grains because there is, in the authors' view, no viable mechanism to remove them. Watson, an earlier contributor to grain chemistry, has recently stated that "except for molecular hydrogen, there is in my opinion still no convincing mechanism for returning molecules from grain surfaces back to the gas." (5) Molecular hydrogen, it is agreed, can thermally evaporate even at interstellar cloud temperatures. There is increasingly strong evidence from infra-red observations that grain mantles do contain an assortment of molecules but it is still unclear whether grain processes contribute to the observed gas phase species.

Another school of thought regards the grains as sites for photochemistry. Greenberg and collaborators (6) have elegantly shown that photon-initiated radical production can stimulate complex molecule synthesis on and in cold, solid materials. Moore and Donn (7) have performed similar experiments with proton bombardment of adsorbed species on cold surfaces. Although these investigators have demonstrated that complex species can be formed on grains, it remains uncertain whether the photon and/or proton flux in dense clouds is sufficient to produce large amounts of complex species.

In addition to grain processes, molecules can be formed via gas phase reactions. The low temperatures and low gas densities of dense clouds restrict gas phase reactions under ambient conditions to two-body exothermic reactions without activation energy (8). Under shock conditions, the activation energy restriction is dramatically eased (9). Although selected exothermic neutral-neutral reactions do not possess activation energy, the more important class of reactions without activation energy are reactions involving at least one charged species. Ion-molecule reactions have been extensively studied in the laboratory (10-13), often possess rate coefficients independent of temperature (if exothermic), and can be understood in terms of simple models such as that due to Langevin. These reactions tend to possess rate coefficients of $\sim 10^{-9}$ cm^3 s^{-1}, a value 1-2 orders of magnitude greater than the fastest neutral-neutral systems possess. In dense interstellar clouds, cosmic ray bombardment provides a source of ionization with a known rate of $\sim 10^{-17}$ s^{-1} (8) although some investigators have argued that internal T Tauri stars and radioactive elements are equally or more important.

To synthesize complex molecules from atoms via gas phase two-body processes, atoms must first combine to form diatomic molecules via radiative association, a process in which the collision complex is stabilized via emission of a photon (14). This is not an efficient process for diatomics, where the lifetime of the collision complex is $\sim 10^{-14}$ s whereas the photon emission time is $>10^{-8}$ s. The actual rate coefficient depends critically on the atoms involved and the diatomic potential energy curves. For molecular hydrogen, the calculated rate coefficient is particularly slow and this dominant interstellar species must be formed on dust grain surfaces. Once diatomics are formed, synthesis can proceed via standard two-body reactions.

A large number of models of dense interstellar clouds have now appeared in the literature which utilize ion-molecule reactions exclusively to form all molecules other than H_2 (8,15-29). In most of these models, molecules are depleted exclusively via gas phase reactions as well although in some models photodissociation processes and/or grain adsorption are also included. It is useful to obtain a sense of the manner in which small molecular species are synthesized in these models. Consider, as examples, the syntheses of H_2O and OH via the processes:

$$H_2 + \text{Cosmic Ray} \longrightarrow H_2^+ + e + \text{Cosmic Ray}, \quad (1)$$

$$H_2 + H_2^+ \longrightarrow H_3^+ + H, \quad (2)*$$

$$H_3^+ + O \longrightarrow OH^+ + H_2 \quad , \qquad (3)*$$

$$OH^+ + H_2 \longrightarrow H_2O^+ + H \quad , \qquad (4)*$$

$$H_2O^+ + H_2 \longrightarrow H_3O^+ + H \quad , \qquad (5)*$$

$$H_3O^+ + e \longrightarrow H_2O + H; \; OH + H_2 \quad . \qquad (6)$$

Starred reactions have been studied in the laboratory. The ion-electron dissociative recombination reaction, process (6), is quite rapid ($k>10^{-7}$ cm^3 s^{-1}) and inversely dependent on temperature, as are all such reactions. Unfortunately, except for $H_3^+ + e$, the branching ratios among the neutral products are not known and must be estimated from crude theory (30). Upon formation, the water molecule is depleted via reaction with ions whereas the hydroxyl radical is also depleted via rapid reactions with the abundant atoms O and N (8).

In most of the models mentioned above, large numbers of "small" (≤ 4 atom) molecules are created and destroyed in analogous reaction schemes. Kinetic equations are written for each species in the model and are normally solved in the steady-state limit (8) or with the "chemical time dependence" approximation (20). In the latter approximation, the abundances of molecules are calculated as a function of time from given initial conditions (normally atomic or diffuse cloud abundances) up through steady-state. Physical conditions such as cloud gas density, temperature, and visual extinction are held constant at dense cloud values and grain adsorption is ignored except for the formation of H_2. The assumed gas density and the gaseous elemental abundances used are important parameters in the solution of the kinetic equations. Normally, the gaseous elemental abundances utilized reflect the so-called normal cosmic abundance values ($H=1$, $He=0.14$, $C=3\times10^{-4}$, $N=9\times10^{-5}$, $O=7\times10^{-4}$, etc.) with heavy elements depleted substantially to form the grains according to diffuse cloud optical studies. Better agreement with observation is obtained, however, if the heaviest elements such as Si, S, and metals are depleted additionally by two orders of magnitude (23).

Although there are differences among models, the general agreement between steady-state calculated abundances of the smaller molecules and observed abundances is good. However, the steady-state and "chemical time dependence" models used are artificial in assuming constant physical conditions. A more realistic picture consists of a diffuse cloud with gas density $\sim 10^2$ cm^{-3} and low visual extinction gradually collapsing gravitationally, cooling slightly, and acquiring a larger visual extinction. The time scale for these physical changes may well be of the same magnitude as the $\sim 10^7$ yrs needed to reach steady-state. A new calculation that includes a realistic variation of physical conditions has been recently accomplished by Prasad, Huntress, and collaborators (31). One major point of disagreement between almost all present models and observation is the large abundance of CI seen in several dense clouds. The standard "chemical time dependence" models show that as time progresses the C^+ present in diffuse clouds first converts itself into C and then (after 10^5-10^6 yrs) to CO. At steady-state conditions, only small amounts of C^+ and C remain. The inclusion of more realistic physical time dependent effects may change this picture. On the other hand, it is possible (but not likely) that dense clouds are "young" (10^5-10^6 yrs past the diffuse stage) and have not yet reached steady-state. Most molecular abundances reach values near or even greater than their steady-state values by $\sim 10^5$ yrs so that by postulating youthful clouds to explain the observed abundance of C, one is not destroying other areas of agreement. In addition, if dense clouds are youthful, the vexing problem of grain adsorption lessens in intensity. However, what seems more likely is that inclusion of low initial gas densities and photodissociation effects, especially in the early stages of cloud collapse, as well as the process of collapse itself will all slow down the C→CO conversion. In addition, grain adsorption and re-ejection effects (32) and internal UV fields (33) may help to maintain a high C abundance even at steady-state. Finally, the recent model of Langer et al. (28) reproduces the large C abundance at steady-state by the simple expedient of setting the C/O elemental gas phase ratio to greater than unity.

In my view, the major drawback of most models has been the virtual exclusion of "large" (>4 atom) molecules. In general, normal ion-molecule reaction pathways fail to produce large species in sufficient abundance because of road blocks in the form of reactions that are endothermic or exothermic but with activation energy barriers, or because the reactions involve species that are of too low concentration. To alleviate this problem, radiative association reactions have been advocated by a variety of investigators (8,34-36). These reactions are always exothermic by their very nature and tend to increase in rate as the reaction partners become larger. The joining of two abundant species to form a larger one is a very efficient synthetic pathway if there is an appreciable rate coefficient. Unfortunately it is difficult to study the rates of radiative association processes under normal laboratory conditions, where the pressures are large enough that ternary (three-body) association processes dominate (34). Therefore, theoretical treatments are of some importance. In addition to their determination via theoretical methods, radiative association rate coefficients can be estimated from ternary data (34).

RADIATIVE ASSOCIATION

In existing theoretical treatments of radiative association, the process is modeled as a two-step one in which reactants A^+ and B collide to form an unstable "complex" AB^{+*} which either redissociates or is stabilized via emission of a photon:

$$A^+ + B \rightleftharpoons AB^{+*} \quad , \quad (7a)$$

$$AB^{+*} \longrightarrow AB^+ + h\nu \quad . \quad (7b)$$

Theories must contain estimates for the complex formation and dissociation steps as well as the complex stabilization rate via photon emission. The first two processes are related by microscopic reversibility, which simplifies the theoretical treatment. In order of increasing complexity, Herbst (36), Bates (37-39), and Bowers and co-workers (40,41) have formulated theories of complex formation and dissociation. The results can differ by up to an order of magnitude (42). Herbst (43) has formulated a theory of radiative stabilization which yields an estimate of $10^{2\pm1}$ s^{-1} for the stabilization rate for most complexes. Overall theoretical estimates of radiative association rates can then be expected to be accurate to only an order of magnitude or so. That such accuracy is attainable has been indicated mainly via comparison of analogous theoretical approaches with the wealth of laboratory data on ternary association reactions (44) in which process (7b) is replaced by collisional complex stabilization:

$$AB^{+*} + C \longrightarrow AB^+ + C \quad . \quad (7c)$$

Recently, however, radiative association reaction rates have been measured in the laboratory for at least three systems of interstellar interest - $C^+ + H_2$, $CH_3^+ + H_2$, and $CH_3^+ + HCN$ (41,45,46). The experimental results tend to be in the expected order-of-magnitude agreement with theory but there is a hint that calculated radiative stabilization rates are too low. In collaboration with D. Noid (Oak Ridge) I am currently working on a new theory of complex stabilization rates that is based on semiclassical techniques.

What all theories of radiative association show is that rates are dependent on several important factors. These are the size and bond energy of the formed molecule, the temperature of the system, and the accessibility of normal exothermic channels. As size and bond energy increase, the lifetime of the complex against redissociation increases dramatically until it can live long enough to be stabilized radiatively with near unit efficiency. Thus, calculated radiative association rates range from $<10^{-20}$ cm^3 s^{-1} for some diatomic systems to $\sim 10^{-9}$ cm^3 s^{-1} (the Langevin collision rate) for "large" (~10 atom) ion-molecule collision partners that form molecules with bond energies of 3-5 eV. Temperature is another critical factor; rates increase as temperature decreases via a power law of the type T^{-n} where n depends on the complexity of the system and the particular theory utilized. Dense interstellar clouds with temperatures in the range 10-100K and large abundances of gaseous ions are excellent sites for rapid radiative association reactions. Unfortunately, interstellar clouds are not normally in thermal equilibrium and theoretical treatments of radiative association reactions under non-equilibrium conditions are in general far more complex.

Some systems show activation energy barriers against complex formation, even if one reactant is an ionic species. These barriers can only be recognized via quantum chemical studies or laboratory (typically ternary) association measurements. If barriers are present, association rates are very small and can be neglected under interstellar conditions. For example, Herbst, Adams, and Smith (47) recently found many hydrocarbon ion-H_2 associations previously thought to be important in interstellar syntheses of complex molecules, to possess significant barriers.

Finally, for some radiative association reactions that might be important in the interstellar medium (35,48) there are competing normal exothermic pathways; that is, the AB^{+*} complex can also dissociate to form products:

$$AB^{+*} \longrightarrow C^+ + D \quad . \qquad (7d)$$

Bates (39) has concluded that the existence of a normal exothermic channel will severely reduce the complex lifetime so that the radiative association rate will be minimal. However, there exists a body of experimental work in which ternary association reactions are surprisingly rapid despite competition from normal two-body channels (e.g. reference 34). I am currently working on a detailed theory of association reaction rates for systems where competitive exothermic channels are present. Present indications are that Bates (39) is correct for most systems but that there are effects which can lead to the opposite conclusion.

ION-MOLECULE SYNTHESES OF COMPLEX MOLECULES

During the past five years, a large number of investigators have considered pathways for the gas phase production of complex interstellar molecules (11,19,24,25,26,27,29,34,35,47,48-51). These pathways include radiative association reactions in two major roles - as a mechanism for hydrogenation and as a producer of complex species containing the CH_3^+ ion. Hydrogenation reactions of the type

$$A^+ + H_2 \longrightarrow AH^+ + H \qquad (8a)$$

are often endothermic or contain activation energy barriers even if exothermic. Therefore, for hydrogenation to proceed, reactions of the type

$$A^+ + H_2 \longrightarrow AH_2^+ + h\nu \qquad (8b)$$

are necessary. Unfortunately, even some association reactions of this type are known to possess activation energy (47). Three-body association studies have demonstrated that the CH_3^+ ion can associate with a variety of neutrals such as H_2O, HCN, and H_2CO to form precursor ions for many complex neutral species. The analogous radiative association reactions will be important in interstellar clouds.

To make some sense of the complex molecule systheses to be presented here, the discussion will be divided into three sections according to type of molecule - hydrocarbons, cyanoacetylenes, and others.

Hydrocarbons

The dominant view of hydrocarbon formation under dense cloud conditions is as follows. The species C^+ and C can undergo "fixation" to form hydrocarbon ions via

$$C^+ + H_2 \longrightarrow CH_2^+ + h\nu \qquad (9)*$$

and

$$C + H_3^+ \longrightarrow CH^+ + H_2 \qquad . \qquad (10)$$

Hydrogenation then proceeds to form one-carbon hydrocarbons; viz.,

$$CH^+ + H_2 \longrightarrow CH_2^+ + H \qquad , \qquad (11)*$$

$$CH_2^+ + H_2 \longrightarrow CH_3^+ + H \qquad , \qquad (12)*$$

$$CH_3^+ + H_2 \longrightarrow CH_5^+ + h\nu \qquad , \qquad (13)*$$

$$CH_5^+ + e \longrightarrow CH_4 + H; \; CH_3 + H_2 \qquad . \qquad (14)$$

Two-carbon hydrocarbons can now be formed via "insertion" reactions involving C^+ and/or C with one-carbon hydrocarbons:

$$C^+ + CH_4 \longrightarrow C_2H_3^+ + H; \; C_2H_2^+ + H_2 \quad , \qquad (15)*$$

$$C + CH_3^+ \longrightarrow C_2H_2^+ + H \qquad , \qquad (16)$$

$$C + CH_2 \longrightarrow C_2H + H \qquad (17$$

followed by

$$C_2H_2^+ + H_2 \longrightarrow C_2H_4^+ + h\nu \qquad , \qquad (18)$$

$$C_2H_2^+ + e \longrightarrow C_2H + H \qquad , \qquad (19)$$

$$C_2H_3^+ + e \longrightarrow C_2H_2 + H; \ C_2H + H_2 \qquad , \qquad (20)$$

$$C_2H_4^+ + e \longrightarrow C_2H_3 + H; \ C_2H_2 + H_2 \ . \qquad (21)$$

An alternative mechanism consists of "condensation" reactions between hydrocarbon ions and hydrocarbon neutrals. As an example, consider the reactions

$$CH_3^+ + CH_4 \longrightarrow C_2H_5^+ + H_2 \qquad , \qquad (22)*$$

$$C_2H_5^+ + e \longrightarrow C_2H_4 + H; \ C_2H_3 + H_2 \ . \qquad (23)$$

The above mechanisms can be used to produce still larger hydrocarbons. For example, the interstellar radical C_4H can be formed via the pathways

$$C^+ + C_2H_2 \longrightarrow C_3H^+ + H_2 \qquad , \qquad (24)*$$

$$C_3H^+ + H_2 \longrightarrow C_3H_3^+ + h\nu \qquad , \qquad (25)$$

$$C_3H_3^+ + C \longrightarrow C_4H_2^+ + H \qquad , \qquad (26)$$

or

$$C_2H_2^+ + C_2H_2 \longrightarrow C_4H_3^+ + H; \ C_4H_2^+ + H_2 \qquad (27)*$$

followed by

$$C_4H_2^+ + e \longrightarrow C_4H + H \qquad , \qquad (28)$$

$$C_4H_3^+ + e \longrightarrow C_4H + H_2 \ . \qquad (29)$$

Still larger species can be produced in an analogous manner but the limited experimental information becomes even more limited for hydrocarbons with more than four carbon atoms. In particular, since hydrogenation reactions of types (8a) and (8b) are often quite slow in the laboratory (47) and are very important in hydrocarbon synthesis, speculation concerning syntheses of larger hydrocarbons should await more experimental data on these and other (e.g. "condensation") reactions.

A unique view of hydrocarbon synthesis somewhat complementary to the above reaction pathways has been advanced by Freed, Oka, and Suzuki (50) and used in a model of carbon chain molecule formation in highly ionized regions by Suzuki (24). In this view, the principal synthetic reactions are radiative associations of the type

$$C^+ + C_n \longrightarrow C_{n+1}^+ + h\nu \qquad . \qquad (30)$$

At first glance the process would appear to be non-synthetic because the newly-produced C_{n+1}^+ ion will dissociate upon recombination via

$$C_{n+1}^+ + e \longrightarrow C_n + C, \quad \text{etc} \qquad . \qquad (31)$$

What is needed for synthesis is at least one hydrogenation step; e.g.,

$$C_{n+1}^+ + H_2 \longrightarrow C_{n+1}H^+ + H \qquad (32)$$

followed by

$$C_{n+1}H^+ + e \longrightarrow C_{n+1} + H \qquad . \qquad (33)$$

Finally, Mitchell (52) has considered the gas phase syntheses of interstellar hydrocarbons following the passage of a shock through an initially low density cloud. Using many of the reactions discussed above in addition to reactions (ion-molecule and neutral-neutral) with low to moderate activation energy "turned on" by the high temperature of the post-shock gas, he is able to write synthetic pathways for species as complex as C_6H_2. In shocks hydrogenation appears to no longer be a problem.

Cyanoacetylenes (Cyanopolyynes)

The initial synthesis of the simplest cyanoacetylenes HC_3N and C_3N to be proposed (11,35) was

$$C_2H_2^+ + HCN \longrightarrow H_2C_3N^+ + H \qquad , \qquad (34)*$$

$$H_2C_3N^+ + e \longrightarrow HC_3N + H;\ C_3N + H_2 \qquad . \qquad (35)$$

Next came the proposed synthesis (48)

$$HCNH^+ + C_2H_2 \longrightarrow H_4C_3N^+ + h\nu \qquad , \qquad (36)$$

$$H_4C_3N^+ + e \longrightarrow HC_3N + H_2 + H;\ C_3N + 2H_2 \ , \quad (37)$$

which was shown to be probably more efficient. More recently, I and other authors (24-27) have preferred syntheses based on abundant nitrogen atoms, such as

$$C_3H_3^+ + N \longrightarrow H_2C_3N^+ + H \qquad (38)$$

followed by (35). At present only one hydrocarbon ion - N atom reaction has been studied in the laboratory; many such studies are needed.

The higher or more complex cyanoacetylenes can be synthesized via analogous reactions to those invoked for HC_3N and C_3N. For example, HC_5N can be produced via the radiative association channel (48)

$$H_2C_3N^+ + C_2H_2 \longrightarrow H_4C_5N^+ + h\nu \qquad , \qquad (39)$$

$$H_4C_5N^+ + e \longrightarrow HC_5N + H_2 + H; \; C_5N + 2H_2. \quad (40)$$

Reaction (39) has been calculated to occur much more efficiently than reaction (36) by Leung, Herbst, and Huebner (29). Other mechanisms for HC_5N production involve N atoms (26), such as

$$C_3H^+ + C_2H_2 \longrightarrow C_5H_2^+ + H \qquad , \qquad (41)*$$

$$C_5H_2^+ + N \longrightarrow HC_5N^+ + H \qquad , \qquad (42)$$

$$HC_5N^+ + H_2 \longrightarrow H_2C_5N^+ + H \qquad , \qquad (43)$$

$$H_2C_5N^+ + e \longrightarrow HC_5N + H; \; C_5N + H_2 \qquad (44)$$

but these are highly conjectural. Lastly, a new scheme of D. Bohme and co-workers (53) builds cyanoacetylenes via C^+ reactions; viz.,

$$C^+ + HC_3N \longrightarrow C_4N^+ + H \qquad , \qquad (45)*$$

$$C_4N^+ + CH_4 \longrightarrow H_2C_5N^+ + H_2 \qquad , \qquad (46)*$$

followed by process (44). This route, which depends critically on the methane abundance, can presumably be extended to larger cyanoacetylenes.

For all of the suggested schemes shown here and their extension to larger cyanoacetylenes, far more laboratory work is required.

Other Molecules

A variety of complex molecules can be formed via radiative association reactions between CH_3^+ and assorted neutrals (12,34). The relative importance of CH_3^+ stems from its ability to associate with many molecules (observed in the laboratory mainly in ternary studies) and its estimated high dense interstellar cloud abundance. Consider the following reaction pathways (12,35) as examples:

$$CH_3^+ + CO \longrightarrow CH_3CO^+ + h\nu \quad , \quad (47)$$

$$CH_3CO^+ + e \longrightarrow CH_2CO(\text{ketene}) + H \quad , \quad (48)$$

$$CH_3^+ + H_2O \longrightarrow CH_3OH_2^+ + h\nu \quad , \quad (49)$$

$$CH_3OH_2^+ + e \longrightarrow CH_3OH(\text{methanol}) + H \quad , \quad (50)$$

$$CH_3^+ + HCN \longrightarrow CH_3CNH^+ + h\nu \quad , \quad (51)*$$

$$CH_3CNH^+ + e \longrightarrow CH_3CN(\text{cyanomethane}) + H, \quad (52)$$

$$CH_3^+ + CH_3OH \longrightarrow CH_3OCH_4^+ + h\nu \quad , \quad (53)$$

$$CH_3OCH_4^+ + e \longrightarrow CH_3OCH_3(\text{ether}) + H \quad . \quad (54)$$

Rates for all of these association reactions have been measured (41) or calculated (29,38).

Other mechanisms for complex molecule production can be viewed in the paper of Huntress and Mitchell (35). Their pathways involve radiative association to a considerable extent. Some examples are

$$HCO^+ + H_2O \longrightarrow HCOOH_2^+ + h\nu \quad , \quad (55)$$

$$HCOOH_2^+ + e \longrightarrow HCOOH(\text{formic acid}) + H \quad , \quad (56)$$

$$C_2H_5^+ + H_2O \longrightarrow C_2H_5OH_2^+ + h\nu \quad , \quad (57)$$

$$C_2H_5OH_2^+ + e \longrightarrow C_2H_5OH(\text{ethanol}) + H \quad . \quad (58)$$

One problem with synthetic schemes based on radiative association is that these reactions have rate coefficients that are inversely dependent on temperature. For mechanisms that are rate limited by radiative association steps, the synthesized molecules should be more abundant in cooler sources assuming all other parameters are held constant. The observational evidence does not strongly support such a correlation. After all, Orion, a warm source, is a source in which many complex molecules are abundant.

MODELS

The past several years have witnessed a spate of activity in the area of interstellar cloud modeling with the inclusion of complex molecules via the reaction networks discussed here. Prior to this effort, only Suzuki (19) had considered large molecules in some detail in the context of a detailed cloud model. I will focus on my own recent activities in this field and, where appropriate, compare our work with those of other investigators.

Rather than attempt a complete model of complex molecule formation, I undertook what was termed a "semi-detailed" calculation (25), in which the abundances of selected atomic and smaller molecular species were fixed at observed or estimated values and the abundances of larger molecules calculated in the steady-state limit. This approach was undertaken to decouple errors in the calculation of the abundances of atoms (e.g. C) and smaller species from those of larger species. The model contains 305 gas phase reactions and is primarily concerned with hydrocarbon and cyanoacetylene production and depletion. It was found that for the cloud TMC-1, the calculated abundances of hydrocarbons (e.g. C_3H_4, C_4H) and, to a lesser extent, cyanoacetylenes were seriously underestimated unless high abundances of C and/or C^+ were assumed. Since the low measured fractional ionization of the cloud (54) seems to preclude a high C^+ abundance in the cloud interior, the high C abundance was preferred, especially considering the detection of C in other sources (55). The C abundance required ($C/H_2 \sim 10^{-5}$) is not reproduced by standard ion-molecule models under steady-state conditions. A glance at process (10) indicates the reason for the high C requirement. Simply stated, C (or C^+) is needed to initiate the process of hydrocarbon neutral and ion formation. Hydrocarbon ions are needed in the syntheses of HC_3N and C_3N via processes such as (38) and of other species via (47)-(58).

The high C^+ possibility has been advocated by Suzuki (24) who has published a model in which carbon chain molecules are formed in regions of low optical depth closer to the dark cloud edge than the central regions normally modeled. The reactions used rely heavily on processes (30)-(33), discussed above. An objection to this approach has been made by Millar and Freeman (26) who think that the high D/H isotopic fractionation observed in selected carbon chain molecules in TMC-1 is inconsistent with Suzuki's chosen physical conditions. A response from Suzuki would be of interest.

Millar and Freeman (26,27) have themselves presented detailed steady-state models of the dark clouds TMC-1 and L183 in which the abundances of complex hydrocarbons and cyanoacetylenes are calculated. Included in their models are small but non-zero photodissociation rates normally excluded from dense cloud calculations. With an assumed A_v of 5 for the central visual extinction in TMC-1 (probably a lower limit to this quantity), Millar and Freeman (26) calculate a carbon (C) fractional abundance of 10^{-7}. This is intermediate between the normal dense cloud result of $\leq 10^{-8}$ (20) and the high abundance needed by Herbst (25). The model appears to be quite satisfactory for many species including C_3N and HC_3N, but (not surprisingly in my view) underestimates the abundances of hydrocarbons such as C_3H_4 (methyl acetylene) and C_4H dramatically. A similar model is available for L183 where changes in cloud geometry and gaseous elemental abundances are utilized to determine different results (27). Detailed reactions and rates are not listed in either article. In my view, this omission should be corrected, because it adds to the difficulty in comparing models. Ion-molecule controlled destruction rates of neutral species can vary widely depending upon whether the products of depletion reactions are assumed to recycle back to the neutral species. In the models that I have calculated alone or in collaboration, recycling rates have been minimized and depletion rates maximized. To compare our results with other models, one must know the depletion rates. Of course, knowledge of formation processes and rates is also useful.

A "chemical time dependence" model including a large number of complex molecules has recently been completed by Leung, Herbst, and Huebner (29). Containing 200 chemical species and over 1800 gas phase reactions, the model presents calculated abundances for species such as HC_3N, HC_5N, CH_3OH, CH_3NH_2, CH_3CHO, C_2H_5OH, CH_3CN, and CH_3OCH_3. Photodissociation is not included in the model. The authors found that calculated steady-state abundances of complex molecules, achieved at times exceeding 10^7 yrs, are normally significantly below observed values in the clouds TMC-1, Orion (ridge), and Sgr B2. However, at times before steady-state is reached (10^5 - 10^6 yrs), about a decade after the CI abundance peaks but while it is still sizable, calculated complex molecule abundances peak and are often as large or even larger than observed abundances. This effect had been previously observed for selected smaller species (21). Clearly, the complex molecule synthetic processes "feed" on CI, in agreement with my "semi-detailed" calculation.

The results of Leung, Herbst, and Huebner (29) can be interpreted as supportive of the "young cloud" hypothesis. However, the same qualifications noted earlier in the discussion of the CI problem are pertinent here. Briefly, inclusion of changes in physical conditions as a cloud evolves from diffuse to dense will serve to delay the C → CO conversion and thereby also preserve the high complex molecule abundances to later times. Inclusion of these variations is planned by the authors.

A shock model of complex hydrocarbon production has been propounded by Mitchell (52) who has included hydrocarbons up to six carbon atoms in size. Containing 108 species and 1252 reactions, this model starts with a pre-shock density relevant to diffuse cloud conditions of 10^2 cm^{-3}. The post-shock density eventually reaches 3×10^4 cm^{-3}. Thus, the cloud evolves from diffuse to dense under shock assistance rather than by the slower process of gravitational contraction. Large abundances of C^+ and C are present at times up to 10^5 yrs after the shock passage. As can be expected, large abundances of hydrocarbons are produced during these times. The key to complex molecule production in shocks appears to be choosing initial conditions carefully. If a shock passes through a dense cloud where much of the carbon is in the form of CO (56), complex hydrocarbons are not formed in high abundance. Shock chemistry seems to mirror normal chemistry in requiring significant C^+ and/or C abundances to power complex molecule syntheses.

FUTURE WORK

Recent activity has proved that gas phase ion-molecule processes can produce large amounts of complex molecules in dense interstellar clouds. However, the models used in calculating molecular abundances are deficient in several respects. First, the need for more laboratory input remains an acute one. Measurements of many normal ion-molecule reactions, preferably at low temperatures, are still needed, sometimes desperately. Photodissociation cross sections are also needed, especially as more realistic models begin to include photodestruction processes. More work - both experimental and theoretical - is needed to determine the rates of radiative association processes, including systems in competition with normal exothermic channels. Secondly, gas phase models must include both complex molecules and variation in physical conditions. It is conceivable that inclusion of photodissociation processes at earlier stages of cloud history when A_v is small will actually aid the process of creating complexity out of simplicity, much as cosmic ray bombardment appears to do. Thirdly, the questions of molecule adsorption onto and reejection from grains must be faced and not ducked. If clouds are older than grain adsorption lifetimes, why does a gas phase exist? Is the ejection of molecules from grains related to grain chemistry?

In conclusion, it is clear that we should not be complacent about past successes, significant as they may be. There remains much work to be accomplished.

ACKNOWLEDGMENTS

I would like to acknowledge the support of my theoretical research program by the National Science Foundation (U.S.) via Grant AST-8312270.

REFERENCES

(1) Lafont, S., Lucas, R., and Omont, A. 1982, Astron. Astrophys. 106, pp. 201-213.
(2) Bell, M. B., Feldman, P. A., and Matthews, H. E. 1983, Astrophys. J. (Letters) 273, pp. L35-L39.

(3) Allen, M. and Robinson, G. W. 1977, Astrophys. J. 212, pp. 396-415.
(4) Tielens. A. G. G. M. and Hagen, W. 1982, Astron. Astrophys. 114, pp. 245-260.
(5) Watson, W. D. 1983, U. of Illinois preprint ILL-(AST)-83-26.
(6) Greenberg, J. M. 1983, "Cosmochemistry and the Origin of Life" (ed. C. Ponnamperuma; Reidel: Dordrecht) pp. 71-112.
(7) Moore, M. H. and Donn, B. 1982, Astrophys. J. (Letters) 257, pp. L47-L50.
(8) Herbst, E. and Klemperer, W. 1973, Astrophys. J. 185, pp. 505-533.
(9) Hartquist, T. W., Oppenheimer, M., and Dalgarno, A. 1980, Astrophys. J. 236, pp. 182-188.
(10) Huntress, W. T., Jr. 1977, Astrophys. J. Supp. Ser. 33, pp. 495-514.
(11) Schiff, H. I. and Bohme, D. K. 1979, Astrophys. J. 232, pp. 740-746.
(12) Smith, D. and Adams, N. G. 1981, Int. Rev. Phys. Chem. 1, pp. 271-308.
(13) Albritton, D. L. 1978, Atomic Data and Nuclear Tables 22, pp. 1-101.
(14) Solomon, P. M. and Klemperer, W. 1972, Astrophys. J. 178, pp. 389-421.
(15) Glassgold, A. E. and Langer, W. D. 1976, Astrophys. J. 206, pp. 85-99.
(16) Suzuki, H., Miki, S., Sato, K., Kiguchi, M., and Nakagawa, Y. 1976, Prog. Theor. Phys. 56, pp. 1111-1125.
(17) Iglesias, E. 1977, Astrophys. J. 218, pp. 697-715.
(18) Mitchell, G. F., Ginsberg, J. L., and Kuntz, P. J. 1978, Astrophys. J. Supp. Ser. 38, pp. 39-68.
(19) Suzuki, H. 1979, Prog. Theor. Phys. $\underline{62}$, pp. 936-956.
(20) Prasad, S. S. and Huntress, W. T., Jr. 1980, Astrophys. J. Supp. Ser. 43, pp. 1-35.
(21) Prasad, S. S. and Huntress, W. T., Jr. 1980, Astrophys. J. 239, pp. 151-165.
(22) Henning, K. 1981, Astron. Astrophys. Supp. 44, pp. 405-435.
(23) Graedel, T. E., Langer, W. D., and Frerking, M. A. 1982, Astrophys. J. Supp. Ser. 48, pp. 321-368.
(24) Suzuki, H. 1983, Astrophys. J. 272, pp. 579-590.
(25) Herbst, E. 1983, Astrophys. J. Supp. Ser. 53, pp. 41-53.
(26) Millar, T. J. and Freeman, A. 1984, M. N. R. A. S. 207, pp. 405-424.
(27) Millar, T. J. and Freeman, A. 1984, M. N. R. A. S. 207, pp. 425-432.
(28) Langer, W. D., Graedel, T. E., Frerking, M. A., and Armentrout, P. B. 1984, Astrophys. J. 277, pp. 581-604.
(29) Leung, C. M., Herbst, E., and Huebner, W. F. 1984, accepted by Astrophys. J.
(30) Herbst, E. 1978, Astrophys. J. 222, pp. 508-516.

(31) Tarafdar, S. P., Prasad, S. S., Huntress, W. T., Villere, K. R., and Black, D. C., pre-print.
(32) Boland, W. and De Jong, T. 1982, Astrophys. J. 261, pp. 110-114.
(33) Prasad, S. S. and Tarafdar, S. P. 1983, Astrophys. J. 267, pp. 603-609.
(34) Smith, D. and Adams, N. G. 1978, Astrophys. J. (Letters) 220, pp. L87-L92.
(35) Huntress, W. T., Jr. and Mitchell, G. F. 1979, Astrophys. J. 231, pp. 456-467.
(36) Herbst, E. 1980, Astrophys. J. 237, pp. 462-470.
(37) Bates, D. R. 1979, J. Phys. B. 12, pp. 4135-4146.
(38) Bates, D. R. 1983, Astrophys. J. 270, pp. 564-577.
(39) Bates, D. R. 1983, Astrophys. J. (Letters) 267, pp. L121-L124.
(40) Bass, L. M., Chesnavich, W. J., and Bowers, M. T. 1979, J. Am. Chem. Soc. 101, pp. 5493-5502.
(41) Bass, L. M., Kemper, P. R., Anicich, V. G., and Bowers, M. T. 1981, J. Am. Chem. Soc. 103, pp. 5283-5292.
(42) Herbst, E. 1981, J. Chem. Phys. 75, pp. 4413-4416.
(43) Herbst, E. 1982, Chem. Phys. 65, pp. 185-195.
(44) Adams, N. G. and Smith, D. 1981, Chem. Phys. Letters 79, pp. 563-567.
(45) Luine, J. A. and Dunn, G. H. 1984, submitted to Phys. Rev. Letters.
(46) Barlow, S. E., Dunn, G. H., and Schauer, M. 1984, Phys. Rev. Letters 52, pp. 902-904.
(47) Herbst, E., Adams, N. G., and Smith, D. 1983, Astrophys. J. 269, pp. 329-333.
(48) Mitchell, G. F., Huntress, W. T., Jr., and Prasad, S. S. 1979, Astrophys. J. 233, pp. 102-108.
(49) Mitchell, G. F. and Huntress, W. T., Jr. 1979, Nature 278, pp. 722-723.
(50) Freed, K. F., Oka, T., and Suzuki, H. 1982, Astrophys. J. 263, pp. 718-722.
(51) Winnewisser, G. and Walmsley, C. M. 1979, Astrophys. Space Sci. 65, pp. 83-93.
(52) Mitchell, G. F. 1983, M. N. R. A. S. 205, pp. 765-772.
(53) Bohme, D. K. 1984, private communication.
(54) Guélin, M., Langer, W. D., and Wilson, R. W. 1982, Astron. Astrophys. 107, pp. 107-127.
(55) Phillips, T. G. and Huggins, P. J. 1982, Astrophys. J. 251, pp. 531-540.
(56) Mitchell, G. F. 1984, Astrophys. J. Supp. Ser. 54, pp. 81-101.

ADDITIONAL NOTE

After completing this manuscript, I have learned that the published articles of Millar and Freeman (26,27) contain a list of their reaction network on microfiche. This list was not contained in the preprint sent to me prior to publication. I regret my statement in the body of the text that "detailed reactions and rates are not listed."

CHEMICAL REACTIONS IN DIFFUSE CLOUDS AND THE CHEMISTRY OF
ISOTOPE FRACTIONATION IN THE INTERSTELLAR GAS

R. M. Crutcher[1] and W. D. Watson[1,2]

Departments of Astronomy[1] and Physics[2]
University of Illinois at Urbana-Champaign
Urbana, IL 61801

ABSTRACT

   The basic processes and reaction networks that cause the formation and destruction of small molecules containing hydrogen, carbon, oxygen and nitrogen in the diffuse interstellar clouds of our galaxy are summarized. Selected, recent advances in the molecular physics/chemistry, in the modelling and interpretation, and in key observational aspects are mentioned. The fractionation of the isotopes $^{13}C/^{12}C$ and D/H in molecules in both the diffuse and dense cloud environments is discussed.

1. INTRODUCTION

   The key processes and basic reaction networks involved in the chemistry of diffuse clouds and in the fractionation of isotopes were largely recognized and delineated some time ago. Most of the recent progress in this area has been either in the laboratory measurement and calculation of the microscopic physical processes, or in the observational measurement of molecular abundances and other relevant properties of the interstellar gas. We will therefore take as our task the reviewing of the key processes and basic reaction networks so that the significance of the new data may better be appreciated -- an endeavor which will presumably be of most value to the laboratory oriented investigators. We will also integrate selected observational and laboratory developments into the discussion.

Since the earliest observations of interstellar molecules (CH, CH^+ and CN in about 1940), the diffuse interstellar clouds have served as the environment in which the most quantitative tests can be made for the reactions which have been proposed to understand the abundances of interstellar molecules. Most chemical elements are primarily in atomic (or ionic) form and the observed molecular species are diatomics. Thus, relatively few reactions tend to be essential in determining the abundance of a molecular species. Compounding of the uncertainties is thus minimized. Because the characteristic time scales for the chemical processes are short (e.g., the time for photodissociation of molecules is typically 100-1000 years), steady-state abundances for <u>in situ</u> reactions are expected to prevail. That is, we do not need to be concerned, for example, that the molecular abundances in diffuse clouds are contaminated by molecules that are ejected in the winds of cool stars as we must in considering dense clouds. Although the strong, ultraviolet transitions of the abundant species (H, H_2, C, N and O) are saturated, abundances can nevertheless be determined to satisfactory accuracy. Uncertainties about the excitation normally are not serious because the observations mostly utilize optical and ultraviolet, electronic transitions in absorption from the ground state.

We consider the following as representative parameters for diffuse interstellar clouds: number density of all particles $\sim$ few hundred cm^{-3}; kinetic temperature $\sim$ 25-100K; total continuous extinction at visual wavelengths $A_v \lesssim 1$ due to dust grains; abundances by number of carbon, nitrogen and oxygen are 10^{-3}-10^{-4} that of hydrogen. Except for He, other elements are significantly less abundant and we will not consider them here. Except for hydrogen, for which the molecular form (H_2) may be comparable in abundance to the atomic form, the chemical elements are in atomic form with the most abundant molecule (CO) being only about 0.01 of the total carbon. The dust grains which cause the extinction of starlight also contribute a non-negligible surface area to which the gas particles can, in principle, stick and be altered by surface reactions. Their temperatures are thought to be about 10-25K depending upon size and composition. The extinction caused by the dust grains varies with wavelength; in the ultraviolet where most of the photodissociation occurs, it is typically about three times that at optical wavelengths. Thus the photodissociation rate at the centers of diffuse clouds is reduced by no more than a factor of about ten from that at the edges of the clouds as a result of the absorption by dust. At particular frequencies, absorption by spectral lines (esp. of H_2) can be much greater. The radiative flux incident upon diffuse clouds can usually be approximated satisfactorily by the galactic background flux.

The environment described above does not contain observable abundances of the large, complex molecules which are of great interest for their involved chemistry, nor are these clouds near the onset of star formation and thus of prime interest for astronomical reasons. It does, however, provide the best environment for testing the basic astrochemistry. The understanding of the chemistry of such regions would seem to be a necessary precondition for having confidence in the more complicated chemistry of the dense, molecular clouds in which the complex molecules are found and which are more closely associated with star formation. Knowledge about molecular processes gained from the study of diffuse clouds has, to varying degrees, provided a basis for subsequent examinations of non – (thermodynamic) equilibrium molecular abundances in other diverse astrophysical contexts, e.g;, in interstellar shock waves, in clouds in the early universe, in the outflowing gas from cool stars. A number of more extensive, but somewhat less up-to-date reviews of this topic are available (e.g., 1-4). Abundances of various molecular species in diffuse clouds have been summarized by Snow (5).

2. BASIC PHYSICAL PROCESSES

Historically the two key processes that have been considered for converting atoms into molecules in the diffuse cloud environment have been surface reactions on dust grains,

$$A + B + grain \rightarrow AB + grain, \tag{1}$$

where A and B are two atomic species, including ions, and radiative association

$$A + B \rightarrow AB + h\nu \tag{2}$$

Coincidentally, the maximum rates for (1) and (2) tend to be comparable. The surface area of dust grains is such that the rate at which a specific atom hits a dust grain is $\alpha_g n \simeq 3 \times 10^{-17}$ n s^{-1} where n(cm^{-3}) is the number density of hydrogen. The most important of the radiative associations is carbon ions with hydrogen. For atomic hydrogen the rate coefficient is also calculated to be about 3×10^{-17} cm^3 s^{-1}; the rate coefficient for molecular hydrogen is considered to be much more uncertain, but seems to be no more than approximately an order-of-magnitude larger than that for atomic hydrogen.

Once molecules are formed, the <u>rate coefficients</u> with which they react to form other molecules,

$$C + AB \rightarrow CA + B \tag{3}$$

can be much greater than for (1) and (2) above. However, the actual rate of creating new molecules tends not to be significantly greater because either (a) the abundance of the molecule AB is low even though the relevant rate coefficient may have the "orbiting" value ($\simeq 2 \times 10^{-9}$ cm^3 s^{-1} for an ion-molecule reaction or $\simeq 4 \times 10^{-11}$ cm^3 s^{-1} for a neutral-neutral reaction), or (b) the atom C cannot react with AB because of an energy barrier. This is the case for reactions of H_2 with the abundant forms of the various chemical elements. In (a), the reactive molecules tend to be no more abundant than about 10^{-8} of hydrogen so that the actual reaction rate for atom C is about the same as for (1) and (2). Although the common forms of the elements do not participate in reactions of type (3), ionized oxygen and nitrogen (but not carbon) do react with molecular hydrogen in this manner. Both have ionization energies greater than that of atomic hydrogen and thus must be ionized (directly or indirectly) by cosmic ray particles and X-rays. The flux of these, and hence the ionization of oxygen and nitrogen, is relatively low. The low abundance of reactive ions thus compensates for the large relative abundance of H_2 and the large rate coefficient ($\simeq 2 \times 10^{-9}$ cm^3 s^{-1}) characteristic of such reactions. In diffuse clouds the direct ionization rate of hydrogen and of species with ionization potentials exceeding that of hydrogen seems to be near the 10^{-17} ionizations per particle per second which is appropriate for the "high energy" (greater than 100 MeV per nucleon) cosmic ray particles. Even if every such ionized atom immediately formed a molecule, the overall rate would still be slower than the maximum rates for (1) and (2) by a factor which is approximately equal to the number density of hydrogen. For elements other than hydrogen, the effective rate of ionization (per particle) may be increased by the transfer of charge from hydrogen. An enhancement factor of approximately [(number density of hydrogen)/(number density of carbon, nitrogen and oxygen)] $\simeq 10^3-10^4$ could ideally be achieved. This enhancement would cause the effective rates for such chemical reactions to be comparable with the maximum rates for (1) and (2).

The primary mechanism for the destruction of molecules is photodissociation. For dissociating transitions at wavelengths in the range 1000-1500 A and with a total oscillator strength of 0.1 (both of which are representative), the photodissociation rate per molecule due to the galactic background radiation is about 10^{-10} s^{-1}. This destruction rate together with the maximum rates for (1) or (2) provide an estimate for the abundance of molecules in diffuse clouds. It is in terms of the abundance ratio,

$$\frac{\text{[molecules containing a particular chemical element]}}{\text{[atoms of the chemical element]}} \simeq 3 \times 10^{-7} n \simeq 10^{-4}\text{-}10^{-5} \qquad (4)$$

Equation (4) is a quite poor estimate for molecular hydrogen as we discuss below.

A few general comments about chemical reactions in diffuse interstellar gas clouds may now be helpful in providing perspective. The low collision rates for particles, together with the relatively rapid destruction rates for molecules, normally makes it necessary that reaction occur at essentially every collision in order for the particular reaction to be an appreciable contributor. At 100K, the kinetic energy in a collision is about 0.01 eV on the average. Due to the long times between collisions, internal excitation (electronic and vibrational) has never been found to alter reaction rates significantly. Thus, to be important a reaction must normally require no energy input for reactants in their ground states and there can be essentially no activation energy barrier even for reactions that are exothermic. Only in the past few years have there been any laboratory measurements at temperatures below 100K, and some 5-10 years ago when the key reaction schemes were delineated, there were few measurements below 300K. Because the rate coefficients for energetically allowed, (positive) ion-molecule reactions have tended to be approximately constant with temperature, normally are unaffected by activation energy barriers and commonly have values near the orbiting rates ($\simeq 2 \times 10^{-9}$ cm^3 s^{-1} when they involve hydrogen), such reactions have been especially relevant in studies of interstellar chemistry. Exothermic reactions between neutrals -- atoms with reactive radicals or radicals with each other -- may also be rapid at low temperatures. In the absence of experimental data, it has seemed (at least to us) that assuming these to be rapid is much more hazardous. Dissociative electron recombinations with the positive molecular ions are another class of reactions that plays an important role in interstellar chemistry. These normally are fast (rate coefficient $\simeq 10^{-6}$ to 10^{-7} cm^3 s^{-1}) and usually limit the abundance of ions which do not react with H or H_2. Most measurements are recent and their application to astrophysical conditions is in some cases quite controversial because of the possible difference in internal excitation. For neutral molecules, photodissociation tends to be a competitive destruction mechanism. Good information on photodissociation cross sections is sparse.

Since the earliest investigations of the formation of interstellar molecules, investigators have been divided about the relative importance of gas phase versus surface reactions. As

noted in the foregoing, there is little doubt that reactive
atoms hit (they also stick to) the surfaces of dust grains at a
rate that is competitive with gas phase reactions. The problem
is in understanding how they are returned to the gas since,
except for H_2 and possibly a few others such as CO and N_2, the
(thermal) vapor pressure should be insufficient due to the low
temperature of the grains. The presence of molecular ions (CH^+,
HCO^+, N_2H^+) indicates that gas phase processes must be at least
somewhat effective. The strong enhancement of deuterium in
molecules (including HD, HCO^+ and N_2H^+) is difficult for us to
imagine as a result of surface reactions, whereas it is nicely
understood through ion-molecule reactions. On the other hand,
H_2 is the most abundant molecule in astronomy and it apparently
must be formed on the surfaces of dust grains under normal
conditions.

3. FORMATION OF MOLECULAR HYDROGEN AND SURFACE REACTIONS

The obvious gas phase processes to examine with regard to
the formation of H_2 are

$$H + H \rightarrow H_2 + h\nu, \tag{5}$$

$$H + H^+ \rightarrow H_2^+ + h\nu, \tag{6}$$

and

$$H + H^- \rightarrow H_2 + e \tag{7}$$

In the case of (5), the two molecular, electronic states (one of
which is the ground state of H_2) that adiabatically connect to
the ground atomic states have different spins. A radiative
transition between them during a collision is then exceedingly
improbable and the rate coefficient for (5) must be many orders
of magnitude less than 10^{-17} cm^3 s^{-1}. Purely vibrational,
radiative transitions are also too slow. Reaction (6) is not
inhibited by spin, but the combined effect of the low abundance
of H^+ and the fact that the upper of the two electronic states
becomes repulsive at large distances causes the overall rate for
(6) to be negligible. Reaction (7) also seems to be significantly less important than formation on grains because of the
low abundances of H^-. The low fractional ionization, together
with the large destruction rates due to photoionization and to
reaction with ions, keep the abundance of the H^- ion small.

The formation of molecular hydrogen has been considered
in detail, especially by Salpeter and collaborators (6,7).
Although the nature of the surface (that is the outer few atomic
layers which determine its catalytic properites) of dust grains

is uncertain, it has been found that even the minimum binding to
the surface is adequate for H-atoms to be retained on the grain
long enough to react with another H-atom. The minimum binding
is considered to be that due to pure physical adsorption plus a
few lattice sites of enhanced binding caused by impurities,
cosmic ray damage, etc. Return of the product H_2 molecules is
no problem because the vapor pressure of H_2 is sufficient to
eject the molecules back into the gas. Enough of the energy of
the reaction itself may be converted to translational energy to
eject the H_2 from the surface. If so it will probably be "hot"
in its vibrational and rotational modes as well (e.g., 8).
These molecules may be the primary contribution to the abundance
of highly, rotationally excited H_2 molecules detected in the
spectra of some diffuse clouds. A semiempirical formation rate
for H_2 can be derived with reasonable confidence by comparing
models with the observations of diffuse clouds. The result is
consistent with the prediction that essentially every H-atom
that hits a dust grain will be converted into an H_2-molecule
(that is, the rate coefficient is near $10^{-17} cm^3 s^{-1}$) [e.g.,
9,10].

We noted in the foregoing that this rate coefficient
together with the photodissociation rate of H_2 in the galactic
background radiation does not yield the observed abundances of
H_2 in diffuse clouds. A further key consideration is the self-
shielding of H_2. Photodissociation of H_2 is a result of absorption in the Lyman and Werner bands near 1000A followed by
radiative decay to continuum vibrational states of the ground
electronic state (11). The absorption lines are, therefore,
intrinsically narrow which together with the large relative
abundance of hydrogen causes the photodissociation rate within
diffuse clouds to be smaller than that at the edges, frequently
by several orders-of-magnitude so that $H_2/H \simeq 1$ when the number
density is only a few hundred per cm^3.

Given the success of surface reactions in converting atomic
to molecular hydrogen, why are they not understood to be more
effective for the formation of other molecules? Unlike H_2,
essentially all other molecules (except perhaps N_2, CO, and O_2)
have low enough vapor pressure that they should freeze out at
the low temperatures of interstellar grains. Atomic C, N, O and
molecules containing these species should readily stick and
remain on the surfaces of the grains. Because of the relative
abundance of hydrogen, most incident C,N,O atoms are expected to
be converted into CH_4, NH_3, and H_2O. Non-thermal mechanisms
have been sought for the ejection of molecules other than H_2
(e.g., 12). They include (i) photodesorption; (ii) evaporation
due to the thermal pulses caused by energy from a cosmic ray or
in the case of small grains, by energy from a single photon or
chemical reaction; (iii) explosion of the grain resulting from

an accumulation of stored energy (13); and (iv) ejection of the
molecule as a result of converting some of its recombination
energy directly into the translational energy of a free
trajectory. The examinations of the thermal pulses utilize the
idea that the grain temperatures are commonly below the Debye
temperatures of the solid so that the specific heat is reduced.

In our opinion, it has not been reliably demonstrated that
any of these processes are sufficiently effective under general
conditions. Even if they were to be effective under the conditions of diffuse clouds, the reduction in starlight and in
reactive atoms together with the increase in the flux of particles hitting grains due to the increased density would seem
to render them insufficient in the dense, molecular clouds.
As far as diffuse clouds go, however, not obviously unreasonable
assumptions about the efficiencies of any one of processes (i),
(ii), or (iv) [and perhaps (iii) as well] would provide an
ejection mechanism that is adequate to prevent substantial
accumulation of the heavy elements (C,N,O, etc.) onto the grain
surfaces. Observationally, the depletion of carbon, nitrogen
and oxygen from the gas in diffuse clouds seems to be small;
that is, no more than about a factor of three. This indicates
that the finite lifetime of the clouds ($\gtrsim 10^7$ yrs), the mixing
of the grains from the inside to the outside of the cloud where
they might be attacked by hot intercloud gas, or one of the
processes (i) – (iv) [or perhaps some, as yet unrecognized,
phenomenon] must be important.

4. FORMATION OF THE HD MOLECULE

Just as the H_2 molecule provides the most clearcut evidence
that surface reactions are important in interstellar chemistry,
the isotopic form HD gives perhaps the most dramatic example of
the impact of gas phase reactions. The observed ratio $[HD]/[H_2]$,
which is about 10^{-6} in representative diffuse clouds, is not on
the face of it surprising since $[D]/[H]$ is thought to be $\simeq 10^{-5}$.
In such clouds, however, the self-shielding of H_2 is substantial
($\simeq 10^4$) whereas there is little if any self-shielding of HD
(also, H_2 does not appreciably shield HD). It then follows
(assuming steady-state) that the rate per D-atom at which
D-atoms are converted into HD can be $\simeq 10^3$ greater than the rate
per H-atom at which H-atoms are converted into H_2 (14). However, an H-atom is converted into H_2 at essentially every
collision with a dust grain. A D-atom can clearly do no better
in producing HD through surface reactions. It then follows that
HD must be produced by gas phase reactions and at a rate which
can be 10^3 or so greater than the rate at which atoms hit
grains.

The reaction

$$D^+ + H_2 \rightleftarrows HD + H^+ \qquad (8)$$

has been found (15,16,17) to do the job because of the "enhancement" of the ionization rate of atomic-D, over that due to direct cosmic ray or X-ray ionization, by the charge-transfer (17)

$$H^+ + D \rightleftarrows H + D^+ \qquad (9)$$

Both (8) and (9) occur at essentially the orbiting rate, though (9) is influenced by threshold effects at the lower temperatures (18).

In addition to being an indicator for the importance of gas-phase reactions, the HD molecule also provides us with the ability to deduce the abundance of ionized hydrogen - a key parameter in assessing the ionizing flux of cosmic rays and X-rays in the galaxy - in a manner that is relatively insensitive to serious uncertainties. To deduce the actual ionizing flux, the neutralizing reactions for H^+ must be brought into the analysis. In diffuse clouds, they are mainly the radiative electron recombination and charge-transfer with atomic-O (equation 25) followed by the reaction of O^+ with H_2 (equations 26-29). Upper limits to the ionizing flux can be derived with reasonable confidence (19,10).

HYDRIDES OF "HEAVY" ELEMENTS

Due to the relative abundance of hydrogen, we expect that the first step in converting the less abundant elements to molecules will be the formation of hydrides.

5. CARBON AND THE CH^+ PROBLEM

Because the ionization energy of atomic carbon is below the cutoff in the galactic radiation caused by the ionization of hydrogen, carbon atoms exist mainly as C^+ in diffuse clouds. Their ionization probably provides most of the free electrons in diffuse clouds, as well.

The exchange reaction

$$C^+ + H_2 \rightarrow CH^+ + H \qquad (10)$$

requires an energy input of 0.4eV for reactants in their ground states and is therefore completely negligible at the standard temperatures of diffuse clouds. Studies have therefore focused on the radiative association of C^+ with hydrogen. Originally the emphasis was on (20)

$$C^+ + H \rightarrow CH^+ + h\nu \tag{11}$$

though more recently attention has centered on (21)

$$C^+ + H_2 \rightarrow CH_2^+ + h\nu \tag{12}$$

It has seemed likely that the rate coefficient for (12) will exceed the maximum rate of 8×10^{-17} cm^3 s^{-1} (22) for (11) by a factor of ten or more. Recent laboratory measurements yield an upper limit for (12) of 2×10^{-15} cm^3 s^{-1} at 10K (23) whereas a semi-empirical calculation gives 10^{-15}–10^{-16} cm^3 s^{-1} for the range 10-100K (24). Although (12) produces neither CH nor CH^+ directly, and (11) does not produce CH, subsequent reactions and photoprocesses do cycle the carbon back into CH and CH^+. These are

$$CH^+ + H_2 \rightarrow CH_2^+ + H, \tag{13}$$

$$CH_2^+ + H_2 \rightarrow CH_3^+ + H, \tag{14}$$

$$CH_3^+ + e \rightarrow CH_2 + H, \; CH + H_2, \tag{15}$$

and the photoprocesses

$$CH + h\nu \rightarrow CH^+ + e, \; C + H, \tag{16}$$

$$CH_2 + h\nu \rightarrow CH + H, \; CH_2^+ + e, \tag{17}$$

$$CH_2^+ + h\nu \rightarrow CH^+ + H, \tag{18}$$

and

$$CH_3^+ + h\nu \rightarrow CH_2^+ + H, \; CH^+ + H_2 \tag{19}$$

The key destruction processes for CH^+ are (13) and the following

$$CH^+ + e \rightarrow C + H, \tag{20}$$

$$CH^+ + h\nu \rightarrow C^+ + H \tag{21}$$

and

$$CH^+ + H \rightarrow C^+ + H_2 \tag{22}$$

of which (21) can be neglected under normal conditions. The above carbon hydrides also react with C^+, N and O atoms though these reactions ordinarily should not be the dominant destruction processes in diffuse clouds. In addition to the radiative associations, attention has also been focused on certain other of the above key reactions. Reaction (20) has been the center of considerable controversy. In a gas of pure atomic-H, the fractional abundance of CH^+ obtained in steady-state by considering only reactions (11) and (20) is

$$[CH^+]/[H] \lesssim \frac{\alpha(11)}{\alpha(20)} \simeq \text{few} \times 10^{-10} \qquad (23)$$

if a typical value is adopted for the rate coefficient $\alpha(20)$ for dissociative recombination ($\simeq$ few $\times 10^{-7}$ cm^3 s^{-1}). Observed values for $[CH^+]/[H]$ range up to perhaps 2×10^{-8}. This dilemma and arguments that the rate for reaction (22) is negligible led investigators to conjecture that $\alpha(20)$ is unusually small ($\simeq 10^{-9}$ cm^3 s^{-1}). A single laboratory experiment has obtained $\alpha(20) \simeq 3 \times 10^{-7}$ cm^3 s^{-1} (25). Because of the difficulty of such measurements and the lack of a confirming study, the issue has not been considered to be completely closed. In the presence of molecular hydrogen, the foregoing reactions indicate that the chemistry is more complicated. The predicted abundance cannot be related so directly to the physical processes as in equation (23). Rates for most of the additional processes are at best only poorly known. Nevertheless, plausible assumptions have been made and detailed calculations for the resulting abundances have been performed. The conclusion is that the presence of H_2 does not alleviate the CH^+ problem (21). In perhaps the best studied case (Zeta Ophiuchi), the calculated abundance of CH^+ is still deficient by a factor of 40 even when favorable assumptions are made about these processes involving H_2. This has been the status of the CH^+ problem under the conditions of diffuse clouds.

Just now, however, a reassessment of earlier data (26) and a laboratory measurement at 300K (27) have been presented for reaction (22). These recent studies find that, in contrast to the prevailing view, there is no reason to expect this reaction to be unusually slow and, in fact, the laboratory measurement shows that almost surely it is quite fast. At 300K the rate coefficient is 7.5×10^{-10} cm^3 s^{-1} and it is increasing with decreasing temperature. If the limiting case of a purely atomic gas is again considered together with only reactions (11) and (22),

$$\frac{[CH^+]}{[H]} \simeq 4 \times 10^{-11} \qquad (24)$$

is the result analogous to equation (23).

The possibility that CH^+ can be understood as occurring under the standard conditions of diffuse interstellar clouds now seems remote. In fact, there is observational evidence that CH^+ is different from other species. There is a tendency for it to occur at different Doppler shifts ($\simeq$ 1-2 km/s) than the other molecules (28,29). In contrast to other molecules, the amount of CH^+ in a given direction is uncorrelated with the total gas (29,31) though see also (30). Its line profile is smooth and symmetric as if it were produced by a hot gas (32). The corresponding temperature ranges up to 4500 K. There appears to be a correlation between this temperature and the amount of CH^+ observed. Detailed calculations (33,34) have shown that adequate CH^+ can reasonably be produced by low velocity shocks (shock velocities < 12 km/s) that are likely to be propagating through diffuse clouds. There is little if any independent evidence for or against the presence of such shocks, however. Shocks are the purview of reviews at this Workshop (A. Dalgarno; B. Draine) and we shall leave further comments to them.

6. HYDRIDES OF HEAVY ELEMENTS - OXYGEN

Hydrides of oxygen are produced primarily in a manner that is quite similar to the formation of HD. Like deuterium, oxygen has an ionization energy that is almost exactly equal to that of hydrogen. It is then expected that the charge transfer

$$H^+ + O \rightleftharpoons H + O^+ \tag{25}$$

may have a rate coefficient near the orbiting value for the exothermic direction and that the abundance of O^+ will be substantially increased by reaction (25). The presence of the fine-structure states introduces complications. Nevertheless, laboratory measurements (35,27) and calculations (36) all indicate a rapid charge-transfer in (25) with a rate coefficient of about 5×10^{-10} cm^3 s^{-1} for the exothermic (reverse) direction.

As in the case of the formation of HD, the O^+ then reacts with H_2 ultimately to produce OH and H_2O (19,37).

$$O^+ + H_2 \rightarrow OH^+ + H \tag{26}$$

In this case further reactions with H_2 are possible and occur rapidly,

$$OH^+ + H_2 \rightarrow H_2O^+ + H, \tag{27}$$

$$H_2O^+ + H_2 \rightarrow H_3O^+ + H \tag{28}$$

followed by

$$H_3O^+ + e \rightarrow H_2O + H \text{ or } OH + H_2 \tag{29}$$

Photodissociation should normally be the primary destruction mode for these species in diffuse clouds (it also converts H_2O into OH), though reaction with C^+, N and O which will be discussed subsequently are also important, potential destruction modes. There have been recent studies for the photodissociation rate of OH (38).

For diffuse clouds with higher densities that are not too cold (T $\gtrsim$ 50K) and which have $H_2/H \simeq 1$, reaction (25) can reasonably be expected to approximate the neutralization rate for H^+. An OH molecule can be assumed to be produced for essentially each forward reaction (25). If the photodissociation rate for OH in diffuse clouds is designated as $\Gamma(OH)$ s^{-1} and the ionization rate per H-atom by cosmic rays (or X-rays) is ξ, we then have

$$[OH]/[H] \simeq \xi/\Gamma(OH) \qquad (30)$$

Although there must, of course, be corrections to (30) for various processes which are ignored, equation (30) does indicate how the H-ionizing flux can be deduced in a manner that is relatively insensitive to the serious uncertainties of the physical conditions, esp. density. Similar information can be found from the abundance of HD as mentioned previously. Application of the OH abundance to determining ξ has been emphasized by Dalgarno and co-workers (19).

As a result of a laboratory measurement for the relevant oscillator strength (39), limits on the abundance of H_2O are considered to be compatible with the foregoing scheme for generating OH.

7. NITROGEN AND SURFACE REACTIONS

Because its ionization energy is 14.5 eV, atomic nitrogen in diffuse clouds is neither ionized appreciably by starlight nor does it acquire appreciable ionization by charge-transfer with H^+. Its radiative association with hydrogen is then expected to be much slower than that of carbon and exchange reactions analogous to (26)-(28) for oxygen are much less effective. The formation of the hydrides of nitrogen can be initiated by

$$H_3^+ + N \rightarrow NH_2^+ + H \qquad (31)$$

Even if the destruction of H_3^+ through dissociative electron recombination is quite small as indicated by the recent measurement of Smith and Adams (40), photodissociation of H_3^+ is still quite rapid [$\simeq 2 \times 10^{-8}$ s^{-1}, (41)]. If the ionization rate of H_2 is $\simeq 10^{-17}$ s^{-1}, the abundance of H_3^+ can be estimated to be

$$[H_3^+]/[H_2] \lesssim 10^{-9} \qquad (32)$$

Species which react with H_3^+ with a rate coefficient that is near the orbiting value then react with an overall rate that is about 10^{-18} [H_2] s^{-1}.

In contrast, the rate at which nitrogen atoms (and other atoms) hit dust grains is at least an order of magnitude greater. The rates for forming hydrides of carbon and oxygen by gas reactions are larger yet.

If the rates for forming CH and OH by gas phase reactions were somewhat slower than given in the foregoing, formation on grains might then be the source of CH and OH. Though the specifics of surface reactions are highly uncertain, a reasonable possibility is that the surfaces are inert and whatever hits simply attaches H-atoms. Oxygen, nitrogen and, if the grains are electrically neutral, carbon will then form hydrides at essentially the same rate per particle. Although loopholes in the argument can be imagined, one would then expect that the NH/OH ratio should not be drastically different from the N/O ratio ($\simeq 7$ for cosmic abundances). However, the observed ratio toward the star Omicron Persei is found to be NH/OH $\lesssim 1/100$ (42). This tends to support the premise that CH and OH are not the result of surface reactions.

8. FORMATION OF THE DIATOMICS OF HEAVY ATOMS -- CO, C_2, CN, O_2, N_2, NO

Once the hydrides are produced as discussed in the foregoing, exchange reactions of the form

$$A + BH \rightarrow AB + H \qquad (33)$$

readily produce the heavy diatomics (43,44) at rates per atom that can be seen as comparable with the rates for attaching H-atoms, or 10^{-17} n(cm^{-3}) s^{-1}. For example, OH and CH are about 10^{-8} of the abundance of hydrogen so that if A represents an ion (C^+ here) which reacts in (33) at near the orbiting rate, the rate per A-atom is just 10^{-17} n. Of course, if both A and BH are neutral, the rate coefficient will tend to be smaller ($\simeq 4 \times 10^{-11}$ cm^3 s^{-1}) and may be much smaller due to a small activation energy barrier that becomes evident only at low temperatures. The presence or absence of such barriers is difficult to predict from calculations or from laboratory measurements at $T \gtrsim 300$ K and essentially no reliable data are available for the relevant temperatures. In addition to diatomics, BH can also represent polyatomic hydrides such as CH_2 which are expected to be abundant but which cannot be detected at the current level of sensitivity. There can thus be several routes to the formation of these species. Thus we have,

$$C^+ + CH \text{ or } OH \to C_2^+ + H, \ CO^+ + H \text{ or } CO + H^+ \qquad (34)$$

which further react with H_2 before being destroyed.

$$C_2^+ \text{ or } CO^+ + H_2 \to C_2H^+ + H \text{ or } HCO^+ + H \qquad (35)$$

and

$$C_2H^+ + H_2 \to C_2H_2^+ + H \qquad (36)$$

before electron recombinations and perhaps photodissociation reduce these species to C_2 and CO. Since CH^+ apparently does not coexist with the other species, formation of CN seems to require a neutral-neutral reaction if nitrogen hydrides are low in abundance as expected in a gas-phase chemistry -- a point emphasized recently by Federman, Danks and Lambert (45). There are many possibilities,

$$CH + N \to CN + H \qquad (37)$$

but also

$$CH_2 + N \to HCN + H \text{ or } CN + H_2 \qquad (38)$$

from which the HCN is photodissociated to produce the CN. Exchange of the heavy atoms is also possible

$$C_2 + N \to CN + C \qquad (39)$$

to produce some and destroy others, e.g.,

$$C_2 \text{ or } CN + O \to CO + C \text{ or } CO + N \qquad (40)$$

Of the above, C_2 and CN are destroyed by (39) and (40) as well as by photodissociation. Due to its strong binding, CO can only be destroyed appreciably by photodissociation. Its large abundance ($\simeq 10^{-6}$ of hydrogen) is a result of its seemingly small cross section for photodissociation, which leads to a rate due to the galactic background radiation of about 10^{-11} s^{-1}, and of numerous reaction channels that terminate in CO. Recent studies of the spectra of CO have been argued to indicate that photodissociation of CO occurs exclusively through sharp lines

(46; see also work presented at this workshop by F. Rostas and by C. R. Vidal.) If so, CO is sufficiently abundant that self shielding can be appreciable. Further, at least one of the (presumed) dissociating lines of CO will be shielded by a line of the H_2 molecule.

The remaining diatomics of C, N and O (O_2, N_2, NO) have not been detected. They should be generated and destroyed by reactions that are analogous to those which are observed, though they do require neutral-neutral reactions for which the presence of an activation energy is always a matter of concern. Molecular oxygen is thought to be formed primarily by

$$O + OH \rightarrow O_2 + H, \tag{41}$$

and destroyed by photodissociation and photoionization at a rate 9×10^{-10} s^{-1} due to the galactic background radiation as well as by reactions such as

$$C^+ + O_2 \rightarrow CO + O^+ \text{ or } CO^+ + O \tag{42}$$

Reaction (41) has been measured down to 250K and shows no evidence of an activation energy barrier. The predicted O_2 abundance for the gas toward Zeta Ophiuchi is [O_2]/[hydrogen] $\simeq 10^{-10}$ [column density of $O_2 = 3 \times 10^{-11}$ cm^{-2}; (47)].

Formation of the NO molecule can proceed through

$$N + OH \rightarrow NO + H \tag{43}$$

for which there is no evidence for an activation energy from measurements at higher temperature, though apparently not through the analogous reaction of O + NH. It can be destroyed by

$$C^+ + NO \rightarrow NO^+ + C \tag{44}$$

followed by dissociative electron recombination as well as by neutral-neutral reactions

$$C + NO \rightarrow CN + O \tag{45}$$

though apparently not through N + NO (nor O + NO due to energetics). It is also destroyed by photodissociation, of course. The model of Black and Dalgarno (41) for Zeta Ophiuchi predicts a column density of NO at $3 \times 10^{11} cm^{-2}$ or [NO]/[hydrogen] $\simeq 10^{-10}$.

There seems to be no evidence against the formation of N_2 through

$$N + NH \rightarrow N_2 + H \qquad (46)$$

or

$$N + NO \rightarrow N_2 + O \qquad (47)$$

though

$$N + CN \rightarrow C + N_2 \qquad (48)$$

is measured to be slow at higher tempertures and is presumeably unimportant in diffuse clouds. The dominant destruction mode is expected to be photodissociation which is estimated to be relatively slow. Again, the classic Black and Dalgarno (41) model for Zeta Ophiuchi predicts the column density for N_2 to be 10^{13} cm^{-2} or $[N_2]/[\text{hydrogen}] \simeq 10^{-8}$. Observational upper limits are somewhat below this value (3.8×10^{12} cm^{-2}, or $[N_2]/[\text{hydrogen}] \simeq 2.6 \times 10^{-9}$) but are in the direction of another star, (δ Sco).

9. ADDITIONAL COMMENTS ON THE COMPARISON WITH OBSERVATION

As emphasized at the outset, one must have (to varying degrees) detailed information about the physical conditions (esp. density, temperature and abundances of the chemical elements) in diffuse interstellar clouds in order to make rigorous tests of the chemistry. Studies of the ultraviolet pumping of the rotational levels of H_2, of the ionization states of the atoms, of the collisional excitation of the low-lying levels of various species are utilized to restrict the physical conditions of the diffuse clouds. The most detailed effort to determine the physical conditions and to test rigorously the chemical reaction networks against observation in a self-consistent manner is the classic model of Black and Dalgarno (41) for the well studied gas toward Zeta Ophiuchi. Except for the abundance of CH^+, agreement with the observed molecular abundances is achieved in this model. We will, however, mention a subsequent study of our own in which we argue that the physical conditions actually are quite different (Section 10).

Another approach to testing observationally the chemistry of diffuse clouds is that taken by Federman and collaborators (48,29,49,45). Molecular abundances obtained from observations of a large number of diffuse clouds are compared in a statistical sense. Extreme cases of high and low density clouds are taken. They find that expected trends of [molecule]/[hydrogen] for CO, CH, C_2, and CN are followed by the observed abundances.

Diffuse clouds and molecular clouds of the interstellar gas have normally been treated as unrelated entities from both the observational and theoretical viewpoints. Observationally, this is probably due largely to the difference in the techniques which are utilized -- optical and ultraviolet wavelengths are used for diffuse clouds whereas observations of molecular clouds have been made primarily at radio wavelengths. Certain recent observations have been directed toward bridging this gap (31) by obtaining the molecular abundances and physical conditions toward stars which are somewhat more heavily obscured than those associated with the traditional diffuse clouds. The stars under consideration (HD 29647 and HD 147889) both have visual extinctions of about 4 magnitudes whereas the "prototypical" diffuse cloud(s) toward Zeta Ophiuchi cause an extinction of about 1 visual magnitude. Both sample the outer envelopes of molecular clouds which have been studied for some time using spectral lines at millimeter wavelengths. In Table 1, we reproduce the physical and chemical data obtained for these regions and contrast it with that for the Zeta Ophiuchi cloud using the temperature and density from (50).

TABLE 1
Comparison of Chemical Properties for Differing Physical Conditions

	Zeta Oph	HD 147889	HD 29647
T_k (K)	65	15	10
A_v (magnitudes)	1	4	4
n_H (cm^{-3})	200	2000	1600
Radiation Field	3	30	1
log X(e$^-$)	-3.6	-3.6	-6.3
log X(CH)	-7.6	-8.0	-7.6
log X(C_2)	-7.9	-8.4	-8.0
log X(OH)	-7.5	-7.5	-7.3
log X(CO)	-5.8	-4.4	-4.2
log X(CN)	-8.7	-9.4	-7.7
log X(CH$^+$)	-8.1	-8.8	<-8.8

Note that log X(A) = log [N(A)/N(H + $2H_2$)] where N represents column densities, n_H is the number density of hydrogen in all forms, T_k is the kinetic temperature and the radiation field in the far-ultraviolet is expressed in terms of the galactic average.

The dark clouds tend to be cooler and more dense. A large number of B stars in the neighborhood of HD 147889 causes the radiation field to be much higher (about a factor of 30) than the galactic average. Its effects are evident in contrasting the fractional electron abundances of the two dark clouds. It is noteworthy that the fractional abundances of CH, C_2 and OH are insensitive to the range of physical conditions sampled in the three clouds. The CN abundance is widely variable, a result that is consistent with the study of Federman, Danks and Lambert (45) toward 15 diffuse clouds. In fact, in a separate study (51) the observed CN abundances vary by a factor of 40 between two nearby positions in the same cloud. Finally, the difference in the entries for CH^+ only reflects the variation in the abundance of hydrogen.

Uncertainties in the penetration of the clouds by ultraviolet radiation become considerably more important for models of the chemistry for clouds with $A_v \simeq 4$. For example, the properties of the grains toward HD 147889 and HD 29647 differ at ultraviolet wavelengths. For the latter, there is only a weak 2200 A feature but a somewhat enhanced extinction in the far-ultraviolet when compared with the norm. Toward HD 147889, the anomalies are just the opposite with the result that the extinction at 1000 A seems to differ by about two (astronomical) magnitudes even though the visual extinction is about 4 magnitudes for both stars. Detailed, theoretical assessments of the influence of the uncertainties in grain properties have been performed (e.g., 52,53).

ISOTOPE FRACTIONATION

In discussing the fractionation of isotopes, we will not restrict ourselves to diffuse clouds. In fact, Section (11) is based exclusively on the observations and physical conditions of molecular clouds.

10. CARBON ISOTOPES AND CONSTRAINTS ON THE NATURE OF DIFFUSE INTERSTELLAR CLOUDS

Other than reactions related to the fractionation of deuterium, the reaction (54,55)

$$^{13}C^+ + {}^{12}CO \rightleftarrows {}^{12}C^+ + {}^{13}CO + \Delta E \qquad (49)$$

($\Delta E/k = 35°$) seems to be the only chemical reaction which has been recognized that is likely to lead to a significant effect. In diffuse clouds and at the edges of denser clouds, where C^+ is the dominant form of carbon, $^{13}CO/^{12}CO$ can be significantly enhanced at the low temperatures ($\simeq$ 20K) that may occur in such regions. Observations (56,57) apparently do demonstrate the effect. As a result of reaction (49) and the nature of the chemical processes which tend to convert C^+ but not CO into carbon-bearing molecules, there is some tendency for molecules to be divided into two groups with regard to their predicted $^{13}C/^{12}C$ ratios - CO (and probably HCO^+) versus everything else. In the CO group, $^{13}C/^{12}C$ will tend to be enhanced whereas in the "everything else" group it will tend to be reduced. Observationally, this separation has not been demonstrated to occur though such a demonstration would be quite difficult. There are many uncertainties at the factor-of-two level in inferring abundances from the observations. The most detailed study which incorporates reaction (49) into an extensive network of chemical reactions (58) does not, however, include the depletion onto grains.

Reaction (49) has been used to restrict the temperature of the gas in the "prototypical" diffuse cloud - that toward Zeta Ophuichi. At the number densities expected for this cloud and for the estimated photodissociation rate for CO, reaction (49) should attain equilibrium. If the gas temperature were as low as has been considered by some investigators ($\simeq$ 20K), a substantial enhancement of $^{13}CO/^{12}CO$ should occur. The upper limit to this ratio obtained by Crutcher and Watson (50) [see also (59)] has been interpreted by these authors to imply that the gas temperature must be greater than about 50K in the region where the bulk of the CO molecules are located.

This apparent conflict with the physical conditions of the classic model by Black and Dalgarno (41) led to a re-examination of the observational constraints on the gas toward Zeta Ophiuchi, and in particular with regard to whether a two component model is necessary for the bulk of the gas. The interpretation of these constraints involves "molecular astrophysics" as well, though the issues are the excitation of the low-lying rotational states of CO, HD and H_2. Excitation of the fine-structure states of atomic carbon and the fractional ionization of carbon are the other key issues. Crutcher and Watson (50) conclude that the chief contraints on the gas toward Zeta Ophiuchi actually are consistent with a single set of physical conditions (density $\simeq$ 200 cm^{-3}, temperature $\simeq$ 65K) that are quite different from the high density, molecule - bearing component of the Black and Dalgarno model. It is unclear precisely how the agreement between the predicted and observed molecular abundances will be affected. Still more recently, observations of the populations of the rotational states of the C_2 molecule have been utilized as a diagnostic for the physical conditions by Danks and Lambert (60) based in part on the analysis of van Dishoeck and Black

(61). These measurements again point toward a major low temperature ($\simeq$ 20K) component of the gas, though as yet there seems to be no definite conflict with the $^{12}CO/^{12}CO$ temperature because of the uncertainties in the data.

In addition to the fractionation of CO resulting from chemical reactions, it has been argued (as mentioned in Section 8) that recent laboratory data suggest that the photodissociation of CO occurs through sharp spectral lines and that preferential self-shielding by the more abundant isotopic forms of CO together with chance coincidences between certain lines of H_2 and CO isotopes will then influence the abundance ratios of the various CO isotopes (46). However, reaction (49) tends to be more much effective than the selective photodissociation and locks together the $^{13}CO/^{12}CO$ ratios (62) at essentially that determined by equation (49). Nevertheless, this selective photodissociation can alter the isotope ratios for oxygen in CO, though the region over which the effect can be significant is quite limited.

11. DEUTERIUM

One of the most remarkable aspects of interstellar chemistry is the large enhancement of deuterium that occurs in the molecules of the molecular clouds (63). The D/H ratio in these molecules typically is larger by factors of 10^2–10^4 than the actual ratio D/H $\simeq 10^{-5}$ for these isotopes in all forms in our neighborhood of the galaxy. Such enhancement seems to be observed in essentially all clouds and molecular species for which there is sufficient sensitivity. Partial summaries of the observations are contained in various sources (e.g., 64).

Most of the deuterium in molecules is presumed to be in the form of HD in molecular clouds (most hydrogen is H_2). Although the binding energy of HD is greater than that for H_2 because of the smaller zero-point energy, this difference is even greater for all other molecules as far as we know. This energy difference (in temperature units) is typically hundreds of degrees Kelvin. At the temperature of molecular clouds ($\simeq$ 10-50K), a large enhancement of deuterium would then be expected to occur if there are exchange reactions for D/H that can proceed appreciably toward thermodynamic equilibrium. It seems that the reaction

$$H_3^+ + HD \rightleftharpoons H_2D^+ + H_2 + \Delta E \qquad (50)$$

($\Delta E/k \simeq 230K$) is the chief cause for the enhancement of deuterium (3). Because the proton affinity is lower than for essentially all (if not all) other molecules, proton (deuteron) transfer to other species passes along the enhancement which results from (50). For example,

$$H_2D^+ + CO \rightarrow DCO^+ + H_2, \tag{51a}$$

$$H_2D^+ + N_2 \rightarrow N_2D^+ + H_2 \tag{51b}$$

but also

$$H_2D^+ + HCN \rightarrow HCND^+ + H_2 \tag{52}$$

followed by

$$HCND^+ + e \rightarrow DNC + H \tag{53}$$

are effective in passing along the enhancement of deuterium. One might expect that all molecular ions will exchange with HD in reactions analogous to (50), but this seems not to be the case (see 65). The ion CH_3^+ seems to be the only other such species for which such exchanges do occur appreciably,

$$CH_3^+ + HD \rightleftarrows CH_2D^+ + H_2 + \Delta E \tag{54}$$

It may be important for molecules with C-H bonds.

Reaction (50) is only effective if it competes with the destruction reactions for H_3^+ which have seemed to be primarily

$$H_3^+ + e \rightarrow H_2 + H \text{ or } 3H \tag{55}$$

and equation (51a) because of the abundance of CO. This limited set of reactions yields for the predicted abundance ratio

$$\left(\frac{H_2D^+}{H_3^+}\right)/(D/H) \simeq \frac{2\,\alpha(50)}{\alpha(50)\,e^{-230/T} + \alpha(55)X(e) + \alpha(51a)X(CO)} \tag{56}$$

The rate coefficients $\alpha(50)$ and $\alpha(51a)$ have been measured at somewhat higher temperatures and are near the orbiting values (see 65). There have also been measurements which indicate that $\alpha(55) \gtrsim 3 \times 10^{-7}$ cm^3 s^{-1} (66,67). If the enhancement of, for example, DCO$^+$ is then ascribed to (50), upper limits to the fractional ionization X(e) and the fractional abundance of CO (and other molecules that react with H_3^+) X(CO) can be obtained. These upper limits are low enough to be useful (X(e) $\lesssim 10^{-7}$) and only depend on the ratios of laboratory quantities in addition to the observed enhancement factor and abundance for deuterium.

There has also been some debate about the value of $\Delta E/k$ in (50) and whether it is large enough to cause the observed fractionation, especially in the warmer clouds. However, agreement now seems to have been reached that the value given here should be satisfactory (c.f. 68,65). Observational studies of the enhancement of DNC as a function of temperature in various clouds gives $\Delta E/k \simeq 240K$ for whatever the exchange reaction is which causes the enhancement if the reaction is assumed to have reached thermodynamic equilibrium and the actual D/H ratio is fixed (69).

Starting from the premise that reaction (50) controls the fractionation and that $(D/H) \simeq 10^{-5}$, Herbst (68) showed that $[DCO^+]/[HCO^+]$ in several molecular clouds is consistent with the temperatures of the gas obtained by independent techniques. Alternatively, this can be taken as evidence that (D/H) does not vary throughout the galaxy.

Despite the two laboratory measurements mentioned above for reaction (55), it has been argued on theoretical grounds that the reaction (55) should be unusually slow at the low temperatures and excitation state of interstellar clouds (70,71). Smith and Adams (40) have now obtained a quite different rate coefficient for (55). In fact, they find only an upper limit of 2×10^{-8} cm^3 s^{-1} at 95K. This will, of course, at least raise the upper limit to X(e) from equation (56) to the level that is of little value. However, another recent measurement by the Pittsburg group gives a value for reaction (55) that is similar to their earlier measurement (72).

Alternative approaches based on the chemistry of molecular clouds do exist such as that examined by Wooten, Snell and Glassgold (73). The sensitivity of this technique will also be reduced by a slower rate for reaction (55). In addition, it depends upon an uncertain rate for the ionization of hydrogen within molecular clouds.

The fractionation of deuterium in molecules may also be influenced by exchange reactions of molecular ions with atomic-D ((74); A. Dalgarno, comments at this Workshop).

12. RECOMMENDATIONS FOR LABORATORY INVESTIGATIONS

Ion-molecule reactions which are slow at 300K often seem to behave unexpectedly at low temperatures (e.g., (75)). Reliable information about this class of reactions is needed for the temperature regime $\simeq 10-60K$ that is appropriate for the conditions of low excitation that prevail in the interstellar gas.

Much less seems to be known about relevant neutral-neutral reactions at 300 K and below. Settling the issue of the dissociative electron recombination of H_3^+ is of highest priority.

ACKNOWLEDGEMENT

Research by the authors is supported in part by US NSF Grants AST81-14887 and AST 82-16723.

REFERENCES

1. Herbst, E. and Klemperer, W. 1976, Phys. Today 29, p. 32.
2. Dalgarno, A. and Black, J. H. 1976, Rep. Prog. Phys. 39, p. 573.
3. Watson, W. D. 1976, Rev. Mod. Phys. 48, p. 513.
4. Watson, W. D. 1983, in Galactic and Extragalactic Infrared Astronomy, M. F. Kessler and J. P. Phillips eds. (Reidel: Dordrecht), p. 69.
5. Snow, T. P. 1980, in Interstellar Molecules, B. H. Andrew ed. (Reidel:Dordrecht), p. 247.
6. Hollenbach, D. and Salpeter, E. E. 1971, Ap. J. 163, p. 155.
7. Hollenbach, D., Werner, M. W. and Salpeter, E. E. 1971, Ap. J. 163, p. 165.
8. Hunter, D. A. and Watson, W. D. 1978, Ap. J. 226, p. 447.
9. Jura, M. 1974, Ap. J. 191, p. 375.
10. O'Donnell, E. J. and Watson, W. D. 1974, Ap. J. 191, p. 89.
11. Stecher, T. P. and Williams, D. A. 1967, Ap. J. Letters 149, p. L29.
12. Watson, W. D. and Salpeter, E. E. 1972, Ap. J. 174, p. 321.
13. Greenberg, J. M., Allamandola, L. J., Hagen, W., van de Bult, C. E. P. and Baas, F. 1980, in Interstellar Molecules, B. H. Andrew ed. (Reidel:Dordrecht), p. 355.
14. Spitzer, L., Drake, J. F., Jenkins, E. B., Morton, D. C., Rogerson, J. B. and York, D. G. 1973, Ap. J. Letters 181, p. L116.
15. Dalgarno, A., Black, J. H. and Weisheit, J. 1973, Ap. Letters 14, p. 77.
16. Fehsenfeld, F. C., Dunkin, D. B., Ferguson, E. E. and Albritton, D. L. 1973, Ap. J. Letters 183, p. L25.
17. Watson, W. D. 1973, Ap. J. Letters 182, p. L73.
18. Watson, W. D., Christensen, R. B. and Deissler, R. 1978, Astron. Ap. 69, p. 159.
19. Black, J. H. and Dalgarno, A. 1973, Ap. J. Lett. 184, p. L101.
20. Bates, D. R. and Spitzer, L. 1951, Ap. J. 113, p. 441.

21. Dalgarno, A. 1976, in Atomic Processes and Applications, P. G. Burke and B. L. Moiseiwitsch eds. (North-Holland: Amsterdam), p. 109.
22. Graff, M. M., Moseley, J. T. and Roueff, E. 1983, Ap. J. 269, p.796.
23. Luine, J. A. and Dunn, G. H. 1981, in Proceedings of the XII ICPEAC, S. Datz ed. (North-Holland:Amsterdam), p. 1035.
24. Herbst, E. 1982, Ap. J. 252, p. 810.
25. Mitchell, J. B. and McGowan, J. W. 1978, Ap. J. Letters 222, p. 77.
26. Chesnavich, W. J., Akin, V. E. and Webb, D. A. 1984, Ap. J. (in press).
27. Federer, W., Villinger, H., Howorka, F., Lindinger, W., Tosi, P., Bassi, D. and Ferguson, E. 1984, Phys. Rev. Letters 52, p. 2084.
28. Chaffee, F. H. 1975, Ap. J. 199, p. 379.
29. Federman, S. 1982, Ap. J. 257, p. 125.
30. Cohen, J. G. 1973, Ap. J. 186, p. 149.
31. Crutcher, R. M. 1984, Ap. J. (in press).
32. Hobbs, L. M. 1973, Ap. J. 181, p. 79.
33. Elitzur, M. and Watson, W. D. 1978, Ap. J. Letters 222, p. L141.
34. Elitzur, M. and Watson, W. D. 1980, Ap. J. 236, p. 172.
35. Fehsenfeld, F. C. and Ferguson, E. E. 1972, J. Chem. Phys. 56, p. 3066.
36. Chambaud, G., Launay, J. M., Levy, B., Millie, P., Roueff, E. and Tran Minh, F. 1980, J. Phys. B 13, p. 4205.
37. Watson, W. D. 1973, Ap. J. Letters 183, p. L17.
38. van Dishoeck, E. and Dalgarno, A. 1984, Ap. J. 277, p. 576.
39. Smith, P. L. and Parkinson, W. H., 1978, Ap. J. Letters 223, p. L127.
40. Smith, D. and Adams, N. 1984, Ap. J. Letters (in press).
41. Black, J. H. and Dalgarno, A. 1977, Ap. J. Suppl. Ser. 34, p. 405.
42. Crutcher, R. M. and Watson, W. D. 1976, Ap. J. 209, p. 778.
43. Solomon, P. and Klemperer, W. 1972, Ap. J. 178, p. 389.
44. Herbst, E. and Klemperer, W. 1973, Ap. J. 185, p. 505.
45. Federman, S. R., Danks, A. C. and Lambert, D. L. 1984, Ap. J. (in press).
46. Glassgold, A. E., Huggins, P. J. and Langer, W. D. 1984, Ap. J. (in press).
47. Black, J. H. and Smith, P. L. 1984, Ap. J. 277, p. 562.
48. Federman, S. R., Glassgold, A. E., Jenkins, E. B. and Shaya, E. J. 1980, Ap. J. 242, p. 525.
49. Danks, A. C., Federman, S. R. and Lambert, D. L. 1984, Astron. Ap. 130, p. 62.
50. Crutcher, R. M. and Watson, W. D. 1981, Ap. J. 244, p. 855.
51. Crutcher, R. M., Churchwell, E. and Ziurys, L. 1984, Ap. J. (in press).

52. Sandell, G. and Mattila, K. 1975, Atron. Ap. 42, p. 357.
53. Roberge, W. G., Dalgarno, A. and Flannery, B. P. 1981, Ap. J. 243, p. 817.
54. Watson, W. D., Anicich, V. G. and Huntress, W. T. 1976, Ap. J. Letters 205, p. L165.
55. Smith, D. and Adams, N. 1980, Ap. J. 242, p. 424.
56. McCutcheon, W. H., Dickman, R. L., Shuter, W. L. H. and Roger, R. S. 1980, in Interstellar Molecules, B. H. Andrew ed. (Reidel:Dordrecht), p. 411.
57. Langer, W. D., Goldsmith, P. F., Carlson, E. R. and Wilson, R. W., 1980, Ap. J. Letters 235, P. L39.
58. Langer, W. D., Graedel, T. E., Frerking, M. A. and Armentrout, P. B. 1984, Ap. J. 277, p. 581.
59. Wannier, P. G., Penzias, A. A. and Jenkins, E. B. 1982, Ap. J. 254, p. 100.
60. Danks, A. C. and Lambert, D. L. 1983, Astron. Ap. 124, p. 188.
61. van Dishoeck, E. and Black, J. H. 1982, Ap. J. 258, p. 533.
62. Chu, Y.-H. and Watson, W. D. 1983, Ap. J. 267, p. 151.
63. Wilson, R. W., Penzias, A. A., Jefferts, K. B. and Solomon, P. M. 1973, Ap. J. Letters 179, p. L107.
64. Brown, R. D. and Rice, E. 1981, Phil. Trans. R. Soc. Lond. A 303, p. 523.
65. Smith, D., Adams, N. and Alge, E. 1982, Ap. J. 263, p. 123.
66. Leu, M. T., Biondi, M. and Johnsen, R. 1973, Phys. Rev. A8, p. 413.
67. Mul, P. M., Mitchell, J. B. A., D'Angelo, V. S., Defrance, P., McGowan, J. W. and Froelich, H. R. 1981, J. Phys. B 14, p. 1353.
68. Herbst, E. 1982, Astron. Ap. 111, p. 76.
69. Snell, R. L. and Wootten, A. 1979, Ap. J. 228, p. 748.
70. Porter, R. 1977, J. Chem. Phys. 66, p. 2756.
71. Michels, H. H. and Hobbs, R. H. 1982, Presented at the 35th Annual Gaseous Electronics Conference, Dallas, Texas.
72. MacDonald, J. A., Biondi, M. and Johnsen, R. 1984, Planetary Sp. Sci. 32, p. 651.
73. Wootten, A., Snell, R. L. and Glassgold, A. E. 1979, Ap. J. 234, p. 876.
74. Watson, W. D. and Walmsley, C. M. 1982, in Regions of Recent Star Formation, R. S. Roger and P. E. Dewdney eds. (Reidel:Dordrecht), p. 357.
75. Luine, J. A. and Dunn, G. H. 1984, in preparation.

THE CHEMISTRY OF SHOCKED REGIONS OF THE INTERSTELLAR GAS

A. Dalgarno

Harvard-Smithsonian Center for Astrophysics
Cambridge, Massachusetts, U.S.A.

I. INTRODUCTION

Shock waves occur in compressible fluids when the pressure gradients are large enough to generate supersonic motion. Because information about the pressure disturbances cannot propagate upstream faster than the velocity of sound the fluid ahead of the shock does not respond dynamically until the shock arrives. The shock then compresses and heats the fluid. The boundary which separates the hot compressed fluid and the upstream fluid is called the shock front. The shocked material will undergo excitation, dissociation and ionization if the shock is rapid. The subsequent recombinations and emissions produce photons which may dissociate and ionize the fluid constituents ahead of and behind the shock. This precursor radiation modifies the effect of the shock and influences its dynamical and thermal evolution. The shock structure is changed in the presence of a magnetic field and if the fractional ionization is small the shock may be preceeded by a magnetic precursor, compressing and heating the material ahead and behind the shock front (1,2,3).

Shocks are a ubiquitous phenomenon in the interstellar medium, occurring during the birth and death of stars and persisting through the life of massive stars. Stars emit ionizing photons and create regions of high pressure which drive shocks into the surrounding interstellar medium. Early type stars undergo mass loss and the stellar winds produce shocks as do collisions of interstellar clouds and supernova explosions.

In the shock front the directed energy of the shock is converted into random thermal energy and the gas is heated to a temperature T_s, which for a strong shock with velocity v_s km s^{-1}, is of the order $40v_s^2$ K (4,5). Thus slow shocks heat the molecular gas and the chemical composition changes. Fast shocks dissociate the molecules and molecules form in new arrangements as the heated post-shock gas cools. Dissociation occurs by collision-induced dissociation

$$H_2 + H_2 \rightarrow H_2 + H + H \tag{1}$$

$$H_2 + H \rightarrow H + H + H . \tag{2}$$

Because of radiative stabilization (6,7,8), the rate coefficients depend upon the density. For heteronuclear molecules radiative stabilization is very efficient (7) and heteronuclear molecules are not destroyed by collision-induced dissociation. They are destroyed by chemical exchange reactions with hydrogen atoms. Thus the sequence

$$H + CO \rightarrow CH + O \tag{3}$$

$$H + CH \rightarrow C + H_2 \tag{4}$$

destroys CO and the reaction

$$H + HD \rightarrow H_2 + D \tag{5}$$

destroys HD. Because of such exchange reactions, any shock incident upon a gas in which the hydrogen is mostly atomic will tend to destroy the molecules that exist in the pre-shock gas.

For high shock velocities, hot electrons are produced and even at low densities will dissociate H_2 by excitation to the repulsive state,

$$e + H_2 \rightarrow e + H_2(b^3\Sigma_u^+) \rightarrow e + H + H . \tag{6}$$

Hollenbach and McKee (9) have calculated the limiting shock velocities above which dissociation is 90% complete as a function of the gas density n_H. At low densities H_2 survives at shock velocities up to about 50km s^{-1}. Because only a small amount of atomic hydrogen suffices to remove CO chemically, CO survives only up to 40km s^{-1}. At high densities, the stabilizing effect of radiative stabilization is suppressed by collisions and H_2 survives only to 25km s^{-1}. Because of its higher dissociation energy, CO is slightly more resilient than H_2 at the lower temperatures created by a 25km s^{-1} shock. The limiting velocities at high densities can be significantly increased to about 45km s^{-1} by magnetic fields (3).

The chemistry of an interstellar molecular gas subjected to a non-dissociative shock is dominated by reactions with H_2,

$$H_nX + H_2 \rightarrow H_{n+1}X + H , \tag{7}$$

both exothermic and endothermic (10). The total abundances of molecules created in the shocked gas depends upon the initial temperature, the endothermicities of the reactions and the time the gas remains warm. The gas is cooled by fine-structure transitions of atomic oxygen and by rotational and vibrational transitions of H_2, CO, H_2O and OH (3,11). Thus the chemical and thermal structures are closely interwoven.

Tabulations of the reaction rate coefficients may be found in references (4), (5) and (12). They should be regarded with caution. In laboratory measurements of chemical reaction rates of neutral species, the ambient density is high and the translational, rotational and vibrational modes are populated in thermal equilibrium at the kinetic temperature. In astrophysical environments, low-lying vibrational levels of homonuclear molecules may be populated thermally at high densities as may low-lying rotational levels of homonuclear and heteronuclear molecules are rapidly depopulated by electric dipole emissions. Thus the OH molecules participating in the reaction

$$OH + H_2 \rightarrow H_2O + H \quad (8)$$

are effectively in their lowest vibrational level. Whether or not a reaction is enhanced by translational energy or by vibrational energy depends upon the details of the reaction mechanism. The subject has been reviewed by Kneba and Wolfrum (13). The particular case of reaction (8) has been studied by Spencer, Endo and Glass (14) and Glass and Chatuverdi (15) who find that at 298K the rate coefficient is enhanced by only 50% if OH is in its v=1 level but by a factor of 155 if H_2 is in its v=1 level. It appears that vibrational excitation of the bond which is broken has a larger influence that vibrational excitation of the spectator bond.

2. OXYGEN CHEMISTRY

In a heated molecular gas, atomic oxygen is rapidly converted to OH and H_2O by the reactions

$$O + H_2 \rightarrow OH + H \quad (9)$$

$$OH + H_2 \rightarrow H_2O + H \quad (10)$$

which have rate coefficients of $4.1 \times 10^{-19} T^{2.44} \exp(-1281/T) \, cm^3 s^{-1}$ (16) and $1.4 \times 10^{-14} T \exp(-3500/T) \, cm^3 s^{-1}$ (3) respectively. Because of the possibility of the reverse reactions, the resulting distribution of oxygen between O, OH and H_2O is sensitive to the ratio of the densities of atomic and molecular hydrogen. For a slow shock in a gas in which the hydrogen is initially molecular a substantial enhancement in the abundance of H_2O is predicted (17).

Elitzur and de Jong (18) invoked reactions (9) and (10) in a model of an OH maser produced behind a shock moving with a velocity of 10km s^{-1} and Elitzur (19) proposed the same sequence as the source of the high H_2O abundance found in the Kleinman-Low region in Orion (20-22).

The deuterated water molecule, HDO, has also been detected in the same vicinity by observations of the 1_{10}-1_{11} (23,24) and 2_{11}-2_{12} (25) transition. Beckman et al. (25) concluded that the ratio for HDO/H_2O is 1×10^{-2} whereas Johansson et al. (24) obtained a value of 1×10^{-4}. The emission region has a complex structure and the two transitions may be excited in different locations. If the high abundances of H_2O and HDO are a consequence of shock-heated gas, exchange reactions

$$H_2O + HD \rightleftarrows HDO + H_2 \tag{11}$$

$$H_2O + D \rightleftarrows HDO + H \tag{12}$$

will tend to eliminate any fractionation. As accurate measure of the HDO/H_2O ratio is a potentially powerful diagnostic of the physical conditions in the shocked material.

Emissions from the excited J=5/2 nebula at 119.23 and 119.44μm have been tentatively identified near the Kleinman-Low nebula by Storey et al. (26), though Stacey et al. (27) did not observe them and gave an upper limit to the total intensity slightly below that reported by Storey et al. (26). However the presence of shock-heated gas is clearly established by observations of emission from high rotational levels of CO (26-30) and from excited vibrational levels of H_2 (31,32).

Using a limited description of the chemistry, Draine et al. (3) have calculated the abundances of OH and H_2O in a molecular gas subjected to shocks with velocities between 5km s^{-1} and 50km s^{-1} in the presence of magnetic fields. Almost all the oxygen that is not locked up in carbon monoxide molecules is converted to H_2O though a modest enhancement in the OH abundance occurs. With some modifications, their model has been applied to a detailed interpretation of the H_2 and CO emission from the shocked region in Orion (33,34). By postulating a shock velocity of 36-38km s^{-1} and a pre-shock magnetic field of 1.5 milligauss (33) or 0.45 milligauss (34), agreement was obtained with the CO and H_2 observations and the calculated OH emission line intensities are close to but below the upper limits of Stacey et al. (27).

A one hundred-fold increase in the OH/CO abundance ratio has been seen in the supernova remnant IC443 in which vibrationally excited H_2 is also detected (35). A shock provides a plausible explanation. The measured ratio is reproduced by model calculations (12) in which a 10km s^{-1} shock is incident upon a diffuse interstellar cloud. Calculations at higher densities are needed to confirm the interpretation.

An elaborate chemistry has been included by Mitchell (5) in calculations of the molecular composition of a dense cloud subjected to shocks with velocities between 5km s^{-1} and 20km s^{-1} but with zero magnetic field. Mitchell predicted a large abundance for HO$_2$ for shocks of 10 and 15km s^{-1} through the reaction

$$O_2 + H_2 \rightarrow HO_2 + H \tag{13}$$

The molecule HO$_2$ had been searched for earlier in Orion (22) but only an upper limit to its abundance could be obtained.

3. SULFUR CHEMISTRY

Hartquist et al. (35) pointed out that SH and H$_2$S are readily produced in heated gas by endothermic reactions similar to (9) and (10)

$$S + H_2 \rightarrow SH + H \tag{14}$$

$$SH + H_2 \rightarrow H_2S + H, \tag{15}$$

the formation of SH and H$_2$S leading to the increased abundances of other sulfur-bearing molecules. In cold gas none of the suggested mechanisms for the production of sulfur compounds is efficient (37-41) and their presence may be an indicator of shock-heated gas. Phosphorous compounds are also difficult to produce in cold gas (42) and if detected may indicate a heated gas.

The abundance ratios of SO and SO$_2$ with respect to CO are much enhanced in the plateau source in Orion compared to the ridge (42,22,24). The enhancements are qualitatively consistent with a shocked gas but the measured SO$_2$/SO ratio of 0.8 is considerably larger than the predicted value of 10^{-3} for an 8km s^{-1} shock (36). The main source of SO$_2$ is the reaction

$$SO + OH \rightarrow SO_2 + H \tag{16}$$

and the discrepancy may result from an underestimate of the OH abundance in the cooling post-shock gas.

The OH abundance is increased by the existence of atomic hydrogen in the pre-shock gas and by the production of atomic hydrogen in the shocked gas. In the model calculations of Hartquist et al. (36), the pre-shock density was 10^6 cm^{-3}, the hydrogen was initially molecular and only small amounts of atomic hydrogen were produced by chemical reactions with H$_2$. The abundance of OH and the distribution of sulfur between SO and SO$_2$ is sensitive to the density of the post shock region and to the shock velocity (36). Lower densities increase the abundance of SO$_2$. Mitchell and Deveau (12) studied the effect of a 10km s^{-1} shock on a diffuse cloud of density 10^2 cm^{-3} and for a gas shielded from ultraviolet photons obtained an SO$_2$/SO ratio of 0.6 which happens to agree with the value for the plateau region in Orion.

Hartquist et al. (36) and Mitchell and Deveau (12) predicted increases in the abundance of CS in shocked gas but little or none is observed in Orion or in IC443 (44,35). A higher initial abundance of atomic oxygen and an increase in shock velocity may resolve the discrepancy by permitting a more effective destruction of CS by the reactions

$$CS + O \rightarrow CO + S \tag{17}$$

$$CS + OH \rightarrow OCS + H . \tag{18}$$

4. SILICON CHEMISTRY

In shocks with enhanced abundances of OH the reaction

$$Si + OH \rightarrow SiO + H \tag{19}$$

should lead to an increased amount of SiO (36). SiO is strongly enhanced in the shocked plateau region of Orion (24,45) but is not in IC433 (35). Because OH is enhanced in IC433, the absence of SiO is unexpected. Perhaps silicon is depleted in the shocked gas cloud.

5. HCO^+ CHEMISTRY

The calculations of Iglesias and Silk (17) and of Mitchell and Deveau (46) for a 10km s^{-1} shock show that the abundance of HCO^+ is reduced by the shock. In constrast, observations of the supernova remnant IC433 (46,35) of Orion (22) and of NGC2071 (47) show HCO^+ to be enhanced by at least one order of magnitude. However HCO^+ is close to normal abundance in the disturbed gas lying towards the FU Orionis star Elias 12 (48).

Mitchell and Deveau (12) have argued that the increase of HCO^+ in IC433 occurs because the pre-shock region is rich in C^+ ions which are converted in the shock by reactions such as

$$C^+ + OH \rightarrow CO^+ + H \tag{20}$$

$$CO^+ + H_2 \rightarrow HCO^+ + H \tag{21}$$

$$C^+ + H_2O \rightarrow HCO^+ + H . \tag{22}$$

Elitzur (49) (see also 50) has put forward the alternative suggestions that the additional HCO^+ is produced by an increase in the ionization rate in the shocked material resulting from its penetration into the remnant cavity where it is exposed to trapped cosmic rays. Other explanations are possible. The diminution of HCO^+ in the shock models (17,45) is mainly due to the reaction with H_2O,

$$HCO^+ + H_2O \rightarrow H_3O^+ + CO . \qquad (23)$$

A model with different initial conditions, producing less H_2O and more OH, would lead to an increase in HCO^+.

6. CARBON CHEMISTRY

In cold clouds the carbon chemistry is initiated by the radiative association of C^+ and H_2 (51),

$$C^+ + H_2 \rightarrow CH_2^+ + h\nu . \qquad (24)$$

An empirical value of the order of $2-5 \times 10^{-16}$ cm^{3}s^{-1} for the rate coefficient of (24) has been derived from the observations of CH in diffuse clouds (54-55). The postulated chemistry fails to reproduce the measured abundances of CH^+. Elitzur and Watson (56) have suggested that CH^+ is produced in a gas heated by shocks driven by expanding HII regions. In a gas with temperatures above about 1000K, the endothermic reaction

$$C^+ + H_2 \rightarrow CH^+ + H \qquad (25)$$

proceeds rapidly (57). The CH^+ ion is converted to CH_2^+ by the reaction

$$CH^+ + H_2 \rightarrow CH_2^+ + H . \qquad (26)$$

It is removed by the reverse of reaction (25),

$$CH^+ + H \rightarrow C^+ + H_2 \qquad (27)$$

and by dissociative recombination

$$CH^+ + e \rightarrow C + H . \qquad (28)$$

The rate coefficients of (27) and of (28) decrease with increasing temperature (58,59).

There is observational support for the hypothesis of Elitzur and Watson (55). Frisch (60) and Federman (61) found that CH^+ has a velocity different from CH, which is presumably formed mostly in the cold compressed gas. Reactions (9) and (10) form OH and H_2O in the same region as CH^+. Crutcher (62) has found that towards ζ Ophiuchi OH exists with two velocity components, one of which he locates in the pre-shock gas and the other in the post-shock gas.

Particularly when radiative stabilization (6) is taken into account, the model tends to produce too much OH and possibly too much H_2O. Photodissociation of OH (63) may limit the production of OH and H_2O. Alternatively magnetic fields may modify the chemistry, because the reaction between C^+ and H_2 will proceed at a rate determined by the streaming velocity of the ions relative to the neutral component of the gas (64). It may be possible to selectively enhance CH^+ formation compared to neutral molecules. Other molecular ions may be created by the same mechanism.

Carbon monoxide is largely unaffected by the passage of a shock (17) unless a significant amount of C or C^+ exists in the pre-shock gas. Most of the carbon ions and atoms are converted to CO though the pathway opened up to reaction (25) leads to a high abundance of CH_4 (17,12).

The response of formaldehyde is sensitive to the H/H_2 ratio. In the 10km s^{-1} shock of Iglesias and Silk (17) sufficient atomic hydrogen was created in the shocked gas that formaldehyde was severely depleted by the reaction

$$H + H_2CO \rightarrow H_2 + HCO ,$$

producing a large increase in HCO. Formaldehyde has been observed in the high velocity gas in Orion with a normal abundance by Wootten et al. (65) who argue that the temperature of the gas must not exceed about 400K. However in the shock calculations of Mitchell (5), the abundance of H_2CO is increased by the shock due to the reactions

$$CH_3 + O \rightarrow H_2CO + H \tag{30}$$

$$CH_3 + O_2 \rightarrow H_2CO + OH . \tag{31}$$

That (29), (30) and (31) result in an increase in H_2CO appears to be due to the assumption of a large abundance of C in the pre-shock gas.

Mitchell (5) found an initial decline in H_2CO after the shock before it increased. The initial decline occurs in the hot gas through collisional dissociation

$$H_2CO + H_2 \rightarrow HCO + H + H_2 . \tag{32}$$

The rate coefficients adopted by Mitchell (5) for the collision-induced dissociation of heteronuclear molecules are all too large (6,7). Only chemical reactions such as (29) are effective in dissociation.

Mitchell (66) has extended his calculations to include hydrocarbons containing from one to six carbon atoms and he finds that provided the initial fraction of C and C^+ is high, substantial amounts of complex hydrocarbon molecules are produced by the passage of a shock.

7. NITROGEN CHEMISTRY

The nitrogen chemistry depends upon the initial distribution of nitrogen between N, N_2 and NH_3. Any atomic nitrogen is processed into N_2 and NH_3. Any CN tends to be removed by

$$CN + H_2 \rightarrow HCN + H \tag{33}$$

but CN (and HCN) can be enhanced if a large fraction of carbon is C or C^+ in the pre-shock gas (12). In the Orion plateau HCN is enhanced (24) but not, compared to CO, in IC433 (35).

8. DISSOCIATIVE SHOCKS

The chemistry of a gas fully dissociated by a fast shock has received less detailed attention. Hollenbach and McKee (4) have summarized the physical and chemical processes. Molecular hydrogen may be reformed by association on the surfaces of grains and at high temperature by the sequences

$$H + e \rightarrow H^- + h\nu \tag{34}$$

$$H + H^- \rightarrow H_2 + e \tag{35}$$

$$H + H^+ \rightarrow H_2^+ + h\nu \tag{36}$$

$$H_2^+ + H \rightarrow H_2 + H^+ \tag{37}$$

The formation processes heat the gas and slow its cooling.

The chemical evolution of the gas following the formation of H_2 is affected by precursor radiation which modifies the gas ahead of and behind the shock. Much of the precursor radiation emerges as Lyman alpha (67). The molecules CH_4, H_2O, OH and NH_3 are readily dissociated by Lyman alpha photons and their abundances in the cooling post-shock gas are diminished by the presursor radiation, allowing the preferential formation of CO. Any oxygen left over after the formation of CO will form OH, H_2O and O_2. Sulfur will be transformed mostly to H_2S and nitrogen to ammonia. Detailed models have yet to be constructed.

The densities of electrons and ions are critical to the behavior of shocks in magnetic fields (2,3). In fast dissociative shocks, ionization is produced by the precursor radiation. In slower non-dissociative shocks, chemical processes of associative ionization such as

$$O + CH \rightarrow HCO^+ + e \tag{38}$$

(68,69) are probably the major source of electrons and ions.

9. SUMMARY

The chemistry of shocked gases is a critical aspect of their evolution and the molecules that are formed dominate the cooling of the shocked gas. The resulting infrared emission can be detected and provides powerful diagnostic probe of the temperature structure. The composition offers additional information on the shock environment. Some species are sensitive to the initial conditions and should be useful in determining the characteristics of the pre-shock gas. Many uncertainties exist in the chemical processes that occur at temperatures of a few thousand degree Kelvin. The efficiencies of dissociation and ionization are perhaps particularly uncertain, but the role of internal excitation in chemical reactions also demands a deeper understanding.

ACKNOWLEDGMENTS

This work was supported in part by the National Science Foundation under Grant AST-81-14718.

REFERENCES

1. Mullan, D.J. 1971, Mon. Not. Roy. Astron. Soc. 153, pp. 145-170.
2. Draine, B.T. 1980, Ap. J. 241, pp. 1021-1038.
3. Draine, B.T., Roberge, W.G. and Dalgarno, A. 1983, Ap. J. 264, pp. 485-507.
4. Hollenbach, D. and McKee, C.F. 1979, Ap. J. Suppl. 41, pp. 555-592.
5. Mitchell, G.F. 1984, Ap. J. Suppl. 54, pp. 81-101.
6. Dalgarno, A. and Roberge, W.G. 1979, Ap. J. Lett. 233, pp. L25-L27.
7. Roberge, W.G. and Dalgarno, A. 1982, Ap. J. 255, pp. 176-180.
8. Lepp, S. and Shull, M.J. 1983, Ap. J. 270, pp. 578-582.
9. Hollenbach, D. and McKee, C.F. 1980, Ap. J. Lett. 241, pp. L47-L50.
10. Aannestad, P. 1973, Ap. J. Suppl. 25, pp. 223-252.
11. McKee, C.F., Chernoff, D.F. and Hollenbach, D.J. 1984, Galactic and Extragalactic Infrared Spectroscopy, Ed. M.F. Kessler and J.P. Phillips (Reidel), pp. 103-131.
12. Mitchell, G.F. and Deveau, T.J. 1983, Ap. J. 266, pp. 646-661.
13. Kneba, M. and Wolfrum, J. 1980, Ann. Rev. Chem. 31, pp. 47-79.
14. Spencer, J., Endo, H. and Glass, G.P. 1977, 16th Int. Symp. Comb. Inst. Pittsburgh, pp. 829-835.
15. Glass, G.P. and Chatuverdi, B.K. 1981, J. Chem. Phys. 75, pp. 2749-2752.
16. Ravishankara, A.R., Nicovich, J.M., Thompson, R.L. and Tully, F.P. 1981, J. Phys. Chem. 85, pp. 2498-2503.
17. Iglesias, F.R. and Silk, J. 1978, Ap. J. 226, pp. 851-857.
18. Elitzur, M. and de Jong, T. 1978, Astron. Ap. 67, pp. 323-332.
19. Elitzur, M. 1979, Ap. J. 229, pp. 560-566.
20. Wannier, P.G. 1978, Ap. J. Lett. 222, pp. L59-L62.
21. Waters, J.W., Bustinic, J.J., Kakar, R.K., Kuiper, T.B.A., Roscoe, H.K., Swanson, P.N., Rodriguez Kuiper, G.N., Kerr, A.R. and Thaddeus, P. 1980, Ap. J. 235, pp. 57-62.
22. Kuiper, T.G.H., Zuckerman, B. and Rodriguez Kuiper, F.N. 1981, Ap. J. 251, pp. 88-102.
23. Turner, B.E., Zuckerman, B., Fourikis, N., Morris, M. and Palmer, P. 1975, Ap. J. Lett. 198, pp. L125-L128.
24. Johansson, L.E.B., Andersson, C., Ellder, J., Friberg, P., Hjalmarson, A., Hoglund, B., Irvine, W.M., Olofsson, H. and Rydbeck, G. 1984, Astron. Ap. 130, pp. 227-256.
25. Beckman, J.E., Watt, G.D., White, G.J., Phillips, J.P., Frost, R.L. and Davis, J.H. 1982, Mon. Not. Roy. Astron. Soc. 201, pp. 357-364.

26. Storey, J.W.V., Watson, D.M. and Townes, C.H. 1981, Ap. J. Lett. 244, pp. L27-L30.
27. Stacey, G.J., Kurtz, N.T., Smyers, S.D. and Hartwit, M.]983, Mon. Not. Roy. Astron. Soc. 202, pp. 25P-29P.
28. Watson, D.M., Storey, J.W.V., Townes, C.H., Haller, E.E. and Hansen, W.L. 1980, Ap. J. Lett. 239, pp. L129-L132.
29. Stacey, G.J., Kurtz, N.T., Smyers, S.D., Harwit, M., Russell, R. and Melnick, G. 1982, Ap. J. Lett. 257, pp. L37-L40.
30. Goldsmith, P.F., Erickson, N.R., Fetterman, H.R., Clifton, B.J., Peck, D.D., Tannenwald, P.F., Koepf, G.A., Buhl, D. and McAvoy, N. 1981, Ap. J. Lett. 243, pp. L79-L82.
31. Beckwith, S., Evans, N.J., Gatley, I., Gull, G. and Russell, R.W. 1983, Ap. J. 264, pp. 152-160.
32. Beck, S.C. and Beckwith, S. 1983, Ap. J. 271, pp. 175-182.
33. Draine, B.T. and Roberge, W.G. 1982, Ap. J. Lett. 259, pp. L91-L96.
34. Chernoff, D.F., Hollenbach, D.J. and McKee, C.F. 1982, Ap. J. Lett. 259, pp. L97-L101.
35. DeNoyer, L.K. and Frerking, M.A. 1981, Ap. J. Lett. 246, pp. L37-L40.
36. Hartquist, T.W., Oppenheimer, M. and Dalgarno, A. 1980, Ap. J. 236, pp. 182-188.
37. Oppenheimer, M. and Dalgarno, A. 1974, Ap. J. 187, pp. 321-332.
38. Duley, W.W., Millar, T.J. and Williams, D.A. 1980, Mon. Not. Roy. Astron. Soc. 192, pp. 945-957.
39. Prasad, S.S. and Huntress, W.T. 1982, Ap. J. 260, pp. 590-598.
40. Millar, T.J. 1982, Mon. Not. Roy. Astron. Soc. 199, pp. 309-319.
41. Millar, T.J. 1983, Mon. Not. Roy. Astron. Soc. 202, pp. 683-689.
42. Thorne, L.R., Anicich, V.G., Prasad, S.S. and Huntress, W.T. 1984, Ap. J. 280, pp. 139-143.
43. Zuckerman, B. and Palmer, P. 1975, Ap. J. Lett. 199, pp. L35-L38.
44. Goldsmith, P.F., Langer, W., Schloerb, F.P. and Scoville, N.Z. 1980, Ap. J. 240, pp. 524-531.
45. Lada, C.J., Oppenheimer, M. and Hartquist, T.W. 1978, Ap. J. Lett. pp. L153-L156.
46. Mitchell, G.F. and Deveau, T.J. 1982, in Regions of Recent Star Formation Ed. R.S. Roger and P.E. Dewdney (Reidel) pp. 107-113.
47. Dickinson, D.F., Rodriguez Kuiper, E.N., Dinger, A.S.C. and Kuiper, T.B.H. 1980, Ap. J. Lett. 237, pp. L43-L45.
48. Wootten, A., Loren, R.B., Sandqvist, A., Friberg, P. and Hjalmarson, A. 1984, Ap. J. 279, pp. 633-649.
49. Levreault, R.M. 1983, Ap. J. 265, pp. 855-863.
50. Elitzur, M. 1983, Ap. J. 267, pp. 174-178.
51. Dalgarno, A. 1981, Phil. Trans. Roy. Soc. A 303, pp. 513-522.

52. Black, J.H. and Dalgarno, A. 1973, Astrophys. Lett. 15, pp. 79-82.
53. Black, J.H. and Dalgarno, A. 1977, Ap. J. Suppl. 34, pp. 405-423.
54. Black, J.H., Hartquist, T.W. and Dalgarno, A. 1978, Ap. J. 224, pp. 448-452.
55. Danks, A.C., Federman, S.R. and Lambert, D.L. 1984, Astron. Ap. 130, pp. 62-66.
56. Elitzur, M. and Watson, W.D. 1980, Ap. J. 236, pp. 172-181.
57. Erwin, K.M. and Armentrout, P.B. 1984, J. Chem. Phys. 80, pp. 2978-2980.
58. Federer, W., Ferguson, E., Tosi, P., Villinger, H., Bassi, D., Howorka, F. and Lindinger, W. 1984, Phys. Rev. Lett. in press.
59. Mul, P.M. and McGowan, J.W. 1980, Ap. J. 237, pp. 749-751.
60. Frisch, P.C. 1979, Ap. J. 227, pp. 474-482.
61. Federman, S.R. 1980, Ap. J. Lett. 241, pp. L109-L112.
62. Crutcher, R.M. 1979, Ap. J. Lett. 231, pp. L151-L153.
63. van Dishoeck, E.F. and Dalgarno, A. 1984, Ap. J. 277, pp. 576-580.
64. Draine, B.T. 1980, Ap. J. 241, pp. 1021-1038.
65. Wootten, A., Loren, R.B. and Bally, J. 1984, Ap. J. 277, pp. 189-195.
66. Mitchell, G.F. 1983, Mon. Not. Roy. Astron. Soc. 205, pp. 765-772.
67. Shull, J.M. and McKee, C.F. 1979, Ap. J. 227, pp. 131-149.
68. Dalgarno, A., Oppenheimer, M. and Berry, R. S. 1973, Ap. J. Lett. 183, L21-L24.
69. Messing, I., Filseth, S. V., Sadowski, C. M., and Carrington, T. 1981, J. Chem. Phys. 74, 3874-3881.

THEORETICAL MODELS OF SHOCK WAVES IN MOLECULAR CLOUDS

B. T. Draine

Princeton University Observatory
Princeton, NJ 08544, U.S.A.

ABSTRACT: The important physical processes acting in shock waves in molecular clouds are discussed. The magnetic field plays an important role in determining the structure of a shock wave in molecular gas. Because of the low fractional ionization, the magnetic field lines, together with the electron-ion plasma, can "slip" through the neutral gas; this "magnetic-ion-slip" can result in shock structures which are qualitatively different from nonmagnetic shock waves. The molecular processes acting in shock waves are reviewed, with attention to molecular excitation and dissociation, and a few important chemical reaction channels. The infrared spectra of C-type shocks are discussed, and theoretical models are compared to observations (in the BN-KL region of the Orion Molecular Cloud) of emission lines from vibrationally and rotationally-excited H_2, and rotationally excited CO and OH. Theoretical shock modelling is currently hindered by uncertainties in various cross sections for collisional excitation; some of these uncertainties are pointed out in hopes of stimulating further theoretical or laboratory work.

I. INTRODUCTION

Radio and infrared spectral line observations over the past decade have made it clear that hypersonic gas motions and shock waves are associated with star formation processes in molecular clouds. Some of these shock waves may occur within material emerging from stellar objects in the form of "winds", when these winds are decelerated as the result of running into the ambient molecular cloud material; other shock waves occur within previously undisturbed molecular cloud material during accretion onto a protostellar condensation (e.g., the solar nebula), or when stellar winds or other forms of energy release from stars drive shock waves outward into the surrounding molecular cloud material.

These shock waves are revealed to us by the line emission from atoms and molecules which are excited within the shock-heated gas. This line emission can provide information regarding the density, temperature, chemical composition, and state of motion of the shock-heated material. Furthermore, it provides an astrophysical "laboratory" which, if "calibrated", could return "experimental" information regarding collisional cross sections which might be difficult to obtain in terrestrial laboratories.

Here we review our present understanding of the nature of shock waves in molecular cloud material, including the likely magnetohydrodynamic structure of the shock, as well as the molecular processes which play an important role in the cooling, chemistry, and emission-line diagnostics of these shock waves.

II. STRUCTURE OF MHD SHOCK WAVES IN MOLECULAR CLOUDS

Recent work [1] on magnetohydrodynamic (MHD) shock waves in predominantly neutral gas has shown that shock waves fall into three broad categories: *(i)* J-type shocks without magnetic precursors; *(ii)* J-type shocks with magnetic precursors; and *(iii)* C-type shocks. J-type shocks (both with and without magnetic precursors) are distinguished by the presence of an abrupt transition layer where the hydrodynamic variables (velocity, density, temperature) change "discontinuously" in a hydrodynamic "jump" (hence the term "J-type") with a thickness of order one molecular mean free path. In J-type shocks *without* magnetic precursors, gas more than a few molecular mean free paths ahead of the hydrodynamic "jump" is essentially undisturbed from its "preshock" state (assuming no appreciable UV radiation from the hot postshock gas); conventional nonmagnetic shock waves fall in this category.

When magnetic fields are present it becomes necessary to treat the magnetized ion-electron plasma as a fluid distinct from the gas of neutral particles; these two fluids interpenetrate, but may locally have different velocities or temperatures (in fact, the ion temperature and electron temperature may also differ from one another). If the magnetic field is sufficiently strong that the Alfven velocity for the ion-electron plasma exceeds the shock speed, then hydromagnetic "signals" can propagate ahead of the "jump" transition, leading to partial compression of the magnetic field together with ion-neutral streaming in a "magnetic precursor" region ahead of the jump transition; such a shock, containing both a "jump" transition *and* a "magnetic precursor", is termed a "J-type shock with a magnetic precursor". The neutral gas within the "precursor" region is both heated and accelerated by collisions with the streaming ions. The spatial extent of the magnetic precursor depends upon the magnetic field strength and the fractional ionization of the gas.

Finally, when the fractional ionization is small and the magnetic field is sufficiently strong, shock structures may exist without a "jump" transition: the hydrodynamic variables (density, velocity, temperature) vary smoothly from the preshock to the postshock values; this is referred to as a "C-type" shock, since the hydrodynamic variables are continuous.

A recent study of shock waves in molecular clouds (ref. 2, hereinafter DRD) has concluded that, for reasonable estimates of fractional ionizations and plausible magnetic field strengths, shock waves in dense ($n_H \gtrsim 10^4 \mathrm{cm}^{-3}$) molecular clouds will be C-type for shock speeds $v_s \lesssim 40 \,\mathrm{km\,s^{-1}}$. Since shocks with shock speeds exceeding this value are apparently uncommon in molecular clouds, the discussion in the present paper will concentrate on C-type MHD shocks, although of course many of the physical processes would also be important in J-type shocks.

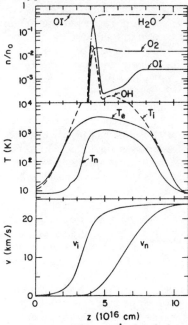

Fig. 1 -- Structure of a $v_s = 25\,\mathrm{km\,s^{-1}}$ shock in a molecular cloud with (preshock) density $n_H = 10^4 \mathrm{cm}^{-3}$, fractional ionization $x_e = 10^{-7}$, and magnetic field $B_0 = 100 \mu\mathrm{G}$. Shown are the velocities, measured relative to the preshock gas, of the neutrals (v_n) and ions (v_i); the temperatures of the neutrals (T_n), ions (T_i), and electrons (T_e), and the concentration, relative to the cosmic oxygen abundance (taken to be $n_O/n_H = 8.5 \times 10^{-4}$), of O, OH, O_2, and H_2O. From ref. [2].

Calculation of the structure of a C-type shock is relatively straightforward. One writes down equations for conservation of mass, energy, and momentum for each of the fluids (neutral and magnetized plasma), including coupling terms for exchange of mass, energy, or momentum between the fluids due to, say, ion-neutral collisions, ionization, recombination, etc., as well as terms representing exchange of energy with the rest of the universe (i.e., radiative cooling, and possible heating by background cosmic rays, X-rays, etc.). If one makes the simplifying assumption that the shock is plane-parallel and steady, and adopts a reference frame moving with the shock, then one obtains a set of coupled ordinary differential equations. The simplest case to examine is the one where the magnetic field $\mathbf{B}_0$ is perpendicular to the shock velocity $\mathbf{v}_s$; all discussion here will refer to this case. One now perturbs the ion velocity by a small amount relative to the preshock conditions, representing the leading edge of the "magnetic precursor", and proceeds to numerically integrate the "equations of motion" for the ion velocity, neutral velocity, temperatures, etc. If the integration proceeds without encountering singularities (this requires that the flow velocity of the neutral gas, in the frame where the shock is stationary, remain everywhere supersonic) then the shock structure is C-type, and one eventually attains a steady-state solution representing quiescent, compressed postshock material. The hydrodynamic structure of a representative C-type shock is shown in Figure 1, which shows a $v_s = 25 \text{ km s}^{-1}$ shock propagating into a molecular cloud of preshock density $n_H = 2n(H_2) = 10^4 \text{cm}^{-3}$, with an assumed magnetic field strength $B_0 = 10^{-4}$ gauss, and an assumed fractional ionization $x_e = n(e)/n_H = 10^{-5}$. It is assumed (for simplicity) that the magnetic field direction is perpendicular to the direction of propagation of the shock. The velocities in Fig. 1 are measured relative to the preshock gas; thus the region to the left of the shock is preshock material (at zero velocity), while the region to the right is postshock material, accelerated to a final velocity of $\sim 23 \text{ km s}^{-1}$. The important feature to note is the extended transition zone (of thickness $\sim 3 \times 10^{16}$ cm) where the ion and neutral velocities are appreciably different. One way of looking at this is as follows: Compressive magnetosonic waves can propagate with a velocity $v_{ms} = (B^2/4\pi\rho_i + 10kT/3m_i)^{1/2}$, where ρ_i is the mass density of the ions (the electron mass being negligible), m_i is the mean ion mass, and the ions are assumed to be singly-charged. For the parameters of the shock in Figure 1, this propagation velocity is very large: $v_{ms} \approx 1500 \text{ km s}^{-1}$ (assuming $m_i \approx 20 m_H$); because of the low ion density, there is relatively little inertia attached to the magnetic field, consequently disturbances in the magnetic field have a large propagation speed. Because $v_{ms} > v_s$, it is possible for compressive magnetosonic waves to propagate *ahead* of the shock and begin to accelerate the magnetized plasma before the neutral gas is disturbed. This results in a region where the ions (and electrons), being accelerated prior to the neutral gas, stream through it. As a result of ion-neutral collisions, the neutral gas then begins to be both accelerated and heated. Thus one develops a hydrodynamic structure (in these C-type shocks) where acceleration (and heating) of the neutrals occurs primarily through collisions with streaming ions (rather than by viscous stresses or by gradients in the neutral gas pressure). In this C-type shock structure, the heating and dissipation associated with the shock takes place over the extended region where ion-neutral streaming occurs; because this region is extended, the local heating rate per volume is limited, and cooling processes (primarily molecular line emission from rotational and vibrational levels of H_2 and H_2O) are able to keep the gas from getting hot enough to appreciably dissociate molecules. By this means it is possible for shock waves to accelerate molecular gas up to velocities of up to $\sim 40 \text{ km s}^{-1}$ without dissociation of the molecules, while producing copious amounts of molecular line emission. **Such shocks have apparently been observed,** the best example being the region of extended molecular line emission in the star-forming dense molecular cloud OMC-1 (see §IV below).

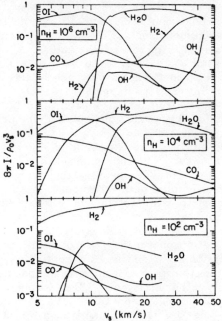

Fig. 2 -- The normalized power radiated by the principal coolants, as a function of shock speed v_s, in molecular gas with preshock $n_H = 10^2 \text{cm}^{-3}$, $x_e = 10^{-4}$, and $B_0 = 10 \mu G$; $n_H = 10^4 \text{cm}^{-3}$, $x_e = 10^{-7}$, and $B_0 = 100 \mu G$; and $n_H = 10^6 \text{cm}^{-3}$, $x_e = 10^{-8}$, and $B_0 = 1 \text{mG}$. From ref. [2].

III. ATOMIC AND MOLECULAR PROCESSES

a) Collisional Excitation

As seen above, the MHD shock models can contain extensive regions where the gas is heated to hundreds or thousands of degrees K. Collisional excitation of atoms and molecules can result in line emission, thereby removing energy from the hot gas. The important cooling processes in C-type shocks have recently been examined [2]. The relative importance of the principal cooling channels in C-type shocks is shown, as a function of shock speed, in Figure 2. From Figure 2 it can be seen that the principal cooling channels are fine structure line emission from atomic oxygen (if present in the preshock gas), and rotational-vibrational emission lines of H_2, and H_2O; emission from OH and CO may provide useful diagnostic information, but is not of major importance for the overall energy budget of the shock.

Rate coefficients for excitation of the fine structure levels of OI are available for collisions with H atoms [3], but rate coefficients for collisions with H_2 are unavailable; DRD arbitrarily adopted H_2-OI collisional rate coefficients equal to 50% of the H-OI rates. Since [OI]63.19μm can be both an important cooling channel and a useful diagnostic, direct calculation of cross sections for excitation of OI($^3P_2-^3P_1$) in collisions with H_2 would be of value.

Cross sections for collisional excitation of H_2 (in collisions with H, He, and H_2) are probably even more uncertain. DRD have attempted to compile a set of empirical formulae for the rate coefficients for $\Delta J = \pm 2, \pm 4, \pm 6$ transitions; rate coefficients for $\Delta J = \pm 2$ have also been estimated by Shull and Beckwith [4]. These rate coefficients, particularly for H_2-H_2 collisions, remain very uncertain, but are very important since the population of very high J levels (e.g., $J=17$) -- from which emission can be detected -- may depend very sensitively on these rate coefficients if the density is not high enough for the levels to be in LTE.

Rate coefficients for collisional excitation of H_2O in collisions with He have been computed by Green [5] for a number of the lower energy levels of H_2O. Because of the many energy levels of H_2O, however, existing shock calculations have not yet attempted detailed calculations of the line emission, but have been restricted to estimates for the total cooling rate [2,6]. In view of the large amounts of energy expected to be radiated in emission lines of H_2O, it is clear that efforts should be directed at computing detailed H_2O emission spectra. Atmospheric H_2O absorption has thus far prevented detection of H_2O emission from shocks, but continuing efforts to obtain spectra from above the atmosphere (from aircraft, balloons, or satellites) should eventually result in the detection of H_2O emission.

As seen from Fig. 2, emission from CO is not a dominant coolant in the shock-heated gas, but the $J \to J-1$ rotational transitions of CO nevertheless radiate enough energy to be detectable. Since for large $J \gtrsim 10$ these transitions tend to be optically thin, measured line intensities can immediately be translated into CO level populations. The observed CO line emission thus provides a powerful diagnostic of the physical conditions in the region where the CO is being collisionally excited, with the utility of this diagnostic being limited primarily by uncertainties in the cross sections for collisional excitation. McKee, Storey, Watson, and Green (ref. 7; hereinafter MSWG) have presented rate coefficients for CO($L \to 0$) transitions induced by collisions with He atoms; the cross sections were computed for $L \leq 32$ using the IOS ("infinite order sudden") scattering approximation. Transitions for $J \to J'$ transitions were obtained from the set of $L \to 0$ rate coefficients using an approximate relation [8]; rate coefficients so obtained may be represented by a convenient empirical formula [9]. MSWG argued that cross sections for H_2-CO collisions should be essentially the same as those for He-CO, giving rate coefficients which are larger by a factor 1.37 .

The IOS approximation for the rate coefficients is thought to be accurate (for a given potential surface) when $|E_J - E_{J'}| \ll kT$. Based on attempts to model the shock in OMC-1 (see §V), Draine and Roberge [10] suggested that perhaps the MSWG rates were too large for large values of $|E_J - E_{J'}|/kT$, and they adopted an *ad-hoc* modification of these rates:

$$<\sigma v>_{J \to J'} = \frac{<\sigma v>_{J \to J'}^{MSWG}}{1 + \alpha(\Delta E/kT)^2}$$

with α a dimensionless parameter of order unity. The MSWG rates are recovered for $\alpha=0$; $\alpha \approx 3$ was indicated by some models of the shock in OMC-1

[10]. Results are shown below for both $\alpha=0$ and $\alpha=3$.

b) Collisional Dissociation

The molecules present in shocked gas are of course subject to dissociation as the result of energetic collsions. Thermal collisions with atoms and molecules are by far the most frequent, in view of the very low fractional ionization in molecular clouds. Because the vibrational-rotational excitation of the molecule (e.g., H_2O) is generally sub-LTE in the shocked gas, "laboratory" dissociation rates are inappropriate [11-13]. It is instead necessary to use state-to-state rate coefficients to compute the (time-dependent) dissociation. Normally it is possible to assume that the *fractional* level populations approach a steady-state, so that the dissociation can be modelled as a time-independent process. Ideally one would perform a detailed state-to-state calculation; in existing calculations, however, the rotational level structure has been ignored, with only vibrational transitions being explicitly considered, although some allowance has been made for sub-LTE excitation of the rotational levels [13]. Unfortunately, at this time the state-to-state collisional rate coefficients are not known sufficiently well to justify the effort of a calculation giving full consideration to the rotational level structure, so H_2 excitation and dissociation in shock waves remains somewhat uncertain.

Dissociation of H_2 can also take place as the result of collisions with streaming ions: $H_2+I^+ \rightarrow H+H+I^+$, where I^+ is a generic ion (e.g., S^+). DRD estimated the rate for this reaction by assuming that all "orbiting" collisions with a center-of-mass energy in excess of the dissociation energy would lead to dissociation; the "Langevin" cross section for such orbiting collisions is proportional to the polarizability of H_2, and to $E^{-1/2}$. In computing the rate coefficient for this process, the fact that the streaming motion of the ions results in a non-Maxwellian distribution of relative velocities between the ions and molecules must be allowed for [2].

Finally, it is also possible for collisions of H_2 with dust grains to result in dissociation of the H_2. The calculations of DRD included this process, but the actual dissociation probability in a grain-H_2 collision is very uncertain.

The above dissociation processes -- collisions with neutrals, and with streaming ions and dust grains -- can also be important for other astrophysically interesting molecules, such as OH, O_2, and H_2O.

c) Collisional Ionization

The MHD structure of the shock is very sensitive to the electron fraction x_e, as the coupling between the magnetized plasma and the neutral gas is proportional to x_e for $x_e \gtrsim 10^{-7}$ (for smaller values of x_e, charged dust grains contribute to the coupling). It is therefore very important to understand those processes which may change the fractional ionization in the gas as it is shocked.

Ionization by electron impact is reasonably well understood for metal atoms such as Na. If the electrons become sufficiently hot, this process could raise the fractional ionization up to $x_e \approx 10^{-6}$ (the fractional abundance of metals with low ionization thresholds).

Ion-neutral collisions are another potential source of ionization, since the ions may be streaming through the gas at large velocities, resulting in large center-of-mass energies in ion-neutral collisions. Neutral-neutral collisions, though less energetic, are much more frequent, and hence may also be important. Unfortunately, ionization cross sections are very poorly known. DRD adopted cross sections based on the estimate by Drawin [14,15] of the cross section for the reaction $A+A \to A^+ +e^- +A$; the cross section formula was also assumed to apply to reactions $A+B \to A^+ +e^- +B$. Drawin's analysis is based on classical arguments, and hence is highly uncertain, particularly near threshold, though it appears to be in fairly good agreement with low energy measurements on $He+He \to He^+ +e^- +He$, $He+Ne \to He^+ +e^- +Ne$ [16] and $H+He \to H^+ +e^- +He$ [17,18]. It would be extremely valuable to have cross sections for ionization in H_2–He, and H_2–I^+ collisions measured at energies near threshold.

Finally, "charge-exchange ionization" events of the form $A+B \to A^+ +B^-$ are also possible, with a lower threshold energy than ordinary ionization; measurements on H_2+H [17,18] indicate $H_2+H \to H_2^+ +H^-$ to be the dominant ionization channel at low energies. Cross sections for reactions $H_2+H_2 \to H_2^+ +H_2^-$ and $H_2+He \to H_2^+ +He^-$ would be of great value for modelling high velocity MHD shocks in which such ionization processes may be important.

d) Chemical Reactions

Reactive collisions can alter the composition of the shocked gas. Detailed chemical reaction networks have been studied for nonmagnetic shock waves in interstellar gas [19-24]; see the review by Dalgarno [25]. Thus far, however, only limited investigations have been made into the chemistry of C-type MHD shock waves in molecular clouds. Because H_2O can be such an important coolant, DRD used a simplified reaction network to follow the abundances of O, OH, H_2O, and O_2. Figure 1 shows the abundances of these oxygen-bearing species in one particular shock model. As discussed below (§V), the observed intensity of line emission from rotationally-excited OH in OMC-1 is in agreement with that expected for OH produced (and collisionally-excited) within the shock wave believed (on other grounds) to be present in OMC-1.

It must also be noted that "ice" mantles on grains may be sputtered away in MHD shocks with $v_s \gtrsim 25 \text{km s}^{-1}$ [2], thereby injecting molecules and radicals into the hot gas, possibly with important consequences for the gas-phase chemistry.

More detailed studies of chemistry in MHD shock waves are clearly called for.

IV. RESULTS OF SHOCK MODELLING

Our information regarding shock waves in molecular clouds is based primarily on observations of infrared and radio emission (and occasionally absorption) lines originating from atoms and molecules in the shocked gas. Aside from H_2O (for which detailed emission spectra have not yet been

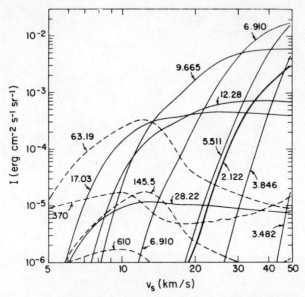

Fig. 3 — Line intensities (normal to the shock front), for plane-parallel shock waves propagating with shock speed v_s into molecular gas with $n_H = 10^4 \text{cm}^{-3}$, $x_e = 10^{-7}$, and $B_0 = 100\,\mu\text{G}$. Lines are labelled by wavelength (μm). Solid lines are rotation-vibration transitions of H_2; broken lines are fine-structure transitions of OI(63.19μm, 145.5μm) and CI(145.5μm, 610μm). From ref. [2].

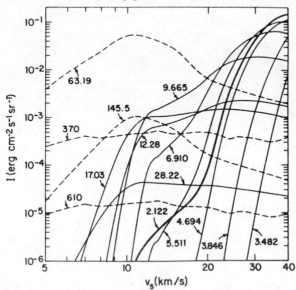

Fig. 4 — Same as Fig. 3, but for $n_H = 10^6 \text{cm}^{-3}$, $x_e = 10^{-8}$, and $B_0 = 1\,\text{mG}$. From ref. [2].

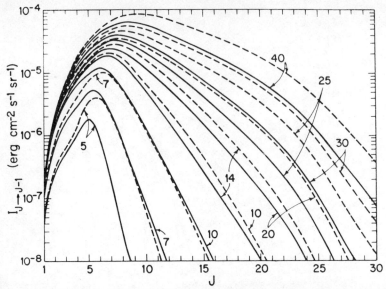

Fig. 5 -- Intensities of rotational transitions $J \to J-1$ of CO, for preshock $n_H = 10^4 \text{cm}^{-3}$, $x_e = 10^{-7}$, and $B_0 = 100\,\mu\text{G}$, and $n(\text{CO})/n_H = 7 \times 10^{-5}$. Curves are labelled by shock speed $v_s (\text{km s}^{-1})$. Broken curves were computed using MSWG rate coefficients; solid curves were computed using $\alpha = 3$, leading to reduced rate coefficients for large ΔJ transitions. From ref. [9].

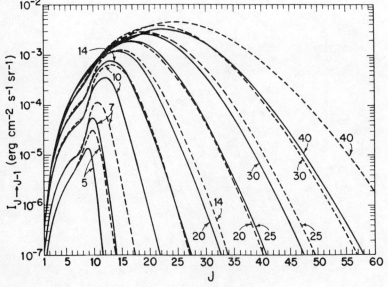

Fig. 6 -- Same as Fig. 5, but for $n_H = 10^6 \text{cm}^{-3}$, $x_e = 10^{-8}$, and $B_0 = 1\,\text{mG}$. From ref. [9].

calculated), the strongest emission lines from such shock waves are due to rotational-vibrational transitions of H_2, CO, and OH, and to fine-structure transitions in C and O. DRD have computed the emission in lines of H_2, O, and C for a large number of models; Figures 3 and 4 show these intensities, as functions of shock speed, for shocks advancing into gas of preshock density $n_H = 10^4$ and $10^6 cm^{-3}$, for assumed preshock magnetic fields of 10^{-4} and 10^{-3} gauss, respectively. As discussed by DRD, lines with $\lambda \lesssim 10 \mu m$ which fall in atmospheric windows can be detected with present-day instrumentation for surface brightnesses $\gtrsim 10^{-4} erg\, cm^{-2} s^{-1} sr^{-1}$, so it is evident from Figures 3 and 4 that C-type shocks with $n_H \gtrsim 10^4 cm^{-3}$ and $v_s \gtrsim 10\, km\, s^{-1}$ are expected to produce detectable emission in a number of different spectral lines of H_2. [OI]63.19μm emission is also expected to be detectable for shocks with $n_H \gtrsim 10^5 cm^{-3}$.

CO line emission from these shocks has also been computed [9]. Intensities of the far-infrared $J \rightarrow J-1$ rotational lines are shown, as functions of J, for a number of different shock speeds and preshock densities $n_H = 10^4 cm^{-3}$ (Fig. 5) and $n_H = 10^6 cm^{-3}$ (Fig. 6). Detection techniques have now been developed which permit detection of emission lines in the $\lambda \approx 100 \mu m$ region with surface brightnesses as small as $10^{-4} erg\, cm^{-2} s^{-1} sr^{-1}$ [26] so it is clear that many of these CO lines, with wavelengths $\lambda \approx 100(26/J) \mu m$, have detectable intensities for $n_H \gtrsim 10^5 cm^{-3}$ and $v_s \gtrsim 30\, km\, s^{-1}$.

V. Comparison with Observation: Molecular Line Emission from OMC-1

The dense condensation of molecular gas known as OMC-1 ("Orion Molecular Cloud 1"), located at a distance of ~500 pc, contains the Becklin-Neugebauer–Kleinmann-Low (BN-KL) cluster of luminous infrared sources -- presumably recently-formed or forming stars together with emission from warm dust heated by radiation from these stars (see ref. 27 for a recent review). This region is also the site of intense infrared line emission from vibrationally and rotationally excited H_2 and from CO in very high rotational states (levels up to $J=34$ have been observed), and shows evidence (from millimeter-wave observations of CO, and high-resolution spectroscopy of 2.1μm emission from H_2) of gas motions of up to $\pm 100\, km\, s^{-1}$ relative to the center-of-mass velocity of OMC-1.

The molecular line emission from this region is quite intense, on the order of ~$10^3 L_\odot$ in the lines of H_2 alone (after correcting for extinction by intervening dust). This emission has been interpreted as originating in shock-heated molecular gas, and detailed shock models have recently been computed [10,28]. Even though highly idealized, the shock models are quite successful in reproducing the observed power emitted in a number of different spectral lines. Figure 7 shows a comparison between column densities in various excited states (v,J) as inferred from observations of emission lines (after correcting for extinction) and as computed [10] for a shock with a shock speed $v_s = 38\, km\, s^{-1}$ advancing into gas with a preshock density $n_H = 7 \times 10^5 cm^{-3}$ and magnetic field $B_0 = 1.5 \times 10^{-3}$ gauss. The agreement is quite good. Note that emission is observed from H_2 levels as little as 1700 K above the ground state $(v=0, J=4)$ and as much as 21000 K above the ground state $(v=0, J=17)$.

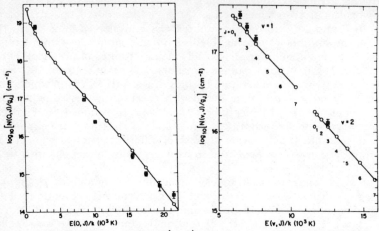

Fig. 7 -- Column densities $N(v,J)$ of rotationally and vibrationally excited levels of H_2, measured normal to a shock front, and divided by the degeneracy g_J, versus the energy of the level relative to the ground state. Open symbols represent the results of theoretical calculations for a $v_s = 38 \mathrm{\,km\,s^{-1}}$ shock propagating into molecular gas with $n_H = 7 \times 10^5 \mathrm{cm^{-3}}$, $x_e \approx 10^{-7}$, and $B_0 = 1.5 \mathrm{\,mG}$. Solid symbols represent column densities inferred from observed line intensities at "Peak 1" [29-31], corrected for extinction (assuming $\tau(2.1\mu m) = 3.5$), and divided by a factor 2.5 to allow for limb-brightening and possible emission from a second shock moving away from the observer. From ref. [10].

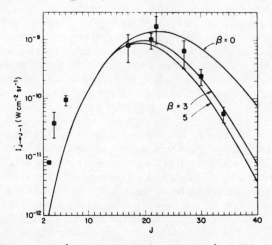

Fig. 8 -- Intensities (averaged over a 60" disk) of CO $J \to J-1$ emission lines, versus J. Solid curves are theoretical values for a $v_s = 38 \mathrm{\,km\,s^{-1}}$ shock with $n_H = 7 \times 10^5 \mathrm{cm^{-3}}$, $x_e \approx 10^{-7}$, $B_0 = 1.5 \mathrm{\,mG}$, and $n(CO)/n_H = 7 \times 10^{-5}$. Shock is assumed to be spherical with an angular diameter of 60". Curves are plotted for three values of the parameter α (here denoted β). Solid symbols are observed intensities [32-38]. From ref. [10].

A number of different rotational lines of CO have now been observed, providing a powerful diagnostic of the conditions in the line-emitting region. Unfortunately, the theoretically-computed emission from high-J levels depends upon the adopted rate coefficients. The observed emission from OMC-1 is compared in Fig. 8 to emission calculated for a shock model with a "normal" CO abundance $[n(CO)/n_H=7\times10^{-5}]$. Three different theoretical curves are shown, differing from one another in the parameter α (see §IIIa); $\alpha=0$ corresponds to the MSWG rate coefficients; the best fit for $J\geq17$ seems to be for $\alpha\approx3$ [the relatively poor agreement between the theoretical curve and $J\leq6$ observations is due at in large part to a crude treatment of radiative transfer in these optically thick lines, and in part because these levels are low enough for emission to be contributed from gas not associated with the shock [9]. Chernoff, Hollenbach, and McKee [28], by using a significantly lower value for the preshock gas density and a higher CO abundance, were able to satisfactorily account for the observed $J\geq17$ CO line emission using the MSWG cross sections (i.e., $\alpha=0$). It is clear that *if* the cross sections for excitation of H_2 by CO were firmly established, then the observed CO line emission could be used to *measure* the density in the shocked gas, thereby allowing the preshock density to be determined.

Emission lines of OH (119.23,119.44μm) have been detected from the shocked region in OMC-1. The observed surface brightness of each of these lines (1.7×10^{-10}erg cm^{-2}s^{-1}sr^{-1}; [26]) is in excellent agreement with the predicted values [10,28], with the OH being produced by chemical reactions within the shock.

On the whole, it appears that this C-type shock model is quite successful in accounting for a variety of observed emission lines, from a number of different species (H_2, CO, OH), and from levels with a broad range of energies. Uncertainties are present in both the theoretical calculation (various rate coefficients) and the interpretation of the observations (especially the corrections for extinction of H_2 line emission by intervening dust), and the model is rather simplistic (e.g., assuming that the observations -- which average over a fairly large area, presumably including a range of physical conditions -- can be modelled by emission from a single shock), so that precise agreement between model and observation would in any case not be expected. The degree of agreement between model and observation does, however, suggest that the basic theory for these shocks is understood, and that such shock modelling can be used to "measure" densities n_H and magnetic fields B_0 in molecular clouds. [It is important to note here that substantial efforts have been made to use non-magnetic shock waves models to reproduce the observations, as originally suggested by several authors [39-41]. Such models for the H_2 and CO line-emitting region in OMC-1 are apparently not possible, given the constraints imposed by observed emission line intensities.]

The two independent efforts to apply MHD shock models to OMC-1 differ in quantitative detail, but reach similar results: Draine and Roberge [10] found $n_H=7\times10^5$cm^{-3} and $B_0=1.5$mG, while Chernoff, Hollenbach, and McKee [28] found $n_H=2\times10^5$cm^{-3} and $B_0=0.45$mG. The differences in B_0 in fact derive primarily from differences in the assumed preshock density; this difference in turn can be traced back to differences in the assumed extinction suffered by the H_2 line emission. It is worth noting that if the collisional rate coefficients for rotational excitation of CO had been known with certainty, then the preshock density could have been determined quite accurately; the differences between the densities derived by these two studies are possible only because the freedom existed to use different rate coefficients for large ΔJ transitions for CO.

VI. Directions for Future Work

The considerable success in modelling the shocked gas in OMC-1 strongly suggests that the theory of MHD shocks described here is correct in its essentials. Further theoretical work can be directed toward various refinements, such as: (1) considering "oblique" shocks where the magnetic field is not precisely perpendicular to the direction of shock propagation; (2) improving the analysis of the motion of the charged dust grains in the shock; (3) studying an expanded chemical network; and (4) obtaining improved cross sections for excitation and dissociation of atoms and molecules. Because molecular astrophysics is the topic of this conference, I shall expand upon (4) and reiterate the most badly-needed cross sections.

The higher rotational levels of both H_2 and CO provide the possibility of excellent density and temperature diagnostics, as optically-thin emission from these levels allows the relative level populations (which are not in LTE for the densities of interest) to be experimentally determined. Accurate cross sections for excitation of the rotational levels of H_2 and CO by collisions with H_2, He, and H are required. It is important that cross sections be calculated or measured for large ΔJ transitions, as these play an important role in excitation of the high J levels.

The relative level populations of $v=1$ and $v=2$ (and, in the future, higher v) levels of H_2 can be measured by means of optically-thin emission lines. Accurate vibrational excitation cross sections would permit an independent estimate of the density in the region where the vibrational line emission occurs.

State-to-state cross sections for excitation of H_2 in collisions with H_2, He, and H are required in order to accurately estimate collisional dissociation rates.

Cross sections for dissociation of molecules in high-velocity ($v \approx 5-30 \, \text{km s}^{-1}$) collisions with ions are required in order to accurately model the chemical network. Such energetic collisions may enhance the production of species which result from endothermic reactions, for example, CH^+ formation by the reaction $C^+ + H_2 + 0.4\text{eV} \rightarrow CH^+ + H$.

Cross sections for ionization in ion-neutral and neutral-neutral collisions are required to study the extent to which the fractional ionization in the shock may be increased by collisional processes. This is important only in high-velocity shocks.

It is important to remember that the infrared and far-infrared observations of OMC-1 -- which have already proven to be so rich -- have been made with either ground-based telescopes (for the H_2 line observations), or with the relatively small (91 cm) telescope aboard the Kuiper Airborne Observatory. In the future we can hope for spectroscopy from orbiting helium-cooled telescopes, such as *ISO* or *SIRTF*, which should provide a wealth of infrared and far-infrared spectral line information for modelling shocks in many molecular clouds other than OMC-1 (the existence of high velocity CO emission and H_2 2.1μm line emission from a number of clouds almost certainly indicates the presence of MHD shocks similar to the shock in OMC-1). If this expected progress on the observational front is accompanied by theoretical efforts, we may anticipate substantial advances in our understanding of the physics, chemistry, and history of molecular clouds.

This research was supported in part by the Alfred P. Sloan Foundation, and in part by grant AST-8341412 from the National Science Foundation.

REFERENCES

1. Draine, B. T. 1980, *Ap. J.*, **241**, pp. 1021-1038 (see also *Ap. J.*, **246**, p. 1045 [1981]).
2. Draine, B. T., Roberge, W. G., and Dalgarno, A. 1983, *Ap. J.*, **264**, pp. 485-507 (DRD).
3. Launay, J. M., and Roueff, E. 1977, *Astr. Ap.*, **56**, pp. 289-292.
4. Shull, J. M, and Beckwith, S. 1982, *Ann. Rev. Astr. Ap.*, **20**, pp. 163-190.
5. Green, S. 1980, *Ap. J. (Suppl.)*, **42**, pp. 103-141.
6. Hollenbach, D., and McKee, C. F. 1979, *Ap. J. (Suppl.)*, **41**, pp. 555-592.
7. McKee, C. F., Storey, J. W. V., Watson, D. M., and Green, S. 1982, *Ap. J.*, **259**, pp. 647-656 (MSWG).
8. DePristo, A. E., Augustin, S. D., Ramaswamy, R., and Rabitz, H. 1979, *J. Chem. Phys.*, **71**, pp. 850-865.
9. Draine, B. T., and Roberge, W. G. 1984, *Ap. J.*, **282**, in press.
10. Draine, B. T., and Roberge, W. G. 1982, *Ap. J. (Letters)*, **259**, pp. L91-L96.
11. Dalgarno, A., and Roberge, W. G. 1979, *Ap. J. (Letters)*, **233**, pp. L25-L27.
12. Roberge, W. G., and Dalgarno, A. 1982, *Ap. J.*, **255**, pp. 176-180.
13. Lepp, S., and Shull, J. M. 1983, *Ap. J.*, **270**, pp. 578-582.
14. Drawin, H. W. 1968, *Z. Phys.*, **211**, pp. 404-417.
15. Drawin, H. W. 1969, *Z. Phys.*, **225**, pp. 483-493.
16. Hayden, H. C., and Utterback, N. G. 1964, *Phys. Rev. A*, **135**, pp. 1575-1597.
17. Van Zyl, B., and Utterback, N. G. 1969, *Sixth Int. Conf. Phys. Electron. Atom. Collisions* (Cambridge: MIT), p. 393.
18. Van Zyl, B., Le, T. Q., and Amme. R. C. 1981, *J. Chem. Phys.*, **74**, pp. 314-323.
19. Iglesias, E. R., and Silk, J. P. 1978, *Ap. J.*, **226**, pp. 851-857.
20. Elitzur, M., and Watson, W. D. 1980, *Ap. J.*, **236**, pp. 172-181.
21. Hartquist, T. W., Oppenheimer, M., and Dalgarno, A. 1980, *Ap. J.*, **236**, pp. 182-188.
22. Mitchell, G. F., and Deveau, T. J. 1982, in *Regions of Recent Star Formation*, ed. R. S. Roger and P. E. Dewdney (Dordrecht: Reidel), pp. 107-116.
23. Mitchell, G. F., and Deveau, T. J. 1983, *Ap. J.*, **266**, pp. 646-661.
24. Mitchell, G. F. 1984, *Ap. J. (Suppl.)*, **54**, pp. 81-101.

25. Dalgarno, A. 1985 (this volume).
26. Watson, D. M. 1982b, Ph. D. thesis, Univ. California, Berkeley.
27. Werner, M. W. 1982, in *Symp. on the Orion Nebula to Honor Henry Draper*, ed. A. E. Glassgold, P. J. Huggins, and E. L. Schucking, pp. 79-98.
28. Chernoff, D. F., Hollenbach, D. J., and McKee, C. F. 1982, *Ap. J. (Letters)*, **259**, pp. L97-L101.
29. Beck, S. C., Lacy, J. H, and Geballe, T. R. 1979, *Ap. J. (Letters)*, **234**, pp. L213-L216.
30. Knacke, R. F., and Young, E. 1981, *Ap. J. (Letters)*, **249**, pp. L65-L69.
31. Beckwith, S. Persson, S. E., Neugebauer, G., and Becklin E. E. 1978, *Ap. J.*, **227**, pp. 436-440.
32. Phillips, T. G., Huggins, P. J., Neugebauer, G., and Werner, M. 1977, *Ap. J. (Letters)*, **217**, pp. L161-L164.
33. van Vliet, A. H. F., de Grauuw, Th., Lee, T. J., Lidholm, S., and v. d. Stadt, H. 1981, *Astr. Ap.*, **101**, pp. L1-L3.
34. Goldsmith, P. F., *et al.* 1981, *Ap. J. (Letters)*, **243**, pp. L79-L82.
35. Stacey, G. J., Kurtz, N. T., Smyers, S. D., Harwit, M., Russell, R. R., and Melnick, G. 1982, *Ap. J. (Letters)*, **257**, pp. L37-L40.
36. Storey, J. W. V., Watson, D. M., and Townes, C. H. 1981, *Ap. J. (Letters)*, **244**, pp. L27-L30.
37. Watson, D. M., Storey, J. W. V., Townes, C. H., Haller, E. E., and Hansen, W. L. 1980, *Ap. J. (Letters)*, **239**., pp. L129-L132.
38. Watson, D. M. 1982a, private communication.
39. Hollenbach, D., and Shull, J. M. 1977, *Ap. J.*, **216**, pp. 419-426.
40. Kwan, J. 1977, *Ap. J.*, **216**, pp. 713-723.
41. London, R., McCray, R., and Chu, S.-I. 1977, *Ap. J.*, **217**, pp. 442-447.

COMETARY COMAE

W. F. Huebner

Los Alamos National Laboratory,
Theoretical Division, T-4
Los Alamos, NM 87545 USA

ABSTRACT

Our knowledge about the structure and composition of a comet nucleus is inferred from observations of the cometary atmosphere (coma). However, photodissociation, photoionization, and chemistry destroy the mother molecules evaporating from the nucleus. To extract the primary information, the chemical kinetics and the physics of the coma are modeled with a computer and the results are compared with coma observations. The physics and chemistry for a dust free coma are described taking into account energy balance, multi-fluid flow for fast atomic and molecular hydrogen and the bulk fluid, and the transition from a collision dominated inner region to the free molecular flow outer region. Special attention is paid to the molecular data requirements for the current models and for extended models which will include solar wind interaction and dust. Such models are an important tool in support of the Giotto mission to Halley's comet, in the analysis and interpretation of coma observations, and in the understanding of the earliest history of the solar system.

Since this is an interdisciplinary meeting, I will first define what a comet is. Comets are small bodies of the solar system that move in elliptic orbits about the sun. "New" comets spend most of their orbiting time in a spherical shell at a distances ~50,000 AU (astronomical units) from the sun (the Oort cloud) such that their orbits are nearly parabolic. When a comet is closer than about 2 AU from the sun, it developes an atmosphere (coma). From Doppler shifts of spectra it has been

G. H. F. Diercksen et al. (eds.), Molecular Astrophysics, 311–330.

determined that the neutral coma gas moves nearly radially outward with speeds of about 1 km/s; i.e., the coma is gravitationally not bound to the central body, the nucleus. This is the distinguishing feature for comets: Planetary atmospheres are gravitationally bound and asteroids have no atmospheres. An additional distinguishing feature for comets is that they may exhibit a dust tail caused by solar radiation pressure acting on the dust grains in the coma and a plasma tail produced by solar wind interaction with coma ions.

The source of the coma, the nucleus, has most probably never been observed; for large comets it is probably only a few kilometers in size. Its chemical composition is not known. It will depend on the place and time of origin and on the formation mechanisms for the nuclei. Unfortunately these are also not known so that intelligent guesses have to be made about the nucleus composition. Chemical equilibrium compositions, i. e., H_2O, CH_4, and NH_3, do not produce the observed radicals CN, C_2, and C_3, nor the ion CO^+ in the observed abundances and heliocentric distance dependences. "Near chemical equilibrium" compositions, i. e., the above equilibrium molecules plus some CO or CO_2, lead to similar difficulties, although more CO^+ is produced. Some interstellar molecules must be included in the models to reproduce observations.

Table I lists the atomic and molecular species that have been identified in comet comae. Not all species are identified in every comet, but CN, C_2, and C_3 are most commonly identified in the nearly circular, visible coma that may also contain varying amounts of dust. Sulfur compounds have been identified only recently, they have a very small range around the nucleus. The hydrogen coma is observed in Lyman-α light; typically it is larger than the sun and has an ellipsoidal extension in the direction away from the sun. Whenever atomic hydrogen is observed in a comet, OH is also detected in about the correct amount to suggest that H and OH are dissociation products of H_2O. The volatile component of comet nuclei is probably dominated by water ice.

TABLE I

Neutral Species Identified in Comet Comae

H	C	O	Na	S	CH	NH	OH	C_2	CN	CO	CS	S_2
NH_2	H_2O	HCN	HCO	C_3	NH_3	CH_3CN						

Comet tails point away from the sun. The dust tail is usually smooth in appearance and somewhat curved. The plasma tail shows structure and is straighter than the dust tail. Table II lists ionic species that are found in plasma tails or in the coma in the direction of the plasma tail. The longest plasma tails contain CO^+. Again, not all ionic species are found in every comet; CO^+ and H_2O^+ are the most common ions in comets.

TABLE II

Ions Detected in Comets Tails

CO^+ H_2O^+ CO_2^+ N_2^+ CH^+ CN^+ OH^+ C^+ H_2S^+

It should be noted that most of the observed species listed in Tables I and II are chemically reactive radicals or ions. They are most likely products of more stable molecules that exist in a frozen state in the nucleus. Whipple's (1950, 1951) icy conglomerate model is the basis for most models of the nucleus. But the details of the physical and chemical composition of the nucleus are not known.

When a comet is within a few astronomical units from the sun, the volatile component, i. e., the frozen gases in the nucleus, evaporate. From the energy of insolation, corrected for reflection and thermal reradiation from the nucleus, one can calculate the amount of material vaporized per unit time. Since the nucleus is small, the atmosphere of this vaporized material is gravitationally not bound, it streams constantly into interplanetary space. This feature distinguishes comets from planets and asteroids. A quantitative description of these processes has been given many times (see, e. g., Huebner et al. 1982, and references therein) and will not be repeated here.

In the neighborhood of the nucleus, the density of the coma gas is about 10^{12} to 10^{14} molecules/cm^3. The temperature of the gas, also obtained from the energy balance for insolation at the surface of the nucleus, is about 150 K and the outstream speed of the gas is close to its sound velocity which is ~300 m/s. These values correspond to a bright comet at r = 1 AU heliocentric distance. For smaller heliocentric distances these values increase somewhat with decreasing r, while for r > 1.5 AU they decrease rapidly with increasing r. The density in the coma decreases approximately with R^{-2}, where R is the distance from the center of the nucleus.

From the above values it is easily seen that chemical reactions between reactive species, in particular ion-molecule reactions, can proceed quite rapidly. But because the coma gas constantly streams radially outward, thereby causing it to become more dilute, the reactions slow down and eventually cease. Thus, some species that could have reacted further if collisions were still of sufficient frequency, will not react and will be "frozen in."

In order to model the coma, two basic processes must be combined with the fluid dynamics of the streaming gas: (1) photo processes that in the solar radiation field, attenuated by the coma gas, cause dissociation and ionization of the molecular constituents and (2) time-dependent chemical kinetics between the so produced reactive species. The temperature of the coma gas can be determined from the energy balance in the coma. It affects the above processes, but especially the fluid dynamics and certain chemical reactions. Although a number of papers have now been published on this subject (see, e. g., Shimizu, 1975, 1976; Houpis and Mendis, 1980; Ip, 1983; Marconi and Mendis, 1982a,b; Huebner and Keady, 1983), the results from modeling are as yet not final. This remains an active field of research that deserves further discussion here.

Modeling a comet coma must take into account the physics and chemistry of the (neutral) coma gas from the point of production on the nucleus surface to the outer hydrogen coma, the production of the plasma component and its streaming until it interacts with the solar wind and moves into the plasma tail, the chemical interaction of the plasma component with the neutral gas, and the entrainment of the dust by the coma gas and the solar radiation pressure that moves the dust into the tail. Such a model involves collision-dominated flow in the inner coma, transition to free molecular flow, multi-fluid flow (the lighter hydrogen decouples from the bulk of the heavier gas, dust and gas flow at different speeds in the innermost coma), counter streaming of the plasma relative to the neutral gas in the solar wind interaction region between the contact surface and the bow shock, heating and cooling, and photochemical reactions. No model exists that takes all these processes into account. In the following, a dust-free model is described without solar wind interaction.

Only one-dimensional coma calculations will be described here. Two-dimensional UV optical depth effects have been modeled; they show deviations from a one-dimensional calculation only in a narrow cone in the tail direction of the comet. Therefore, for the inner coma a one-dimensional calculation is sufficient, but in the outer coma, where solar wind interactions occur, two-dimensional calculations will be absolutely essential for the ions. The inner coma models are based on the processes occurring in a shell of coma gas as it expands and moves radially outward with velocity v.

To obtain the density, temperature, and flow speed profiles the fluid dynamic conservation equations must be solved. Ideally they should be solved for each species with collisional coupling between the species and the chemical kinetics as part of the source and sink terms. But the approach taken here is to solve the chemical reaction network with a stiff differential equation solver method on a fine mesh and to solve the conservation equations for mass (rather than individual elements), momentum, and energy for the bulk motion of all molecules in a Lagrangian coordinate system, but corrected for the fast nonthermalized hydrogen species:

$$d(\rho v R^2) = \mu R^2 dR \text{ , mass conservation,} \quad (1)$$

$$d(\rho v^2 R^2) + R^2 dp = \Pi R^2 dR \text{ , momentum conservation,} \quad (2)$$

$$d[\rho v^3 R^2/2 + \gamma p v R^2/(\gamma - 1)] = \varepsilon R^2 dR \text{ ,}$$
$$\text{energy conservation} \quad (3)$$

In these equations ρ, v, and p are the mass density, velocity, and pressure of the gas at a distance R from the center of the nucleus. The terms μ, Π, and ε are the sources and sinks for mass, momentum, and energy, each quantity per unit volume and per unit time. Contributions to these terms from fast hydrogens are discussed below. Effects on these terms from gas-dust collisions have been investigated by Marconi and Mendis (1982a,b). But effects from gas-dust collisions, solar wind interaction, and radiation pressure on the chemistry have never been investigated in a chemically comprehensive model.

Endothermic reactions and escaping radiation, emitted from excited states of molecular species, are energy sinks. In addition, expansion of the coma gas into near vacuum is a cooling process. Excess energy imparted to products of exothermic chemical reactions or of photodissociations are energy sources if the product particles can be thermalized collisionally. If such a product particle has a speed much higher than the bulk speed of the fluid, then this fast particle is likely to leave the shell (energy sink for that shell) but can be collisionally thermalized in another shell (energy source for that shell). The probability of leaving one shell and being thermalized in another depends on the thermalization mean free path and the direction of motion of the fast particle.

The excess photon energy above the dissociation threshold, if it is not too large, goes mostly into kinetic energy of the dissociation products; usually only a small fraction goes into vibrational and rotational excitation. Excess kinetic energy in dissociation and kinetic energy of exothermic reaction products are the main source for the relative pressure over density increase and the related temperature rise in the region $R \cong 10^2$ km to 10^4 km, illustrated in Fig. 1.

Because of momentum conservation, the particle with smaller mass m_s of two photofragments receives the larger fraction of the excess energy in the ratio $m_\ell/(m_\ell + m_s)$, where m_ℓ is the fragment with the larger mass. In photoionization the electron receives practically all of the excess energy.

Product molecules with molecular weight nearly equal to that of the bulk fluid, in which they are immersed, are almost thermalized in a single collision; this means that their thermalization mean free path is about equal to their collisional mean free path, $\Lambda_{th} \cong \Lambda_{coll}$. Molecules much lighter than the bulk fluid's mean value require many collisions before they are thermalized. For example, fast hydrogen atoms in water vapor require about 9 or 10 random walk collisions to lose half their energy (we shall call that thermalized), $\Lambda_{th} \cong 9.5^{1/2} \Lambda_{coll}$.

In the photodissociation of H_2O, hydrogen receives most of the excess energy, but tends to carry it away instead of sharing it with the bulk of the coma gas. The coma gas will therefore not be heated effectively by this process. This also holds true for any dissociation or reaction producing H or H_2, and to a lesser degree it also holds for C, N, and O products and their hydrides.

We shall assume that the fluid is composed of three components: one for fast atomic hydrogen H(f), one for fast molecular hydrogen $H_2(f)$, and one for the rest of the bulk fluid, the latter also includes the thermalized atomic and molecular hydrogen. If the cross section radius a_b = 4.4 Å for collisions between molecules of the bulk fluid and a_f = 3.3 Å for both of the fast hydrogen components H(f) and $H_2(f)$ colliding with fluid molecules, then the collision mean free path between bulk fluid molecules is $\Lambda_b = (\sqrt{2}\, n\, \pi\, a_b^2)^{-1} \cong 1.16 \times 10^{14}/n$ cm and for fast hydrogens [H(f) and $H_2(f)$] it is $\Lambda_f \cong 2.07 \times 10^{14}/n$ cm, where n is the fluid number density in cm^{-3}. We define the thermalized mean free path for the fast hydrogens $\Lambda = 9.5^{1/2} \Lambda_f$ and for the bulk fluid $\Lambda' = \Lambda_b$.

Mean free paths depend on the local number density n and therefore on the radial distance R. The mean free path concept is most important near the diffuse collision zone boundary,

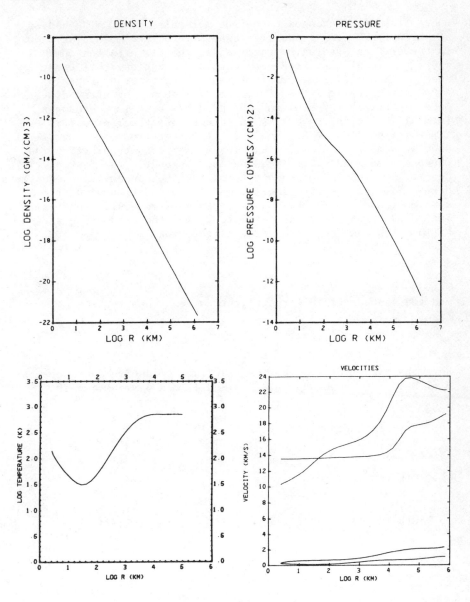

Fig. 1. Model results for density ρ, pressure p, temperature T, and velocities plotted as function of distance into the coma, log R. In the velocity panel the bottom curve represents the sound speed v_s, the next curve up is the bulk speed v, this is followed by the fast molecular hydrogen speed $v_{H_2(f)}$, and the fast atomic hydrogen speed $v_{H(f)}$ at top right.

where fluid flow changes into free molecular flow. The radius at which fluid flow gradually changes into free molecular flow is the critical radius R_c, defined as the place in the coma where the local mean free path equals the radius. Different species have different critical radii. For fast hydrogen $R_c^H = \Lambda(R=R_c)$; typically $R_c^H \sim 3 \times 10^3$ km at 1 AU heliocentric distance. Once the critical radius and the production of fast hydrogen from all processes $P_H = \Sigma_i k_i^H n_i^H$ are known, the fast particle first flight flux, $\Phi(R)$, and first flight current density (first angular moment of flux), $j(R)$, for a R^{-2} density distribution (Huebner and Keady, 1984) can be used to obtain the fast particle density n_H and the inhomogeneous terms for the conservation equations. Here k_i^H means the rate coefficient for production of H from species i and n_i is the number density of molecular species i that produces H. Similar terms apply for the production of $H_2(f)$. The fast particle first flight flux is

$$\Phi(R) = P_f R^2 \Phi_1(R/R_c) / (R_c) \quad , \tag{4}$$

and the fast particle first flight current density is

$$j(R) = P_f R^2 j_1(R/R_c) / (R_c) \quad . \tag{5}$$

Here the subscript f signifies reactions producing fast species H or H_2. The functions $\Phi_1(R/R_c)$ and $j_1(R/R_c)$ are shown in Fig. 2.

The inhomogenous terms of the fluid conservation equations can be written:

$$\mu(R) R^2 dR = -d[R^2 j(R)] m \times [v_f / (v_f + v)] \quad , \tag{6}$$

$$\Pi(R) R^2 dR = d[R^2 j(R)] |mv_f| \times [v_f / (v_f + v)] \quad , \tag{7}$$

$$\varepsilon(R) R^2 dR = E \{P_{fv} R^2 dR - d[R^2 j(R)]\} \times [v_f / (v_f + v)]$$

$$+ E' \{P_v R^2 dR - d[R^2 j'(R)]\}$$

$$- \varepsilon_{rad} R^2 dR \quad . \tag{8}$$

Here the right hand sides are summed over particles, E is the fraction of excess energy imparted to fast hydrogens and E' is the fraction of excess energy imparted to products other than

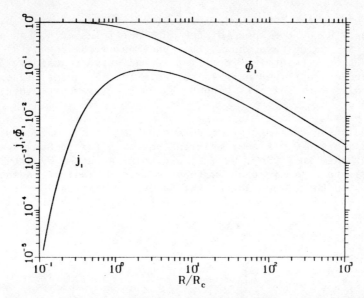

Fig. 2. Normalized first flight emitted particle flux Φ_1 and current density j_1 for fast particles as function of $\log(R/R_c)$.

fast hydrogesn, $(E_x = E + E')$, P_{fv} and P_v are the photo production rates [cm^{-3} s^{-1}] of fast hydrogens and of all other products, respectively, $j'(R)$ is the first flight current density of particles other than fast hydrogens with $R'_c = \Lambda'(R=R'_c)$ for the bulk fluid, v stands for the speed of the bulk fluid and v_f for that of the fast hydrogens relative to the bulk fluid. Since the function $j(R)$ considers the birth and death of the fast particles, birth corresponding to a mass loss of the bulk fluid in Eq. (6), negative values of $j(R)$ correspond to the reverse, namely addition of mass to the bulk fluid. The energy gain of the fast particles in Eq. (8) is not completely at cost to the bulk fluid, but a large part (EP_{fv} and EP_v) comes from the UV photon field and must therefore be added separately. The additional factor, involving only velocities, corrects for the motion of the fluid in which the fast particles are produced. The ratio of transit time of fast hydrogen to that of the bulk fluid through a given distance is equal to the reciprocal of the corresponding speeds: $v / (v_f + v)$. One minus this fraction, namely $v_f / (v_f + v)$, determines the fractional loss of fast hydrogen from the moving shell. If the fast hydrogen were produced with $v_f = 0$, then this "fast" hydrogen would never leave the shell, regardless of how large its mean free path is. On the other hand, if $v_f \gg v$, then the loss of fast hydrogen from the fluid shell is determined entirely by its mean free path.

The density of fast particles at point R is obtained by dividing $\Phi(R)$ by the speed of the fast particles at R. As Huebner and Keady have shown, the functions $\Phi_1(R)$ and $j_1(R)$ are related to the Feautrier formalism in radiative transfer. Thus the procedure outlined here is suitable for including higher order collisional effects without using the more costly Monte Carlo procedure.

Radiative cooling is designated by ε_{rad}; the main contributor to it is H_2O. A semiempirical equation for it has been given by Shimizu (1976), but, as was pointed out by J. Crovisier in a private communication, without consideration for radiation trapping in the coma. An optical depth calculation is sufficient to approximate the radiation trapping. The optical depth is

$$\tau' = \int_R^\infty n\,\sigma\,dR \cong n_o v_o R_o^2 \sigma \int_R^\infty \frac{dR}{v R^2}, \qquad (9)$$

From this it can be shown that radiative cooling is not important before the onset of dissociative and chemical heating, unless the comet is very small or is at a large heliocentric distance r so that its gas production and therefore its density n_o are very low. Shimizu's formula for radiative cooling by H_2O, modified by the above optical depth, is

$$\varepsilon_{rad} = \frac{8.5 \times 10^{-19} T^2 n^2}{n + 2.7 \times 10^7 T} \times \exp \tau', \quad \text{erg cm}^{-3}\,\text{s}^{-1}, \quad (10)$$

where n is the R-dependent number density of water molecules.

Eqs. (1) through (3) can now be solved for ρ, p, and v. The local number density n can then be obtained using the mean molecular weight for the local gas composition and from this and the ideal gas law the local temperature T can be obtained. Model results for density ρ, pressure p, temperature T, and velocities v_s, v, $v_{H(f)}$, and $v_{H_2(f)}$ are shown in Fig. 1. These results are influenced by chemistry that will be discussed below.

Chemical reactions between evaporated mother molecules are very slow and of secondary importance only in large comets, but dissociation and ionization by solar UV radiation produce highly reactive radicals and ions. While it is mostly the visual part of the solar spectrum that causes production of gas from the nucleus, it is the UV spectrum that is responsible for initiating chemical reactions. However, near the nucleus surface, the coma is opaque to UV radiation except in small comets and comets at large heliocentric distance. From the relationship between density n and distance R in the coma, from the photo cross sections of the coma constituents, and from their relative abundances, the attenuation of the solar UV radiation can be calculated. This attenuation depends not only on the gas density and distance above the nucleus, but also on wavelength and angle of incidence of the sunlight. However, calculations show that the attenuation varies only weakly with angle as measured from the comet-sun axis except for a narrow cone in the antisolar direction.

Mother molecules dominate the composition in the region where attenuation is largest. Assuming an R^{-2} dependence for the density n, which is a good approximation except very close to the nucleus, the UV optical depth is

$$\tau''(\lambda) = -\int_\infty^R \sum_i n_i \sigma_i(\lambda) \, dR \cong -\sum_i n_i \sigma_i(\lambda) R \quad . \tag{11}$$

The summation over i is essentially only for mother molecules. The contribution from radicals is very small and can only be calculated after a complete chemistry calculation of the coma. A second iteration of the coma calculation can be carried out in which the contribution from the radicals and a more exact density profile are considered.

The effective photo rate coefficients can be calculated from

$$k_i = \sum_j k_i(\lambda_j) \times e^{-\tau''(\lambda_j)} , \tag{12}$$

here the $k_i(\lambda_j)$ are the photo rate coefficients for species i in the wavelength bin λ_j.

An alternative to producing radicals by dissociation of mother molecules in the solar UV radiation field is the decomposition in visible sunlight of clathrates containing radicals, should they exist in the comet nucleus. In that case chemical reactions could take place even in the innermost coma where UV radiation is strongy attenuated. So far clathrates have not been identified in comets.

Another very important process is photodissociative ionization (PDI). PDI is a process in which an extreme UV photon causes a molecule to be ionized into an excited state from which the molecular ion decays by dissociation. An example is

$$CO_2 + h\nu \rightarrow O + CO^+ + e \quad . \tag{13}$$

Even though the rate coefficients for PDI are small, the hard photons necessary for PDI penetrate deeply into the coma where they are deposited in a relatively thin shell with high density n_+. Therefore, this process is important for the production of CO^+ deep in the coma, at about 10^3 km from the nucleus, where CO^+ is observed. PDI of other CO-bearing molecules gives similar results.

Photon energies in excess of the ionization potential produce energetic electrons (often 10 eV or more) that can cause further ionization and dissociation by impact. This is a secondary process with effects on the coma that have not been investigated in detail, but are expected to increase the fraction of ions.

Electron dissociative recombination is another very important process in which an electron recombines with a molecular ion and simultaneously dissociates that ion. Because of the Coulomb interaction between the electron and the ion, this process has a very large cross section and therefore a large rate coefficient. In the temperature region of interest in comet comae, the cross section increases approximately inversely proportional to the electron velocity relative to the ion, or proportional to $T^{-1/2}$.

Ion gas phase reactions including positive ion-atom interchange and positive ion charge transfer compose a large part of the chemical reaction network that must be considered. Examples of these two reactions are

$$CO^+ + H_2O \rightarrow HCO^+ + OH \quad , \tag{14}$$

$$CO^+ + H_2O \rightarrow H_2O^+ + CO \quad . \tag{15}$$

Also important is radiative electronic state deexcitation, such as

$$O(^1D) \rightarrow O(^3P) + h\nu \quad , \tag{16}$$

which is one of the forbidden O transitions.

Of lesser importance are neutral rearrangements, radiation stabilized neutral recombinations, radiation stabilized positive ion-neutral associations, neutral-neutral associative ionizations, radiative recombinations, three-body neutral recombinations, and three-body positive ion-neutral associations. Negative ions play no role in comet comae because the densities are too low. Heterogeneous (grain surface) reactions may be important, but have not been considered because virtually no rate coefficients are available. Methane and ethane may be produced by catalytic reactions if iron or sodium are present on grain surfaces.

Because of the expansion of the coma gas as it moves radially outward, chemical reactions take place in a continually diluting gas exposed to ever increasing solar UV radiation. For this reason chemical kinetics must be considered instead of a steady state. Some chemical reactions proceed very fast compared to the fluid dynamic dilution, while others are slow. In the practical application of solving a network with several hundred photo-chemical reactions, the time step for the chemistry needs to be much smaller than the fluid dynamic time step at which the density, temperature, and attenuated solar flux are recalculated. Fluid dynamic time steps are spaced approximately logarithmically. Only at large distances from the nucleus does the chemical time step approach the fluid dynamic time step. A stiff differential equation solver technique is used to solve the chemical reaction network. Although some very fast chemical reactions may reach a chemical steady state in any one fluid dynamic time step, many reactions will not, and some species will "freeze-in" as the gas expands and reactions terminate.

Since different reactions dominate in the same comet at different heliocentric distances and at different distances into the coma, a network of several hundred reactions is required. competing processes leading to the same product or destroying the same reactant can change their relative importance. For example, at heliocentric distance $r \cong 1$ AU, in the innermost coma, CO^+ is produced by photodissociative ionization of CO-bearing molecules, while in the intermediate coma it is produced by photoionization of CO. The crossover of these two processes occurs within $\sim 10^3$ km, while at $r \cong 3$ AU this crossover is at $\sim 5 \times 10^3$ km. At $r \cong 1$ AU destruction of CO^+ in the inner coma occurs by positive ion-atom interchange (e. g., reaction 14) and charge exchange (e. g., reaction 15), while in the outer coma electron dissociative recombination is the dominant destruction mechanism of CO^+. The crossover occurs within $\sim 10^4$ km, while ar $r \cong 3$ AU it is at $\sim 2 \times 10^4$ km. Another example is the production of H_3O^+. Different initial compositions may cause different reactions to dominate. For some particular composition, at heliocentric distance $r \cong 3$ AU, the process

$$H_2O^+ + H_2O \rightarrow H_3O^+ + OH \quad , \tag{17}$$

may dominate in the entire coma out to 10^5 km. At $r \cong 1$ AU, it may dominate out to $\sim 3 \times 10^4$ km; beyond this the reaction

$$H_2^+ + H_2O \rightarrow H_3O^+ + H \quad , \tag{18}$$

may dominate. At $r \cong 0.6$ AU reaction (17) dominates out to $\sim 10^4$ km, at $\sim 3 \times 10^4$ km the reaction

$$HCO^+ + H_2O \rightarrow H_3O^+ + CO \quad , \tag{19}$$

dominates, while at $\sim 10^5$ km

$$H_2O^+ + H_2 \rightarrow H_3O^+ + H \quad , \tag{20}$$

dominates. Competing, parallel processes also lead to an unexpected benefit: they tend to minimize the effects from uncertainties in the rate coefficients.

Figure 3 illustrates results for number densities from a typical calculation and composition (see Table III). Predicted, but as yet not detected species include H_3O^+ with certainty, HCO^+, C_2H_2, C_2H_4, and possibly HCO_2^+ and CH_2OH^+ if a hydrocarbon precursor exists in the nucleus with sufficient abundance, NH_4^+ if NH_3 exists even with small abundance, and NO and H_2CN^+. (Even though C_2H_2 is a common result of coma chemistry, in some compositions it is also assumed to be mother molecule.)

TABLE III

An Assumed Composition for the Frozen Gases in the Nucleus

Species	Percent	Species	Percent	Species	Percent
H_2O	43.004	N_2	5.225	NH_2CH_3	0.201
H_2CO	22.105	CO	2.814	C_2H_2	0.161
CH_4	13.464	HCN	0.482	NH_3	0.080
CO_2	12.057	CH_3CN	0.201	$H_2C_3H_2$	0.005

For $R > R_c$ collisions become rare, chemical reactions (except those involving Coulomb interactions) cease, dissociation and chemistry products stop to share their excess energy and continue on their path as determined by their initial velocity. (We ignore here further complications caused by radiation pressure on neutrals, solar wind interaction, and gyration of ions and electrons in magnetic fields that cause counter streaming and two-stream instabilities.)

Of particular interest and importance in the free molecular flow region is the H-coma. The velocities of atomic hydrogen can be determined from the Doppler profile of solar Lyman-α

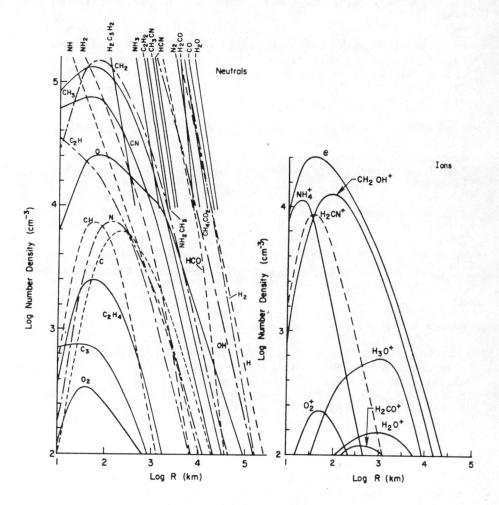

Fig 3. Model results for number densities of selected species plotted as function of distance into the coma, log R, for r = 1 AU. Mother molecules are not shown below a density of 10^4 cm^{-3} in order to keep the graph readable. Also the ion species are plotted separately for the purpose of clarity.

radiation scattered by the H-coma. Analysis has shown that at least two major velocity components exist: one at about 20 km/s, the other at about 8 km/s. It is believed that the first of these components results from the excess energy imparted to H in the dissociation of OH, which itself is a dissociation product of water. The dissociation cross section of OH is only approximately known; predissociation and superposition of solar Fraunhofer lines on the predissociation lines play an important role (Jackson, 1980).

From the column density profile of H and OH a production rate of H_2O can be determined. For a large number of bright comets this water production rate is $\sim 10^{29}$ molecule/s. For the typical value $Z \cong 10^{17}$ molecules/(cm^2 s), this yields a radius of a few kilometers for the nucleus.

Radiation pressure of solar Lyman-α on the H-coma causes deviations from spherical symmetry. On the subsolar side the atomic hydrogen density decreases more rapidly than in the direction perpendicular toward the sun and it extends further in the tail direction.

The neutral component of the coma gas streams almost radially outward, essentially unimpeded by the solar wind. Small deviations from this are caused by radiation pressure on the neutrals and near the contact surface (the boundary for solar wind penetration into the coma) where some collisions occur between neutrals and nonradially moving ion streams. Radiation pressure is most effective on species with strong resonance lines in the visible region of the spectrum. Just as in the H-coma, radiation pressure causes an asymmetry of the coma in the tail direction.

On the other hand, coma ions move radially outward until they encounter the contact surface. At that point they experience strong interactions with the solar wind ions. Between the contact surface (probably several times 10^3 km from the nucleus in the subsolar direction) and the bow shock ($\sim 10^6$ km) electrons, solar wind protons, and photons impacting with coma neutrals may produce some of these ions. The ion-loaded solar wind slows down as it penetrates inward from the bow shock, and finally stagnates at the subsolar point on the contact surface. In this region cometary ions couple to the solar wind and are swept tailwards. The solar wind interaction with the coma has been modeled extensively by Schmidt and Wegmann (1982). They have also taken into account the Alfvén critical velocity effect. This effect is a collective process, caused by plasma instabilities, in which electrons produced by collisional ionization are heated so that they cause further ionization when a plasma with a magnetic field collides with a neutral gas and the relative kinetic energy between plasma (ion) and gas (molecule) is larger than the ionization energy of a gas molecule.

In the region between contact surface and outer shock the outward streaming coma plasma and the solar wind can be treated as fluids since Coulomb interaction results in large collision cross sections. Photolytic processes continue in this region and further complicate the situation. Fortunately, at least in first approximation, this region can be treated semi-independently from the region inside the contact surface. But multi-fluid flow must be considered.

No direct comparisons of the model density, pressure, and temperature profiles shown in Fig. 1 are possible. However, comparisons with the model velocity profiles can be made. Observations indicate a mean bulk velocity in the visual coma (out to a few 10^4 km) of about 1 km/s. Model results are in good agreement with these. The velocity profile for the fast atomic hydrogen H(f) in Fig. 1 is an average from many different dissociations, but H_2O dissociation is an important contributor to it. In the outer coma, beyond several 10^4 km, the model indicates a velocity of ~20 km/s or slightly higher, in agreement with the observed high velocity component. The low velocity (~8 km/s) component has not been separated out in the model calculations. $H_2(f)$ cannot be observed; the abundance in the coma is too low for present instrumentation.

Model column densities for selected species are shown in Fig. 4. The atomic hydrogen density is split into two parts: the thermalized hydrogen, labeled H, and the fast hydrogen, labeled H(f). The total column density is the sum of these two components and gives a profile (not shown) that increases somewhat towards the center of the coma. This is a feature that qualitatively agrees with observations.

The C_2 column density in Fig. 4 is much flatter than that of any other neutral species. This is in agreement with observations. Many competing reaction paths lead to the production of C_2, giving it this profile. Comparison of ranges for various species, taking into account their relative oscillator strengths (e. g., a factor of ~40 for C_3/C_2), is in qualitative agreement with observed ranges.

Reactions (14) and (15) are main reasons to keep the abundance ratio of CO^+/H_2O^+ in model calculations from reaching values much larger than one. This is in disagreement with the only measured value which is ~30.

Observations of comets usually yield column densities over a small region of the coma or averaged over the coma subtended by the instrument aperture. Some observers convert these to a production rate Q, assuming a steady state,

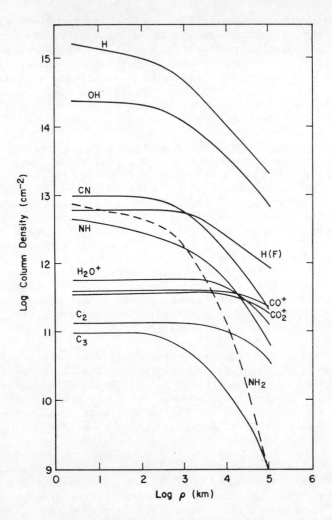

Fig. 4. Model results for column densities of selected species plotted as function of projected distance on the sky, log ρ. Thermal hydrogen is indicated by H, while fast hydrogen is labeled H(F). Note that the column density for C_2 is flat compared to that of the other neutral species.

$$Q = N/\tau \, , \tag{21}$$

where N is the number of emitters in the column. The life time τ of the species depends not only on its radiative destruction time, but also on chemical reactions (including solar wind charge exchange for ions) and therefore on a coma model, usually the Haser model. For comparing with the model presented here, production rates are less desirable than column densities at a given distance in the coma or averaged over a given distance.

In summary, dust-free models for parabolic comets give reasonable agreement with overall, averaged observations. Although some models have been developed with dust or solar wind interaction, they are preliminary and lack internal consistency. Sulfur chemistry and sulfur compound observations may become probes for the most central part of the coma. We believe to understand the overall features in the physics and chemistry, but lack most details.

Model calculations for any particular initial composition of the nucleus are only as good as the atomic and molecular data that are used in the model. Cross sections for photodissociation, photoionization, and photodissociative ionization for a number of molecules are not known. For other molecules cross sections are only known over very limited ranges of photon energies. For still others, the cross sections may be known, but the branching ratios for the various decay channels are not known. To these groups of molecules or radicals belong: OH, CN, C_2, NH, C_2H, HCO, CH_3CN, NH_2, C_3, HCN, and various more complex hydrocarbons.

Also not well known are some of the rate coefficients for chemical reactions in the temperature range of interest (25 K < T < 600K) or the branching ratios for their products. Although the total rates for electron dissociative recombination are reasonably well established for comet chemistry, the branching ratios for the products are very often not known at all.

ACKNOWLEDGMENTS

I want to thank Dr. J. J. Keady for valuable discussions and for the preparation of the figures. I am grateful to the North Atlantic Treaty Organization and to the National Science Foundation for supporting presentation of this research. The research was supported by funds from the National Aeronautics and Space Administration Planetary Atmospheres Program, and performed under the auspices of the Department of Energy.

REFERENCES

Houpis, H. L. F. and Mendis, D. A. 1980, Astrophys. J. 239, pp. 1107-1118.

Huebner, W. F. and Keady, J. J. 1983, "Energy balance and photochemical processes in the inner coma, in International Conference on Cometary Exploration, Hungarian Academy of Sciences, Vol. I, pp. 165-183.

Huebner, W. F., Giguere, P. T. and Slattery, W. L. 1982, "Photochemical processes in the inner Coma," in Comets, L. L. Wilkening, Ed., Univ. Arizona Press, pp. 496-515.

Huebner, W. F. and Keady, J. J. 1984, Astron. Astrophys. 135, pp. 177-180.

Ip, W. H. 1983, Astrophys. J. 264, pp. 726-732.

Jackson, W. M. 1980, Icarus 41, pp. 147-152.

Marconi, M. L. and Mendis, D. A. 1982a, Moon Planets 27, pp. 27-46.

Marconi, M. L. and Mendis, D. A. 1982b, Moon Planets 27, pp. 431-452.

Schmidt, H.-U. and Wegmann, R. 1982, "Plasma flow and magnetic fields in comets," in Comets. L. L. Wilkening, Ed., Univ. Arizona Press, pp. 538-560.

Shimizu, M. 1975, Astrophys. Space Sci. 36, pp. 353-361.

Shimizu, M. 1976, Astrophys. Space Sci. 40, pp. 149-155.

Whipple, F. L. 1950, Astrophys. J. 111, pp. 375-394.

Whipple, F. L. 1951, Astrophys. J. 113, pp. 464-474.

MOLECULAR SPECTRA OF THE OUTER PLANETS AND SATELLITES

Catherine de Bergh

Observatoire de Paris-Meudon

Methods and results of analyses of the most recent molecular spectra of the outer planets and the satellites Titan and Triton obtained from ground-based, plane, satellite or spacecraft observations are reviewed. These spectra cover a large part of the electromagnetic spectrum. They have largely contributed to our present understanding of the nature, formation and evolution of these outer solar system objects. In order to better interpret existing and forthcoming data, it is essential to pursue or initiate appropriate laboratory and theoretical studies of the molecules detected in these atmospheres.

INTRODUCTION

Although several space probes - Pioneer 10 and 11, Voyager 1 and 2 - have visited the two closest outer planets of our solar system, Jupiter and Saturn, and the satellite Titan, no probe has been sent into their atmospheres. Our knowledge on the atmospheres of planets and satellites of the outer solar system therefore only relies on remote-sensing observations, either from the ground or from space. The different approaches which can be used to probe a planetary atmosphere from a distance are : photographic, photometric, polarimetric, radiometric and spectroscopic observations, stellar occultations and, during a fly-by, occultations by the planetary atmosphere of the radio signals emitted by the spacecraft.

The studies which can be made by molecular spectroscopy, combined or not with other types of measurements, pertain to different fields of planetology. The knowledge on the bulk atmospheric composition of a planet obtained by molecular spectroscopy can provide

G. H. F. Diercksen et al. (eds.), Molecular Astrophysics, 331–349.

informations on the physico-chemical processes active in the atmosphere, on the internal structure of the planet, on the formation and evolution of the planet and its atmosphere and on the composition of the primitive solar nebula from which the planets have originated. Concerning this last point, the study of the giant planets which, contrarily to the terrestrial planets, have retained their primordial atmosphere, is particularly interesting.
The measurements of the abundance and vertical distribution of photodissociation products inferred from the analysis of their spectral signatures give indications on the major photochemical processes occurring in the upper planetary atmospheres. Through the study by molecular spectroscopy of the vertical distribution of condensable species and of their local and seasonal variations, some informations on meteorology can be obtained. Finally, the nature of the surface of planets or satellites can also be probed by molecular spectroscopy.

The present review includes all the outer planets and only the satellites with an atmosphere or likely to possess an atmosphere. After a brief introduction on the transfer of radiation in a planetary atmosphere, whose understanding is essential in order to interpret planetary spectra, we give for each planet or satellite the most recent results obtained from the study of their molecular spectra with some indications on the techniques used and the remaining causes of uncertainties on the obtained measurements. Finally we enumerate some of the expected new developments in the exploration of the outer solar system by molecular spectroscopy.

1. RADIATION TRANSFER THROUGH A PLANETARY ATMOSPHERE

For a planet with an atmosphere, molecular spectra are the result of transfer of planetary radiation through the atmosphere. Some characteristics of planetary radiation and its transfer through a planetary atmosphere are given here.
The most obvious source of planetary radiation is solar radiation, which can be approximated as the radiation of a blackbody at 6500°K and, therefore, with a maximum near 0.5 micron. The amount of solar radiation received by a planet or a satellite depends on its distance to the Sun and its diameter. The molecules and cloud particles present in the planetary atmosphere and, in the case of planets or satellites with a thin atmosphere, the planetary surface itself, absorb and scatter solar radiation. Part of the solar radiation not absorbed is reflected back to space at the same wavelengths. The radiation which is not reflected back to space heats the atmosphere which then emits thermal radiation, essentially in the infrared because of the low temperatures of the planets. However, measurements of the global thermal radiation emitted by the giant planets have indicated that three of the giant planets, Jupiter, Saturn and Uranus, radiate more energy than they receive from

the Sun (1). This additional energy is probably related to processes occurring in the interiors of these planets which do not have a solid surface. The bulk of planetary radiation therefore consists of reflected solar radiation which peaks near 0.5 micron and thermal infrared radiation (from solar and thermal origins) whose maximum depends on the planetary temperature.

In the thermal infrared range, as a result of the transfer of radiation in the atmosphere, the planetary radiation emitted at a given wavelength comes preferentially from atmospheric levels above which the optical thickness is about unity. Because of the variation of atmospheric opacity with wavelength, different altitudes are probed at different wavelengths. Molecular signatures can be detected in planetary spectra because the temperature varies with altitude. If the temperature increases with altitude - as in the stratosphere of a planet - a molecular absorption line will appear in emission in the planetary spectrum, as the level probed at the wavelength corresponding to the line center (maximum of absorption) is higher and therefore warmer than the level probed at the wavelengths corresponding to the wings of the line (minimum of absorption). On the contrary, if the temperature decreases with altitude - as in the troposphere of a planet - the spectral signature will appear in absorption. As a consequence, spectra in the thermal infrared can be used to probe the vertical thermal profile if the atmospheric composition - and therefore opacity - is known or to probe the atmospheric opacity - and therefore composition - if the thermal profile is known.

In the reflected sunlight range, the situation is completely different. The part of the atmosphere probed at a given wavelength is

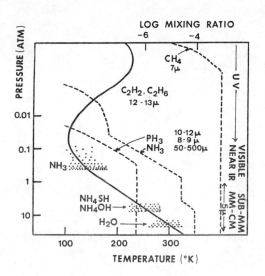

Figure 1.
Vertical structure of the atmosphere of Jupiter with probable cloud locations and compositions. The levels probed in different parts of the electromagnetic spectrum are indicated. (1)

essentially limited at the bottom by the cloud opacities which, if too high, prevent penetration of solar radiation (figure 1). The detectability of molecular signatures therefore depends essentially on the column abundance of molecules present in the atmospheric region which can be probed. Signatures always appear in absorption. An accurate knowledge of the atmospheric thermal structure is not required to interpret planetary spectra. However, at these short wavelengths, scattering by gaseous molecules and -or- cloud particles may have a strong influence on the tranfer of radiation and this a major difficulty in the analyses of spectra in the reflected sunlight radiation range.

2. SPECTRA OF JUPITER AND SATURN

A large number of molecules have been detected in spectra of Jupiter and Saturn, mostly from ground-based spectroscopy. These molecules are, with some approximate mixing ratios :
H_2 (90 %), He (10 %), CH_4 (2×10^{-3}), $^{13}CH_4$, NH_3 (4×10^{-4}), HD, CH_3D (3×10^{-7}), C_2H_2 (3×10^{-8}), C_2H_6 (5×10^{-6}), GeH_4 (7×10^{-10}), H_2O (3×10^{-5}), CO (10^{-9}), PH_3 (6×10^{-7}) and HCN (2×10^{-9}) for Jupiter and H_2 (96 %), He (4 %), CH_4 (5×10^{-3}), $^{13}CH_4$, NH_3 (10^{-4}), HD, CH_3D (4×10^{-7}), C_2H_2 (2×10^{-7}), C_2H_6 (3×10^{-6}), PH_3 (10^{-6}), HCN (2×10^{-9}), C_3H_4 and C_3H_8 for Saturn (references can be found in the review papers 1 and 2). The detections of H_2O and GeH_4 in Jupiter were obtained from a plane. The detections of C_3H_4 and C_3H_8 in Saturn were made in Voyager spectra. Because of the very high quality and extended spectral range - from 4.2 to 55 microns- of the spectra recorded by the IRIS interferometer on-board the Voyager spacecrafts, and because of the very large number of spectra obtained with a good spatial resolution, the Voyager IRIS investigation has strongly contributed to the recent improvements in gaseous abundance measurements and in the determination of the temperature and cloud structures of these two planets.

2.1. The collision-induced H_2 absorptions

The absorption path lengths of H_2 and He in the atmospheres of the giant planets are such that signatures due to collision-induced H_2-H_2 and H_2-He dipole transitions can be detected. The spectral range between 200 and 700 cm^{-1} where the H_2-H_2 collision-induced opacity dominates and where there is some contribution, although very weak, of the collision-induced H_2-He absorption to the opacity, have been used extensively for two purposes : temperature measurements in the troposphere and measurements of the H_2/He ratio. The best analyses have been made from the Voyager-IRIS spectra.

Temperature measurements were obtained by direct inversion of the equation of radiative transfer at several frequencies between 280 and 600 cm^{-1} (3, 4) by selecting spectral ranges where the H_2-H_2

Figure 2. Atmospheric pressure levels in Jupiter where the optical depth equals 1 at the spectral resolution of the Voyager-IRIS spectra. At each wavenumber, the maximum contribution to the emergent radiation originates from this level. (20)

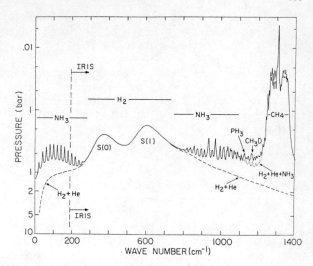

absorption coefficient is very sensitive to temperature. Because of the significant variation of the H_2 opacity in this spectral range, the temperature is obtained over an extended altitude range-between 200 and 1000 mb in the case of Jupiter (figure 2).

Determinations of the H_2/He ratio were obtained by two different methods (5). The first method utilized only the two IRIS spectra corresponding to the locations of ingress and egress of the Voyager spacecraft along with the radio occultation measurements at the time of ingress and egress which provide temperature profiles as a function of the mean atmospheric molecular weight. By combining these two sets of measurements, the mean molecular weight of the atmosphere and, therefore the percentage of He, is obtained. The second method takes advantage of the different shapes of the H_2-H_2 and H_2-He collision-induced absorptions to extract the very small contribution of He to the absorption in the IRIS spectra. Many measurements are possible since a very large number of IRIS spectra have been recorded, but this method depends in a very critical way on the accuracy of the H_2-H_2 and H_2-He laboratory measurements and theoretical data. In the case of Jupiter these two methods have given essentially the same results. In table 1 is given the average He/H_2 ratio. The same techniques have been applied to Saturn spectra (6) and it has been found that, as predicted by Stevenson and Salpeter (7, 8), there is proportionally less helium in Saturn than in Jupiter (table 1). This difference would be due to differences in the present internal structures of the two planets. There would be in Saturn, but not in Jupiter, condensation of helium in the interior of the planet. This He condensation would have started at some stage of the planetary evolution as the planet cooled and helium became immiscible with metallic hydrogen. The condensation of helium at the center of the planet would have induced depletion of He in its atmosphere. If this scenario is the right one, then the sources of internal energy would not be the same for Jupiter and Saturn (7, 8).

Two features have been recently detected in the region of the H_2-H_2 and H_2-He pressure-induced absorptions at 354.4 cm^{-1} and 587.1 cm^{-1} in IRIS spectra of Jupiter and Saturn (9). These features are currently being interpreted as due to rotational transitions in the $(H_2)_2$ dimer (10, 11).

2.2. The $^{12}C/^{13}C$ ratio

Measurements of the $^{12}C/^{13}C$ ratio in Jupiter have been made from a study or $^{13}CH_4$ signatures in two different spectral ranges : at 1.1 micron from the ground (12) and at 7 microns from Voyager spectra (13). These two sets of measurements give different results (table 1). The value obtained from Voyager spectra is surprisingly high compared to measurements in the terrestrial planets which all indicate values close to the terrestrial ratio, except in the case of radio measurements which probe the mesosphere of Venus (14) and lead to very high values for the $^{12}C/^{13}C$ ratio (185 ± 69). However, this last measurement is made from CO which is a minor carbon compound in the atmosphere of Venus and some ^{13}C fractionation between ^{13}CO and $^{13}CO_2$ may have occurred at high altitudes (14). The Jupiter measurement at 1.1 micron in the troposphere of the planet was made in a spectrum recorded at a resolution of 0.1 cm^{-1} in which three manifolds of the $3\nu_3$ band of $^{13}CH_4$ could be detected (12). The main source of uncertainty in this measurement is due to a lack of spectroscopic data for the very weak $^{12}CH_4$ absorptions which are mixed with the $^{13}CH_4$ absorptions. The measurement of Courtin et al. (13) in the stratosphere of the planet (figure 2) is based on the analysis of an average of 1260 Voyager spectra at a resolution of 4.3 cm^{-1} in which the contribution of $^{13}CH_4$ to the total radiance can be clearly detected, especially in the region of the Q-branch of the ν_4 band, near 1298 cm^{-1}. Uncertainties on this measurement are due to uncertainties on line intensities and line shapes for the lines of methane. There is at present no obvious explanation for the discrepancy in the $^{12}C/^{13}C$ ratios obtained by these two independent methods.

In the case of Saturn, the only measurement of the $^{12}C/^{13}C$ ratio (table 1) comes from measurements of the $^{12}CH_4/^{13}CH_4$ ratio in spectra at 1.1 micron recorded for a resolution of 0.2 cm^{-1} (12). New laboratory data and improved planetary spectra have been recently obtained which should help to improve the ground-based measurements at 1.1 micron for the two planets (15 and de Bergh, private communication).

2.3. The D/H ratio

As deuterium is destroyed in stars, the D/H ratio decreases with time in the interstellar medium. The measurement of this ratio in planets like Jupiter and Saturn which have kept their primordial atmosphere should therefore provide a measurement of D/H in the interstellar medium 4.5 billion years ago (16). The most direct measurements of the D/H ratio in Jupiter and Saturn are obtained from a comparaison between HD and H_2 abundances (D/H = 1/2 x HD/H_2). Several extremely weak dipole lines of HD in the 4-0 and 5-0 bands have been detected in visible spectra of Jupiter and Saturn. The measurements of D/H have evolved with time as more accurate HD line intensities became available and also as the effects of scattering on the line equivalent widths were better handled, but there is still a large scatter in the measurements. Only the latest ones are quoted in table 1.

Deuterium has also been detected under the form of CH_3D in Jupiter and Saturn. In principle, measurements of the D/H ratio can be obtained through measurements of the CH_3D/CH_4 ratio, but in a more indirect way. Indeed, the relation between these two abundance ratios is : D/H = (1/4f) x (CH_3D / CH_4), where f, the fractionation factor (which is higher or equal to 1), accounts for differences in binding energies between HD and CH_3D molecules. This factor increases as temperature decreases. Uncertainties on the values of f appropriate for Jupiter and Saturn introduce additional uncertainties on the D/H ratio itself. For Jupiter and Saturn, measurements of the CH_3D/CH_4 ratio have been obtained from 5 microns and 8.6 microns ground-based and Voyager spectra (20, 21, 22, 23). An average of measurements for Jupiter is given in table 1 (23). In the case of Saturn, the $3\nu_2$ band of CH_3D has been detected in ground-based spectra at 1.6 micron and is currently beeing used to provide an additional measurement (24). The analysis of Voyager IRIS spectra at 8.6 microns (23) indicates (in agreement with preliminary results at 1.6 micron) that the CH_3D/CH_4 ratio is smaller in Saturn than in Jupiter (table 1), which was totally unexpected, but the errors are still large. Improved measurements are required before drawing firm conclusions on the D/H ratios in these two planets.

2.4. The $^{14}N/^{15}N$ ratio

Several lines of $^{15}NH_3$ have been detected in ground-based spectra of Jupiter near 12 microns at spectral resolutions better than 1 cm^{-1} (25, 26, 27). However, because of the low spatial resolution of these ground-based observations and as both temperature profiles and NH_3 vertical distributions are very inhomogeneous over extended parts of the disk of Jupiter, the accuracy on the measurements of the $^{14}N/^{15}N$ is not very good. Nevertheless, it has been suggested recently that the $^{14}N/^{15}N$ ratio may be higher on Jupiter than on Earth (terrestrial ratio : 270) from a preliminary analysis of several new ground-based measurements (28).

Table 1. Some abundance ratios in Jupiter and Saturn atmospheres

abundance ratio	JUPITER	SATURN
$Y = He/H_2$ (per mass) (solar ratio : 0.22 - 0.27)	0.118 ± 0.04 (5)	0.06 ± 0.05 (6)
C/H (solar ratio : 4.7×10^{-4})	$1.09 \pm 0.08 \times 10^{-3}$ (29)	$2.2 ^{+1.2}_{-1} \times 10^{-3}$ (23) $2 \pm 1 \times 10^{-3}$ (31)
D/H from HD : from CH_3D :	$5.1 \pm 0.7 \times 10^{-5}$ (17) $2.0 ^{+1.1}_{-0.8} \times 10^{-5}$ (18) $3.3 \pm 1.1 \times 10^{-5}$ (23)	$5.5 \pm 2.9 \times 10^{-5}$ (19) $1.6 ^{+1.3}_{-1.2} \times 10^{-5}$ (23)
$^{12}C/\ ^{13}C$ (solar and terrestrial ratios : 89)	at 1.1 micron : $89 ^{+12}_{-10}$ (12) at 7 microns : $160 ^{+40}_{-55}$ (13)	$89 ^{+25}_{-18}$ (12)

2.5. The C/H ratio

The most accurate measurements of this ratio in the atmosphere of Jupiter has been obtained from the Voyager IRIS spectra (29). The measurements were made in the region of the ν_4 band of methane. Three spectral ranges were used between 1215 and 1455 cm^{-1} which correspond to spectral ranges with minimum contamination by absorptions due to other molecules and to spectral ranges where the uncertainty on the exact shape factor of absorption lines far in the wings did not have too much influence. Indeed, at the spectral resolution of the IRIS spectra (4.3 cm^{-1}) far wings of strong absorption lines can strongly affect the planetary radiance in nearby regions and, unfortunately, little is known about the exact shapes of lines such as CH_4, CH_3D, PH_3 and NH_3 far in the wings when these lines are broadened by H_2. A selection of 94 IRIS spectra was made for which the influence of clouds should be weak and these spectra were co-added. The temperature profile in the probed atmospheric regions was obtained from radiance measurements at a few selected frequencies between 200 and 1400 cm^{-1}. The uncertainty on the value of the obtained C/H ratio (table 1, 30) includes an uncertainty of $\pm$ 10 % on the ν_4 band intensity itself, about which there has been a long controversy (29).

Similar measurements were made for Saturn from the IRIS spectra (23). The enrichment in C/H compared to the solar C/H ratio appears to be even more important for Saturn than for Jupiter. The result obtained for Saturn is much less accurate than the result for Jupiter because of the colder temperatures of Saturn. The obtained values for Saturn from the IRIS measurements are in good agreement with earlier measurements made from the ground at 1.1 micron (31). The interpretation of the 1.1 micron spectra - at a resolution of 0.2 cm^{-1} - was made by using scattering models and line-by-line techniques for the methane absorptions.

Now that the enrichment in carbon on Jupiter and Saturn compared to the Sun seems to be fairly well established, its causes remain to be found. Several suggestions have been made. One of them is that the enrichment in carbon could be due to the evaporation of some of the ices initially contained in the core of the planets, as the planets warm up, which would later induce some carbon enrichment in their atmosphere compared to an original solar composition. Another suggestion is that this enrichment would be the result of impacts of planetesimals and meteorites which would bring additional volatile elements into the planetary atmospheres after the planets were formed (30). In this respect, it would be interesting to measure other elemental abundance ratios such as N/H, O/H or P/H but, as shown in the next paragraph, these measurements are difficult to obtain.

2.6. Molecules with a variable mixing ratio

Ammonia condenses in the atmospheres of Jupiter and Saturn at pressures lower than 1 atmosphere. In addition, thermochemical models predict the existence of clouds of NH_4SH or NH_3-H_2O in the deep tropospheres of these two planets and the presence of such clouds seems to be corroborated by abrupt changes which are observed in the ammonia mixing ratios at pressures of a few bars (32, 33). Consequently, only measurements which probe below these clouds can provide N/H measurements representative of the global planetary compositions. Microwave spectra, which probe down to pressures of about 10 bars, are the most appropriate for this purpose. The analyses of such measurements give values of the N/H ratio which are about twice higher than the solar ratio in the case of Jupiter and 3 times higher than the solar ratio in the case of Saturn (23, 32, 34). However, there are still some ambiguities on these measurements because of possible additional opacity by water vapor clouds - whose existence and location are debatable - and because of uncertainties on the line shape of ammonia lines broadened by hydrogen at millimeter wavelengths (32, 35).

No oxygen molecule has been detected in Saturn. The most abundant oxygen molecule in Jupiter, by far, is H_2O and therefore, measurements of the ratio O/H can be inferred from H_2O/H_2 measurements. Water vapor has been detected at 5 microns. As none of the dominant absorbers in Jupiter - CH_4, NH_3, H_2 - has strong absorption bands near 5 microns, measurements at 5 microns allow to probe deep atmospheric layers (figure 1). The values of H_2O/H_2 obtained from such measurements are very low. They correspond to a ratio O/H 30 to 100 times less than the solar ratio (20, 36). The Jupiter atmosphere is therefore surprisingly deficient in oxygen. However, as H_2O may condense at lower altitudes, the measured O/H ratio probably provides only a lower limit to the actual O/H ratio corresponding to the bulk atmospheric composition.

The PH_3/H_2 mixing ratio varies with altitude in both Jupiter and Saturn. From measurements at 5 microns, the ones which allow to probe the deepest into the armosphere, it si found that the P/H ratio may be solar in Jupiter and 2 to 3 times solar in Saturn, with the restriction that, at least for Jupiter, local variations of the PH_3 abundance by as much as a factor of 2 have been detected (23, 37, 38).

The mixing ratio of a molecule can vary with altitude because of condensation, convection, chemical reactions or photodissociation. Ammonia signatures, which are present in numerous spectral ranges from the ultraviolet to the microwave, have been used extensively to probe photodissociation in the stratosphere, altitudes and opacities of the cloud layers and cloud distributions over the disk of the planets (20, 23, 32, 33, 39). Some correlations have also been established between the clouds detected by molecular spectroscopy and the Voyager visual images (33). Concerning PH_3, as its presence in the upper atmospheres of Jupiter and Saturn is difficult to explain by thermochemical models (40), the fact that PH_3 is detected at high altitudes in Jupiter and Saturn (20, 23) may be due to the very turbulent nature of these atmospheres. Consequently, PH_3 is probably a very good tracer of atmospheric circulation and the study of its local variations is particularly interesting for the study of atmospheric dynamics. The vertical distributions of molecules such as C_2H_2 and C_2H_6 have been inferred from signatures in the thermal infrared and ultraviolet. All these measurements help to a better understanding of the complex photochemistry which takes place in the stratospheres of these two planets (23, 41, 42).

3. SPECTRA OF TITAN

3.1. The atmosphere of Titan

The presence of an atmosphere around Titan had been suspected since 1908. The first detection of an atmosphere of methane was made by Kuiper in 1944 (43). In 1975, several molecules were discovered in ground-based spectra near 10 microns : C_2H_2, C_2H_4, C_2H_6, CH_3D (44). The Voyager exploration of Titan in 1980 has been extremely rewarding. It has established the bulk composition of this satellite and has revealed a number of other molecules present in very small amounts in the stratosphere of Titan including : C_3H_4, C_3H_8, C_4H_2, HCN, C_2N_2, HC_3N and CO_2, all detected in IRIS spectra (figure 3). In addition, a weak contribution of the H_2-N_2 collision-induced absorption to the tropospheric opacity was detected in IRIS spectra, corresponding to a very small amount of H_2. N_2 emissions were also detected in Voyager spectra in the extreme ultraviolet (see 1 for references).

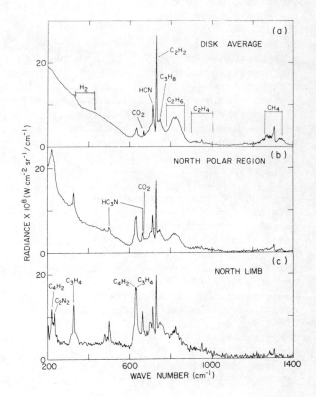

Figure 3.
Detection of molecules in the stratosphere of Titan from Voyager-IRIS spectra as a function of latitude.
a) average of 346 low latitude spectra
b) average of 30 north polar spectra
c) average of 3 north limb spectra.
(50)

For many years, it had been thought thas methane was the major atmospheric constituent of this satellite. However, in the seventies, it was repeatedly suggested that methane might not be the major atmospheric constituent. The evidence that methane is only a minor atmospheric constituent in an atmosphere of N_2 comes from two sets of Voyager measurements : the detection of N_2 very high in the atmosphere of the satellite and the determination of the mean molecular weight of the atmosphere near the surface which was obtained by combining radio occultation measurements and infrared observations. The approximate value obtained for the mean molecular weight is indeed 28. Since the Voyager exploration, another molecules, CO, has been detected in Titan from ground-based observations at 1.6 microns with a spectral resolution of 1.2 cm^{-1} (45) and at 1.3 mm (46). Besides, the ground-based observations at 1.6 microns also provide a measurement of the CH_3D/CH_4 ratio - difficult to obtain from the Voyager spectra (47) - and, consequently, of the D/H ratio, in Titan (24).

The major constituents of the atmosphere of Titan detected up to now are, with their approximate mixing ratios : N_2 (94 %), CH_4 (2-6 x 10^{-2}), H_2 (2 x 10^{-3}), CO (6 x 10^{-5}). Given the low temperatures in the atmosphere of Titan, all the molecules detected, except N_2, H_2 and CO, condense in the atmosphere of Titan or at the surface.

Analogies between the bulk compositions of Titan and the Earth atmospheres, detections in Titan of the HCN, C_2N_2 and HC_3N nitriles and of oxygen molecules have strongly enhanced the interest in the study of this satellite. A very detailed model of possible Titan photochemistry which includes the last discoveries - and especially the oxygen molecules which had not been included in previous models - has been recently published (48). It provides some insights on the nature of the primitive atmosphere of Titan.

3.2. The surface of Titan

The vertical thermal structure in the atmosphere of Titan has been obtained by combining radio-occultation and infrared Voyager measurements, and it has been found that, at the surface of Titan, the temperature is about 94°K and the pressure 1.5 atm. The CH_4 mixing ratio, although still uncertain since most of the measurements which have been made correspond to the stratosphere, is of the order of 2 to 6 %. With such a low surface temperature, CH_4 could condense at the surface. However, the measured temperature gradient near the surface (1.4 °K/ km) is lower than the wet adiabatic gradient for CH_4, and, therefore, excludes this possibility and another suggestion has been made concerning the nature of the surface which includes all the observational evidences which exist at the present time. It is argued that, because of the efficient photolysis of CH_4 in the upper atmosphere and easy conversion into C_2H_6

which is itself very stable against photolysis, a kilometer deep ocean composed of about 75 % of C_2H_6 and 25 % of CH_4 could have accumulated at the surface of Titan over the age of the solar system. C_2H_6, which is much less volatile than CH_4 at 94 °K, would help to maintain the observed amount of CH_4 in the atmosphere (49).

4. SPECTRA OF URANUS AND NEPTUNE

Our knowledge on these two planets is still very limited. The molecules which have been detected in Uranus are : CH_4, H_2, HD and CH_3D (see 2 for references), and, possibly, very small amounts of C_2H_6 in the stratosphere (51). The molecules detected in Neptune are CH_4, H_2 and C_2H_6 (2). The major constituents of these two planets are almost certainly H_2 and He but there is at present no detection of He. It is hoped that the Voyager explorations of Uranus in 1986 and Neptune in 1989 will provide measurements of the H_2/He ratio. Such measurements may help to understand why Uranus, which has about the same size and mass as Neptune, does not however radiate more energy than it receives from the Sun.

The masses of the cores of Uranus and Neptune are expected to be about the same as those of the cores of Jupiter and Saturn, whereas their atmospheres would be relatively less important. In addition, the cores of Uranus and Neptune should contain a higher proportion of ices since they were formed further away from the Sun. For these two reasons, if the bulk atmospheric compositions of the giant planets indeed reflect their internal structures, major differences may be expected in the relative proportions of the atmospheric constituents of Uranus and Neptune, compared to those of Jupiter and Saturn, and, more particularly, much higher C/H ratios in Uranus and Neptune than in Jupiter and Saturn. The D/H ratios in Uranus and Neptune should also be strongly enhanced compared to the D/H ratios in Jupiter and Saturn which would have remained primordial (52, 53, 54).

Many attempts have been made to obtain measurements of the CH_4/H_2 ratio in these two planets. Analyses have been done in the visible, near infrared and thermal infrared ranges. They all conclude to a C/H ratio higher than the solar ratio but the values are still very imprecise as CH_4 condenses in the atmospheres of these planets. The most recent ones range between 2 and 100 times the solar value in the case of Uranus and 10 to 100 times the solar value in the case of Neptune (51, 55).

Two deuterated molecules have been detected in Uranus. The most recent measurements of an HD absorption in spectra of Uranus indicate that the D/H ratio must be of the order of 10^{-4} (56) but the uncertainties are probably quite large because of the difficulty to account properly for the effects of scattering in visible spectra.

CH$_3$D has been detected in spectra at 1.6 microns with a resolution of 1.2 cm^{-1} (24). If we assume a fractionation factor of 1, the obtained value for D/H is $9^{+9}_{-4.5} \times 10^{-5}$. So, if there is indeed an enrichment of deuterium in Uranus compared to Jupiter, this enrichment is smaller than predicted (52). This may very well mean that the equilibrium between the core and the atmosphere is not reached (54), unless fractionation is not as efficient as assumed for the ices initially contained in the core of the planets. This result seems to be anyway in contradiction with the CH$_4$ measurements. Improved measurements of C/H and D/H are certainly needed. A measurement of D/H in the atmosphere of Neptune would also be extremely useful.

5. SPECTRA OF TRITON

Triton which may be slightly bigger than Pluto has a mass about 10 times higher than that of Pluto. It had been thought for some time that Triton might have the right mass and temperature to retain an atmosphere. However, because of the extreme faintness of this satellite, it is only recently (1979) that some indication of an atmosphere around Triton was obtained (57). An absorption feature attributed to gaseous methane was seen near 2.3 microns in spectra of Triton between 1.4 and 2.6 microns. More recent observations, still at low spectral resolution, between 0.8 and 2.5 microns reveal a number of other absorption bands which, for most of them, can be identified as due to methane (58). However, a comparison between the observed spectra, synthetic spectra for gaseous methane at 55°K (the expected temperature at the surface of Triton) based on band model techniques and laboratory spectra of methane frost indicate that the methane must be preferentially in the ice phase (58). An absorption which cannot be due to CH$_4$ and which occurs at 2.16 microns has been repeatedly seen in Triton spectra and has been tentativelly explained by the 2-0 pressure-induced band of liquid N$_2$ (59). A layer a few tens of centimeter thick at the surface of Triton would explain the observed absorption. If indeed there is some liquid N$_2$ at the surface of Triton, then the atmosphere itself may very well consist primarily of gaseous N$_2$ (59).

6. SPECTRA OF PLUTO

Pluto with its extremely small size (radius : 1500 km) and very eccentric orbit is a planet completely different from the other outer planets. The first detection of a signature in spectra of Pluto was obtained in 1976 by spectrophotometric studies in the near infrared (60). The absorption observed near 1.6 microns was attributed to methane frost. More recent studies at this wavelength seem to confirm this attribution (61). However, in 1980, Fink et al. (62) recorded with a CCD a spectrum of Pluto between 0.58 and

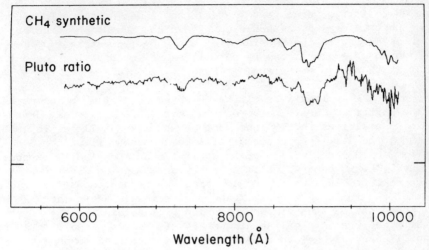

Figure 4. Comparison between a CCD spectrum of Pluto - divided by the spectrum of a standard star - and a synthetic spectrum of gaseous CH_4 computed with band models for the CH_4 absorption. (61)

1.06 microns at a resolution of 25 Å (figure 4) in which they have detected several CH_4 absorption bands which are better characterized by CH_4 in the gaseous state. This is the first evidence for an atmosphere around Pluto. Given the low temperatures expected at the surface of Pluto - between 50 and 65°K - the amount of CH_4 present in its atmosphere is probably limited by the surface temperature (63, 64). In the case of Pluto also, it is quite possible that CH_4 is only a minor atmospheric constituent. Gases such as N_2, Ar or CO may very well be present in the atmosphere of Pluto (63, 64).

FUTURE PROSPECTS

Continuous progress in detector sensitivies ans the availability of larger collectors will allow to still improve the spectroscopic ground-based studies of the faintest objects of the outer solar system : Uranus, Neptune, Titan and Triton, studies which, for at least three of these objects, are still at an early stage. At the same time, it will be possible to get spectra of Jupiter and Saturn at higher spectral and spatial resolutions. With the construction of new millimetric large antennae and interferometers, a number of observations of the giant planets and Titan in the millimetric will become possible.

The launch of the Space Telescope with its first generation of instruments which includes a high resolution ultraviolet spectrograph and, above all, several years later, the launch of the Infrared Space Observatory which should carry a high-resolution interferometer operating in the far-infrared and sub-millimetric ranges are the two major explorations by satellites of interest for the outer solar system which are planned for the next future.

Several space explorations of the outer solar system are expected in the next few years. The Voyager 2 spacecraft will visit Uranus and its satellites in 1986 and Neptune and its satellites in 1989. In 1986, the Galileo spacecraft will be launched. It will reach Jupiter in 1989 and will stay in orbit around the planet for about 2 years. This spacecraft will be carrying a mapping spectrometer which will provide spectra at low spectral resolution but very high spatial resolution between 0.7 and 5 microns. In addition, a probe will be sent into the atmosphere which will accomplish the first in-situ measurements of the atmosphere of a giant planet. This probe will carry, among other instruments, a mass spectrometer and a special device to measure the H_2/He ratio. Another space project, but for the nineties, is the Cassini ESA-NASA mission. It would consist of an orbiter around the Saturn-Titan system and a probe which would be sent into the atmosphere of Titan.

Better quality of ground-based observations, the opening of additional spectral ranges from the ground or from satellites, the first space explorations of Uranus, Neptune and Triton and the Galileo orbiter around Jupiter will increase the possibilities of atmospheric probing by molecular spectroscopy. New signatures of molecules already discovered will be identified which will allow to sound different altitude levels and improve vertical structure and abundance determinations. It will be possible to search for molecules such as H_2S, HCP, CH_3, N_2H_4 ... in Jupiter and Saturn or CH_3CN, CH_3NH_2 ... in Titan, which all have signatures in the far infrared, sub-millimetric or millimetric ranges. As the quality of the spectra improves, more accurate spectroscopic parameters will be needed and the lack of data on line intensities at room and low temperatures, line assignments, line shapes or line broadening parameters which is too often a limitation to the present analyses will become even more critical. If we want to improve this situation, it is therefore absolutely essential to prepare future observations of planets by acquiring ahead of time, as much as possible, the required spectroscopic parameters.

REFERENCES

(1) Hunt, G.E. 1983, Ann. Rev. Earth Planet. Sci. 11, pp. 415-459.
(2) Encrenaz, T. 1984, Space Sci. Rev. 38, pp. 35-87.
(3) Gautier, D., Marten, A., Baluteau, J.P. and Lacombe, A. 1979, Icarus 37, pp. 214-235.
(4) Gautier, D., Lacombe, A. and Revah, I. 1977, J. of the Atm. Sci. 34, pp. 1130-1137.
(5) Gautier, D., Conrath, B., Flasar, M., Hanel, R., Kunde, V., Chedin, A. and Scott, N. 1981, J. of Geophys. Res. 86, pp. 8713-8720.
(6) Conrath, B.J., Gautier, D., Hanel, R.A. and Hornstein, J.S. 1984, Astrophys. J., in press.
(7) Stevenson, D.J. and Salpeter, E.E. 1977, Astrophys. J. Suppl. 35, pp. 221-237.
(8) Stevenson, D.J. and Salpeter, E.E. 1977, Astrophys. J. Suppl. 35, pp. 239-261.
(9) Gautier, D., Marten, A., Baluteau, J.P. and Bachet, G. 1983, Can. J. Phys. 61, pp. 1455-1461.
(10) McKellar, A.R.W. 1984, Can. J. Phys. 62, pp. 760-763.
(11) Fromhold, L., Samuelson, R. and Birnbaum, G. 1984, Astrophys. J. Letters, in press.
(12) Combes, M., Maillard, J.P. and de Bergh, C. 1977, Astron. Astrophys. 61, pp. 531-537.
(13) Courtin, R., Gautier, D., Marten, A. and Kunde, V. 1983, Icarus 53, pp. 121-132.
(14) Clancy, R.T. and Muhleman, D.O. 1983, Astrophys. J. 273, pp. 829-836.
(15) de Bergh, C. 1984, Colloque CNES - Rapports isotopiques dans le système solaire - Paris, in press.
(16) Gautier, D. and Owen, T. 1983, Nature 302, pp. 215-218.
(17) Trauger, J.T., Roesler, F.L. and Mickelson, M.E. 1977, Bull. Am. Astron. Soc. 9, p. 516.
(18) Encrenaz, T. and Combes, M. 1982, Icarus 52, pp. 54-61.
(19) Macy, W. and Smith, W.H. 1978, Astrophys. J. 222, L73-L75.
(20) Kunde, V., Hanel, R., Maguire, W., Gautier, D., Baluteau, J.P., Marten, A., Chedin, A., Husson, N. and Scott, N. 1982, Astrophys. J. 263, pp. 443-467.
(21) Knacke, R., Kim, S.J., Ridgway, S.T. and Tokunaga, A.T. 1982, Astrophys. J. 262, pp. 388 et sq.
(22) Drossart, P., Encrenaz, T., Kunde, V., Hanel, R. and Combes, M. 1982, Icarus 49, pp. 416-426.
(23) Courtin, R., Gautier, D., Marten, A. and Bezard, B. 1984, Astrophys. J., in press.
(24) de Bergh, C., Lutz, B.L. and Owen, T. 1984, Colloque CNES - Rapports isotopiques dans le système solaire - Paris, in press.
(25) Encrenaz, T., Combes, M. and Zeau, Y. 1978, Astron. Astrophys. 70, pp. 29-36.
(26) Tokunaga, A.T., Knacke, R.F., Ridgway, S.T. and Wallace, L. 1979, Astrophys. J. 232, pp. 603-615.

(27) Tokunaga, A.T., Knacke, R.F. and Ridgway, S.T. 1980, Icarus 44, pp. 93-101.
(28) Drossart, P., Encrenaz, T. and Combes, M. 1984, Colloque CNES - Rapports isotopiques dans le système solaire - Paris, in press.
(29) Gautier, D., Bezard, B., Marten, A., Baluteau, J.P., Scott, N., Chedin, A., Kunde, V. and Hanel, R. 1982, Astrophys. J. 257, pp. 901-912.
(30) Gautier, D. and Owen, T. 1983, Nature 304, pp. 691-694.
(31) Buriez, J.C. and de Bergh, C. 1981, Astron. Astrophys. 94, pp. 382-390.
(32) Marten, A., Courtin, R., Gautier, D. and Lacombe, A. 1980, Icarus 41, pp. 410-422.
(33) Bezard, B., Baluteau, J.P. and Marten, A. 1983, Icarus 54, pp. 434-455.
(34) de Pater, I., Kenderdine, S. and Dickel, J.R. 1982, Icarus 51, pp. 25-38.
(35) de Pater, I. and Massie, S.T. 1984, preprint.
(36) Drossart, P. and Encrenaz, T. 1982, Icarus 52, pp. 483-491.
(37) Drossart, P., Encrenaz, T., Kunde, V., Hanel, R. and Combes, M. 1982, Icarus 49, pp. 416-426.
(38) Drossart, P.,Encrenaz, T. and Tokunaga, A. 1984, preprint.
(39) Combes, M., Courtin, R., Caldwell, J., Encrenaz, T., Fricke, K.H., Moore, V., Owen, T., Butterworth, P.S. 1981, Adv. Space Res. 1, pp. 169-175.
(40) Barshay, S.T. and Lewis, J.S. 1978, Icarus 33, pp. 593-611.
(41) Strobel, D.F. 1983, Int. Rev. Phys. Chem. 3, pp. 145-176.
(42) Atreya, S. and Romani, P.N. 1984, in Planetary Meteorology, ed. G.E. Hunt (Cambridge University Press), in press.
(43) Kuiper, G. 1944, Astrophys. J. 100, pp. 378-391.
(44) Gillett, F.C. 1975, Astrophys. J. Lett. 201, L41-L43.
(45) Lutz, B.L., de Bergh, C. and Owen, T. 1983, Science 220, pp. 1374-1375.
(46) Muhleman, D.O., Berge, G.L. and Clancy, R.T. 1984, Science 223, pp. 393-396.
(47) Kim, S.J. and Caldwell, J. 1982, Icarus 52, pp. 473-482.
(48) Yung, Y.L., Allen, M. and Pinto, J.P. 1984, Astrophys. J. Suppl. Ser. 55, pp. 465-506.
(49) Lunine, J.L., Stevenson, D.J. and Yung, Y.L. 1983, Science 222, pp. 1229-1230.
(50) Samuelson, R.E., Maguire, W.C., Hanel, R.A., Kunde, V.G., Jennings, D.E., Yung, Y.L. and Aikin, A.C. 1983, J. of Geophys. Res. 88, pp. 8709-8715.
(51) Orton, G.S., Tokunaga, A.T. and Caldwell, J. 1983, Icarus 56, pp. 147-164.
(52) Hubbard, W.B. and Mac Farlane, J.J. 1980, Icarus 44, pp. 676-682.
(53) Hubbard, W.B. and Mac Farlane, J.J. 1980, J. Geophys. Res. 85, pp. 225-234.

(54) Mac Farlane, J.J. and Hubbard, W.B. 1982, in Uranus and the Outer Planets, ed. by G. Hunt, Cambridge Univ. Press, pp. 111-124.
(55) Bergstralh, J.T. and Neff, J.H. 1983, Icarus 55, pp. 40-49.
(56) Cochran, W.D. and Smith, W.H. 1983, Astrophys. J. 271, pp. 859-864.
(57) Cruikshank, D.P. and Silvaggio, P.M. 1979, Astrophys. J. 233, pp. 1016-1020.
(58) Cruikshank, D.P. and Apt, J. 1984, Icarus 58, pp. 306-311.
(59) Cruikshank, D.P., Brown, R.H. and Clark, R.N. 1984, Icarus 58, pp. 293-305.
(60) Cruikshank, D.P., Pilcher, C.B. and Morrison, D. 1976, Science 194, pp. 835-837.
(61) Cruikshank, D.P. and Silvaggio, P.M. 1980, Icarus 41, pp. 96-102.
(62) Fink, U., Smith, B.A., Benner, D.C., Johnson, J.R. and Reitsema, H.J. 1980, Icarus 44, pp. 62-71.
(63) Trafton, L. and Stern, S.A. 1983, Astrophys. J. 267, pp. 872-881.
(64) Stern, S.A. and Trafton, L. 1984, Icarus 57, pp. 231-240.

PART II

THEORETICAL AND EXPERIMENTAL LABORATORY STUDIES

MOLECULAR IONS IN ASTROPHYSICS AND IN THE LABORATORY

Sydney Leach

Laboratoire de Photophysique Moléculaire, Université de Paris-Sud, 91405 Orsay Cédex, France and Département d'Astrophysique Fondamentale, Observatoire de Paris-Meudon, 92190 Meudon, France

ABSTRACT

Singly-charged molecular cations are present in stellar and planetary atmospheres, in comets, and in the interstellar medium. Laboratory studies of molecular ions are required to support astrophysical observation and interpretation. The range of such investigations is discussed and techniques are presented for the study of radiative and nonradiative transitions in molecular ions. Astrophysical applications are made of the results of several different types of laboratory studies. A discussion is also given concerning the possible formation and direct or indirect observation of doubly-charged molecular cations in astrophysical situations.

I. INTRODUCTION

Molecular ions are ubiquitous in astrophysics. Their spectroscopic observation has been made from the Earth, from satellites and from space probes. Mass spectrometric techniques adapted to satellite and space probe carriers have provided other direct evidence of molecular ions within the solar system. The presence in astrophysical situations of various molecular ions, not directly observed, has been inferred from physico-chemical schemes devised to model existing observations.

Three types of molecular ion have, or may have, astrophysical existence. The most certain are singly-charged molecular cations, which are present in the Sun (e.g. SiH^+), in planetary atmospheres/ionospheres (e.g. CO^+, N_2^+, NO^+, O_2^+, CO_2^+, ..., ion clusters), in comets (e.g. CH^+, CO^+, N_2^+, OH^+, H_2O^+, CO_2^+, ...) and in the

G. H. F. Diercksen et al. (eds.), Molecular Astrophysics, 353–374.

gas clouds of the interstellar medium (e.g. CH^+, SiH^+, HCO^+, N_2H^+, ...). I also conjecture that molecular cations are trapped in interstellar dust grains. Singly-charged molecular anions play an important role in the Earth's ionosphere (e.g. O_2^-, O_3^-, NO_2^-, CO_3^-, NO_3^-, ..., and corresponding ion clusters). Suggestions have been made that molecular anions could take part in interstellar gas phase chemistry (1), that S_2^- and S_3^- are trapped in interstellar grains (2), and that C_2^- could be abundant in carbon stars (3). Their observational possibilities have also been considered (4, 5). The possible astrophysical occurrence of doubly-charged molecular cations will be discussed in this report.

The types of laboratory studies of molecular ions, (mainly cations) and the techniques used, will first be discussed. Present and possible astrophysical applications will be mentioned. Recent gas phase ion studies carried out at Orsay will be described within the context of their application to astrophysical problems.

II. THE RANGE OF LABORATORY STUDIES OF MOLECULAR CATIONS

Laboratory studies of gas phase molecular ions involve the following themes (6, 7) : (i) physics of molecular ion formation, (ii) spectroscopy and structure, (iii) intramolecular nonradiative transitions, (iv) intramolecular chemistry, (v) collisionally induced physics, (vi) collisionally induced chemistry.

The physics of molecular ion formation concerns direct ionization and autoionization processes which occur by interaction between neutral molecules and particles of sufficient energy : photons, electrons, fast atoms or molecules, atoms or molecules in metastable excited states, charged atoms or molecules. Table 1 lists the principal ionization processes and indicates the corresponding energy conversion.

TABLE 1 : Ionization processes

Mechanism	Process [a]	Energy conversion
Photoionization	$AB + h\nu \rightarrow AB^+$	Electromagnetic $\rightarrow$ electronic
Multiphoton ionization	$AB + nh\nu \rightarrow AB^+$	Electromagnetic $\rightarrow$ electronic
(Fast) collisions	$AB + M(v_1) \rightarrow AB^+ + M(v_2)$	Translational $\rightarrow$ electronic
Penning ionization	$AB + M^* \rightarrow AB^+ + M$	Electronic $\rightarrow$ electronic
Charge transfer	$AB + M^+ \rightarrow AB^+ + M$	Electronic $\rightarrow$ electronic
Electric field (E) ionization	$AB + \vec{E} \rightarrow AB^+$	Electric field $\rightarrow$ electronic

[a] AB and/or AB^+ can be in ground or excited states

Recently, the formation of ions has been studied where the initial state is a neutral species in a highly excited metastable Rydberg state. Ionization then proceeds by thermal collisions, infrared or microwave absorption, or through weak applied electric fields (8). Strong electric field (> 10^8 V cm^{-1}) ionization of unexcited neutrals, and ionization by multiphoton absorption via real and/or virtual intermediate states are increasingly being used as methods of molecular cation formation. Although all of the processes listed in Table 1 are useful for laboratory production of molecular ions, only single-photon ionization, fast-collision ionization and charge-transfer ionization are likely to be of major importance in astrophysical situations. Particle collisions with Rydberg state atoms are known to be important to astrophysics, but would be less so for molecules. Excited state product ions play a role in the ion chemistry of the Earth's atmosphere (9).

Theoretical studies in the area of molecular ion formation have mainly been on photoionization, electron impact ionization and fast collision ionization ; attention is increasingly being paid to multiphoton ionization and to Penning ionization. Calculation of total and partial cross sections can be helpful in providing physical insight into these processes, in particular when autoionization, multiple ionization and core effects occur.

The spectroscopy of molecular ions is a basic source of information on their geometrical structure, internal dynamics and electronic configurations in various states of excitation. Classical spectroscopic techniques successful with neutral species are rarely of use for molecular ions. This is due to the difficulty of creating ion densities sufficiently high and of sufficient duration for spectroscopic measurements to be made by traditional methods. The processes which limit the densities of a specific ion are (i) ion-electron recombination, (ii) ion-molecule reactions and (iii) spontaneous dissociation of excited electronic states. Carefully designed special ion sources are therefore necessary for spectroscopic studies.

Molecular ions are increasingly being studied by microwave spectroscopy (10), both in the laboratory and in the interstellar medium. Infrared spectra of a few gas phase ions have been obtained in the laboratory (11). Direct application to astrophysical observation is increasingly being sought in this rapidly developing field. The infrared spectra of several molecular cations (and anions) trapped in low temperature matrices have been measured (12, 13).

The electronic spectroscopy (visible, ultraviolet) of molecular ions has developed considerably in the last few years. The emission spectra of many gas phase ions have been obtained using various electron impact and discharge sources (14, 15). More recently, molecular ions in beams (16), ion traps (17) and in plasmas produced by Penning ionization (18) or by time-gated electron impact (19) have been photoexcited by lasers to obtain fluorescence excitation spectra and dispersed fluorescence spectra (20). An ancillary technique of much value beginning to be applied to molecular ions (20) is jet expansion cooling resulting in low vibrational, and very low rotational, temperatures of gas phase species.

Low (21) and high (22) energy electron excitation, laser excitation (23), and coincidence studies (24) of molecular photoions have been used to measure the lifetimes of electronic excited ion states. Techniques in which relatively high ion densities are formed (21-23) suffer from space charge problems which can create distortions in measurement of optical lifetimes of ions (21, 25).

Spectroscopic information on molecular ions has also been obtained through the study of the photodissociation spectra of beams of positive ions (26), including the doubly-charged ion N_2^{++} (27), and of negative ions (28), in which fragment ions are monitored. In these experiments one can study threshold phenomena by selecting and measuring the zero kinetic energy products as a function of excitation wavelength. Alternatively one can measure the kinetic energy spectrum of the photoejected species at a fixed excitation wavelength. In similar fashion, ion and electron spectroscopies can be used to provide information about the internal energy states of molecular ions formed by inelastic collisions of neutral molecules with photons (29) or electrons (30). Photoelectron spectroscopy can indeed provide much information on molecular ion states but is essentially limited, by low energy resolution, to electronic and vibrational levels.

Absorption spectra of a few molecular ions have been obtained using flash discharge and pulsed electron (Febetron) techniques (31) and, more recently, using matrix isolation methods (32).

Electronic spectra are known for several hundred molecular ions. Rotational analysis of electronic spectra has been carried out for many diatomic ions, but for only a handful of polyatomic ions. Indeed, in the last nine years not one ion has been added to the meagre list of rotationally analyzed polyatomic ions known in 1975 : N_2O^+, CO_2^+, CS_2^+, H_2O^+, H_2S^+, $C_4H_2^+$ and the deuterated variants of the last three ions (33) although new information has been obtained for most of the listed ions since then. Rotational band contour analysis has been initiated for a few large ions (34).

It should be stressed that detailed spectroscopic information obtained in any one spectral region can be of use for estimating molecular properties and predicting spectra to be observed in other spectral regions. This is often particularly important for astrophysics, where even low resolution or approximate energy prediction can help narrow the bounds of observational search. Quantum-chemical calculations of molecular ion energy levels and oscillator strengths are also of value in this respect.

Intramolecular nonradiative transitions, involving the coupling of two or more electronic states have rarely been studied explicitly in molecular ions. Nevertheless, such transitions have often been invoked, in particular to account for ionic fragmentation. For example, in the quasi-equilibrium theory of mass spectra (35) it is assumed that molecular ions in excited electronic states rapidly convert their energy into vibrational energy of the ground electronic state. This can subsequently lead to Herzberg case II (vibrational) predissociation (36) provided that sufficient vibrational energy becomes available ; a further proviso is that the intramolecular dynamics leads to a dissociative nuclear configuration at a rate faster than any other competitive process which tends to deactivate the newly attained high vibrational levels of the ground state. Similar processes can obviously occur for excited neutral molecules.

Although some theoretical work has been done on unimolecular dissociation from the standpoint of radiationless transitions (37), most research on radiationless transition theory has been concerned with non-dissociative situations (38), in which the electronic excitation energy is implicitly or explicitly considered to be less than that necessary to achieve fragmentation. Radiationless transition models have been applied mainly to neutral molecules but applications to ions have been increasing in the past few years (39). The extension of radiationless transition models to include specific problems arising in molecular ion formation and relaxation is underway, and appropriate tests are being devised (see e.g. refs. 20, 24 and 39).

Another area in which particular examples of radiationless transitions occur is that of intramolecular chemistry i.e. isomerization and unimolecular fragmentation. The theoretical approach in dealing with these problems has mainly been cast in a statistical framework (40) but more recent work includes studies of the interaction (nonadiabatic coupling) of potential surfaces and specific trajectory calculations (41). The question of dissociation to three products has recently arisen in connection with the fragmentation of SO_2^+ (42) and SO_2^{++} (43). Experimental work in the area of intramolecular ion chemistry relies on a wide range of mass spectrometric techniques and of methods of molecular ion preparation. The most specific data on intramolecular ion chemistry comes from the use of state selection techniques, mainly using coincidence methods, as well as measurement of product internal and kinetic energies (44).

Collisionally induced physics and chemistry of molecular ions involve a wide range of techniques and objectives. The techniques include the use of ion traps, flow-tube methods, molecular beams, and a large range of coincidence methods and particle and/or photon detection devices. Relatively recent developments in ion-molecule reaction studies include the use of low temperature techniques (45, 46), the study of chemical (47, 48) and Penning ionization, improved state-selection methods (49) and the application of collisional activation mass spectrometries (50).

III. RADIATIVE AND NONRADIATIVE TRANSITIONS IN MOLECULAR IONS : TECHNIQUES

Our studies at Orsay have been concerned with the electronic spectroscopy and intramolecular relaxation of molecular ions. As a starting point it is necessary to determine whether a particular ion fluoresces and for this we use a number of coincidence techniques which enable us to measure ion fluorescence quantum yields ϕ_F and lifetimes τ_m (24, 39). When $\phi_F = 0$, only intramolecular relaxation (internal conversion and/or dissociation) occurs. If $\phi_F > 0$ then radiative processes occur and will be the only relaxation processes if $\phi_F = 1$. These will exist in competition with intramolecular relaxation if $0 < \phi_F < 1$.

The measured ϕ_F and τ_m are related to the nonradiative, k_{nr}, and radiative, k_r, rates by the expressions :

$$k_r = \phi_F \tau_m^{-1} \qquad (1)$$

$$k_{nr} = (1 - \phi_F)\tau_m^{-1} \qquad (2)$$

In these equations, τ_m is assumed to correspond to a single monoexponential decay process. A more complex treatment is necessary if the measured lifetime is bi- or multi-exponential (39).

The measured lifetime and quantum yield can be used to obtain an approximate value of the electronic transition oscillator strength f via the following expression (51) :

$$f = 1.5 \, \phi_F / \tilde{\nu}^2 \tau_m \qquad (3)$$

where $\tilde{\nu}$ is an average transition frequency in cm^{-1} and the measured lifetime τ_m is in seconds. Excited state lifetimes and transition oscillator strengths so determined can have many applications in modelling excitation and de-excitation processes in astrophysics.

A number of special ion sources have been developed at Orsay to obtain high resolution emission spectra of molecular ions that fluoresce. They are of the following types : electron impact sources (linear electron beam), including crossed molecule-electron beams (52, 53) with extension to multiple beams (54) ; hot cathode discharges ; Penning ionization sources ; low pressure discharges with crossed electric and magnetic fields (14) ; hollow cathode discharges. Reviews of some of these and other molecular ion emission sources have been given by Cossart (15, 55).

Coincidence techniques used in studies of intramolecular relaxation of molecular ions (24, 56) are tabulated in Table 2.

The PIFCO technique is used to measure the average fluorescence lifetime $\bar{\tau}_m$ and quantum yield $\bar{\phi}_F$, the average being over occupied vibrational levels of a particular emitting electronic state of the ion. PEFCO and T-PEFCO allow measurements to be made at specific excitation energy levels within the resolution of the particular methods (30-100 meV). Thus one can derive $\tau_m(v)$ and $\phi_F(v)$ values

Table 2 : Coincidence techniques used for the study of intramolecular relaxation processes in molecular ions.

Acronym	Coincidence between	Excitation Source (a)	Principal Ionization Mechanism : Direct (D) Autoionization (A)	Channel Studied
PIFCO (57,58)	PhotoIon-Fluorescence photon	RGL	D	Ion-selected parent or fragment-ion fluorescence
PEFCO (59,60)	PhotoElectron-Fluorescence photon	RGL	D	Energy-selected parent or fragment-ion fluorescence
T-PEFCO (61)	Threshold PhotoElectron-Fluorescence photon	MSR	D,A	Energy-selected parent or fragment-ion fluorescence
PEPICO (44)	PhotoElectron-PhotoIon	RGL	D	Energy-selected parent or fragment-ion formation
T-PEPICO (44)	Threshold PhotoElectron-PhotoIon	MSR	D,A	Energy-selected parent or fragment-ion formation
PIPICO (43)	PhotoIon-PhotoIon	MSR	D,(A)	Energy selected fragment-ion formation from doubly charged ions.

(a) RGL = Rare-gas lamp

MSR = Monochromatized synchrotron radiation

for assigned vibrational levels v of an excited electronic state. The PEPICO and T-PEPICO techniques used in ion fragmentation studies also have an energy resolution in the range 30-100 meV.

When the excitation source is a rare-gas lamp whose photon energy is at least 1 or 2 eV greater than the energy of the ion states studied (i.e. in the PIFCO, PEFCO and PEPICO techniques) information is obtained on processes resulting from direct ionization. In the T-PEFCO and T-PEPICO techniques, where coincidences are counted between threshold photoelectrons and photon or ion products respectively, the data obtained relate not only to direct ionization processes and products but also to those of superexcited autoionizing states. Further useful information on autoionization processes can be obtained from measurement of the photoelectron energy distribution at specific energies corresponding to superexcited state formation.

In the PIPICO technique, doubly charged ions AB^{++} are produced by photoionization ($AB + h\nu \rightarrow AB^{++} + 2e^-$). We have used monochromatized synchrotron radiation as an excitation source. The AB^{++} ions are unstable or metastable. Because of Coulombic repulsion forces they tend to dissociate into two singly-charged ionic fragments ($AB^{++} \rightarrow A^+ + B^+$) rather than into a doubly charged and a neutral fragment ($AB^{++} \rightarrow A^{++} + B$). The PIPICO method enables one to study the $AB^{++} \rightarrow A^+ + B^+$ channels. Both ionic fragments are detected and measurement is made of $\Delta t = t(A^+) - t(B^+)$, the difference in the times of flight of A^+ and B^+. This is carried out for different excitation energies. The fragment masses and internal energies as well as the kinetic energy released in fragmentation can be deduced from the spectrum of Δt values at particular excitation energies. The PIPICO technique has been used to determine the energy, structure and relaxation properties of electronic states of doubly charged molecular cations and to measure the partial and total cross sections for double photoionization (43).

The six coincidence techniques are described in the references cited in Table 2. A compilation of references to PIFCO, PEFCO and T-PEFCO studies (up to June 1982) can be found in Table V of reference 24.

IV. RADIATIVE AND NONRADIATIVE TRANSITIONS IN MOLECULAR IONS : SOME ASTROPHYSICAL APPLICATIONS

Many molecular ions of astrophysical interest have been studied in our laboratory by optical spectroscopy and/or by coincidence techniques. These include the following species :

diatomics : OH^+, SH^+, N_2^+, O_2^+, CS^+, SO^+

triatomics : CO_2^+, COS^+, CS_2^+, N_2O^+, H_2S^+, SO_2^+, $ClCN^+$

polyatomics : $C_2H_2^+$, $C_2N_2^+$, NH_3^+, CH_3CN^+, $Cl-C\equiv C-H^+$, $Cl-C\equiv C-Cl^+$,

$H-(C\equiv C)_2-H^+$, $CH_3-(C\equiv C)_2-CH_3^+$, $C_6H_6^+$

doubly charged ions : CO_2^{++}, SO_2^{++}, NH_3^{++}, CH_4^{++}

A few of these species are singled out for presentation in this report.

1. CS^+

The emission spectrum of CS^+ was observed in the laboratory at high resolution by Gauyacq and Horani (62). They carried out rotational analysis of the $A^2\Pi \rightarrow X^2\Sigma^+$ bands in the 6000-7000 Å region. The spectrum consists of a double-headed progression of five red-degraded bands assigned to $A^2\Pi(v') \rightarrow X^2\Sigma^+(v''=0)$ where $v' = 1$ to 5. Each band exhibits two double-headed subbands whose interval of 300 cm^{-1} is approximately equal to the spin-orbit coupling constant of $A^2\Pi$. One of the diffuse interstellar cloud absorption doublets observed in the lines of sight of ζ Oph and δ Sco closely coincides with the $^RQ_{21}$ heads of the $A^2\Pi_{1/2} - X^2\Sigma^+$ component of the (3-0) emission band, at 6701.6 Å (63). If confirmed by observation of the corresponding $A^2\Pi_{3/2} - X^2\Sigma^+$ (3-0) band features at 6840 Å, CS^+ would be the first sulphur-bearing molecule detected in diffuse clouds.

2. SH^+

The laboratory emission spectrum of SH^+ was first observed, between 3170 and 4000 Å, and vibrationally analyzed, by Horani et al in 1963 (64) ; rotational analysis of the 0,0 band of the $A^3\Pi - X^3\Sigma^-$ transition of SH^+ and SD^+ was also carried out (65). More recently, the (1-0), (0-0), (0-1) and (0-2) bands of both isotopic species have been rotationally analyzed and precise molecular constants obtained for the $A^3\Pi$ and $X^3\Sigma^-$ states (66). Predissociation of the $A^3\Pi$ state for $v' > 1$ allows only about 15 lines of the (1-0) band of SH^+ to be detected in emission. The ion dissociation spectrum has been obtained by laser excitation of an SH^+ ion beam, by Edwards et al (67). Radiative and nonradiative properties of the $A^3\Pi$ state have been discussed (66, 67). Relevant spectroscopic properties in the UV, IR and microwave domains have been determined or estimated, and assessed, in order to aid detection of SH^+ in astrophysical objects (68). In particular A-X v', v" transition probabilities, oscillator strengths, and the equivalent widths for predicted absorption lines have been estimated. Predictions of the ultraviolet absorption and emission spectra have been made for various temperature conditions. To aid in the detection of SH^+ in the mm, near and far infrared regions, a set of rotational term values has been determined for the $X^3\Sigma^-$, v = 0, 1 and 2 levels. The rotational lines involving N = 0, i.e. $1_0 - 0_1$ and $1_2 - 0_2$, are predicted around 340 and 530 GHz respectively. These lines are expected to be split into doublets by hyperfine interactions. Using the hyperfine interaction constants derived from the predissociation spectrum (67), estimated splittings of 12 MHz for the $1_0 - 0_1$ line and 107 MHz for the $1_2 - 0_2$ line have been obtained (68).

The chemistry of SH^+ formation in interstellar clouds has been discussed (69). The SH^+ ion may be present in shock-heated regions of the interstellar medium ; its detection is potentially an indicator of the applicability of the shock model (70). Another likely domain of astrophysical occurence of SH^+ is comet tails (71).

3. H_2S^+

The emission spectrum of H_2S^+ occurs between 4000 and 5000 Å ; it was first observed (64) by electron excitation, using a crossed beam technique (53). Vibrational (72) and rotational (73) analyses of the $\tilde{A}^2A_1 - \tilde{X}^2B_1$ transition have been carried out. The main band system consists of a progression in the excited state bending vibration ν'_2. Subsidiary progressions in ν'_2 with one and two quanta of ν''_2 excited are also observed. Both ground and upper states are bent, the bond angles being 92.9° ($\tilde{X}^2B_1$) and 127° ($\tilde{A}^2A_1$) respectively. The barrier to linearity in the $\tilde{A}^2A_1$ state is ca. 4400 cm^{-1}. In the longer wavelength bands of H_2S^+, corresponding to $v'_2 < 4$, the most prominent features are two violet degraded heads, assigned to $^RQ_{0,N}$ and $^PQ_{1,N}$ branches. Bands with $v'_2 > 4$ have only one head, assigned to $^PQ_{1,N}$ branches.

Neutral H_2S has been observed in interstellar clouds and so it is of interest to search for the positive ion. In a review of cometary spectra (74), following the observation of H_2O^+ in comet Kohoutek (75), Herbig suggested that a cometary search be made for the isosteric species H_2S^+. A tentative assignment of a few emission bands of H_2S in spectra of Comet Bradfield 1980t was made by Cosmovici et al (76). Very recently Cosmovici and Ortolani (77) have assigned nine emission bands in the visible spectrum of Comet IRAS-Araki-Alcock 1983d to H_2S^+ (Table 3). The published spectra, taken at a resolution of 6 Å, exhibit a low signal-to-noise ratio for the H_2S^+ bands assigned on the microdensitometer tracing. Spectra of higher resolution and contrast are desirable.

4. Molecular Ions of Low Fluorescence Quantum Yield

The ions listed in Table 4 are all of present or potential astrophysical interest for studies of the interstellar medium, comets or planetary atmospheres. It is therefore important to know whether these ions fluoresce, in order to observe them directly or through the use of electronic emission spectra to predict the corresponding absorption as well as transitions in other spectral regions.

PIFCO coincidence measurements have been carried out in order to determine the fluorescence yields of these ions. Only in SO_2^+ was fluorescence detected positively (78), albeit with a very low

TABLE 3 : H_2S^+ emission bands ($\tilde{A}^2A_1 \to \tilde{X}^2B_1$) observed in the laboratory[72] and assigned in Comet IRAS-Araki-Alcock 1983d[77]

Band centre (Å)	Intensity [a]	Assignment	Assigned in Comet IRAS-Aracki-Alcock 1983d
4962.39	m	2_1^3	+
4898.84	w	2_0^2	
4818.48	m	2_2^5	+
4755.75	m	2_1^4	+
4692.39	s	2_0^3	+ b
4616.93	m	2_2^6	+
4566.07	vw	2_1^5	
4507.20	s	2_0^4	+
4421.33	m	2_2^7	+
4384.91	vw	2_1^6	
4336.38	m	2_0^5	+
4207.79	w	2_1^7	
4172.46	m	2_0^6	
4012.03	w	2_0^7	+ c

a : s = strong, m = moderate, w = weak, vw = very weak
b : overlapped by C_2 band
c : overlapped by C_3 band

TABLE 4 : Fluorescence yield $\bar{\phi}_F$ of some molecular ions of astrophysical interest (low or zero $\bar{\phi}_F$ cases)

Ion	Excited State Energy (eV)	Dissociation Energy (eV)	$\bar{\phi}_F$	Relaxation Process [a]	References
SO_2^+	3.69	3.63	6×10^{-5}	F ; D	(78)
CH_3CN^+	2.92	1.80	$< 10^{-4}$	D	(79)
NH_3^+	4.83	5.58	$< 5 \times 10^{-5}$	I.C.	(80)
$C_2H_2^+$	4.97	5.82	$< 10^{-4}$	I.C.	(79)
$C_2N_2^+$	2.11	4.10	$< 10^{-4}$	I.C.	(79)
$C_6H_6^+$	2.24	3.69	$< 2.5 \times 10^{-4}$	I.C.	(81)

a : F = Fluorescence, D = Dissociation, I.C. = Internal Conversion

yield. In all other cases listed in Table 4 no fluorescence was observed above the experimental noise level and only an upper limit to the yield could be given.

The low or zero fluorescence yield is readily understandable for SO_2^+ and CH_3CN^+ where the excited state energy lies above the lowest dissociation limit, so that the dominant relaxation process is fragmentation. However in the other ions the excited state lies <u>below</u> the first dissociation limit. In these cases, interpretation of effective intramolecular fluorescence quenching involves postulating that internal conversion processes occur, in which the excited state couples with high vibrational levels of lower-lying (generally the ground) states. Such processes have been shown to be highly probable for NH_3^+ (80) and $C_6H_6^+$ (81). The result is to effectively lengthen the lifetime of the excited state so that it is eventually deactivated by a (rare) collisional process under our experimental conditions or by vibrational radiative relaxation in the infrared (82). Radiative relaxation processes of such long-lived states might be observable in many astrophysical situations where low particle densities occur. However, the low effective fluorescence quantum yields render exceedingly difficult the observation in the laboratory of dispersed fluorescence spectra of these ions. Thus the emission band near 2338 Å observed in an electron impact experiment on ammonia and tentatively assigned to NH_3^+ (83), is unlikely to be due to this ion.

5. CO_2^+

Spectral emission of the CO_2^+ ion has been assigned to two electronic transitions, $\tilde{B}^2\Sigma_u^+ \rightarrow \tilde{X}^2\Pi_g$ around 2900 Å and $\tilde{A}^2\Pi_u \rightarrow \tilde{X}^2\Pi_g$

between 2800 and 5000 Å (84, 85). Bands corresponding to these transitions are among the important features observed by Mariners 6, 7 and 9 in the atmosphere of Mars. For Mariners 6 and 7 spectroscopic observations of the atmosphere were each made twice, tangentially to the bright disc as the spacecraft flew by the planet. Spectral resolution was 10-20 Å (86-88). Intensity measurements were made with Mariner 9 on a number of the planetary limb crossings. Published results on CO_2^+ features (89) concern observations made on four occasions. Highest wavelength cut-off was 3400 Å. The upper atmosphere spectra of Mariner 9 exhibit the same spectral features observed in Mariners 6 and 7 but with higher signal to noise ratio.

Interpretation of the CO_2^+ emissions in the Martian dayglow requires a knowledge of the $\tilde{B}^2\Sigma_u^+$ and $\tilde{A}^2\Pi_u$ state excitation and subsequent emission processes. The original models considered photoionization, fluorescent scattering and photoelectron impact excitation mechanisms (90, 91). Dalgarno et al (90, 91) calculated the contributions expected from these three mechanisms to the mean subsolar zenith intensities for a set of solar zenith angles. The three excitation mechanisms were also used in calculations of limb or slant intensities as a function of altitude by Stewart (88) for the $\tilde{B} \rightarrow \tilde{X}$ and $\tilde{A} \rightarrow \tilde{X}$ emissions and by McConnell (92) for the $\tilde{A}$-$\tilde{X}$ emission. Comparison of model calculation results and spacecraft observations of the Martian dayglow were used for determining the physical composition and dynamic processes in the Martian upper atmosphere.

In this earlier work, the following explicit or implicit assumptions were made : (i) the electronic state branching ratios are independent of excitation wavelengths, (ii) the $\tilde{C}^2\Sigma_g^+$ state of CO_2^+ decays to the $\tilde{A}^2\Pi_u$ and $\tilde{B}^2\Sigma_u^+$ states with a cascade branching ratio of 4.1, (iii) collisional deactivation of $\tilde{A}^2\Pi_u$ and $\tilde{B}^2\Sigma_u^+$ states is negligible, (iv) no intramolecular coupling between the $\tilde{B}^2\Sigma_u^+$ and $\tilde{A}^2\Pi_u$ states.

It was shown elsewhere (93) that assumptions (i), (ii) and (iv) are invalid. Furthermore, insufficiently accurate data on vibrational distributions was used in the modelling studies. A critical examination of earlier modelling assumptions and data, taking also into account new information available from the Viking I probe (94), led to revised values for the excited state production rates (93). In particular, the emission rates are notably affected by intramolecular coupling which was shown to exist between the $\tilde{B}^2\Sigma_u^+$ and $\tilde{A}^2\Pi_u$ states of CO_2^+ (95, 96). It was found that emission from $\tilde{B}^2\Sigma_u^+$ rotational levels occurs not only in the "normal" 2900 Å region but also at $\lambda > 3000$ Å. Thus the Mariner observations must include considerable redshifted emission to the $\tilde{A}$-$\tilde{X}$ spectral region from $\tilde{B}^2\Sigma_u^+$ rovibronic levels populated in the original photoionization, fluorescent scattering or photoelectron impact excitation. PIFCO measurements on CO_2^+ have shown that the $\tilde{B}^2\Sigma_u^+ \rightsquigarrow \tilde{A}^2\Pi_u$ radiationless transition yield is 0.41 ± 0.08 (94). This value has recently been confirmed in a laser excitation experiment on time-gated electron impact created CO_2^+ (20) in which $\phi_F(J')$, the $\tilde{B} \rightsquigarrow \tilde{A}$ radiationless yield for individual rotational levels J' of the $\tilde{B}^2\Sigma_u^+$ state, was measured and integrated over a rotational distribution corresponding to that resulting in a photoionization experiment.

Fox and Dalgarno (97) have recently made a revised model of the upper atmosphere of Mars, also based on Viking I data (94) and taking into account the changes in basic parameter values resulting from the absence of cascade processes from the $\tilde{C}^2\Sigma_g^+$ state as well as the existence of intramolecular coupling between the $\tilde{B}$ and $\tilde{A}$ states.

Thus analysis of the Martian dayglow observations has depended critically on (i) the results of detailed laboratory high resolution spectroscopic studies on CO_2^+ to determine the relevant energy levels and transition probabilities, (ii) knowledge of CO_2^+ spectral perturbations and (iii) radiationless transition yields, the latter being determined from coincidence studies and from laser excitation experiments.

Spectral emission of the CO_2^+ ion is expected in the atmosphere of Venus (98) and indeed may have been observed (99). Cometary spectra, especially from comet-tails of the plasma type, also show CO_2^+ emission (71). Understanding the excitation mechanisms in these cases requires information of the type discussed here (100).

V. DOUBLY CHARGED MOLECULAR CATIONS

Doubly charged molecular cations have not previously been considered, to my knowledge, as taking part in astrophysical processes. This may be due to insufficient laboratory-based information on these species, and to the difficulty of distinguishing between processes originating in doubly and in singly charged molecular ions. One must also consider whether astrophysical conditions exist which are propitious for the formation of non-negligible amounts of doubly charged cations. The properties of doubly charged molecular cations will be discussed and speculation made on the possibility of their formation and detection in astrophysical situations.

For many molecules the ratio of double to single ionization potential is ≈ 2.8 (101). Thus doubly charged molecular ions can be formed at energies of the order of 20-40 eV. Methods of inducing double ionization include electron impact, ion impact, double charge transfer (e.g. $H^+ + AB \rightarrow H^- + AB^{++}$), charge stripping and photon impact (43). Nascent doubly charged cations are relatively short-lived. The thermodynamic limit for dissociation of doubly charged cations lies <u>below</u> the double ionization energy, due to Coulomb repulsion forces (Fig. 1), thus making the ion unstable with respect to dissociation to singly charged products. Furthermore, the doubly charged cation may have a different equilibrium structure from that of the parent neutral molecule, so that a vertical transition can lead directly to the dissociative part of the doubly charged ion potential surface. However, as shown in Fig. 1, a local minimum can occur on this surface giving rise to a certain metastability of the cation. Indeed, doubly charged cations that are stable on the time scale of the experiment, i.e. that persist for times longer than 1 µs, have been detected by various mass-spectrometric techniques (102).

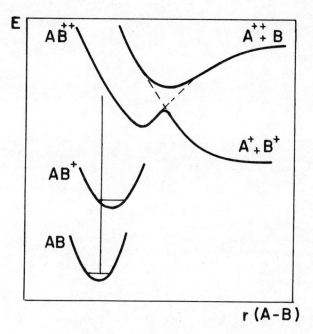

Fig. 1 : Schematic representation of potential surfaces of a neutral (AB) singly ionized (AB^+) and doubly ionized (AB^{++}) diatomic molecule.

In addition to $AB^{++} \rightarrow A^+ + B^+$ dissociation, a second fragmentation pathway is also possible, $AB^{++} \rightarrow A^{++} + B$. Both types of reaction have been observed but dissociation to two singly charged products is expected to be more probable than to one doubly charged fragment, at least for the lower electronic states of AB^{++} (43).

A number of events can give rise to observations which signal the involvement of doubly charged molecular cations :

a) Because of strong Coulomb repulsion forces, the singly charged ion products of the dissociation $AB^{++} \rightarrow A^+ + B^+$ will tend to have high total kinetic energy ($\gtrsim 4$ eV) (43). This would be unusual in the case of singly charged ions produced as parent or fragment ions in a single ionization process. Thus certain ion-molecule reactions which are endothermic at thermal energies could occur via fast fragment ion impact when the latter is produced by dissociation of a doubly charged molecular ion. This may be of some interest in cosmochemical schemes involving ion-molecule reactions.

b) If the A^+ (or B^+) dissociation product is a singly charged <u>molecular</u> ion fragment, it could be formed with an internal vibrational and/or rotational energy distribution that differs markedly from that occuring in a single ionization process. This would be relatively easy to monitor if the products formed are in excited electronic or vibrational states that fluoresce.

c) Doubly charged ions that live long enough in their excited states could exhibit characteristic emission spectra. (Only one example is known so far, that of N_2^{++} (103). A systematic search for other fluorescent doubly charged cations is planned, using a variant of the PIFCO technique (see Table 2)).

d) Specific ion-molecule reactions can occur : $AB^{++} + CD \rightarrow AB + CD^{++} / AB^+ + CD^+ / AB + C^+ + D^+$, etc, perhaps producing products that are chemically unusual or are in unusual states of internal energy (104, 105).

Finally the question must be asked as to where and how doubly charged molecular ions could occur in astrophysical situations. Consider first of all the <u>solar system</u>. Both solar wind particles (electrons, protons, other atomic ions) and solar photons can have energies sufficient for the collisional production of doubly charged molecular cations. With respect to solar photons, special mention should be made to HeII emission at 303 Å (40 eV) which is particularly important in the far ultraviolet region of the solar spectrum (106). Thus one can imagine the following processes as giving rise to doubly charged molecular ions by interaction with molecules in planetary and cometary atmospheres :

$$AB + h\nu \rightarrow AB^{++} + 2e^- \qquad (4a)$$

$$AB^+ + h\nu \rightarrow AB^{++} + e^- \qquad (4b)$$

$$AB + H^+ \rightarrow AB^{++} + H^- \qquad (4c)$$

$$AB^+ + H^+ \rightarrow AB^{++} + H \qquad (4d)$$

$$AB + e^- \rightarrow AB^{++} + 3e^- \qquad (4e)$$

$$AB^+ + e^- \rightarrow AB^{++} + 2e^- \qquad (4f)$$

The cross-sections for these processes have been relatively little studied but representative values can be given for certain ranges of the incident energy : $\sigma(4a) \approx 0.1 - 1 Mb$ (43), $\sigma(4b) \approx$

100 Mb (107), $\sigma(4c) \approx 0.01 - 1$ Mb (108), $\sigma(4d) \approx 0.1 - 1$ Mb (108), $\sigma(4e) \approx 1 - 10$ Mb (109), $\sigma(4f) \approx 0.1 - 10$ Mb (110, 111). The values given for $\sigma(4b)$ and $\sigma(4f)$ are for studies on atoms since these processes have not yet been investigated for molecules.

Apart from the solar system, doubly-charged ions could also be formed in the <u>interstellar medium</u> by interaction of molecules with high energy photons and/or cosmic particles. Ultraviolet radiation emitted by a star embedded in a dense cloud is absorbed in the star's immediate neighbourhood by the H atom ionization continuum at $\lambda < 912$ Å (13.6 eV). However, the H atom ionization cross-section at $\lambda < 100$ Å (124 eV) is sufficiently small for these photons to be able to penetrate great depths of interstellar space. These soft X-rays, as well as the low energy components of cosmic rays, could play a role in creating doubly-charged molecular ions in the interstellar medium, as they do in the formation of multiply-charged atoms (112).

VI. CONCLUSION

Molecular ions are observed in three principal astrophysical situations : planetary atmospheres and ionospheres, comet tails, interstellar space. Molecular ions in atmospheres and ionospheres of planetary objects have been studied mainly by deep space missions. Measurement has generally concerned identification of molecular ions but more rarely energy states of these species. However, it is important to study such energy states, and as a function of time, in order to understand the internal dynamics of molecular ions and the chemistry in which they are involved. Thus optical spectroscopy at the highest possible resolution is required. Both emission and absorption studies should be made since intramolecular radiationless transitions and dissociative processes may effectively quench spectral emission of many molecular ions. A further area of interest in planetary atmospheres concerns the formation and dynamics of molecular ions complexed to simple molecules (113). These processes have been mainly investigated by mass spectrometry (114) but they can be studied advantageously by optical spectroscopy. Molecular ions in comets have mainly been investigated from the Earth, with limitations due to air absorption. Although the emission spectra of most molecular ions are expected to lie at $\lambda > 3300$ Å (39), the more intense absorption spectra of these species should lie at $\lambda < 3300$ Å. Molecular ions in the interstellar medium have principally been observed by microwave spectroscopy, but more optical spectroscopic studies are likely in the future.

The observation of molecular ions, and the interpretation of their widespread role, in astrophysics, requires laboratory studies of the electronic transitions and excited state relaxation properties of these species. A range of appropriate structural and dynamic studies has been presented and the relevant experimental techniques discussed, especially concerning radiative and nonradiative transitions of molecular ions. A number of astrophysical applications (planetary atmospheres, comets, interstellar medium) of laboratory results obtained at Orsay have been presented. The experimental data have been obtained by the use, variously, of spectroscopic, laser and coincidence techniques. Considerations and speculations have also been given on the possible existence and detection of doubly-charged molecular ions in astrophysical situations. Finally a considerable extension of laboratory work on molecular ions is timely in relation to future observations by satellites and space probes and the extension of ground-based spectroscopic astrophysical observations (115, 116). New, and it is hoped higher resolution, information on astrophysical molecular ions is certain to result.

REFERENCES

1. Dalgarno, A. and McCray, R.A., 1973, Ap. J. 181, 95.
2. Exarhos, G.H., Mayer, J. and Klemperer, W., 1981, Phil. Trans. R. Soc. Lond. A303, 503.
3. Vardya, M.S., 1967, Mem. Roy. Astron. Soc. 71, 249.
4. Sarre, P.J., 1980, J. Chim. Phys. 77, 769.
5. Vardya, M.S. and Krishna Swamy, K.S. 1980, Chem. Phys. Lett. 73, 616.
6. Leach, S., 1979, J. Chim. Phys. 76, 1043.
7. Leach, S., 1980, J. Chim. Phys. 77, 585.
8. Stebbings, R.F., and Dunning, F.B. (editors), "Rydberg states of atoms and molecules", 1983, Cambridge Univ. Press, Cambridge, U.K.
9. Ferguson, E.E., Fehsenfeld, F.C. and Albritton, D.L., 1979 in "Gas Phase Ion Chemistry" (ed. M.T. Bowers), Academic, N.Y. Vol. 1, pp. 45-82.
10. Woods, R.C., 1983 in "Molecular Ions : Spectroscopy, Structure and Chemistry" (eds. T.A. Miller and V.E. Bondybey), North-Holland, Amsterdam, pp. 11-47.
11. Saykally, R.J., 1984, this volume.
12. Jacox, M., 1978, Rev. Chem. Intermed. 2, pp. 1-36.
13. Andrews, L., 1983, in "Molecular Ions : Geometric and Electronic Structures" (eds. J. Berkowitz and K.O. Groenveld), Plenum, N.Y., pp. 153-182 and 183-215.
14. Cossart, D., J. Chim. Phys. 76, 1045.
15. Cossart, D., J. Chim. Phys. 78, 703.
16. Brown, R.D., Godfrey, P.D., McGilvery, D.C., and Crofts, J.G., 1981, Chem. Phys. Lett. 84, 437.
17. Winn, J.S., 1983, in "Molecular Ions : Geometric and Electronic Structures" (eds. J. Berkowitz and K.O. Groenveld), Plenum, pp. 53-67.
18. Miller, T.A., and Bondybey, V.E., 1980, J. Chim. Phys. 77, 695.
19. Allison, J., Kondow, T., and Zare, R.N., 1979, Chem. Phys. Lett. 64, 202.

20. Johnson, M.A., Zare, R.N., Rostas, J., and Leach, S., 1984, J. Chem. Phys. 80, 2407.
21. Mohlmann, G.R., and de Heer, F.J., 1977, Phys. Scripta 16, 51.
22. Erman, P., 1984, this volume.
23. Maier, J.P., Ochsner, M., and Thommen, F., 1983, Faraday Disc. Chem. Soc. 75, 77.
24. Leach, S., **1985**, in "Photophysics and Photochemistry in the Vacuum Ultraviolet", (S. McGlynn, G. Findley and R. Huebner, eds.) Reidel, Dordrecht, **p. 297.**
25. Curtis, L.J., and Erman, P., 1977, J. Opt. Soc. Am. 67, 1218.
26. Moseley, J., and Durup, J., 1980, J. Chim. Phys. 77, 673.
27. Cosby, P.C., Möller, R., and Helm, H., 1983, Phys. Rev. A28, 766.
28. Cosby, P.C., Moseley, J.T., Peterson, J.R., and Ling, J.H., 1978, J. Chem. Phys. 69, 2771.
29. Rabelais, J.W., 1977, "Principles of Ultraviolet Photoelectron Spectroscopy", Wiley, N.Y.
30. Brion, C.E., and Hamnett, A., 1981, Adv. Chem. Phys. 45, 2.
31. Herzberg, G., 1971, Quart. Rev. Chem. Soc. 25, 201.
32. Bondybey, V.E., English, J.H., and Miller, T.A., 1980, J. Chem. Phys. 81, 455.
33. Leach, S., 1976, in "Spectroscopy of the Excited State", (ed. B. di Bartolo), Plenum, N.Y., pp. 369-375.
34. Klapstein, D., Leutwyler, S., Maier, J.P., Cossart-Magos, C., Cossart, D., and Leach, S., 1984, Mol. Phys. 51, 413.
35. Rosenstock, H.M., 1968, Adv. Mass Spectrom. 4, 523.
36. Herzberg, G., 1966, "Electronic Spectra of Polyatomic Molecules", Van Nostrand, N.Y.
37. Gelbart, W.M., 1977, Ann. Rev. Phys. Chem. 28, 323..
38. Freed, K.F., 1976, Top. Appl. Phys. 15, 23 ;
    Avouris, P., Gelbart, W.M., and El Sayed, M.A., Chem. Revs. 77, 793.
39. Leach, S., Dujardin, G., and Taieb, G., 1980, J. Chim. Phys. 77, 705.
40. Chesnavich, W.J., and Bowers, M.T., 1979, in "Gas Phase Ion Chemistry" Vol. 1 (M.T. Bowers, ed.) Academic, N.Y., pp. 119-151.
41. Lorquet, J.C., Dehareng, D., Sannen, C., and Raseev, G., 1980, J. Chim. Phys. 77, 719.
42. Dujardin, G., Govers, T., and Leach, S., to be published.
43. Dujardin, G., Leach, S., Dutuit, O., Guyon, P-M., and Richard-Viard, M., 1984, Chem. Phys. 88, 339.
44. Baer, T., 1979, in "Gas Phase Ion Chemistry", Vol. 1 (M.T. Bowers, ed.), Academic, N.Y. pp. 153-196.
45. Smith, D., and Adams, N.G., **1985, this volume.**
46. Rowe, B., **1985, this volume.**
47. Field, F.H., 1968, Adv. Mass Spectrom. 4, 645.
48. Harrison, A.G., 1983, "Chemical Ionization Mass Spectrometry", C.R.C. Press, Boca Raton.
49. Govers, T.R., Guyon, P-M., Baer, T., Cole, K., Frohlich, H.,

and Lavollée, M., 1984, Chem. Phys. $\underline{87}$, 373.
50. McLafferty, F.W., 1980, "Interpretation of Mass Spectra", 3rd edition, Univ. Sci. Books, Calif.
51. Calvert, J.G., and Pitts, J.N., Jr., 1966, "Photochemistry", Wiley, N.Y.
52. Horani, M., and Leach, S., 1959, Comptes Rendus Ac. Sci. (Paris) $\underline{248}$, 2196.
53. Horani, M., and Leach, S., 1961, J. Chim. Phys. $\underline{58}$, 825.
54. Horani, M., 1967, J. Chim. Phys. $\underline{64}$, 331.
55. Cossart, D., 1974, Thèse d'Etat, Université Paris-Sud, Orsay.
56. Leach, S., and Dujardin, G., 1983, Laser Chem. $\underline{2}$, 285.
57. Eland, J.H.D., Devoret, M., and Leach, S., 1976, Chem. Phys. Lett. $\underline{43}$, 97.
58. Dujardin, G., Leach, S., and Taieb, G., 1980, Chem. Phys. $\underline{46}$, 407.
59. Maier, J.P., and Thommen, F., 1980, Chem. Phys. $\underline{51}$, 319.
60. Winkoun, D., Dujardin, G., and Leach, S., 1984, Can. J. Phys., in press.
61. Dujardin, G., Leach, S., Dutuit, O., Govers, T., and Guyon, P.-M., 1983, J. Chem. Phys. $\underline{79}$, 644.
62. Gauyacq, D., and Horani, M., 1978, Can. J. Phys. $\underline{56}$, 587.
63. Ferlet, R., Roueff, E., Horani, M., and Rostas, J., 1983, Astron. Astrophys. $\underline{125}$, L5.
64. Horani, M., Leach, S., and Rostas, J., 1963, VIe Conf. Int. Phenom. Ionisation dans les Gaz, Vol. 1 (S.E.R.M.A.) p. 45.
65. Horani, M., Leach, S., and Rostas, J., 1967, J. Molec. Spectrosc. $\underline{23}$, 115.
66. Rostas, J., Horani, M., Brion, J., Daumont, D., and Malicet, J., 1984, Mol. Phys. $\underline{52}$, 1431.
67. Edwards, C.P., Maclean, C.S., and Sarre, P.J., 1984, Mol. Phys. $\underline{52}$, 1453.
68. Horani, M., Rostas, J., and Roueff, E., 1984, Astron. Astrophys., submitted.
69. Oppenheimer, M., and Dalgarno, A., 1974, Ap. J. $\underline{187}$, 231.
70. Dalgarno, A., 1981, Phil. Trans. Roy. Soc. (London) $\underline{A303}$, 513.
71. Huebner, W.F., **1985, this volume.**
72. Dixon, R.N., Duxbury, G., Horani, M., and Rostas, J., 1971, Mol. Phys. $\underline{22}$, 977.
73. Duxbury, G., Horani, M., and Rostas, J., 1972, Proc. Roy. Soc. Lond. $\underline{A331}$, 109.
74. Herbig, G.H., 1976, in "The Study of Comets", I.A.U. Colloq. N° 25 (ed. B. Donn, M. Mumma, W. Jackson, M. A'Hearn and R. Harrington) NASA SP-393, p. 136.
75. Wehinger, P.A., Wyckoff, S., Herbig, G.H., Herzberg, G., and Lew, H., 1974, Ap. J. $\underline{190}$, L43.
76. Cosmovici, C.B., Barbieri, C., Bonoli, C., Bortoletto, F., and Hamzaoglu, E., 1982, Astron. Astrophys. $\underline{114}$, 373.
77. Cosmovici, C.B., and Ortolani, S., 1984, Nature $\underline{310}$, 122.
78. Dujardin, G., and Leach, S., 1981, J. Chem. Phys. $\underline{75}$, 2521.
79. Devoret, M., 1976, Thèse Dr. 3e cycle, Univ. Paris-Sud, Orsay.

80. Dujardin, G., and Leach, S., 1984, Can. J. Chem., submitted.
81. Braitbart, O., Castellucci, E., Dujardin, G., and Leach, S., 1983, J. Phys. Chem. 87, 4799.
82. Dujardin, G., Leach, S., Taieb, G., Maier, J.P., and Gelbart, W.M., 1980, J. Chem. Phys. 73, 4987.
83. Herzberg, G., 1980, in "Interstellar Molecules", I.A.U. Symposium 87 (ed. B.H. Andrews) p. 231.
84. Gauyacq, D., Horani, M., Leach, S., and Rostas, J., 1975, Can. J. Phys. 53, 2040.
85. Gauyacq, D., Larcher, C., and Rostas, J., 1979, Can. J. Phys. 57, 1634.
86. Barth, C.A., Fastie, W.G., Hord, C.W., Pearce, J.B., Kelly, K.K., Stewart, A.I., Thomas, G.E., Anderson, G.P., and Raper, O.F., 1969, Science 165, 1004.
87. Barth, C.A., Hord, C.W., Pearce, J.B., Kelly, K.K., Anderson, G.P., and Stewart, A.I., 1971, J. Geophys. Res. 76, 2213.
88. Stewart, A.I., 1972, J. Geophys. Res. 77, 54.
89. Stewart, A.I., Barth, C.A., Hord, C.W., and Lane, A.L., 1972, Icarus 17, 469.
90. Dalgarno, A., Degges, T.C., and Stewart, A.I., 1970, Science 167, 1490.
91. Dalgarno, A., and Degges, T.C., 1971, in "Planetary Atmospheres", I.A.U. Symposium N° 40 (eds. C. Sagan, T.C. Owen and H.J. Smith), Springer Verlag, N.Y., pp. 337-345.
92. McConnell, J.C., 1973, in "Physics and Chemistry of Upper Atmospheres" (ed. B.M. McCormac), Reidel, Dordrecht, pp. 309-334.
93. Leach, S., 1977, 1980, "Les Spectres des Molécules Simples au Laboratoire et en Astrophysique", 21st Int. Colloq. Astrophys. Liège, pp. 403-421.
94. Nier, A.O., Hanson, W.B., Seiff, A., McElroy, M.B., Spencer, N.W., Duckett, R.J., Knight, T.C.D., and Cook, W.S., 1976, Science 193, 786.
95. Leach, S., Devoret, M., and Eland, J.H.D., 1978, Chem. Phys. 33, 113.
96. Leach, S., Stannard, P.R., and Gelbart, W.M., 1978, Mol. Phys. 36, 1119.
97. Fox, J.L., and Dalgarno, A., 1979, J. Geophys. Res. 84, 7315.
98. Swings, P., 1969, Proc. Am. Philos. Soc. 113, 229.
99. Polyakova, G.N., Fogel, Ya.M., and Chiu, Y.M., 1963, Sov. Astron. AJ 7, 276.
100. Festou, M.C., Feldman, P.D., and Weaver, H.A. 1982, Ap. J. 256, 331.
101. Tsai, B.P., and Eland, J.H.D., 1980, Int. J. Mass Spectrom. Ion Phys. 36, 143.
102. See references 14-18 cited in reference 43.
103. Carroll, P.K., 1958, Can. J. Phys. 36, 1585.
104. Bearman, G.H., Ranjbar, F., Harris, H.H., and Leventhal, J.J., 1976, Chem. Phys. Lett. 42, 335.
105. Kemp, D.L., and Cooks, R.G., 1978 in "Collision Spectroscopy" (ed. R.G. Cooks), Plenum, N.Y., pp. 257-288.

106. Nikol'skij, G.M., in Handbuch der Physik (S. Flugge ed.), 1982, Vol. 49/6, Geophysik III, Part 6, pp. 309-377.
107. Lyon, I.C., Peart, B., West, J.B., Kingston, A.E., and Dolder, K., 1984, J. Phys. B17, L345.
108. Appell, J., 1978 in "Collision Spectroscopy" (ed. R.G. Cooks) Plenum, N.Y., pp. 227-256.
109. Daly, N.R., and Powell, R.E., 1966, Proc. Phys. Soc. (London) 89, 273.
110. Kieffer, L.J., and Dunn, G.H., 1966, Rev. Mod. Phys. 38, 1.
111. Falk, R.A., and Dunn, G.H., 1983, Phys. Rev. A27, 754.
112. Watson, W.D., and Kunz, A.B., 1975, Ap. J. 201, 165.
113. Capone, L.A., Prasad, S.S., Huntress, W.T., Whitten, R.C., Dubach, J., and Santhanam, K., 1981, Nature 293, 45.
114. Speller, C.V., Fitaire, M., and Pointu, A.M., 1982, Nature 300, 507.
115. Cayrel, R., and Felenbock, P., (eds.), "Haute Résolution Spectrale en Astrophysique : Applications au Télescope Spatial", 2e Colloque National du Conseil Français du Télescope Spatial, Orsay 10-12 March, 1981.
116. Field, G.B., (guest editor), "Special Issue : New Instruments for Astronomy", 1982, Physics Today 35, N° 11, pp. 25-67.

MILLIMETER AND SUBMILLIMETER WAVE SPECTROSCOPY IN THE LABORATORY
AND IN THE INTERSTELLAR MEDIUM

Manfred Winnewisser and Brenda P. Winnewisser

Physikalisch-Chemisches Institut,
Justus-Liebig-Universität
Heinrich-Buff-Ring 58, D-6300 Giessen

Gisbert Winnewisser

I. Physikalisches Institut, Universität zu Köln
Universtätsstrasse 14, D-5000 Köln 41

Since 1968 radio astronomers have succeeded in identifying more than 50 interstellar molecules in the interstellar gas. Millimeter and submillimeter wave laboratory spectroscopy rapidly became crucial to this success. All of the recent advances in this field exemplify the interaction between developments in laboratory and interstellar measurements. The analysis of molecular dynamics is shown to be necessary for understanding the frequencies and intensities of the spectra observed in both mediums and for extrapolating to other parts of the spectra. Methods of high spectral and spatial resolution in molecular spectroscopy have greatly contributed to our knowledge of the physics, dynamics and chemistry of interstellar molecular clouds.

I. INTRODUCTION

Of the 60 or so molecules observed so far in interstellar clouds, the majority have been either identified or observed in the millimeter (MM) or submillimeter (SMM) wave region. The reason for this is seen in Fig. 1, in which the peak absorption coefficient of pure rotational transitions of OCS are plotted as a function of frequency at room temperature (1). Although a

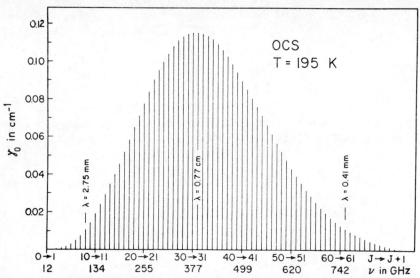

Fig. 1. Peak absorption intensity of pure rotational lines of OCS. From Ref. (1).

corresponding plot for low interstellar temperatures would have its maximum at lower frequency, this shift is largely compensated for by the smaller mass of many interstellar species and the faster frequency dependence (ν^4) of spontaneous emission. Finally, the small important molecules such as CO, HCN, HNC, H_2O can only be studied in the MM and SMM region.

Although many species were first identified from their microwave spectrum, the MM and SMM regions are found to be sprinkled with lines from molecules which are known to be there, but for which only low frequency measurements were available and extrapolation to high frequency was unsatisfactory (2). (See also G. Blake, this meeting). It is therefore mandatory to extend laboratory measurements on even rather pedestrian molecules well into the MM region. We have followed this policy for independent reasons, and have measured the spectra of many species such as HNCO (3-5), vinyl cyanide, CH_2CHCN (6), vinyl isocyanide, CH_2CHNC (7), and acrolein, CH_2CHCHO (8), up to 300 GHz already ten years ago. Vinyl cyanide was first observed on the basis of these measurements in the microwave region (9), but much stronger interstellar transitions were observed in the MM region (10). The possible identification of acrolein in the 72-80 GHz region was recently reported by Turner (11) on the basis of our measurements.

The stimulation provided by interstellar observations should not be overlooked. The laboratory search for HNC was clearly provoked by the suggested identification of the interstellar line (12, 13, 14). Measurement of the rotational spectra of ions,

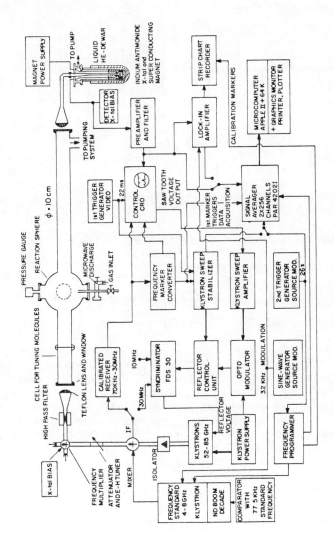

Fig. 2. Block diagram of the submillimeter wave spectrometer showing reaction sphere and free space absorption cell as used for the spectroscopy of unstable molecular species. The flow chart represents an overview of the frequency control and data acquisition system. The spectrometer has two modes of operation, selected by the switches indicated. Mode 1 provides fast sweeping of the microwave source and video display of the absorption signal. Mode 2 utilizes the phase-lock capability, source modulation and phase-sensitive detection of the absorption signal. From Ref. 20.

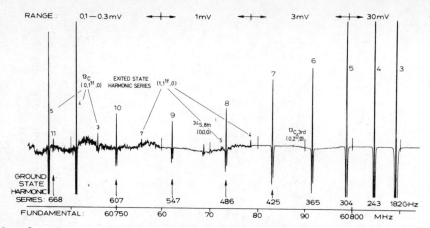

Fig. 3. Source modulation scan of the rotational spectrum of OCS observed within a 50 MHz region of the klystron fundamental frequency. The integers in the upper part of the figure indicate the klystron harmonic of each observed line. From Ref. (20).

pioneered by Claude Woods (15) and now pursued in several laboratories with very encouraging results (16, 17, 18, 19) is also motivated by interstellar observations.

II. LABORATORY SPECTROSCOPY

All of the MM and SMM laboratories currently studying molecules of interstellar interest use spectrometers basically equivalent to that used in the Giessen laboratory and shown in Fig. 2 (20). A monochromatic radiation source produces an approximately collimated beam of radiation, a (usually) single-pass absorption cell is filled to a pressure of a few mTorr (0.2 Pa) with the gas to be investigated, and the signal is focussed on a liquid He-cooled InSb detector. Such signals are processed either by averaging a large number of fast sweeps through a small frequency range, or by a slow sweep using a modulation technique and phase sensitive detection. The radiation source in Fig. 2 is a silicon Zener-diode frequency multiplier driven by a klystron in the range 30-120 GHz. This technique was developed by Gordy and coworkers over 30 years ago (21) and various improved versions thereof are now used routinely in the laboratories of DeLucia (22), Woods (23), Cohen and Pickett (24), Bogey et al. (18) and others in addition to our own laboratories. Recent efforts to exploit alternative sources of monochromatic MM and especially SMM radiation will be mentioned later in this report.

The capability of the Giessen SMM spectrometer is illustrated in Fig. 3, which shows the lines of OCS observed within a fairly narrow region of the tuning range of the klystron, due to the

Table 1. Rotational Transition Frequencies of the Isotopic species of CO.

Isotopic Species	$J_l - J_u$	no. of measurements	av.obs.freq./MHz (std.dev.)	(obs.-calc.)/MHz
$^{12}C^{16}O$	0 - 1	44	115 271.2051 (52)	0.0007
	1 - 2	20	230 538.0016 (50)	-0.0016
	2 - 3	6	345 795.9906 (36)	-0.0006
	3 - 4	5	461 040.7652 (38)	0.0022
	4 - 5	9	576 267.9131(463)	-0.0009
$^{12}C^{17}O$ [a]	0 - 1	3	112 359.2837 (5)	0.0017
	1 - 2	7	224 714.3850 (33)	-0.0001
	2 - 3	3	337 061.1298(100)	-0.0011
	3 - 4	5	449 395.3412(141)	0.0004
$^{12}C^{18}O$	0 - 1	29	109 782.1734 (63)	-0.0032
	1 - 2	22	219 560.3568 (81)	-0.0011
	2 - 3	7	329 330.5453 (40)	-0.0037
	3 - 4	9	439 088.7631 (79)	0.0083
	4 - 5	4	548 830.9775(329)	-0.0034
$^{13}C^{16}O$	0 - 1	24	110 201.3541 (51)	0.0051
	1 - 2	16	220 398.6765 (53)	0.0005
	2 - 3	21	330 587.9601 (49)	0.0020
	3 - 4	12	440 765.1668 (94)	-0.0065
	4 - 5	6	550 926.3029(304)	0.0029
$^{13}C^{17}O$ [a]	0 - 1	4	107 288.9525(153)	0.0025
	1 - 2	7	214 574.0812(117)	-0.0053
	2 - 3	1	321 851.5036 (?)	-0.0924 [b]
$^{13}C^{18}O$	0 - 1	14	104 711.4035 (57)	0.0047
	1 - 2	26	209 419.1594 (44)	-0.0043
	2 - 3	10	314 119.6632(115)	0.0025
	3 - 4	13	418 809.2549 (61)	-0.0014
	4 - 5	11	523 484.3172(225)	0.0004

[a] Unperturbed frequencies, preliminary data reported.

[b] Not used in fit.

simultaneous generation of successive harmonics of the fundamental frequency. The decrease in intensity for higher harmonics reflects the decreasing energy emitted at the succesive harmonics, but selective tuning and signal enhancement allow the measurement of lines up to the 10th harmonic. The line shape is Doppler-limited, the line centers are reproducible to within about 5 kHz per 100 GHz, and we have been able to measure the spectrum of a molecule with a dipole moment as low as 0.02 Debye, as was shown to be the case for HCCl (25). An example of the performance of the spectrometer and its application for work relevant to interstellar spectrosopy is provided by recent measurements by J. Reinstädtler of the stable isotopic species of CO, the preliminary results of which are shown in Table 1. The calculated frequencies referred to in the table are obtained from the adjusted constants

Table 2. Rotational Constants of the Isotopic Species of Carbon Monoxide, CO.

	$^{12}C^{16}O$	$^{12}C^{17}O$	$^{12}C^{18}O$	$^{13}C^{16}O$	$^{13}C^{17}O$	$^{13}C^{18}O$	Ref.
B_0 /MHz	57635.9693(3)	56179.9892(4)	54891.4212(10)	55101.0098(9)	53644.7928(13)	52356.0022(7)	(a)
	.9679	.993	.4263	.0181	.793	.0066	(b)
	.9687(26)	.9828(252)	.4239(122)	.0205(122)	--	.0108(122)	(c)
D_0 /kHz	183.565(8)	174.117(14)	166.468(27)	167.602(23)	158.896^b	151.416(18)	(a)
	.482	.308	.385	.661	.896	.336	(b)
	.567(70)	.388(100)^a	.462(95)^a	.745(96)^a	--	.405(87)^a	(c)
H_0 /mHz	164.3	152.1	141.9	143.5	132.4	123.1	(b)
st.dev. /kHz	1.7 / 35.3	1.5	5.9	5.2		4.0	(a) (c)

Ref.: (a) present work (MMW).

(b) Guelachvili, G., DeVilleneuve, D., Farreng, R., Urban, W., and Verges, J., J. Mol. Spec., 98, 64-79 (1983), (IR-data).

(c) Lovas, F.J., and Krupenie, P.H., J. Phys. Chem. Ref. Data, 3, 245 (1974); Original data were taken from W. Gordy et al. ($^{12}C^{16}O$, 1950-70) and B. Rosenblum et al. (Isotopes, 1957-58), (MMW-data); References see there.

^{a} D_0 was obtained from the $^{12}C^{16}O$ value via the isotope relation.

^{b} D_0 was taken from Ref. (b).

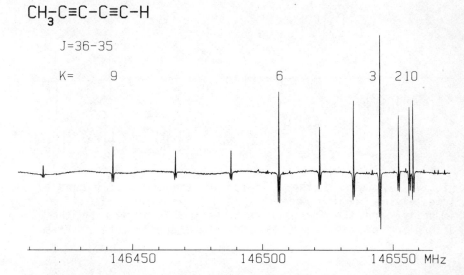

Fig. 4. Laboratory spectrum of one rotational transition of methyldiacetylene with associated K components. From Ref. (27).

given in Table 2. The analysis of the hyperfine structure in the spectrum of $C^{17}O$ must still be completed.

The new Cologne spectrometer for laboratory measurements uses a new, efficient multiplier construction (26), specifically designed for use as a frequency multiplier in the receiver system of the 3m radiotelescope. The spectrometer also employs a phase-locked millimeter wave reflex klystron or solid state Gunn oscillator, the frequency of which is controlled by a digital synthesizer, both in the video and lock-in mode. Signal averaging follows in both modes (27).

In this report we will describe the results of the study of groups of molecules which have occupied us in recent years. In each case, the motivation is a combination of the possibility of interstellar observation and unusual properties which entice molecular phsicists or chemists.

A. Long Carbon Chain Molecules

As is now well known, the cyanopolyines, HC_nN, have been observed with astounding abundance in some interstellar clouds (see H. W. Kroto, this meeting). They have been observed both in the microwave and MM regions. Laboratory measurements for both HCCCN and HCCCCCN made in the Giessen laboratory allowed the determination of correct rest frequencies of the interstellar MM lines (28-30).

Not only the CN group, but also the methyl group can provide

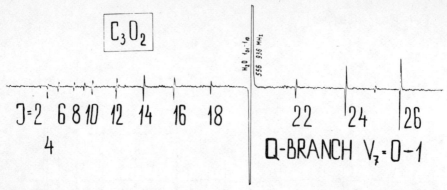

Fig. 5. Origin of the Q branch of the fundmental band of the lowest-lying bending mode of OCCCO. The strong line is a water impurity, and the intensity progression is masked by the broad absorption of atmospheric water in the spectrometer. From Ref. (36).

a dipole moment as a substituent on acetylene or a polyacetylene, and indeed, CH_3CCH has been observed in the interstellar medium (31,32). Its MM spectrum was already studied extensively (33). The next obvious candidate in this series, CH_3C_4H, first studied in the laboratory by Heath et al. (34), has now been studied in the MM region using the new laboratory spectrometer in Cologne. Fig. 4 shows the full K structure of one of the observed transitions (26). In order to determine interstellar column densities, the transition moments and thus the dipole moment are needed. The dipole moment of this molecule was measured in Giessen, with the Hewlett-Packard microwave spectrometer provided by the Max Planck Institute of Radioastronomy. The interstellar search for this species is described below.

Another group of molecules with a long carbon chain is the series OCCCO, OCCCS and SCCCS. Our initial interest in these molecules was aroused because of the quasilinear bending potential function determined from spectroscopic studies of OCCCO (35). Although it has no permanent axial dipole moment, its lowest bending mode falls at 18.2 cm^{-1}. The SMM group of Krupnov, in Gorki, succeeded in measuring this band and its hot bands by microwave techniques between 0.4 and 1.2 THz. In contrast to the spectrometers described above, this spectrometer uses the fundamental frequency output of a high power backward-wave oscillator and acoustic detection of the power absorbed by the gas (36-38). This spectrometer can measure over a wide range of sample pressures. Part of the Q branch of the low-lying bending fundamental of OCCCO is displayed in Fig. 5.

The analogous species OCCCS was first synthesized by Winnewisser and Christiansen, and its microwave spectrum **studied**

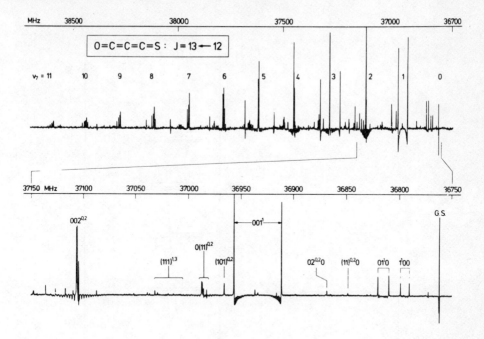

Fig. 6. Microwave spectrum of one rotational transition of OCCCS, showing absorption of molecules in various excited states $v_5 v_6 v_7^\ell$. From Ref. (40).

thoroughly in Giessen (39 - 42). An excerpt from this spectrum is shown in Fig. 6. It is much more rigid than OCCCO, but still has a rather low-lying bending mode at 83 cm^{-1} (41). From the combined analysis of the frequencies and intensities of the microwave absorptions, the vibrational energy level scheme shown in Fig. 7 was derived. The contrast with the energy level scheme of the lighter species OCCCO emphasizes the unusual bending behavior of OCCCO.

In the meantime, the molecules CO, CS, and OCS were observed in the interstellar medium some years ago (2). Just last year, the longer molecule CCCO was observed both in the laboratory (43) and in Sag B2 (44). Also last year, Turner (11) published the tentative identification of OCCCS in the interstellar medium. This means that OCCCO is almost certainly also present, and from our experience with the cyanopolyines, there may well be more molecules in this series present in the interstellar medium. We have therefore reopened the study of OCCCS in the MM region and have also started work on the infrared spectrum of SCCCS using the diode laser spectrometer in the Cologne laboratory.

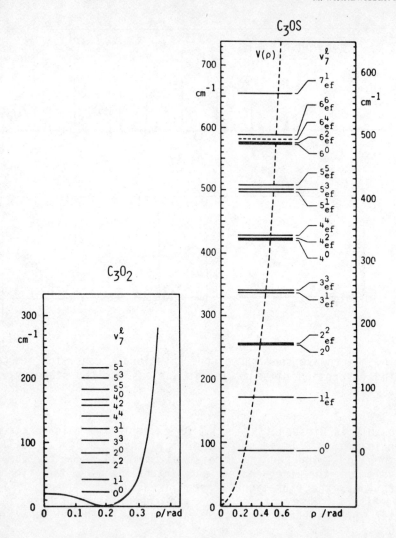

Fig. 7. Bending energy levels and potential functions for the lowest bending mode of OCCCO and OCCCS plotted on the same scale.

B. Molecules with Inversion

The most prominent interstellar molecule with large amplitude inversion is, of course, ammonia. Fig. 8 shows the range of potential functions possible for species of the type NH_2X.

An interstellar search for the deuterated ammonia species NH_2D is described below. The MM spectrum and inversion potential of this species have been studied by Cohen and Pickett (24), and

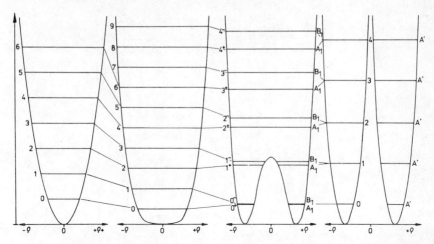

Fig. 8. Correlation of the energy levels and potential functions of molecules with the possibility of inversion, such as ammonia and its derivatives. The left-hand example represents a rigid planar configuration, the right-hand example the two equivalent forms of a rigidly bent invertor. The two middle examples illustrate nonrigid intermediate cases.

the group theory carefully outlined by Papoušek and Špirko (45). The lowest rotational levels of NH_2D are shown in Fig. 9, indicating the splittings due to asymmetry, inversion and the ^{14}N-nuclear quadrupole hyperfine interaction. Since the H atoms are fermions, NH_2D shows ortho and para states, with nuclear spin weights of 3 to 1. Thus for the levels of the $1_{11} - 1_{01}$ transition the 85 GHz transition is favoured by spin statistics over the 110.2 transition. The details of recent laboratory measurements of these lines are reported by Bester et al. (46).

Other substituted ammonias are cyanamide, NH_2CN, and isocyanamide, NH_2NC. Both show large amplitude inversion, but each has a different potential function; cyanamide is farther to the left in Fig. 8 (more planar) than isocyanamide. These species are discussed in the next section.

C. Cyanamide Isomers

Of the group of molecules shown in Fig. 10, the most stable is cyanamide. It is also the only one which has been found so far in the interstellar medium (2). Due to its astrophysical importance its MM and SMM spectrum has been measured and its inversion potential analyzed by Read et al. (47). Although both diazirine and diazomethane are explosive, both are relatively stable in a glass system at low pressure and their spectra have been studied in the past.

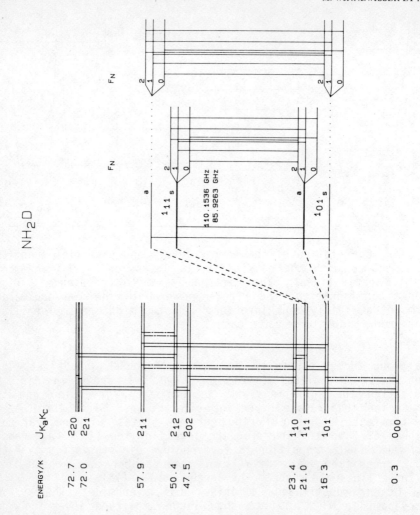

Fig. 9. Energy level diagram of NH$_2$D showing the lowest rotational levels and a magnification of the 1_{11} and 1_{01} levels with their associated hyperfine structure. The two detected interstellar transitons carry different spin statistical weights (3 to 1) as indicated by the thickness of the horizontal bars of the energy levels. Broken vertical lines indicate a-type rotational transitions, whereas solid lines represent the stronger c-type rotation-inversion transitions.

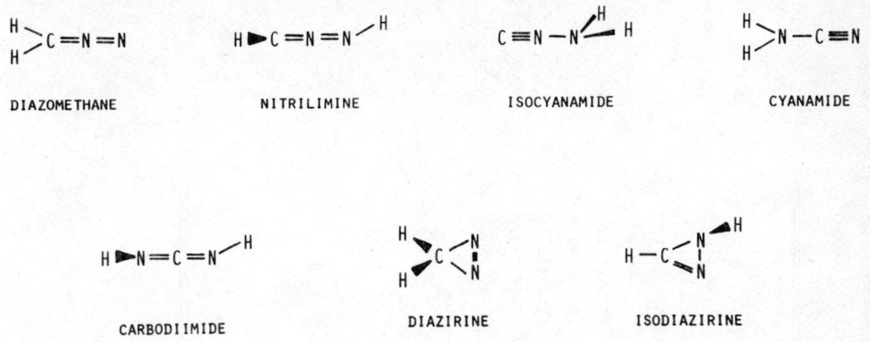

Fig. 10. Structural isomers of cyanamide.

Primarily out of interest in the inversion motion, we began the study of isocyanamide (48). The observed MM and SMM spectrum allowed us to assemble the energy level diagram shown in Fig. 11. As in asymmetrically deuterated ammonia, there are no pure inversion transitions, but there are c-type rotation-inversiton transitions as well as pure rotational a-type transitions. A Fortrat diagram, shown in Fig. 12, indicates how the spectrum of such a molecule looks. The splitting between the two c-type branches is just twice the inversion splitting of the ground state (48). This interval turns out to be nearly the same in isocyanamide as in NH_2D (24). As can be seen from the Fortrat diagram, the full expanse of the MM and SMM regions is necessary to sample the various branches in the rotation and rotation-inversion spectrum of such molecules.

Our work with isocyanamide has led us to MM and SMM as well as infrared work on diazomethane (49-51) and diazirine (52-54), and an infrared study of cyanamide. Diazirine is discussed here even though it is a cyclic molecule, and according to present experience not a likely candidate for the interstellar medium. We have found it a challenging species, however. It is almost totally asymmetric, so that the pure rotational spectrum consists not only of $\Delta K_a = 0$ transitions, but also of almost equally strong $\Delta K_a = 2$ transitions, as indicated in Fig. 13. Many of these transitions are split by hyperfine interactions, which, if observed in the interstellar medium would allow immediate identification. It was found that not only the quadrupole interaction with the two equivalent nitrogen atoms, but also the spin-rotation interaction had to be considered. The contribution of the latter is demonstrated in Fig. 14. We have now determined rotational constants, centrifugal distortion constants and quadrupole coupling constants for all of the isotopic species of diazirine listed in Table 3.

Table 3. Rotational and Centrifugal Distortion Constants in the A-Reduced Hamiltonian for Diazirine Isotopomers in the Ground Vibrational State.

	$H_2{}^{13}C^{14}N_2$	$H_2{}^{12}C^{14}N_2$	$H_2{}^{12}C^{14}N^{15}N$	$H_2{}^{12}C^{15}N_2$	$D_2{}^{12}C^{14}N_2$ [b]
A / MHz	40950.3414(23)	40951.3526(11)	39798.966(79)	38604.89635(99)	35821.697(10)
B / MHz	22869.8200(10)	23667.98345(46)	23416.7922(91)	23200.85858(61)	18760.7153(41)
C / MHz	16322.66702(93)	16726.11565(35)	16406.788(11)	16095.59679(63)	14854.6298(40)
Δ_J / kHz	30.116(13)	31.5651(22)	30.29(33)	30.101(10)	18.374(44)
Δ_{JK} / kHz	103.145(52)	107.940(12)	102.6(41)	100.242(19)	76.905(90)
Δ_K / kHz	-17.40(16)	-23.456(50)	-23.3(119)	-26.084(50)	-15.39(32)
δ_J / kHz	8.8737(18)	9.4407(12)	9.33(25)	9.28800(99)	3.941(29)
δ_K / kHz	74.975(44)	77.179(13)	75.7(37)	73.7025(97)	16.866(40)
H_{JK} / Hz	-0.551(70)	-0.56(12)	-0.54(16)	-0.794(14)	–
H_{KJ} / Hz	2.59(34)	2.77(16)	–	2.247(58)	–
h_J / Hz	-0.0108 [c]	-0.0108(33)	–	–	–
h_{JK} / Hz	3.31 [c]	3.31(49)	–	–	–
σ / kHz	12.6	16.0	23.6	9.6	71.4
κ	-0.4683	-0.4269	-0.4007	-0.3687	-.6274

[a] Figures in parentheses are standard deviations in units of the least significant figures
[b] preliminary fit
[c] held fixed during the fit

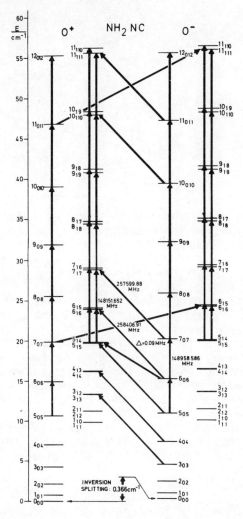

Fig. 11. Energy level scheme of the rotational structure in the two lowest inversion energy levels of isocyanamide. Vertical arrows show observed a-type rotational transitions, and oblique arrows show observed c-type, rotation-inversion transitons. From Ref. (48).

D. HCN Dimer Molecules

In addition to the HCN van der Waals dimer, there are several chemical dimers of HCN which are predicted to be physically stable. The only one which has been identified in the gas phase is N-cyanomethanimine, H_2CNCN. We have just completed a study of

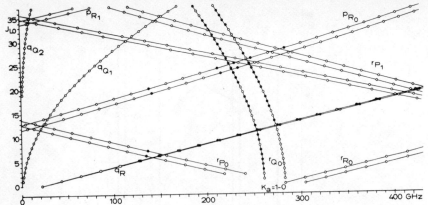

Fig. 12. Fortrat diagram showing the simple branch structure of the observed rotation and rotation-inversion spectrum of isocyanamide. Filled circles indicate observed transitions. From Ref. (48).

the MM spectrum of this molecule, which has a halflife of about ten minutes in a glass cell at 2 mTorr, including measurement of both a and b type transitions and an analysis of the quadrupole coupling due to the two nitrogen atoms (55). This molecule is similar in many ways to vinyl cyanide, and we therefore consider it a likely candidate for the interstellar medium.

Although with rotational spectroscopy we are probing essentially time independent phenomena in the laboratory and in the interstellar medium, we hope by investigating various isomeric systems to explore neighboring portions of the same energy hypersurface, and thus obtain information which contributes to the rapidly increasing understanding of interstellar chemistry.

III. INTERSTELLAR SPECTROSCOPY

During the past 15 years interstellar radio spectroscopy has developed into a powerful tool for probing the physical conditions, chemical composition and isotope abundance ratios in molecular gas by means of molecular and atomic transitions (2). They are now being used to investigate the molecular clouds from their tenuous outer surroundings to their dense interiors, where star formation can take place. Whereas these latter confined areas can be studied profitably only with high angular resolution, in the arc sec region (HPBW 40 arc sec, using large single dish telescopes or interferometric techniques), it is the domain of the smaller radiotelescopes with their larger beamwidth to measure the total sizes and masses of the molecular clouds and their galactic distribution. In addition small radiotelescopes with high surface accuracies are ideal instruments for extending molecular line observations into the SMM region.

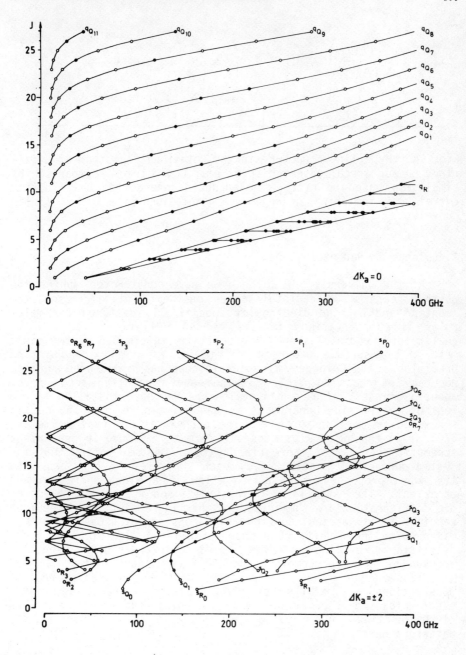

Fig. 13. Fortrat diagrams for diazirine, which has only a-type transitions, but for which the asymmetry leads to a spectrum which appears unstructured. Full circles indicate observed transitions.

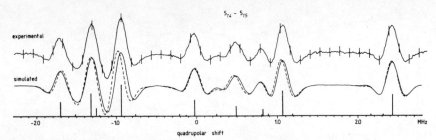

Fig. 14. Hyperfine structure in the rotational spectrum of diazirine. Upper trace: observed spectrum. Lower trace: comparison of simulation based on nitrogen quadrupole interaction only (dotted curve) and including hydrogen spin-rotation interaction (solid curve).

A. Cologne 3m Radiotelescope

At the University of Cologne we have constructed during the past 3 years a 3m radiotelescope together with its associated frontend and backend electronics specifically designed for molecular line observations in the MM and SMM region (56). The quality of a radiotelescope is essentially determined by two parameters: the surface accuracy of the reflector and the pointing precision. The Cologne 3m radiotelescope has a measured surface accuracy of about 30 μm (rms value) and a pointing accuracy of about 3 arc sec. The 3m radiotelescope has been installed on the roof of the Physics Department as can be seen in Fig. 15, but is scheduled to be transported to the Gornergrat Observatory (elev. 3200 m) in the Swiss Alps close to Zermatt. Figure 16 presents a schematic overview of the entire system. One may notice that the system concept is modular, i.e. each subsystem is provided with its own microprocessor so as to allow stand-alone operation and check out. The essential technical data and its present electronic configuration are summarized in the figure captions. The line receiver in its present configuration consists of a GaAs-Schottky-mixer, with the novelty of a phase-stabilized Gunn-diode as local oscillator. For line work three spectrometers are being used, two acousto-optical spectrometers and one filter spectrometer.

Since January 1984 the telescope has been in use for routine observations. Details of the telescope, its associated electronics and observational mode are being described in a series of forthcoming papers to be published in the Zeitschrift für Naturforschung.

B. Results with the 3m Radiotelescope.

The bandwidth (72 - 87.5 GHz) of the line receiver used with the 3m radiotelescope allows observation of a large number of

Fig. 15. The new Cologne 3m radiotelescope on the roof of the physics building. The main reflector consists of four glass fiber panels with a measured total surface accuracy of 30 μm (rms). The telescope is presenty equipped with a cooled Schottky-barrier mixer-receiver covering 72-90 GHz with a double-side-band noise temperature of 230-270 K. The telescope is transportable and has a total weight of ca. 3.5 t.

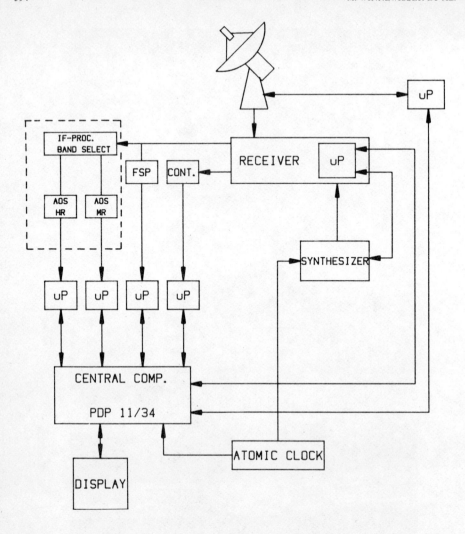

Fig. 16. Schematic lay-out of the electronic system of the 3m radiotelescope. The most important functional groups are indicated by boxes, i. e. receiver, acousto-optical spectrometers (in dashed frame), the filter spectrometer, the 500 MHz continuum backend, and the various microprocessors (uP). The bandwidth and resolution of each of the spectrometers are: filter spectrometer, 256 MHz, 1 MHz/channel; medium resolution acousto-optical spectrometer, 250 MHz, 200 kHz/channel; high reslution spectrometer, 64 MHz, 31.2 kHz/channel.

important molecular transitions such as those of HCN, HCO$^+$, CCH, SO, SiO. We have started a mapping program using essentially HCN, HCO$^+$ and CCH in several types of molecular clouds; cold dark clouds like TMC1 and L183, and warm clouds including those associated with HII regions, such as Orion, W51, and S140. Figure 17 presents some typical sample spectra obtained with the high resolution acousto-optical spectrometer. Maps of several molecular clouds, including isotopic species, are now becoming available.

Aside from these detailed mapping programs with rather abundant species, we have used the telescope for observation of molecules with lower interstellar abundance. It is a surprising and encouraging result that molecules like NH$_2$D and more complicated molecules like CH$_3$CCH can be detected with the 3m radiotelescope.

C. Detection of NH$_2$D

We have recently conducted an interstellar search for the two rotation-inversion components of the $1_{11} - 1_{01}$ transition of NH$_2$D, discussed above and shown in Fig. 9, towards four selected molecular clouds DR 21(OH), S140, L183 and TMC1. However, the interstellar detection of both transitions is beset with difficulties. The lower of the two transitions at 85.9 GHz is blended with a transition of methyl formate (HCOOCH$_3$) which for all clouds where methyl formate has been found (i.e. Orion A, Sgr B2) makes an identification of NH$_2$D very uncertain. The upper of the two transitions at 110.2 GHz is considerably weaker (~1/5) and thus extremely difficult to detect. The first claimed detections of interstellar NH$_2$D (57, 58) suffer heavily from these difficulties. Olberg et al. (59) have now unambiguously detected NH$_2$D by high resolution observations in molecular clouds where it is possible to resolve the associated hyperfine structure. NH$_2$D has been observed in DR21(OH), S140 and L183. A sample of the spectra obtained with the Onsala radiotelescope is presented in Fig. 18. Two results of Olberg et al.(59) have important consequences: (i) the striking intensity difference between the two rotation-inversion transitions of NH$_2$D in favour of the 85.9 GHz line, is partly due to the nuclear spin statistics, but in part may have to be explained by radiative transfer calculations, and (ii) the negative results for TMC1 in comparison to the other sources suggest that differences in chemical fractionation effects between the carbon-deuterium and the nitrogen-deuterium bond could cause this apparent deficiency in NH$_2$D abundance.

D. CH$_3$CCCCH - A New Interstellar Molecule.

With the Effelsberg 100m radiotelescope we have recently detected the K = 0 and K = 1 components of the J = 6 → 5 and J = 5 → 4 transitions of CH$_3$CCCCH, methyldiacetylene, in the nearby dust

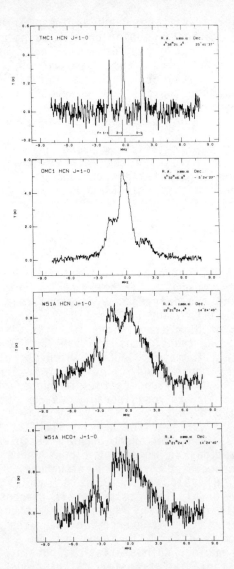

Fig. 17. Interstellar spectra of the lowest rotational transitions of HCN and HCO$^+$ in several molecular clouds. The spectra were taken with the 3m radiotelescope with the high resolution acousto-optical spectrometer by the Cologne spectroscopy group. The hyperfine structure is clearly resolved for TMC1, barely in OMC1, and blended with velocity components in W51. For comparison we present for W51 also the HCO$^+$ spectrum; both show self-absorption.

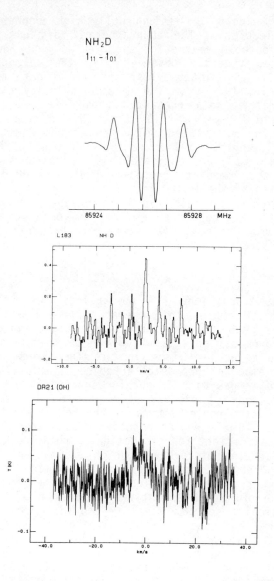

Fig. 18. Comparison of the laboratory spectrum of NH_2D with the interstellar spectrum of the dark cloud L183 (Onsala 20m telescope). The spectrum was taken with an autocorrelator spectrometer set at a resolution of 29.3 kHz/channel, sufficient to resolve the ^{14}N-nuclear hyperfine splitting. The DR21(OH) spectrum was taken with the Cologne 3m radiotelescope using the 2048 channel acousto-optical spectrometer with a channel spacing of 31.2 kHz.

cloud TMC1 (60) known for its unusually high abundance of carbon containing molecules. The abundance of this new interstellar species is comparable to that of the cyanopolyynes and to that of methylacetylene, recently observed with the Cologne 3m radiotelescope (61). The interstellar spectra of both molecules are presented in Fig. 19. From the data of Walmsley et al.(60) and the new laboratory data presented above (26) one obtains the abundance ratio $[CH_3C_4H]/[CH_3C_2H]$ ~ 0.25, which suggests that abundance of the methylated polyynes decreases with increasing carbon chain length. The ratios $[CH_3C_4H]/[C_4H]$ ~ 0.1 and $[CH_3C_4H]/[HC_5N]$ ~ 0.3 (60) indicate that hydrocarbons such as C_4H_x and C_5H_y are the immediate progenitors of both HC_5N and CH_3C_4H. Thus the problem of the formation of long carbon chain molcules is that of forming long-chain hydrocarbons and than of attaching various functional groups such as CH_3, CN, or NH_2, to the hydrocarbon skeleton. One may, therefore, expect to find more related molecules and species with even longer carbon chains.

IV. FUTURE PROSPECTS

Molecular transitions are now being used to investigate interstellar gas ranging in temperature from about 10K in cold dark clouds to 5000 K in molecular shock fronts and circumstellar shells. Astronomical objects which can be investigated by means of atomic and molecular lines include the quiescent cold dark clouds, young embedded protostellar objects, molecular masers and shocked molecular gas with temperatures above 3000 K, as well as molecular envelopes of evolved stars and their associated massloss rates, and the replenishment of the interstellar medium with heavy elements. The energies associated with these astrophysical phenomena are sufficient for excitation of rotational and rovibrational levels. The discovery of high excitation lines such as those of H_2, CO or NH_3 (see for example Ungerechts and Winnewisser (62) for a summary) clearly indicate that the SMM and far infrared region will be of superb scientific interest both for laboratory and astrophysical research. The SMM and far infrared wavelength region has remained - due to technical problems and the absorption in the earth´s atmosphere - the last major unexplored part of the electromagnetic spectrum, in astrophysics even more than in the laboratory. Present plans both by ESA and NASA for future space astronomy missions within the scientific program survey for 1985-2004 call for an opening of infrared and SMM astronomy by launching high-throughput heterodyne spectroscopy missions (Space Science Horizon, ESA, Report of the Survey Committee, July 1984).

Laboratory spectroscopy in the SMM region in the immediate future will be most actively pursued for the study of molecular ions, and for probing the behavior of molecules exhibiting large amplitude motions, which may be internal rotation, quasilinear bending, ring puckering, the loose bonding of van der Waals molecules or clusters, or combinations of these.

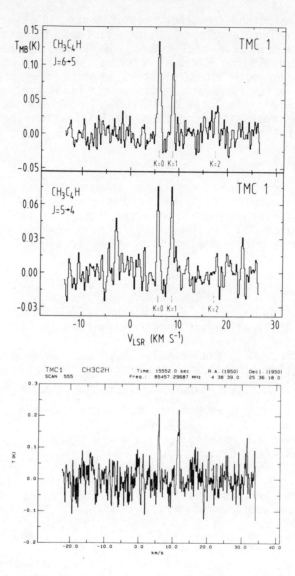

Fig. 19. Insterstellar spectra of methylacetylenes. This first detection of CH_3CCCCH was recorded with the Effelsberg 100m radiotelescope using a 1024 channel autocorrelator set at a resolution of 12.2 and 6.1 kHz/channel for the two transitions, respectively (55), whereas the CH_3CCH spectrum was recorded with the Cologne 3m radiotelescope using the high resolution acousto-optical spectrometer with 31.25 kHz/channel (56).

High resolution frequency measurements in the terahertz region are being developed using various techniques: high frequency backward wave oscillators, difference frequency laser systems, laser-klystron sideband systems, and Fourier transform techniques. None of these systems appears now to be truly efficient for extensive measurements over a broad frequency range, or for use as reliable local oscillators. However, the molecular physics and astrophysics accessible in this region induces us to exploit all of these methods.

ACKNOWLEDGEMENTS

The work reported here has been supported by the Deutsche Forschungsgemeinschaft in both laboratories. The Cologne 3m radiotelescope was constructed with the support of the DFG Sonderforschungsbereich 131, Radioastronomy. The work of the Giessen laboratory was also supported by the Fonds der Chemischen Industrie.

REFERENCES

(1) Gordy, W., and Cook, R. L., "Microwave Molecular Spectra" in Chemical Applications of Spectroscopy Part II (W. West, ed.), Interscience, (Wiley) New York, 1970.
(2) Winnewisser, G., Churchwell, E., and Walmsley, C. M., "Astrophysics of Interstellar Molecules" in Modern Aspects of Microwave Spectroscopy (G. W. Chantry, ed.), Academic Press, London, 1979.
(3) Hocking, W. H., Gerry, M. C. L., and Winnewisser, G., 1975, Can. J. Phys. 53, pp. 1869-1901.
(4) Winnewisser, G., Hocking, W. H., and Gerry, M. C. L., 1976, J. Phys. Chem. Ref. Data 5, pp. 79-101.
(5) Yamada, K., 1980, J. Mol. Spectrosc. 79, pp- 323-344.
(6) Gerry, M. C. L., Yamada, K., and Winnewisser G., 1979, J. Phys. Chem. Ref. Data 8, pp. 107-123.
(7) Yamada K., and Winnewisser, M., 1975, Z. Naturforsch. 30a, pp. 672-689.
(8) Winnewisser, M., Winnewisser, G., Honda, T., and Hirota, E., 1975, Z. Naturforsch. 30a, pp. 1001-1014.
(9) Gardner, F. F., and Winnewisser, G., 1975, Astrophys. J. 195, pp. L127-L130.
(10) Matthews, H.E., and Sears, T. J., 1983, Astrophys. J. 272, pp. 149-153.
(11) Turner, B. E., 1983, Astrophys. Lett. 23, pp. 217-224.
(12) Creswell, R. A., Pearson, E. F., Winnewisser, M., and Winnewisser, G., 1976, Z. Naturforsch. 31a, pp. 221-224.
(13) Saykally, R. J., Szanto, P. G., Anderson, T. G., and Woods, R. C., 1976, Astrophys. J. 204, pp. L143-L145.
(14) Blackman, G. L., Brown, R. D., Godfrey, P. D. and Gunn, H. I., 1976, Nature 261, pp. 395-396.

(15) Woods, R. C., 1983, in "Molecular Ions: Spectroscopy, Structure and Chemistry" (T. A. Miller and V. E. Bondybey, eds.), North-Holland, Amsterdam.
(16) De Lucia, F. C., Herbst, E., Plummer, G. M., and Blake, G. A., 1983, J. Chem. Phys. 78, pp. 2312-2316.
(17) Molecular Ions: Spectroscopy, Structure and Chemistry (T. A. Miller and V. E. Bondybey, eds.) North-Holland, Amsterdam, 1983.
(18) Bogey, M., Demuynck, C., and Destombes, J. L., 1984, Astron. Astrophys., in press; Destombes, J. L., this meeting.
(19) Gudemann, C. S. and Saykally, R. J., 1984, Ann. Rev, Phys. Chem. 35, pp. 387-418.
(20) Schäfer E., and Winnewisser, M., 1983, Ber. Bunsenges. Phys. Chem. 87, pp. 327-334.
(21) Gordy, W., 1983, J. Mol. Struct. 97, pp. 17-32.
(22) Helminger, P., Messer, J. K., and De Lucia, F. C., 1983, Appl. Phys. Lett. 42, pp. 309-310.
(23) Warner, H. E., Conner, W. T., Petrmichl, R. H., and Woods, R. C., 1984, in press.
(24) Cohen, E. A., and Pickett, H. M., 1982, J. Mol. Spectrosc. 93, pp. 83-100.
(25) Schäfer, E., and Christiansen, J. J., 1983, J. Mol. Struct. 97, pp. 101-114.
(26) Bester, M., Yamada, K., Winnewisser, G., Joentgen, W., Altenbach H.-J., and Vogel E., 1984, Astron. Astrophys. Lett. in press.
(27) Bester, M., 1984, Dissertation, University of Cologne.
(28) Creswell, R. A., Winnewisser, G., and Gerry, M. C. L., 1977, J. Mol. Spectrosc. 65, pp. 420-429.
(29) Winnewisser, G., Creswell, R. A., and Winnewisser, M., 1978, Z. Naturforsch. 33a, pp. 1169-1182.
(30) Winnewisser, M., 1981, Faraday Disc. Chem. Soc. 71, pp. 31-55.
(31) Hollis, J. M., Snyder, L. E., Blake, D. H., Lovas, F. J., Suenram, R. D., Ulich, B. L., 1981, Astrophys. J. 251, pp. 541-548.
(32) Irvine, W. M., Höglund, B., Friberg, P., Askne, J., Ellder, J., 1981, Astrophys. J. 248, pp. L113-L117.
(33) Bauer, A., and Burie, J., 1969, Compt. rend. 268B, pp. 800.
(34) Heath, G. A., Thomas, L. F., Sherrard, E. I., and Sheridan, J., 1955, Disc. Faraday Soc. 19, pp.38.
(35) Winnewisser, B. P., 1985, "Molecular Spectroscopy: Modern Research Vol. III" (K. N. Rao, ed.), Academic Press, New York, in press.
(36) Burenin, A. V., Karyakin, E. N., Krupnov, A. F., and Shapin, S. M., 1979, J. Mol. Spectrosc. 78, pp. 181-184.
(37) Karyakin, E. N., Krupnov, A. F., and Shapin, S. M., 1982, J. Mol. Spectrosc. 94, pp. 283-301.
(38) Krupnov, A. F., in "Modern Aspects of Microwave Spectroscopy" (G. W. Chantry, ed.), Academic Press, London 1979,

pp. 217-256.
(39) Winnewisser, M., and Christiansen, J. J., 1976, Chem. Phys. Letters, 37, pp. 270-275.
(40) Winnewisser, M., Peau, E. W., Yamada, K., and Christiansen, J. J., 1981, Z. Naturforsch. 36a, pp. 819-830.
(41) Winnewisser, M., and Peau, E. W., 1982, Chem. Phys. 71, pp. 377-387.
(42) Winnewisser, M., and Peau, E. W., 1984, Acta Phys. Hungarica 55, pp. 33-44.
(43) Brown, R. D., Eastwood, F. W., Elmes, P. S. and Godfrey, P. D., 1983, J. Am. Chem. Soc. 105, pp. 6496-6497.
(44) Matthews, H. E., Irvine, W. M., Friberg, P., Brown, R. D., and Godfrey, P. D., 1984, Nature, in press.
(45) Papoušek, D., and Špirko, V., 1982, Top. Curr. Chem. 68, pp. 59.
(46) Bester, M., Urban, S., Yamada, K., and Winnewisser, G., 1983, Astron. Astrophys. 121, pp. L13-L14.
(47) Read, W. G., Cohen, E. A., and Pickett, H. M., 1985, in press.
(48) Schäfer, E. and Winnewisser, M., 1982, Ber. Bunsenges. Phys. Chem. 86, pp. 780-790.
(49) Schäfer, E. and Winnewisser, M., 1983, J. Mol. Spectrosc. 97, pp. 154-164.
(50) Vogt, J., and Winnewisser, M., 1983, Z. Naturforsch. 38a, pp. 1138-1145.
(51) Vogt, J., Winnewisser, M., Yamada, K., and Winnewisser, G., 1984, Chem. Phys. 83, pp. 309-318.
(52) Vogt, J., Winnewisser M., and Christiansen, J. J., 1984, J. Mol. Spectrosc. 103, pp. 95-104.
(53) Bogey, M., Winnewisser, M., and Christiansen, J. J., 1984, Can. J. Phys., in press.
(54) Möller, K., Vogt, J., Winnewisser, M., and Christiansen, J. J., 1984, Can. J. Phys., in press.
(55) Winnewisser, M., Winnewisser B. P., and Wentrup, C., 1984, J. Mol. Spectrosc. 105, pp. 193-205.
(56) Winnewisser, G., Vowinkel, B., 1984, Sterne u. Weltraum pp. 132-137.
(57) Turner, B. E., Zuckerman, B., Morris, M., and Palmer, P., 1978, Astrophys. J. 219, pp. L43.
(58) Rodriguez-Kuiper, E. N., Zuckerman, B., Kuiper, T. B. H., 1978, Astrophys. J. 219, pp. L49-L53.
(59) Olberg, M., Bester, M., Rau, G., Pauls, T., Winnewisser, G., Johansson, L. E. B., Hjalmarson, A., 1984, Astron. Astrophys. (submitted).
(60) Walmsley, C. M., Jewell, P. R., Snyder, L. E., and Winnewisser, G., 1984, Astron. Astrophys. 134, pp. L11-L14.
(61) Zensen, U., 1984, Dissertation, Cologne University.
(62) Ungerechts, H., and Winnewisser, G., 1984, in "Galactic and Extragalactic Infrared Spectroscopy" (M. F. Kessler and J. P. Phillips, eds.), Reidel, pp. 177-191.

STUDIES OF ASTROPHYSICALLY IMPORTANT MOLECULAR IONS WITH ULTRASENSITIVE INFRARED LASER TECHNIQUES

Richard J. Saykally
Department of Chemistry
UC Berkeley
Berkeley, CA 94720

Over the last decade, modeling of the chemistry occurring in interstellar gas clouds has evolved dramatically. The early qualitative versions of Solomon and Klemperer[1] and Herbst and Klemperer,[2] which first predicted the preeminence of ion-molecular reactions in these cold, diffuse environments, are now supplanted by the modern sophisticated computer models of Mitchell, Ginzberg and Kuntz,[3] Prasad and Huntress,[4] Watson,[5] and others. Quantitative predictions of molecular abundances are now given for an impressive number of species, including some two dozen of the most important molecular ions. In Table 1, the molecular ions expected to be most abundant in cold (50°K), dense ($n \approx 10^6$ cm^{-3}) clouds are listed, along with abundances (relative to H_2) predicted by Mitchell, Ginzberg, and Kuntz,[3] appropriate for a cloud density of 10^6 cm^{-3}.

Generally, the stable, closed-shell "proton adducts" are the most abundant ions in cold, dense clouds, because open-shell species generally react rapidly with H_2 until these thermodynamically favored products are finally obtained. Actually, the synthesis of these ions is most likely to proceed from the direct proton transfer reaction of the principal reactant ion, H_3^+, with abundant stable molecules (e.g., CO, H_2O, CO_2, NH_3, etc.) to yield the respective proton adducts. Until the present, only three such protonated species have been definitively identified in interstellar clouds--HOC^+, HNN^+ and HCS^+. In addition, CH^+ (the unstable isomer of HCO^+)[6] and HCO_2^+ [7] have been tentatively identified. This situation is simply the result of a general lack of high-resolution laboratory spectroscopic information on polyatomic ions. While approximately 45 diatomic ions have now been studied at high resolution in the laboratory,[8,9] only 15 polyatomic species[10] have thus far been detected by spectroscopic techniques capable of resolving their rotational energy level structures. Definitive identification of molecular transitions observed by microwave, millimeter, or infrared astronomical techniques clearly requires comparison with such high-precision laboratory data for the general cases of nonlinear molecules, even though the linear molecules HCO^+, HNN^+, and HCS^+ were, in fact, originally identified quite reliably from millimeter observations alone.[8,9]

TABLE 1
Molecular Ions Predicted to be Abundant in Cold (50°K), Dense ($n \approx 10^6$ cm^{-3}) Interstellar Clouds

Ion (x)		Log n_x/n_H
H_3^+		-10.3
CH_3^+		-11.99
CH_5^+		-14.3
O_2^+	P	-11.4
OH^+	P	-14.8
H_2O^+	P	-14.2
H_3O^+		-11.0
HCO^+	*	-9.8
CO^+	P	-14.8
HO_2^+	P	-13.8
H_2CO^+	P	-11.9
HCO_2^+		-14.6
H_3CO^+		-14.1
NH_2^+		-14.8
NH_3^+	P	-11.9
NH_4^+		-13.4
NO^+		-12.7
HNN^+	*	-11.2
H_2CN^+		-13.1
CCN^+		-14.6
SH^+	P	-11.3
H_2S^+	P	-14.8
SO^+	P	-10.8
S_2^+	P	-14.8
HCS^+	*	-14.5
H_2CS^+	P	-13.8
$HSiO^+$		-12.9

P → paramagnetic

* → detected by radioastronomy

A number of interesting open-shell polyatomic ions are also predicted by current models to be abundant in dense clouds (n=10^6); for example, Mitchell, Ginzberg, and Kuntz,[3] predict high abundances of O_2^+ HO_2^+, SH^+, H_2S^+, S_2^+, NH_3^+, H_2CO^+, and HCS^+ in such regions. For these cases as well, there exist very little spectroscopic data which can be of use in identifying spectra observed in astronomical sources. Indeed, while several open-shell polyatomic ions have, in fact, been studied, both in the laboratory and in comet tails,[8,9] principally by optical emission spectroscopy, these open-shell species are likely to be even more difficult to study by high-resolution techniques than closed-shell ions because of their extreme chemical reactivity.

In this paper we discuss the development of two infrared laser techniques--far-infrared laser magnetic resonance and velocity-modulation infrared laser spectroscopy--which have demonstrated the capability to study molecular ions with sufficient precision to produce a positive identification of their corresponding astronomical spectra. We anticipate that in the near future these methods will make it possible to detect, identify, and make detailed studies of spectra of a variety of astrophysically important molecular ions. The extension of astronomical measurements to new ionic species is of obvious importance to understanding the chemistry of the interstellar medium, particularly in view of the facts that most of the ion-molecule reaction data used as the input for modern chemical models have rarely been obtained at the low temperatures appropriate for the ISM, and that many of the critically important dissociative recombination and radiative association reactions have not been studied at all.

The first infrared spectrum of a gaseous molecular ion (NO^+) was observed in connection with atmospheric nuclear weapons tests conducted in the 1960s.[11] In 1979, Schwartz developed a method for observing broadband IR absorption spectra of ions generated by pulsed radiolysis at atmospheric pressure, and applied it to $H_3O^+(H_2O)_n$ (n=0-6) in 1975[12] and to $NH_4^+(NH_3)_n$ (n=0-4) in 1979.[13] The first laboratory IR spectra of ions to be obtained with high resolution were those of HD^+, observed by the Doppler-tuned fast ion beam technique of Wing and coworkers[14] in 1975. Subsequently, his group was able to study transitions in HeH^+,[15] D_3^+,[16] and H_2D^+[17] by this method. Carrington has incorporated a second laser with the fast ion beam experiment to study HD^+,[18] HeH^+,[19] CH^+,[20] H_3^+,[21] and other ions at internal energies near or even exceeding their lowest dissociation limits. In 1979, K.M. Evenson and I[22] reported the application of far-infrared laser magnetic resonance spectroscopy for the measurement of rotational transitions in the ground $^2\Pi$ state of HBr^+. This work has been continued at Berkeley, and has been extended to the ground states of HCl^+,[23] HF^+,[24] OH^+,[25] and H_2O^+.[26] In 1980, Oka[27] opened the door to the investigation of vibrational spectra of molecular ions with tunable infrared lasers with his work on H_3^+, subsequently extended to HeH^+ [28] and NeH^+ [29] by Amano and coworkers at Ottawa. Brault and Davis[30] observed the vibration-rotation spectrum of ArH^+ and KrH^+ as well. Van den Huevel and Dymanus[32] pioneered the application of tunable far-infrared laser sources to rotational spectroscopy of ions in 1983 with their study of CO^+, HCO^+, and HNN^+. In 1983, Gudeman,

Begemann, Pfaff, and myself reported the development of a new method for observing molecular ion vibration-rotation spectra, called "velocity-modulation laser spectroscopy," and its described application to HCO^+,[33] HNN^+,[34] and H_3O^+.[35] This approach has now been extended to include NH_4^+,[36,37] H_2F^+,[38,39] D_3O^+,[40] and DNN^+ [41] in studies at Berkeley, to NH_4^+,[42] H_3O^+,[43] $HCNH^+$,[44] and OH^+ [45] in studies by Oka's group at Chicago, and to HCO^+ [46] studied by Davies at Cambridge. Recently Amano and coworkers at Ottawa have reported studies of HCO^+,[47] H_2D^+,[48] HD_2^+,[48] and $DCNH^+$ [49] with infrared difference frequency spectroscopy of a discharge, and McKellar and coworkers have studied HCO^+,[50] HNN^+, DCO^+ and DNN^+ [51] with a diode laser. The molecular ions that have been studied with high-resolution infrared methods are listed in Table 2. C.S. Gudeman and myself[10] have reviewed most of this work very recently; consequently, only a brief discussion of the results of laser magnetic resonance spectroscopy and infrared laser spectroscopy experiments that are directly relevant to astrophysical studies is given here.

TABLE 2
Molecular Ions Studied by High-Resolution Infrared Spectroscopy

Ion	References
HD^+	14,18
HeH^+	5,19,28
NeH^+	29
KrH^+	31
$ArH^+(ArD^+)$	31,31,71
HF^+	24
HCl^+	23,55
HBr^+	22,55
CH^+	20
OH^+	25,45
$HCO^+(DCO^+)$	33,46,47,50
$HNN^+(DNN^+)$	34,41,51,64
$H_3^+(H_2D^+,HD_2^+,D_3^+)$	27,16,17,48
H_2F^+	38,39
H_2O^+	26
$H_3O^+(D_3O^+)$	35,40,43
NH_4^+	36,37,42,70
$HCNH^+(DCNH^+)$	44,49

Characteristics of Infrared Spectroscopy

Unlike in the microwave and millimeter regime, where the spectral resolution is typically limited by collisional line broadening, infrared spectra are usually Doppler-limited. For typical light molecules, this corresponds to a linewidth (hwhm) which is $\sim 1 \times 10^{-6}$ of the transition frequency. At 3000 cm^{-1}, this is roughly 100 MHz, whereas at 30 cm^{-1} (in the far infrared) it is 1 MHz. Considering that one can generally measure a line center ~ 10 times more accurately than the fractional linewidth, this implies a conservative Doppler-limited measurement precision ranging from ~ 10 MHz in the IR to ~ 100 kHz in the FIR. Typical measurement precision for the microwave or millimeter region is 10 kHz. The loss of precision in the IR is partially offset by the fact that often 50 to 100 rotational lines can be measured for a given molecule. When a small number of parameters are required to fit the spectra, as in the cases of HCO^+, HNN^+, or $HCNH^+$, knowledge of the astrophysically interesting lowest transition frequency essentially increases like the standard deviation of a mean--i.e., as the square root of the number of measurements. Hence, one can, in principle, regain an order of magnitude in precision from infrared studies, yielding ~ 1 MHz measurement precision at 3000 cm^{-1}. Moreover, sub-Doppler experiments can sometimes be done to increase this precision even further. In view of these facts, studies of molecular ions at frequencies ranging from the infrared to the FIR can yield predictions of either vibrational or rotational transitions of molecular ions with sufficient precision to afford a definitive identification of astronomical rotational and vibrational spectra--particularly if several rotational transitions are observed. A disadvantage of infrared spectroscopy is that the small closed-shell quadrupole hyperfine splittings, often measured in microwave studies, are not normally observable; hence this extremely useful diagnostic tool is lost. As we shall see, infrared spectroscopy often has greater capability for studying a given polyatomic ion than does microwave spectroscopy, particularly because of the difficulty involved with searching suitably large frequency regions by microwave techniques. In any case, microwave measurements are highly complementary to infrared vibration-rotation spectroscopy, and clearly both should be employed whenever possible.

Our approach has been to study the very reactive open-shell ions by the most sensitive absorption method known--far infrared laser magnetic resonance (LMR)--and to study the more stable closed-shell species by velocity-modulation spectroscopy with a narrowband tunable infrared laser. These two approaches are contrasted in the following sections.

Laser Magnetic Resonance Rotational Spectroscopy

The LMR experiment, initially developed by K.M. Evenson and his coworkers, has been described in detail several times;[52,53] its application to molecular ions generated in discharge plasmas has been discussed as well,[54,23,24] hence no further experimental details will be given here. The two principal advantages of LMR are its extremely high sensitivity (~10^6 ions/cm^3 for HCl$^+$) and its high resolution (~0.1 MHz). These features make LMR a powerful tool for studying spectra of open-shell ions, which usually react rapidly with even trace amounts of water or other hydrogen-containing molecules to yield more stable ions, and hence generally occur in very low concentrations (~10^8 cm^{-3}) in laboratory plasmas.

We have given a preliminary account[23] of our study of the ground state of HCl$^+$, diagrammed in Figure 1, by LMR. In a subsequent and more detailed study, we have measured numerous hyperfine-Zeeman transitions and analyzed[55] them with both the effective Hamiltonian of Brown[56] and with the Hund's Case C formalism of Veseth.[57] A typical LMR spectrum

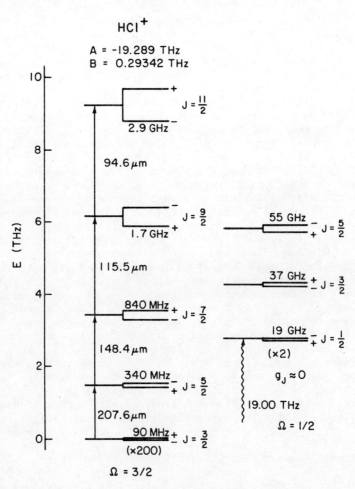

Figure 1. Energy level diagram of the X$^2\Pi$ ground state of HCl$^+$.

HCl^+ is shown in Figure 2, in which the hyperfine splitting from the
$I=3/2$ ^{35}Cl nucleus is revealed, as well as the larger lamda-doublet
splitting. A similar spectrum, exhibiting the same hyperfine effects
for $H^{37}Cl^+$ is shown in Figure 3. A total of 80 lines were measured
for $H^{35}Cl^+$ and for $H^{37}Cl^+$, and analyzed with a 10-parameter fit for
each isotope to give an average deviation of ~2 MHz. Unfortunately,

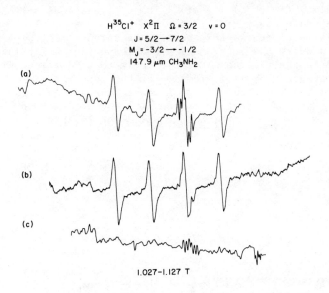

Figure 2. LMR rotational spectra of $H^{35}Cl^+$.
a) 2 mTorr HCl in 1 Torr He.
b) 6 mTorr HCl in 1 Torr He.
c) 20 mTorr HCl in 1 Torr He. The small pair of triplets is due to NH.

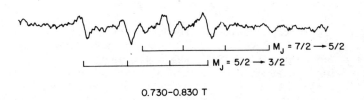

Figure 3. LMR rotational spectra of $H^{37}Cl^+$.

parameter correlation is a problem in such complex analyses as these, impairing our ability to make accurate determinations of all eigenvalues of the $X^2\Pi$ state (in particular, our poor determination of the d hyperfine parameter results in a relatively large uncertaintly for the $\Omega=\frac{1}{2}$ eigenvalues). Nevertheless, it is possible to predict both the lambda-doubling transitions as well as the pure rotational transitions for the astronomically relevant lower spin state $\Omega=3/2$ with ±2 MHz uncertainty (at the 1σ level). These predictions will be presented in a separate paper.[55]

In Figure 4 we present a LMR spectrum measured[24] for the isovalent species HF^+, in which the very small proton hyperfine splitting is resolved; unfortunately we have not been able to accomplish this for HCl^+ so far, but Figure 4 is indicative of the capability of LMR to make such high precision measurements in favorable cases. In Figure 5 is shown a recent LMR detection of the H_2O^+ ion by my group.[26] While a detailed analysis has not yet been carried out for the three rotational transitions observed thus far, we certainly expect to obtain a precise determination of the pure rotational frequencies for this important ion shortly.

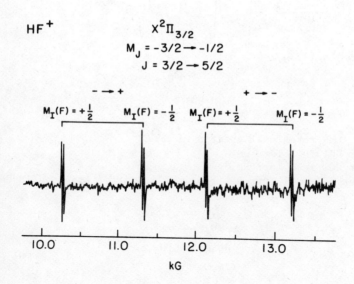

Figure 4. LMR rotational spectra of HF^+.

While LMR clearly possesses the sensitivity and resolution to produce a great deal of valuable information about molecular ions, it possesses one major liability; the major problem now encountered when attempting to detect LMR spectra of ions generated in discharge plasmas is simply the obfuscation of weak ion transitions by the much stronger and more numerous lines of neutral species also formed in these environments. It is just this difficulty which must be transcended if LMR is to become a truly general tool for molecular spectroscopy.

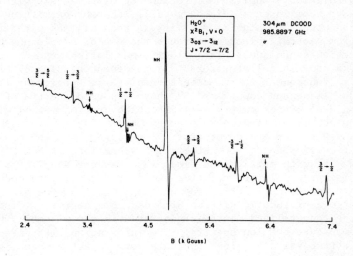

Figure 5. LMR rotational spectra of H_2O^+. A singlet hyperfine pattern is observed for this rotational transition.

The same difficulty -- that of distinguishing the spectra of ions from those of more abundant neutrals--also limited vibration-rotation studies of ions to the cases of very simple species which could be generated in chemically uncomplicated environments. Until nearly 1983, only HD^+, HeH^+, H_3^+, H_2D^+, and NeH^+, HrH^+, and ArH^+ had been studied by high-resolution infrared techniques.[10] On Christmas Eve of 1982, Gudeman, Begemann, and Pfaff performed the first velocity-modulation detection of a molecular ion spectrum (the v_1 band of HCO^+).[33]

Subsequently, HNN^+, DNN^+, H_3O^+, D_3O^+, NH_4^+, and H_2F^+ were studied at Berkeley by this method,[34-41] in which the acceleration of an ion to drift velocities near 300 m/sec. in a discharge plasma and the resulting Doppler shift in its transition frequencies is exploited to switch vibration-rotation transitions into and out of resonance with a color-center laser beam in an AC discharge cell, thereby permitting ions to be detected with very high sensitivity in such plasmas without interference from neutral molecules through the use of simple phase sensitive detection techniques. The velocity-modulation method was also used by Oka and his collaborators[42-45] in recent studies of NH_4^+, H_3O^+, and $HCNH^+$. Subsequent to the detection of these species by velocity modulation, Amano and coworkers[47-49] have reported spectra of HCO^+, $DCNH^+$, and H_2D^+ by the multipassed direct absorption method with a difference frequency laser first developed by Oka[27] in his original study of H_3^+, and McKellar and coworkers[50,51] have reported the measurement of lower frequency heavy atom stretching vibrations of HCO^+, DCC^+, HNN^+, and DNN^+ with multipassed diode laser spectroscopy of a hollow cathode discharge. In these latter two methods, velocity modulation discrimination against neutral was not employed, but the high frequency discharge amplitude modulation afforded some degree of discrimination against the longer-lived neutrals generated in the plasmas, which might otherwise obscure the ion spectra. The information with direct relevance to astrophysics obtained in these very recent studies is summarized in the following paragraphs.

Astrophysical Implications

A. HCl^+

The cosmic elemental abundance[58] of chlorine is $\sim 5 \times 10^{-6}$, that of H_2, about the same as that of phosphorous. The ionization potential of Cl is 12.7 ev, 0.9 ev less than that of hydrogen, implying that a substantial fraction of interstellar chlorine may be present in ionized form. Neutral HCl has been detected by Blake et al.[59] in the ISM. HCl^+ can be produced by the fast reaction $H_2 + Cl^+ \to HCl^+ + H$. Hence HCl^+ may be expected in regions with low density and high excitation. A search for the lambda-doubling transitions in HCl^+, analogous to those already detected for OH and CH, must be made at RF frequencies; this constitutes a difficult endeavor, requiring a very large antenna. The lowest rotational transition ($J=3/2 \to 5/2$) occurs near 207.6 μm, making a search at FIR frequencies possible with modern FIR techniques. Both are viable possibilities, given the large predicted dipole moment of HCl^+ (1.8D).[60] Because HCl^+ will react rapidly with H_2 to give H_2Cl^+, much of the ionized chlorine is likely to be present as the triatomic ion in denser sources. Although this species has yet to be detected by spectroscopic methods, the recent study of the ν_1 and ν_3 vibrations of the isovalent ion H_2F^+ made at Berkeley[38,39] indicates that the infrared spectrum of H_2Cl^+ will very likely be observed in the near future, providing predictions of rotational transitions which could be detectable by microwave or sub-millimeter astronomy.

B. $HCO^+(DCO^+)$

The HCO^+ ion has been studied extensively in the millimeter region by the R.C. Woods group at Wisconsin.[61] In fact, HCO^+ is the first polyatomic ion to yield a microwave equilibrium structure. Gudeman and Woods[62] have observed the $J=0 \rightarrow 1$ rotational transition in all three singly excited vibrational modes, thus providing the capability to study vibrationally excited ions in interstellar space. In the case of HCO^+, the recent vibrational spectroscopy studies provide a precise determination of the ν_1(CH and CD) and ν_3(CO) stretching frequencies, but do not improve on the microwave determinations of rotatioanl transition frequencies.

C. $HNN^+(DNN^+)$

Curiously, the microwave studies[63] made for ground state HNN^+ and DNN^+ have not been extended to excited vibrational states. Hence, the Berkeley studies of the ν_1 fundamentals and bend-excited hot bands of both species[44,41,64] and the Ottawa studies[51] of the ν_3(N-N) vibrations, now produce reliable predictions for the astrophysically important $J=0 \rightarrow 1$ rotational transition in all singly excited fundamentals, and in the doubly excited (1 1 0) mode. Moreover, the ℓ-doubling frequencies in the $\nu_2=1$ modes of both ions are accurately predicted to in the region near 30 GHz.[41,64]

D. $HCNH^+$

The Oka group's recent study[44] of protonated hydrogen cyanide, followed by Amano's work[49] on $DCNH^+$, provide good predictions of the astrophysically accessible rotational transitions for these important ions. The difficulty with detecting these species is that their dipole moments are very small--approximately 0.3 D.[65-67] This situation is improved somewhat for the $HCND^+$, in which the center of mass is shifted away from the center of change,[67] increasing the dipole moment somewhat. Because of this small polarity of $HCNH^+$, searches made for its laboratory microwave spectrum have not yet been successful.[68]

E. H_3O^+

While the extremely important hydronium ion escaped detection by high-resolution spectroscopy until 1983, measurements on three of its four allowed normal vibrations have now been carried out by velocity-modulation spectroscopy. The ν_1 and ν_3 bands (symmetric and antisymmetric stretches, respectively) have been observed for H_3O^+ and D_3O^+ at Berkeley,[40,69] while the ν_2 (nondegenerate bend) mode has been studied at Chicago.[43] Experiments on the asymmetric deutrated forms are also in progress at Berkeley.[69]

Astronomical detection of H_3O^+, or one of its isomeric forms, could be accomplished by far-infrared observations of its rotation-inversion transitions, by microwave or FIR observation of rotational transitions in the asymmetric isomers, or by infrared detection of vibrational transitions, most likely the ν_2 band, near 1000 cm^{-1}.

F. NH_4^+

The ν_3 mode (triply degenerate asymmetric stretch) of   ammonium was detected initially by both my group[36] and by Oka's group.[42] An extensive study of this complicated spectrum has been carried out by Schaefer, Robiette, and myself,[37] leading to accurate predictions of the forbidden transitions, necessary to determine the structure of this molecule. Croften and Oka[70] have recently reported the observation of these forbidden transitions.

Detection of NH_4^+ in the ISM would probably best be accomplished by FIR observation of the rotation-inversion spectrum of the symmetric top isotopomer NH_3D^+, which will be quite strongly allowed by the shift of the center-of-mass away from the the center of charge.

G. OH^+

The ground state of the hydroxylion is $X^3\Sigma^-$. It has been detected only very recently by high-resolution instruments, in discharges of He and H_2O,[25] and in H_2 and O_2.[25,45] Oka[45] has reported observation of its vibrational fundamental by velocity modulation spectroscopy, and Gruebele, Muller and myself[25] have measured three different rotational transitions with six FIR laser lines by laser magnetic resonance. A typical LMR rotational spectrum measured at Berkeley is shown in Figure 6. Unfortunately, a detailed analysis of these LMR spectra is still in progress, but clearly these measurements of rotational spectra of OH^+ will provide precise predictions of the FIR transitions from the lowest rotational state, as well as of the fine-structure and hyperfine splitting patterns, which will serve to unambiguously identify astrophysical spectra of this species.

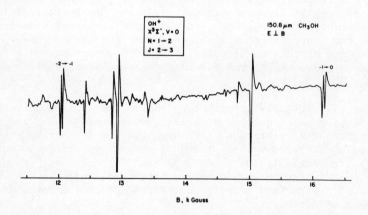

Figure 6. A typical LMR spectrum of OH^+, observed in a H_2/O_2 plasma.

Acknowledgements

The LMR effort at Berkeley is supported by the Director, Office of Basic Research, U.S. Department of Energy, with equipment provided by the National Science Foundation. The infrared velocity modulation spectroscopy work is supported by the National Science Foundation (CHE 84-02861) and the Petroleum Research Fund. Postdoctoral support was provided by the California Space Institute. I thank all of my students and postdocs at Berkeley who did the work described in this paper.

REFERENCES

1. P.M. Solomon and W. Klemperer, Astrophys. J. 178, 389 (1972).

2. E. Herbst and W. Klemperer, Astrophys. J. 185, 505 (1973).

3. G.F. Mitchell, J.L. Ginsberg, and P.J. Kuntz, Astrophys. J. Suppl. 38, 39 (1978).

4. S.S. Prasad and W.T. Huntress, Jr., Astrophys. J. Suppl. 43, 1 (1980).

5. W.D. Watson, Astrophys. J. 188, 35 (1974).; Astrophys. J. 189, 22 (1974).

6. R.C. Woods, Private Communication.

7. P. Thaddeus. M. Guelin, and R.A. Linke, Astrophys. J. 246, L41 (1981); J.L. Destombes, C. Demuynck, and M. Bogey, preprint.

8. R.J. Saykally and R.C. Woods, Ann. Rev. Phys. Chem. 32, 403 (1981).

9. T.A. Miller and V. Bondybey, editors, Molecular Ions: Structure and Chemistry, North-Holland, Amsterdam (1983).

10. C.S. Gudeman and R.J. Saykally, Ann. Rev. Phys. Chem. 35, 387 (1984).

11. Billingsley, Chem. Phys. Lett. 23, 160 (1973).

12. H.A. Schwarz, J. Chem. Phys. 67, 5525 (1977).

13. H.A. Schwarz, J. Chem. Phys. 72, 284 (1980).

14. W.H. Wing, G.A. Ruff, W.E. Lamb, and J.J. Spezeski, Phys. Rev. Lett. 36, 1488 (1976).

15. D.E. Tolliver, G.A. Kyrala, and W.H. Wing, Phys. Rev. Lett. 43, 1719 (1979).

16. J.T. Shy, J.W. Farley, W.E. Lamb, and W.H. Wing, Phys. Rev. Lett. 45, 535 (1980).

17. J.T. Shy, J.W. Farley, and W.H. Wing, Phys. Rev. A 24, 1146 (1981).

18. A. Carrington and J. Buttenshaw, Mol. Phys. 44, 267 (1981); A. Carrington, J. Buttenshaw, and R. Kennedy, Mol. Phys. 48 775 (1983); J. Mol. Structure 80, 47 (1982).

19. A. Carrington, J. Buttenshaw, R. Kennedy, and T.P. Softley, Mol. Phys. 44, 1233 (1981); A. Carrington, R. Kennedy, T. Softley, P. Fournier, and E.G. Richard, Chem. Phys. 81, 25 (1983).

20. A. Carrington, J. Buttenshaw, and R. Kennedy, Mol. Phys. 45, 747 (1982).

21. A. Carrington, J. Buttenshaw, and R. Kennedy, Mol. Phys. 45, 753 (1982); A. Carrington and R. Kennedy, in Ions and Light, ed. M. Bowers (1982); A. Carrington, Proc. Roy. Soc. London A367, 433 (1979); A. Carrington and R. Kennedy, J. Chem. Phys. 81, 91 (1984).

22. R.J. Saykally and K.M. Evenson, Phys. Rev. Lett. 43, 515 (1979).

23. K.G. Lubic, D. Ray, and R.J. Saykally, Mol. Phys. 46, 217 (1982).

24. D.C. Hovde, E. Schafer, S.E. Strahan, C.A. Ferrari, D. Ray, K.G. Lubic, and R.J. Saykally, Mol. Phys. 52, 245 (1984).

25. M. Grueble, R.P. Mueller, and R.J. Saykally, in preparation.

26. S. Strahan, R.P. Mueller, and R.J. Saykally, in preparation.

27. T. Oka, Phys. Rev. Lett. 45, 53 (1980).

28. P. Bernath and T. Amano, Phys. Rev. Lett. 48, 20 (1982).

29. M. Wong, P. Bernath, and T. Amano, J. Chem. Phys. 77, 693 (1982).

30. J.W. Brault and S.P. Davis, Physica Scripta 25, 268 (1982).

31. J.W.C. Johns, J. Mol. Spec. 106, 124 (1984).

32. F.C. Van den Heuvel and A. Dymanus, Chem. Phys. Lett. 92, 219 (1982).

33. C.S. Gudeman, M.H. Begemann, J. Pfaff, and R.J. Saykally, Phys. Rev. Lett. 50, 727 (1983).

34. C.S. Gudeman, M.H. Begemann, J. Pfaff, and R.J. Saykally, J. Chem. Phys. 78, 5837 (1983).

35. M.H. Begemann, C.S. Gudeman, J. Pfaff, and R.J. Saykally, Phys. Rev. Lett. 51, 554 (1983).

36. E. Schafer, M.H. Begemann, C.S. Gudeman, and R.J. Saykally, J. Chem. Phys. 79, 3159 (1983).

37. E. Schafer, R.J. Saykally, and A.G. Robiette, J. Chem. Phys. 80, 3969 (1984).

38. E. Schafer and R.J. Saykally, J. Chem. Phys. 80, 2973 (1984).

39. E. Schafer and R.J. Saykally, J. Chem. Phys., in press.

40. D.J. Nesbitt, H. Petek, C.S. Gudeman, R.J. Saykally, and C.B. Moore, in preparation.

41. D.J. Nesbitt, H. Petek, C.S. Gudeman, R.J. Saykally, and C.B. Moore, J. Chem. Phys., in preparation.

42. M.W. Crofton and T. Oka, J. Chem. Phys. 79, 3157 (1983).

43. N.N. Haese and T. Oka, J. Chem. Phys. 80, 572 (1984).

44. R.S. Altman, M.W. Crofton, and T. Oka, J. Chem. Phys. 80, 3911 (1984).

45. T. Oka, 39th Symposium on Molecular Spectroscopy, Columbus, Ohio; June, 1984.

46. P.B. Davies and W.J. Rothwell, J. Chem. Phys., in press.

47. T. Amano, J. Chem. Phys. 79, 3595 (1983).

48. T. Amano and J.K.G. Watson, 39th Symposium on Molecular Spectroscopy, Columbus, Ohio; June, 1984, paper TE-3. K.G. Lubic and T. Amano, same conference, paper TE-13.

49. T. Amano, 39th Symposium on Molecular Spectroscopy, Columbus, Ohio; June, 1984, paper TE-7.

50. S.C. Foster, A.R.W. McKellar, and T.J. Sears, J. Chem. Phys. 81, 579 (1984).

51. S.C. Foster and A.R.W. McKellar, J. Chem. Phys., in press.

52. K.M. Evenson, R.J. Saykally, D.A. Jennings, R.F. Curl, Jr., and J.M. Brown, "Chemical and Biochemical Applications of Lasers, Vol. V," C.B. Moore, Editor (Academic Press, New York, 1980).

53. K.M. Evenson, Faraday Discussions Chem. Soc. 71, 7 (1981).

54. R.J. Saykally, K.G. Lubic, and K.M. Evenson, in Molecular Ions: Geometric and Electronic Structures, J. Berkowitz and K. Groeneveld, eds. (Plenum Inc., 1980, New York).

55. K.G. Lubic, D. Ray, L. Veseth. and R.J. Saykally, J. Chem. Phys., submitted.

56. J.M. Brown and J.K.G. Watson, J. Mol. Spec. 65, 65 (1977); J.M. Brown, E.A. Colbourn, J.K.G. Watson, and F.D. Wayne, J. Mol. Spec.

<u>74</u>, 294 (1979).

57. L. Veseth, J. Mol. Spec. <u>59</u>, 51 (±976); <u>63</u>, 180 (1976).

58. V. Trimble, Rev. Mod. Phys. <u>47</u>, 877 (1975).

59. G. Blake, private communication.

60. S. Green, private communication.

61. R.C. Woods, T.A. Dixon, R.J. Saykally, and P.G. Szanto, Phys. Rev. Lett. <u>35</u>, 1269 (1975); R.C. Woods, R.J. Saykally, T.G. Anderson, T.A. Dixon, and P.G. Szanto, J. Chem. Phys. <u>75</u>, 4256 (1981).

62. C.S. Gudeman and R.C. Woods, in preparation.

63. R.J. Saykally, T.A. Dixon, T.G. Anderson, P.G. Szanto, and R.C. Woods, Astrophys. J. <u>205</u>, L101 (1976); P.G. Szanto, T.G. Anderson, R.J. Saykally, N.D. Piltch, T.A. Dixon, and R.C. Woods, J. Chem. Phys. <u>75</u>, 4261 (1981).

64. J. Owrutsky, C.S. Gudeman, C.C. Martner, M.H. Begemann, N.H. Rosenbaum, L. Tack, and R.J. Saykally, in preparation.

65. N.N. Haese and R.C. Woods, Chem. Phys. Lett. <u>61</u>, 396 (1979).

66. T.T. Lee and H.F. Schaefer, preprint.

67. P. Bottschwina, private communication.

68. R.C. Woods, private communication.

69. M.H. Begemann and R.J. Saykally, in preparation.

70. M. Crofton and T. Oka, private communication.

71. N.N. Haese, F.S. Pan, and T. Oka, Phys. Rev. Lett. <u>50</u>, 1575 (1983).

TIME RESOLVED PROPERTIES OF SMALL ASTROPHYSICAL MOLECULES

Peter Erman

Royal Institute of Technology
S 100 44 Stockholm, Sweden

ABSTRACT

   Periodic excitations of molecular levels makes possible not only the determination of resulting emission lines but also their individual lifetimes. Today most measurements of this kind are performed either using high power electron excitation, such as applied in the High Frequency Deflection (HFD) technique developed at our laboratory, or pulsed laser excitation. Completed by for instance supersonic jet targets, time resolved spectroscopy should now be considered as an important branch of molecular physics with a number of astrophysical applications.
   While most astrophysical observations of molecules have recently been carried out at radio wavelengths, the most accurate abundance estimates are generally performed from optical absorption spectra provided that the associated f-values are known. Thanks to the development of time resolved spectroscopy, f-values are now available for transitions in most of the important diatomic molecules, which combined with equivalent widths may yield abundances with uncertainties as small as 15 %.
   Other important applications of molecular lifetime investigations are found in studies of various kinds of radiationless transitions, for which purpose it is much more sensitive than classical spectroscopic tools. Thus the inverse of our recently discovered new kind of predissociations through direct interaction between bound-continuum levels of two attractive states in the carbon group of hydrides could be important at molecular formation at low temperatures. Other applications are found in determinations of, for instance thermal collision cross sections and various kinds of rate coefficients.

1. INTRODUCTION

The number of small molecules observed at optical wavelengths in the atmospheres of the stars and other cosmological object is continously increasing. In addition refined studies of the optically observed interstellar molecules are carried out. One of the prime reasons for performing these optical observations is that they enable quantitative estimates of molecular abundances with an accuracy which frequently surpass analyses from radiospectra. For this purpose the oscillatorstrength f of the studied optical transition must be known with as high precision as possible. Accordingly the greatly increased accuracy in estimates of molecular abundances during the past decade is partially due to refined laboratory techniques for measuring f-values. Early f-value estimates were frequently performed using indirect methods like the "Hook-method" or shocktube absorption, which unfortunately may involve parameters which are hard to estimate. In most cases a direct measurement of the lifetime τ of the associated transition and a subsequent conversion $\tau \rightarrow f$ forms a much more accurate and convenient technique. Therefore the rapid progress in recent estimates of molecular abundances from optical spectra is intimately connected with the development of laboratory techniques for time resolved spectroscopy (cf ref (1)). In addition, it is also possible today to perform reliable ab initio calculations using various approaches. However, the astrophysical applications of time resolved spectroscopy are not only restricted to f-values and abundance estimates. As will be discussed below, this approach is also very convenient for instance in studies of radiationless transitions and interactions between excited molecules and the environment.

2. LABORATORY TECHNIQUES

Several review articles have discussed various techniques for time resolved molecular spectroscopy (cf ref (2)), and in this context we will just mention some newer refinements. Thus the introduction of supersonic jet targets have implied a great simplification in time resolved spectroscopy (cf refs (3), (4)). Fig 1 shows an example of such an arrangement installed in the target chamber of a high power electron accelerator used for lifetime investigations using the High Frequency Deflection (HFD) technique developed at our laboratory.

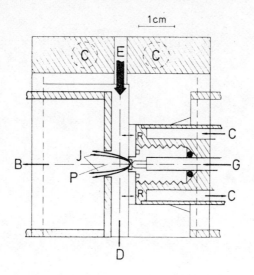

Fig 1. Design of a supersonic jet arrangement (4).

The target gas G (pressure 3-5 atm) is expanded through a 100 μm diameter nozzle forming a jet J which is directed into a high capacity booster pump P. The exciting electron beam (20 keV, 20-40 mA) enters periodically downwards. This arrangement yields a strong cooling effect following a supersonic expansion and we have observed rotational temperatures as low as 15 K for states excited directly from the ground state in neutral molecules. However, when electron exchange is involved, (for instance when a singlet ground state is excited to triplet levels) or ionization processes (which causes a pick up of angular momentum), the observed rotational temperatures become higher.

As seen from Fig 2 the use of supersonic jet targets yield a considerable simplification of recorded molecular spectra in view of the exclusion of lines from higher rotational levels. We have also found (4) that collision effects (at a given target pressure) decrease hundredfold in lifetime studies using the jet as compared to an ordinary static gas target.

Other recent improvements in lifetime measuring techniques include the introduction of ion-traps in studies of molecular ions. Earlier published lifetimes of this kind of species have frequently been too short in view of distortions from Coulomb repulsion effects (cf ref (5)). Probably the most satisfactory way of reducing this effect is the use of radio-frequency ion-traps, which has recently been done by various workers (cf ref (6)).

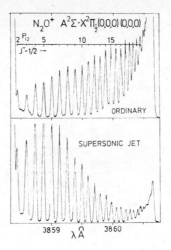

Fig 2. Spectrum of N_2O^+ recorded using an ordinary and a supersonic jet target (4).

3. AB INITIO CALCULATIONS OF TRANSITION PROBABILITIES

Quantummechanical ab initio methods have been extended to include transition probabilities and radiative lifetimes with increasing accuracy during recent years. Several different ab initio approaches have been tried such as multi-reference single and double-excitation configuration interaction (MRD-CI) (cf ref (7)), complete active space SCF (CASSCF) and propagator methods. In particular the CASSCF method (cf ref (8)) has been used by our group for a number of calculations correlated to our experimental results. Very good agreement between CASSCF and accurate experimental results have been obtained in a number of cases. For instance in the important case of the CH A state, our experiments give τ = 534±5 ns (9) (which is confirmed by an independent laser excitation work yielding 537±5 ns (10)) while CASSCF gives τ = 525 ns (11). In particular reliable ab initio methods are of high value where, for different reasons, experimental results are hard to collect or if they afflicted with systematic errors. To the latter category belong longlived ionic species as mentioned in the Introduction. As an example we could here mention the astrophysically important CH^+ radical for which published values of oscillator strength f(A-X) has been gradually decreasing during the past decade (cf ref (12)).

However, very recent experimental results using ion-traps (13) nearly coincide with extended CASSCF calculations (14). Since the amount of reliable ab initio calculations is steadily increasing, they should hopefully be included in the next extended compilation of molecular radiative lifetimes.

4. ABUNDANCE ESTIMATES FROM MOLECULAR LIFETIMES

Recent compilations of measured molecular lifetimes are found for instance in refs (15) and (16). However, in converting τ to f or other dynamical variables, the customer has to be careful in applying the statistical factors in formulas given in the literature in order to avoid "factor of two errors". Since different authors sometimes use dissimilar definitions of f, linestrengths etc, the review compilation (17) has been written in order to clarify the use of τ to f conversion formulas.

If the electronic transition moment R_e does not vary strongly with the internuclear distance $\bar{r}$, the application of the conversion formulas are straight forward. On the other hand, if R_e vary strongly with $\bar{r}$, this might introduce a considerable uncertainty in the τ to f conversions. Different methods for determining the $R_e = f(\bar{r})$ variation experimentally are discussed for instance in ref (1). Reasonable accuracy can be obtained in these procedures for instance in cases where the relative intensities $I_{v',0}$, $I_{v',1}$ ... can be measured for transitions emitted from a common upper level v'. However, with the increased accuracy of ab initio calculations, a more general procedure is to use the theoretical $R_e = f(\bar{r})$ relative variation and normalise to the experimentally measured lifetimes.

Still considerable variations in f-values remain in the literature which partially could be due to uncertainties in the R_e variation. One such example is the C_2 Phillips system which has recently been used as an indicator of interstellar C_2 column densities (cf refs (18) and (19) and references given therein). The lowest f-values follow from analyses of solar spectra (19) and the largest from early shocktube measurements (20). Recent MRD-CI calculations (21) support the measurements (20), while laser excitation studies (22) support the experiments (18) (and (19)). Since it is known that ab initio calculations frequently tend to give too low τ-values unless very large basis-sets are applied (compare the above mentioned CH case) probably the most accurate f-value in this case may be deduced from the calculated R_e variation normalized to the experimental lifetimes (18), (22) which are almost a factor of two longer at lower v'-levels.

5. RADIATIONLESS TRANSITIONS, MOLECULAR FORMATION, DISSOCIATION ENERGIES

As pointed out earlier in numerous connections (cf ref (2)), time resolved spectroscopy is a very convenient way of tracing and classifying predissociations in molecules. A relative change of the order of 1 % can be detected say in τ-measurements of a sequency of rotational levels, which corresponds to a 10^4-10^5 higher sensitivity as compared to absorption line width determinations. Accordingly many new cases of predissociations have been discovered for instance using the HFD technique in our laboratory. The astrophysical interest in processes of this kind is manyfold. For instance their inverse, i.e. molecular formation by inverse predissociation, may be an important process at the low pressures and temperatures which exist in many celestial environments. Of particular interest is the case reproduced in Fig 3, i.e. predissociations caused by the continuum of a (stable) ground state. This process has been discovered for the first time in our laboratory in the $A^2\Delta$ state of the carbon group of hydrides (refs (23), (24) and (25)).

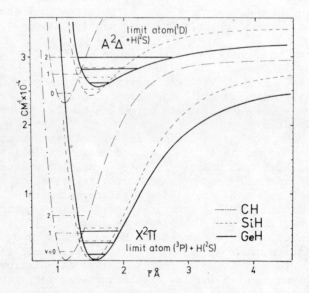

Fig 3. Potential curves of the isovalent molecules CH, SiH and GeH (25).

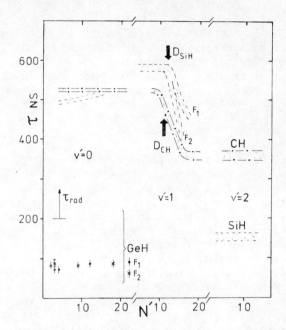

Fig 4. Experimentally measured lifetimes of the A state in CH, SiH and GeH (25).

The potential curves given in Fig 3 indicate that the A state levels above $v'=1$, $N'=11$ and $v'=1$, $N'=12$ in CH and SiH, respectively, lie higher than their corresponding ground state dissociation limits. As seen from Fig 4, a sudden decrease in the measured lifetimes at these limits indicate the onset of very weak predissociations which are interpreted as interactions with the continuum of the $X^2\Pi$ ground state. In the case of GeH, all the $A^2\Delta$ state levels are above the dissociation limit and subjected to predissociations of the same kind. This conclusion follows from a comparison between the experimental lifetimes (75-85 ns) and the result of our ab initio calculations which give a lower limit of 200 ns for the radiative lifetime. The measured nonradiative transition probabilities can be converted to rate coefficients for molecular formation through the inverse of this (indirect) process and they are of the order of $10^{-20} - 10^{-18}$ cm^3 s^{-1}. Accordingly they could play an important role in molecular formation at low temperatures in the universe.

Time resolved studies of predissociations also yield accurate estimates of dissociation energies. For instance for SiH is found $D_0^0(\text{SiH}) = 26950 \pm 200$ cm^{-1}(24), a result which differs drastically from the "older" upper limit ($D_0^0(\text{SiH}) \leq 24680$ cm^{-1}). Even more accurate dissociation energies can be measured in the cases of predissociation by rotation as may be examplified by the B state in CH (23).

In the more common case of predissociations caused by a crossing repulsive state, lifetime studies may be used for instance in deducing molecular constants for this repulsive state. Recent examples of studies of this kind are found in the OH and OD radicals (cf ref (26)).

6. COLLISION PROCESSES

If an excited molecule collides with external particles (molecules, atoms or electrons), the radiative decay is distorted. Accordingly time resolved spectroscopy may also be used for studies of various kinds of collision processes. On the other hand collision processes might also affect astronomical recordings of (for instance) equivalent line widths. A particular kind of collision processes is transfer transitions between adjacent levels caused by collisions with particles in the environment (see Fig 5). More specifically, if a molecule is excited to level Y, collisions with a foreign molecule M may catalyze a transfer to level A via a radiationless transition from Y´ which coincides with A. This give rise to an increased intensity of the A to X emission and a lengthening of the A state lifetime which is readily observable in the decay curve.

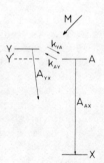

Fig. 5. Simplified diagram of collisional transfer (1).

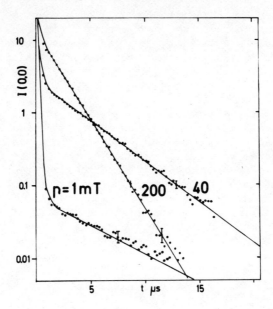

Fig 6. Longlived transfer components in the decay of the C_2 Swan bands (27).

Figure 6 shows experimentally observed decay curves of the C_2 Swan bands (27) where the longlived tails have different pressure dependent lifetimes which probably originate from close lying levels of the C_2 c state. The presence of these transfer effects in the Swan bands could be the reason why several early lifetime recordings of this system have yielded too high values. From the slope of the second (transfer) component it is possible to deduce the transfer rate coefficient. In the case of the C_2 Swan bands, this rate coefficient is about $6 \cdot 10^{-11}$ cm^3 s^{-1}, but in some cases values as high as 10^{-8} cm^3 s^{-1} have been found (cf ref (28)). Very large effects have also been found in radicals formed via dissociative collision transfer as for instance in the case of OH formed from H_2O or ethylene glycol (29).

Various kinds of transfer processes should occur at the astrophysical conditions which exist for instance in the star atmospheres and they could possibly distort measurements of absorption strengths. A possible example of this occurs in the CN radical, which is used for nitrogen abundance estimates in giant stars. Very strong transfer effect has been found (30) in the red CN bands and they were confirmed in subsequent laser flourescence measurements (31). Our measured lifetimes (30) are about 4.2 μs for the CN A state levels independent or v', which is more than a factor of two lower result (for low v') as compared to results from an analysis of the solar spectrum (32) and ab initio calculations (33). Very recent CASSCF calculations (34), however, tend to support the experimental values (30) except for low v'-levels.

This "CN-dilemma" is not solved yet and better laboratory experiments should be carried out as well as investigations of the influence of transfer effects on the astronomical data.

It is a well known fact that chemical processes may proceed at very high rates when excited states of the participating species are involved. Accordingly time resolved spectroscopy may be a convenient tool also for studying collision processes of this kind. As an example we could mention studies of electron activated N_2 gas in the presence of trace amounts of H_2O and CO_2 (35). It was found that such reactions from a very efficient way for producing NO, which has relevance to the auroral NO production.

REFERENCES

1. Erman, P. 1979, Physica Scripta 20, pp. 575-581.

2. Erman, P. 1979, Specialists Periodical Reports, Molecular Spectroscopy Vol 6, pp. 174-231.

3. Levy, D.H., Scientific Am., Feb. 84.

4. Erman, P., Gustafsson, O. and Larsson, M. 1983, Physica Scripta 27, pp. 192-200.

5. Curtis, L.J. and Erman, P. 1977, J. Opt. Soc 67, pp. 1218-1230.

6. Wineland, D.J., Itano, W.M. and van Dyck, R.S. 1983, Adv. in at. and mol. physics 19, pp. 135-186.

7. Bruna, P.J. and Peyerimhoff, S. 1983, Bull. Soc. Chim. Belg. 92, pp. 525-546.

8. Siegbahn, P.E.M., Almlöf, J., Heiberg, A. and Roos, B.O. 1981, J. Chem. Phys. 74, pp. 2384-2396.

9. Brzozowski, J., Bunker, P., Elander, N. and Erman, P. 1976, Astrophys. J. 207, pp. 414-424.

10. Becker, K.H., Brenig, H.H. and Tatarczyk, T. 1980, Chem. Phys. Lett. 71, pp. 242-245.

11. Larsson, M. and Siegbahn, P.E.M. 1983, Physica Scripta 79, 2270-2277.

12. Erman, P. 1977, Astrophys. J. 213, pp. L89-L81.

13. Mahan, B.H. and O'Keefe, A. 1981, Astrophys. J. 248, pp. 1209-1216.

14. Larsson, M. and Siegbahn, P.E.M. 1983, Chem. Phys. 76, pp. 175-184.

15. Kuzmenko, N.E., Kuznetsova, L.A., Monyakin, A.P., Kuzyaka, Yu Ya. and Plastinin, Yu. A. 1979, Sov. Phys. Usp. 22, pp. 160-

16. Dumont, M.N. and Remy, F. 1982, Spectroscopy Letters 15, pp. 699-772.

17. Larsson, M. 1983, Astron. Astrophys. 128, pp. 291-298.

18. Erman, P., Lambert, D.L., Larsson, M. and Mannfors, B. 1982, Astrophys. J. 253, pp. 983-988.

19. Brault, J.W., Delbouille, L., Grevesse, N., Roland,G., Sauval, A.J. and Testerman, L. 1982, Astron, Astrophys. 198, pp. 201-206.

20. Cooper, D. and Nicholls, R.W. 1975, J. Quant. Spectr. Rad. Transf. 15, pp. 139-150.

21. van Dishoek, E. 1983, Chem. Phys. 77, 277-286.

22. Hubrich, C., Wildt, J., Bauer, W. and Becker, K.H. (to be published).

23. Brzozwski, J., Bunker, P., Elander, N. and Erman P. 1976, Astrophys. J. 207, pp. 414-424.

24. Carlson, T.A., Duric, N., Erman, P. and Larsson, M. 1978, J. Phys. B11, pp. 3667-3675.

25. Erman, P., Gustafsson, O, and Larsson, M. 1983, Physica Scripta 27, pp. 256-260.

26. Bergeman, T., Erman, P., Haratym, Z. and Larsson, M. 1981, Physica Scripta 23, pp. 45-53.

27. Erman, P. 1980, Physica Scripta 22, pp. 108-113.

28. Carlsson, T.A., Duric, N., Erman, P. and Larsson, M. 1978, Z. Phys. A 287, pp. 123-136.

29. Erman, P. and Larsson, M. 1980, Physica Scripta 22, pp. 348-352.

30. Duric, N., Erman, P. and Larsson, M. 1978, Physica Scripta 18, pp. 39-46.

31. Katayama, D.R., Miller, T.A. and Bondybey, V.E. 1979, J. Chem. Phys. 71, pp. 1662-1669.

32. Sneden, C. and Lambert, D.L. 1982, Astrphys. J. 259, pp. 381-391.

33. Cartwright, D.C. and Hay, P.J. 1982, Astrophys. J. 257, pp. 383-387.

34. Larsson, M., Siegbahn, P.E.M. and Ågren, H. 1983, Astrophys. J. 272, pp 369-376.

35. Pendleton Jr., W., Erman, P., Larsson, M. and Witt, G. 1983, Physica Scripta 28, pp. 532-538.

ROTATIONAL EXCITATION IN MOLECULAR BEAM EXPERIMENTS

U. Buck

Max-Planck-Institut für Strömungsforschung,
D3400 Göttingen, Federal Republic of Germany

ABSTRACT

The general behaviour of differential cross sections for the rotational excitation in molecular collisions is discussed. The sensitivity of these cross sections to the non-spherical potential surface and experimental methods to measure state-to-state cross sections are reviewed. Based on this input information reliable potential surfaces are determined. Examples which are relevant for astrophysical applications are given for the systems H_2-H_2, H_2-CO and NH_3-He.

1. INTRODUCTION

Rotational energy transfer is one of the most efficient processes in molecular dynamics. The astrophysical interest in this type of collisions comes from the fact that the intensity of the observed radiation of rotational lines from molecules in dense interstellar clouds depends strongly on the rates for rotational population and depopulation. In principle, this information could be obtained by directly measuring these rates in the laboratory under exactly the same conditions as they are needed for the simulation. In practise, however, this can be a difficult problem, since the experimental requirements are not easy to fulfil in non-beam experiments. Therefore it has been common practise to solve the problem by determining the underlying non-spherical intermolecular potential surface which causes rotational energy

G. H. F. Diercksen et al. (eds.), Molecular Astrophysics, 433–451.

transfer. Once the potential surface is known to a sufficient accuracy, all the necessary data can be calculated from it.

There are essentially two approaches in the literature which have been used to solve this problem. In the first approach, the intermolecular potential is completely calculated by ab initio methods, if possible with correlation included. According to

$$V(R,\gamma) \rightarrow Q(j \rightarrow j',E) \rightarrow k(j \rightarrow j',T) \qquad (1)$$

state selective integral cross sections $Q(j \rightarrow j',E)$ as well as temperature dependent rates $k(j \rightarrow j',T)$ are calculated. The bottle-neck of this procedure is the calcuation of accurate potential surfaces. While a hierachy of quantum and semiclassical methods [1-3] is available for the calculation of cross sections, it is still a big problem to calculate very accurate potential surfaces even for few electron systems, since the usually involved van der Waals interaction is very weak.

The second approach starts from a precise measurement of state-resolved differential cross sections $\sigma(j \rightarrow j',\theta,E)$ and determines the potential surface either by direct inversion techniques [4] or by comparison of measured and calculated data based on model potentials [5]:

$$\sigma(j \rightarrow j',\theta,E) \Leftrightarrow V(R,\gamma) \qquad . \qquad (2)$$

The problem of this approach is the preparation of the input information closest to the intermolecular potential: the state-to-state differential cross section, which is very difficult to measure for neutral systems. Nevertheless there has been great progress in the last five years in measuring these cross sections and determining the interaction potentials for a selected number of systems, namely the hydrogen molecule-rare gas [6,7] and the hydrogen molecule dimer systems and their isotope variations [8-11], the sodium dimer-rare gas systems [12,13] and the N_2,O_2-He systems [14,15]. However, very often problems arise for a unique determination of the potential surface, since the measured data are incomplete with respect to the angular range and/or the resolution so that only differential energy loss spectra $\sigma(\Delta E,\theta,E)$ or integral state selected cross sections $Q(j \rightarrow j',E)$ are available.

Therefore, the most powerful approach to solve the problem is to use a combination of both methods. The experiments are performed under optimal conditions with regard to experimental requirements and the resulting cross sections are compared with calculations based on interaction potentials that contain as much theoretical input information as possible. For the experimental input differential energy loss spectra over a large angular range are used. This type of cross sections can now be measured for a large class of systems, in contrast to state-to-state differential cross sections, which are only available for a small number of systems. For the theoretical input large basis set SCF-calculations and long range dispersion coefficients from perturbation theory are used, which both can be obtained for a large number of systems. By adjusting one or two parameters very reliable potential surfaces result which can then be used to calculate all the desired data. This combination of a multiproperty analysis of experimental data with the so-called Hartree-Fock-dispersion potential model, which both proved to be very successful in spherical potential analysis [16,17], seems to be the most promising procedure for analysing the much more complicated non-spherical potential surfaces, which are responsible for rotational energy transfer.

The present paper is organized as follows. The next section will summarize the sensitivity of scattering experiments to potential features with special emphasis on the interpretation of rotationally inelastic scattering in terms of the anisotropy of the potential. A short discussion on the experimental set up to measure differential cross sections for rotational energy transfer will follow. Finally we will present results for 3 systems which are relevant for astrophysical applications and which illustrate the state of the art of crossed molecular beam studies of inelastic scattering and the evaluation of potential surfaces. We start with the hydrogen molecular dimer for which differential cross sections have been measured at a state resolved level and which have been analysed by full close-coupling calculations based on a potential surface with configuration interaction included. For D_2+CO and NH_3+He, we analysed differential energy loss spectra by performing centrifugal sudden calculations based on a potential at SCF-level for the repulsion and damped dispersion coefficients for the attraction.

2. SENSITIVITY OF SCATTERING EXPERIMENTS

The great advantage of analyzing differential scattering experiments is that it is very easy to relate the measured angular dependence to the underlying forces. Generally speaking small angle scattering is caused by attractive and large angle scattering by repulsive forces. In elastic scattering the classical rainbow angle separates these two regions very clearly [1]. Systems with small reduced masses and potential well depths are dominated by diffraction oscillations. The angular positions of these oscillations give a direct measure of the diameter of the repulsive wall R_0 to better than 1 %. The amplitudes are sensitive to the form of the potential between the minimum and the distance where it passes through zero. The large angle scattering data are determined by the steepness of the repulsive wall and allow its determination to within 20 %.

According to the anisotropy of the system we distinguish between weak and strong coupling. In the weak coupling case the rotationally inelastic cross sections are much smaller than the elastic ones. The potential surface is well represented by a few terms in the usual expansion in spherical harmonics and there is a direct correspondence of single state-to-state differential cross sections to single potential terms. A schematic drawing of the isotropic (V_o) and the anisotropic (V_a) part of the potential is shown in Fig. 1. The system can be described by the distorted wave approximation, which connects the differential cross section $\sigma(j \to j', \theta, E)$ for the rotational transition from $j \to j'$ with the anisotropy of the potential V_a:

$$\sigma(j \to j', \theta) \propto \left[\int \psi_{j\ell}^{(o)}(R) \, V_a(R) \, \psi_{j'\ell'}^{(o)} \, dR \right]^2 \quad . \tag{3}$$

θ is the deflection angle, R the internuclear distance and $\psi_{j\ell}^{(o)}$ are the zero-order channel wave functions calculated from the isotropic part of the potential $V_o(R)$ only. The inelastic cross sections are directly proportional to V_a^2. However, the range

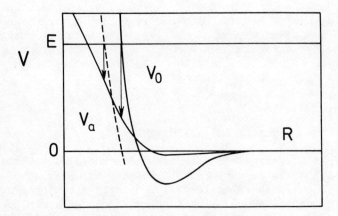

Figure 1: Isotropic (V_o) and anisotropic (V_a) potentials for the weak coupling case. If the classical turning point determined by V_o is shifted, a different part of V_a is probed.

which is probed is given by the "elastic wave functions" near the classical turning point. If V_o is shifted inwards, a larger value of V_a is probed. Therefore the complete analysis always requires information on both V_a (from inelastic scattering) and V_o (from elastic scattering).

In the strong coupling case the anisotropy is much larger than the isotropic part of the potential and many potential terms of the expansion in spherical harmonics contribute so that a simple relationship between cross sections and potential terms is not anymore valid. However, in this case, the cross sections are dominated by the rotational rainbow effect, which manifests itself in a maximum of the differential cross section as a function of the final angular momentum j´ at a fixed deflection angle θ or vice versa [2,18]. The reason for this behaviour is a maximum of the classical excitation function J(γ), which connects the final rotational angular momentum with the orientation angle of the molecule also shown in Fig. 2 for the simple case of the scattering of an atom from a homonuclear diatom. For a hard ellipsoid potential model with the long and short semiaxis A and B the rotational rainbow angular momentum is given by

$$j_r = 2k\ (A-B)\ \sin\frac{\theta}{2} \qquad (4)$$

where k is the wavenumber and Θ the deflection angle. This for-

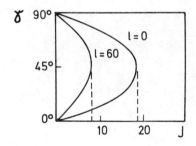
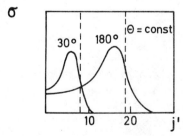
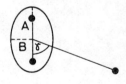

Figure 2: The origin of the rotational rainbow structure in the differential cross sections σ demonstrated by the final angular momentum J(γ) as function of the orientation angle of the molecule. ℓ is the orbital angular momentum which corresponds to the deflection angle θ.

mula connects in a simple way the measured cross section features with the anisotropy of the potential surface and enables us to predict the general behaviour of the inelastic cross sections. Although derived under special conditions (hard ellipsoid, sudden approximation) Eq. (2) gives often a good description of the rainbow curve for realistic potential surfaces. Again, as in the weak coupling case, only the relative anisotropy is obtained from the inelastic cross sections. To determine the absolute scale of the interaction potential additional information is necessary, for instance, from the measurement of the diffraction oscillations.

3. EXPERIMENTAL

The general aim of the experiment is to measure state selected differential cross section $\sigma(j \rightarrow j',\theta,E)$ for rotational transitions. The critical point of such an experiment is the preparation of a single state before and the detection after the collision. There are essentially two possibilities to solve this problem for crossed beam experiments.

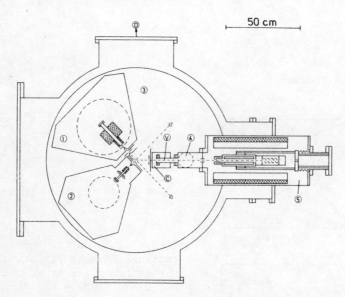

Figure 3: Schematic view of the cross molecular beam apparatus. (1) primary beam source chamber, (2) secondary beam source chamber, (3) scattering chamber, (4) detector buffer chamber, (5) detector chamber, (C) pseudorandom chopper, (D) diffusion pump, (V) UHV-valve.

In the velocity change method two well collimated and nearly monochromatic beams are scattered. The inelastic transition is identified by measuring the energy loss after the collision. A typical example of such an apparatus used in our laboratory is shown in Fig. 3 [6,7]. The two colliding beams, which cross at 90°, are produced as nozzle beams from two differentially pumped chambers (1) and (2). The angular dependence is measured by rotating the source unit with respect to the fixed detector assembly (4,5). The scattered particles are detected by a double differentially pumped mass spectrometer with electron impact ionization operating at pressures $< 10^{-8}$ Pa. Elastic and inelastic events are separated by time-of-flight analysis of the scattered particles using the pseudo-random chopping technique (C). The flight path varies between 449.5 or 593 mm. In this way the velocity distribution of the scattered particles were measured with a resolution of ca. 2 % where effects of the finite ionization volume, the shutter function and the channel width have been induced. Collection and processing of the data were executed by a minicomputer. The pumping facilities for the two beam sources permit the use of pressures up to 15 MPa with 18 μm diameter nozzles. Therefore, the two intersecting beams not only have velocity distributions with a full width at half maximum of better than 5 % but also expand in such a manner that all molecules are in their lowest available rotational states. The actual beam conditions for some of the scattering species are given in Table 1.

Parameter	D_2	H_2	D_2	CO	NH_3[a]	He
Nozzle diameter/μm	18	20	10	100	40	30
Source pressure/10^5 Pa	150	70	191	6	2.5	30
Source temperature/K	303	306	300	303	305	305
Peak velocity /ms^{-1}	2085	2780	2042	780	1623	1785
Speed ratio S	40.2	44.2	30.5	16.8	22.3	75.5
Rotational temperature/K	50	119	70	-	10	-

Table 1: Beam data. [a]An 8 % mixture of NH_3 in He is used.

For the CO secondary beam the pressure has been kept at low values in order to avoid condensation. For NH_3 the seeded beam technique has been used to increase the speed ratio and therefore the velocity resolution of the experiment.

If the experimental resolution is good enough to separate the time-of-flight distributions for different rotational transitions from each other, these intensity peaks are easily converted to state selected differential cross section. If this is not the case, the TOF-spectra are converted to differential energy loss spectra $\sigma(\Delta E,\theta,E)$ where the energy loss $\Delta E = E(j´)-E(j)$ is considered to be a continuous variable [19,20]. The big advantage of this experimental arrangement is its universal applicability to nearly any system provided we are able to produce resonably good nozzle beams. The disadvantage is the unspecificity and therefore to a certain extend also the unsensitivity of the detection mechanism.

This problem is overcome in the second experimental method the state-changing method, where state-specific detection mechanisms are used. The state change is detected by laser induced fluorescence without or in combination with state selectors based on inhomogeneous electric or magnetic fields. In contrast to the first method, this method is restricted to a certain class of molecules depending on the availability of laser sources for the excitation or permanent moments to interact with. State-to-state differential cross sections have, up to now, only been measured for Na_2 [12,13,21]. Therefore it is also interesting to look for integral state-to-state cross section $Q(j \rightarrow j´,E)$ which have been measured for an increasing number of systems such as LiH [22], NO [23] and NH_3 [24]. Of special interest for astrophysical application is the measurement of these cross section for the radical OH scattered from H_2 [25].

4. RESULTS

In this section we will present results for three systems which are important for astrophysical applications. In all cases the measured detailed differential cross sections have been analysed in terms of complete angular dependent potential functions.

4.1 H_2-H_2

In the last 15 years this system has attracted a great deal of interest. It is of principle interest because it represents the simplest diatom-diatom system which is open to accurate quantum chemical structure calculations. A wide range of experimental data is available ranging from the equation of state in solids [26], liquids [27] and gases [28] over transport and relaxation phenomena [29] to the spectroscopy of the van der Waals dimer [30] and molecular beam experiments for the different isotope combinations. In our group we have measured differential cross sections for elastic and rotationally inelastic transitions in $HD+D_2$ collisions at 45.2 meV [8] and 70.3 meV [9]. The experimental resolution was sufficient to resolve the single rotational transitions 0 → 0, 0 → 1 and 0 → 2 for HD. Although this system is interesting in itself because of astrophysical applications and the dominating 0 → 1 inelastic transition [31] we tried to relate the measured cross sections to the H_2-H_2 potential surface which was transformed to the HD+D_2 coordinate system. As a result of such an investigation we found that the cross sections for 0 → 0 an 0 → 1 transitions are essentailly sensitive to the isotropic part of the H_2-H_2 surface, since the anisotropic coupling term for the letter transition is essentially due to the shift of the center-of-mass of HD compared to the symmetric species H_2 and D_2. In order to have a direct test of the original H_2-H_2 surface we measured, in addition, the system D_2+H_2 at E = 84.1 meV. The results are displayed in Fig. 4. The system was measured in two different nuclear spin modifications ortho(o)·D_2 + para(p)·H_2 where both molecules are initially in j=0 and oD_2+oH_2 where D_2 is in j=0 and H_2 in j=1. Besides the elastic cross sections the 0 → 2 rotational transitions for D_2 and H_2 have been resolved although they were found be only a few percent of the elastic cross section. The most striking feature is the fact that the cross sections for the 01 → 21 transition, where the scattering partner H_2 rotates before and after the collision, is about a factor of two larger than the same cross section 00 → 20 with a non-rotating H_2 as scattering partner [10]. The oscillatory part of the elastic cross section is shown in an enlarged view in Fig. 5.

The hydrogen dimer system is to a certain extent an extraordinary case. Not only did we resolve single rotational transitions but also exact close coupling calculations were feasible since only few rotational states are open at the present collision energy. In addition, the system is small enough to be tackled in ab initio calculations for a number of geometries based on configuration interaction [32]. Only very few potential terms of the usual expansion in spherical harmonics contribute. Thus this system is a typical example for a weak coupling case. The results of the ab initio calculation are shown in Fig. 6. The leading term is V_{000} the isotropic part. The anisotropy is represented by V_{202}, which is responsible for the excitation of one hydrogen molecule, and V_{224}, which couples two rotating molecules and which is asymptotically due to the long range interaction of the quadrupole moments of the two molecules. A direct comparison of an exact close coupling calculation with the measured values is shown in Fig. 4 by the solid line. The overall agreement especially for the inelastic transitions is excellent.

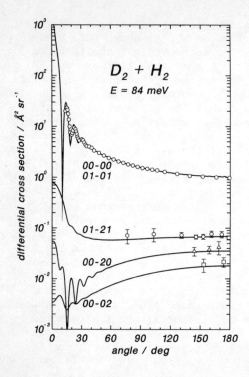

Figure 4: Measured differential cross sections for D_2+H_2 collisions compared with close coupling calculations based on the ab initio potential of Ref.[32]. The rotational transitions are marked by $j(D_2)j(H_2) \rightarrow j'(D_2)j'(H_2)$.

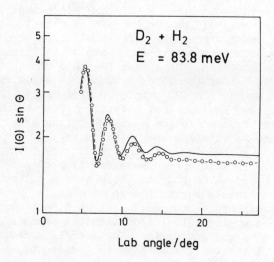

Figure 5: Measured elastic differential cross sections weighted by $\sin\theta$ as a function of the deflection angle. The curves are calculated using different potential function: (———) ab initio calculation, ‒‒‒ experimental fit.

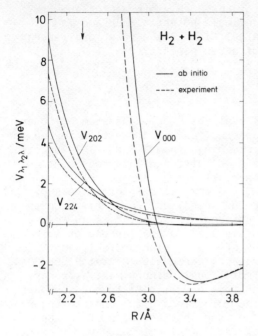

Figure 6: Comparison of the best fit potential surface (---) for H_2-H_2 with the ab initio calculation (—) of Ref. [32].

Note that no adjustable parameter is used. Detailed investigations show that the 00 → 02 and 00 → 20 cross sections are directly correleted with V_{202} whereas both V_{202} and V_{224} are responsible for the 01 → 21 transitions, which explains the much larger magnitude of these cross sections. The careful inspection of the oscillatory part of the elastic cross section, as shown in Fig. 5, reveals a small but significant deviation of the calculation in the position and the amplitude of the diffraction oscillations. The corrected isotropic potential V_{000} which is in agreement with the differential cross section is also shown in Fig. 6 by the dashed line. The potential well depth has to be increased by 7 % and the distance where the potential goes through zero has to be shifted inwards by about 0.08 Å. These changes have been confirmed in other experimental investigations ranging from molecular beam scattering data [8,10,11], the spectroscopy of van der Waals complexes [30] to equilibrium data of the solid [26,11]. Once we know that the isotropic part of the potential V_{000} has to be shifted, also the anisotropic parts V_{202} and V_{224} have to be changed according to what has been said in Sec. 2 (see Fig. 1). The final version of the potential surface which results from the comparison with the beam data is shown in Fig. 6 by dashed lines. A parametrization of this result is given in Ref. [11].

We conclude that in the case of the H_2-H_2 interaction and its isotope combinations a very reliable potential surface is avialable which allows the calculation of the rotational energy transfer within an accuracy of a few percent.

4.2 H_2-CO

The experimental results available for the system D_2+CO are as follows: (1) Energy loss spectra for six deflection angles in the center-of-mass system in the range from 36° to 105° and total (summed over all rotational states) differential cross section with well resolved diffraction oscillations at the collision energy of E = 87.2 meV [19]. The energy loss spectra are shown in Fig. 7. Single transitions are not resolved. However, the data show a pronounced rotational rainbow structure, a large peak at small energy transfer (about 5 meV) and a smaller one at larger energy transfers (about 25 meV at 105°) which are shifted to large energy losses with increasing deflection angle. Since CO is a heteronuclear molecule two rotational rainbow maxima appear, the one at smaller energy transfers corresponds to the approach of the H_2 from the O-end of the molecule and the one at larger energy transfers corresponds to the approach of H_2 from the C-end of CO. Any excitation of the D_2 molecule appears at energy losses larger than 22 meV. It should be noted that in the backward direction more than half of the collision energy is transferred which corresponds to final rotational states of $j' \approx 12$ for CO. Therefore this system is a typical case of a strong coupling system.

These data have been analysed in terms of an interaction potential expanded in Legendre polynomials

$$V(R,\gamma_1,\gamma_2) = \sum_{\lambda_1 \lambda_2} [V^{SCF}_{\lambda_1 \lambda_2}(R) + V^{dis}(R) \cdot f(R)] P_{\lambda_1}(\cos\gamma_1) P_{\lambda_2}(\cos\gamma_2) \quad (5)$$

where R is the internuclear separation of the two molecular centers-of-mass, γ_1 and γ_2 are the orientation angles of the CO and H_2, respectively, with respect to $\underline{R}$. The potential is averaged over the small dependence on the twist angle ϕ. V^{SCF} is calculated using a large basis set to reduce the so-called superposition error. The long range attraction V^{dis} is taken from perturbation theory calculations and $f(R) = \exp[-\gamma(D/R-1)^2]$ is a suitable damping function for small distances [33]. Besides the particular form of $f(R)$ γ and D are the only adjustable parameters in the model which have to be fitted to the experimental data.

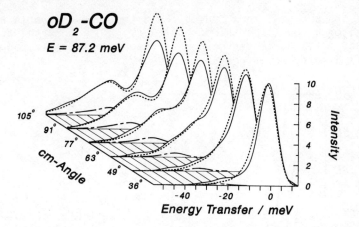

Figure 7: Comparison of measured (———) and calculated (---) energy transfer distribution for D_2+CO. The theoretical contribution for the D_2 excitation is separately shown (-.-).

The dynamical calculations are performed within the quantal centrifugal sudden (CS) approximation for both molecules. Treating the CO rotation in the energy sudden approximation or by classical mechanics as previously done in Ref. [19] and Ref. [34], respectively, leads to errors in the position of the rotational rainbow peak which are outside the experimental errors. By adjusting the two parameters of f(R) a complete potential surface is determined. The comparison of full CS calculations based on this potential with the experimental results is also shown in Fig. 7 by the dashed line. The contributions from the D_2 excitation are separately marked by the dashed-dotted line. Theory and experiment are normalized to each other via the fit to the total differential cross sections which are in good agreement with the calculations. The analysis shows that both quantities must be fitted simultaneously for a unique determination of the potential. The rotational rainbows obtained from the energy loss spectra are most sensitive to the relative anisotropy, namely the constants A-B of Eq. (4). For the determination of the absolute range the diffraction oscillations of the total differential cross sections are used.

The resulting potential surface is shown in Fig. 8. As expected, the repulsive wall is shifted to smaller distances going from the C-approach ($\gamma_1 = 0$) over the O-approach ($\gamma_1 = 180°$) to the perpendicular approach ($\gamma_1 = 90°$). The parallel configuration of the two molecules is the most stable one with a minimum distance of 4.23 Å and a well depth of 9.2 meV. Other potential surfaces available for this system could not satisfactorily reproduce the total differential cross sections [19] or the energy loss spectra [35] thus indicating a two deep isotropic well (Ref. [36]) or a too large anisotropy (Ref. [37]). Only the potential of Ref. [38] comes very close to the present one, although some small discrepancies remain.

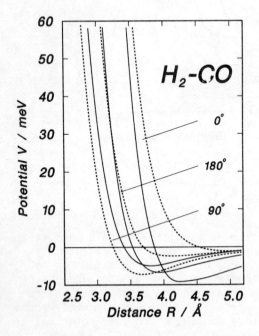

Figure 8: H_2-CO interaction potential vs internuclear distance R for CO orientation angles $\gamma_1 = 0°$, $90°$, and $180°$ and H_2 orientation angles $\gamma_2 = 0°$ (——) and $90°$ (- - -).

We summarize that the presented potential gives the best overall description of the molecular beam data. The preliminary calculation of integral inelastic cross sections showed much better agreement with measured values than obtained previously.

4.3 NH_3-He

The measurement of inelastic differential cross sections discussed so far rely on the fact that very effective nozzle beams of the molecules can be generated. Experiments with ammonia suffer from the fact that this molecule tends to built large clusters based on hydrogen bonding. We therefore decided to measure these cross sections by scattering the heavier NH_3 from the light He. If in addition, the laboratory angular range is restricted to certain values, only monomer NH_3 molecules are detected [39]. Because of this restriction it is difficult to generate measured time-of-flight distributions in the c.m. system, so that we have to compare the experimental results with calculations in the laboratory system. Measurements have been performed for NH_3+He at E = 97.7 meV and a lab angular range from

21.5° to 31.5° which covers a c.m. angular range form 60° to 160° [40]. The expansion conditions of the primary beam are such (see Table 1) that only the lowest rotational states 0,0 and 1,1 are populated in the nuclear spin modifications o-NH_3 and p-NH_3, respectively. A typical time-of-flight spectrum is shown in Fig. 9. The fast peak contains information on c.m. scattering angles from 96.7° to 160.7° whereas the slow peak probes c.m. angles from 140° to 180°. In both regions the maximum of the energy transfer is shifted from the elastic transition which is marked by arrows. However, in contrast to D_2+CO, the maximal energy transfer never exceeds 10 to 20 meV.

The potential model used in this case is similar to the one used in Eq. (5). We get

$$V(R,\theta,\phi) = \sum_{\lambda,\mu} [v^{SCF}_{\lambda\mu}(R) + v^{dis}_{\lambda\mu}(R)f(R)] T_{\lambda\mu}(\theta,\phi) \quad . \tag{6}$$

The angular dependent part is a suitable linear combination of spherical harmonics, which takes into account the symmetry of the system [41]. R is the nuclear distance between the c.m. of the molecule and the He atom, θ and ϕ are angles of a spherical coodinate system chosen in such a way that $\theta = 0$ coincides with the C_{3V}-axis of NH_3 (z-axis). The SCF-part and the long range dispersion part have been calculated as described above [42]. The final potential is determined by adjusting the parameter in the damping function in a simultaneous fit to the measured time-of-flight spectra and the diffraction oscillations of the total differential cross section [43].

The calculations for the rotationally inelastic cross sections have been performed in the CS-approximation [40]. After summing up all the different cross sections which are measured at one lab angle, these quantities are transformed into the lab system and suitably averaged over the experimental distribution functions. The result is shown in Fig. 9 as a solid line. The agreement is very satisfactory. The potential on which this calculation is based is shown in Fig. 10. For a fixed aximutal angle $\phi = 0$ the distance dependence for different θ angles is shown. $\theta = 66°$ marks the approach along the H-N bond and is found to be the most repulsive part while the approach between the three H-atoms ($\theta = 0$) leads to the innermost potential values. Note that the largest anisotropy in the respulsive part is about 0.3 Å which is much smaller than the values found for CO+H_2 (see Fig. 8). The results show that also for NH_3 a reliable potential function can be derived. Since SCF-calculations of similar quality are available for NH_3-H_2 [44] also this system can be evaluated in the same way.

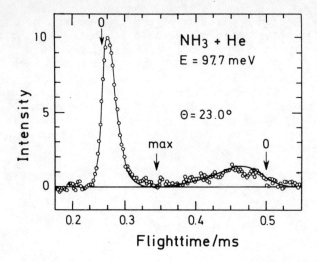

Figure 9: Measured time-of-flight spectrum for NH_3+He at the collision energy E = 97.7 meV. The arrows mark the position of the elastic contribution. The solid line is a calculation based on the fit potential of Fig. 10.

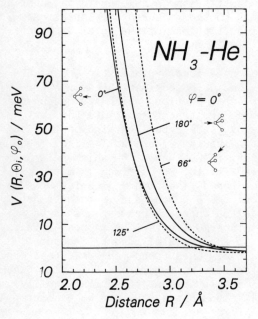

Figure 10: NH_3-He interaction potential vs internuclear distance for $\phi = 0$ and different θ values which correspond to different approaches of the He atom.

5. SUMMARY

Due to a combined effort of experimental and theoretical developments the rotational energy transfer in molecular dynamics is well understood. These results can be used to derive reliable non-spherical potential surfaces from the measurement of differential cross section for rotational transitions. If data is available on a state selected level it is no problem to determine the potential surface as it has been demonstrated for the H_2-H_2 system. But even for unresolved differential energy loss spectra reliable potential functions can be obtained if additional information from large basis set SCF-calculations and damped dispersion coefficients is used. Examples for this method, which is generally applicable, are given for H_2-CO and NH_3-He.

REFERENCES

[1] Atom-Molecule Collision Theory 1979, ed. Bernstein, R.B. (Plenum, New York)

[2] Schinke, R. and Bowman, J.M. 1983, in Molecular Collision Dynamics, ed. Bowman, J.M. (Springer, Berlin), Chap. 4

[3] Billing, G.D. this volume

[4] Gerber, R.B., Buch, V., Buck, U., Maneke, G. and Schleusener, J. 1980, Phys.Rev.Lett. 44, 1397-1400

[5] Buck, U. 1982, Faraday Discuss. Chem.Soc. 73, 187-203

[6] Buck, U., Huisken F., Schleusener, J. and Schaefer, J. 1980, J.Chem.Phys. 73, 1512-1523

[7] Andres, J., Buck, U., Huisken, F., Schleusener, J. and Torello, F. 1980, J.Chem.Phys. 73, 5620-5630

[8] Buck, U., Huisken, F., Schleusener, J. and Schaefer, J. 1981, J.Chem.Phys. 74, 535-544

[9] Buck, U., Huisken, F., Maneke, G., and Schaefer J. 1983, J.Chem.Phys. 78, 4430-4438

[10] Buck U., Huisken, F., Kohlhase, A., Otten, D. and Schaefer, J. 1983, J.Chem.Phys. 78, 4439-4450

[11] Norman, M.J., Watts, R.O. and Buck, U., J.Chem.Phys. in press

[12] Hefter, U. Jones, P.L., Mattheus, A., Witt, J., Bergmann, K. and Schinke R. 1981, Phys.Rev.Lett. 46, 915-918

[13] Jones, P.L., Hefter, U., Mattheus, A., Witt, J., Bergmann, K., Müller, W., Meyer, W. and Schinke, R. 1982, Phys.Rev. A26, 1283-1301

[14] Faubel, M., Kohl, K.H., Toennies, J.P., Tang, K.T. and Yung, Y.Y. 1982, Faraday Discuss. Chem.Soc. 73, 205-220

[15] Faubel, M., Kohl, K.H., Toennies, J.P. and Gianturco, F.A. 1983, J.Chem.Phys. 78, 5629-5636

[16] Ahlrichs, R., Penco, R. and Scoles, G. 1977, Chem.Phys. 19, 119-130

[17] Tang, K.T. and Toennies, J.P. 1977, J.Chem.Phys. 66, 1496-1506

[18] Beck, D. 1982, Physics of Electronic and Atomic Collisions, ed. Datz, S. (North Holland, Amsterdam), pp. 331-342

[19] Andres, J., Buck, U., Meyer, H. and Launay, J.M. 1982, J.Chem.Phys. 76, 1417-1429

[20] Buck, U., Otten, D., Schinke, R. and Poppe, D. 1984, J.Chem.Phys. submitted

[21] Serri, J.A., Becker, C.H., Elbel, M.B., Kinsey, J.L., Moskowitz, W. and Pritchard, D.E. 1981, J.Chem.Phys. 74, 5116-5119

[22] Dagdigian, P.J., Wilcomb, B.E., Alexander, M.A. 1979, J.Chem.Phys. 71, 2726-2728

[23] Andresen, P., Joswig, H., Pauly, H. and Schinke, R. 1982, J.Chem.Phys. 77, 2204-2205

[24] Klaasen, D.B.M. 1982, PhD thesis, University of Nijmegen

[25] Andresen, P., Häusler, D. and Lülf, H.W. 1984, J.Chem.Phys. in press

[26] Silvera, I.F. and Goldmann, V.V. 1978, J.Chem.Phys. 69, 4209-4213

[27] Schaefer, J. and Watts, R.O. 1982, Mol.Phys. 47, 933-944

[28] McConville, G.T. 1981, J.Chem.Phys. 74, 2201-2205

[29] Köhler, W.E. and Schaefer, J. 1983, J.Chem.Phys. 78, 4862-4874; J.Chem.Phys. 78, 6602-6610

[30] Waaijer, M., Jakobs, M. and Reuss, J. 1980, Chem.Phys. 63, 257-261

[31] Schaefer, J., this volume

[32] Schaefer, J. and Meyer, W. 1979, J.Chem.Phys. 70, 344-360. We used the corrected values by Meyer and Schaefer (unpublished) and Schaefer and Liu (unpublished).

[33] Schinke, R., Meyer, H., Buck, U. and Diercksen, G.H.F. 1984, J.Chem.Phys., in press

[34] Billing, G.D. and Poulsen, L.L. 1982, Chem.Phys. 70, 119-126

[35] Billing, G.D. and Poulsen, L.L. 1983, Chem.Phys. 99, 368-371

[36] Flower, D.R., Launay, J.M., Kochanski, E. and Prisette, J. 1979, Chem.Phys. 37, 355-362

[37] van Hemert, M.C. 1983, J.Chem.Phys. 78, 2345-2354

[38] Poulsen, L.L. 1982, Chem.Phys. 68, 29-40

[39] Buck, U. and Meyer, H. 1984, Ber.Bunsenges. Phys.Chem. 88, 254-256

[40] Meyer, H., PhD thesis, Göttingen, in preparation

[41] Green, S. 1980, J.Chem.Phys. 78, 2740-2750

[42] Meyer, H., Buck, U., Schinke, R. and Diercksen, G.H.F., to be published

[43] Slankas, J.T., Keil, M. and Kuppermann, A. 1979, J.Chem.Phys. 70, 1482-1491

[44] Diercksen, G.H.F. 1983, private communication

ION-MOLECULE REACTIONS AT LOW TEMPERATURES

D. Smith and N.G. Adams

Department of Space Research, University of Birmingham,
Birmingham B15 2TT, England

Laboratory studies of ion-molecule reactions relevant to interstellar clouds (ISC) are discussed. It is stressed that when such reactions are exoergic and fast at room temperature, they may be assumed to be so at ISC temperatures. However, when they are slow at room temperature, then assumptions concerning their rates at ISC temperatures are dangerous. Some reactions relevant to interstellar chemistry are discussed to illustrate this point. Isotope exchange in ion-molecule reactions is also discussed and a simple kinetic rule which facilitates the estimation of the rate coefficients for such reactions at ISC temperatures is given. Some ternary association reactions are considered in relation to the analogous radiative association reactions which are now seen to be important in the synthesis of molecules in ISC. Data are presented which show how kinetic excitation in the ions can increase the rates of some important interstellar reactions and the possible rôle of kinetically excited ions in the chemistry of shocked regions of ISC is briefly discussed.

1. INTRODUCTION

The agreement between the observed abundances and those predicted by gas phase ion-chemical models for several interstellar molecular species leaves little room for doubt that gas phase ion chemistry plays a major rôle in the synthesis of interstellar molecules [1-5] as was proposed more than a decade ago [6,7]. The laboratory kinetic data (i.e. rate coefficients and product distributions) for the numerous ion-atom and ion-molecule reactions included in the models have mostly been obtained at room temperature and above using flow tube and ion cyclotron resonance experiments [8-10]. Fortunately – as it turns out – most of the reactions included in the models are fast at room temperature and above, i.e. the rate coefficients, k, are close to the respective collisional rate coefficients, k_c, and a growing body of laboratory data indicates that when $k \approx k_c$ at 300K then k is also close to k_c at lower temperatures [11]. However, the k for some of the reactions which are important in interstellar chemistry are much less than k_c at room temperature and it is also apparent from the laboratory data that when $k < k_c$ at 300K, then k almost invariably changes with temperature, the sense of the temperature variation being unpredictable in most cases [11]. Notably exceptions to this are isotope exchange reactions which we discuss later.

These features of ion-molecule reactions lead one to wonder if some of the outstanding problems in gas phase interstellar chemistry, such as the failure of current models to predict the relative abundances of NH_3 and H_2S, are due to the inclusion of grossly incorrect rate coefficients in the quantitative models. Some evidence for this is emerging, for example, in relation to NH_3 synthesis (discussed below).

It is clearly desirable in the light of the above comments to study reactions of potential importance to interstellar molecular synthesis at suitably low temperatures, ideally down to ~5K. This is impractical in most types of experiments, especially since many of the reactant neutral species have little vapour pressure at such low temperatures. However, it is possible to study the reactions of almost any ion species with several of the permanent gases present in ISC (H_2, CO, N_2, O_2, CH_4 etc) down to a temperature of 80K using the variable-temperature selected ion flow tube (VT - SIFT) developed in our laboratory [11,12]. Indeed, using our VT - SIFT, we have now studied several hundred positive ion-molecule reactions relevant to interstellar ion chemistry at 80K, including exoergic and near-thermoneutral two-body reactions, isotope exchange reactions and three-body reactions [11,13]. We summarise the more relevant results in this paper. There is however no inherent reason why the VT - SIFT could not be operated at much lower temperatures (e.g. using liquid Ne or H_2 as coolants) and there is obviously some need for this. Meanwhile, other low temperature experiments have been developed including an expanding jet experiment [14] (called CRESU, which can be operated down to ~10K) and an ion trap experiment [15] (down to ~5K). Some of the data obtained using these experiments will be referred to in this paper.

An important question relevant to interstellar chemistry relates to the influence on reaction kinetics of ion-molecule centre-of-mass interaction energies, KE_{cm}, greater than those KE_{cm} appropriate to the thermalised species. Clearly, KE_{cm} is elevated in shocked regions of the ISC (due either to gas dynamic or MHD shocks; refs. 16 and 17) and it is well known that elevated KE_{cm} can influence the rates of slow ion-molecule reactions and also initiate endothermic reactions [18]. Suprathermal ions (and hence elevated KE_{cm}) are also produced in exoergic ion-molecule reactions. We discuss this kinetic phenomenon in relation to several reactions of potential interstellar importance and present some data on the influence of KE_{cm} on the rate coefficients for some of these reactions. These data have been obtained using our recently developed variable-temperature selected ion flow drift tube (VT-SIFDT) [19]. In this apparatus, ion-molecule reactions can be studied at any temperature within the 80-600K range and for KE_{cm} up to about one eV, and it is therefore a suitable apparatus with which to study interstellar shock chemistry.

2. SOME RECENT LABORATORY DATA RELEVANT TO INTERSTELLAR ION CHEMISTRY

Both binary (two-body) and ternary (three-body) ion-molecule reactions are common in laboratory plasmas. Ternary reactions can only proceed at significant rates when the ambient pressure is sufficiently high and the temperature is suitably low. In interstellar clouds, the pressures are never sufficiently high for ternary reactions to be important. Nevertheless, laboratory studies of ternary reactions are proving to be of great value in providing data relating to the rates of binary radiative association reactions which are now considered to be important in interstellar molecular synthesis. So in this paper, we discuss some ternary association reactions, but the majority of the paper will be concerned with binary ion-molecule reactions. The order of discussion in this section is: 2.1. Binary reactions including fast exoergic reactions, exoergic reactions which are slow at room temperature, slightly endoergic reactions and appreciably endoergic reactions, 2.2. Isotope exchange reactions and 2.3. Ternary association and radiative association reactions.

2.1. Binary Reactions

Fast exoergic binary reactions. As has already been stated, when ion-molecule reactions are rapid, i.e. when $k \approx k_c$, then the k do not usually vary with temperature, at least over the temperature range accessible to laboratory experiments. This principle has been well demonstrated in numerous flowing afterglow (FA) and selected ion flow tube (SIFT) studies during the last decade or so [11]. Since the large majority of the k for exoergic ion-molecule reactions, including many reactions of importance in interstellar chemistry, are observed to be close to k_c at room temperature, then it is reasonable to make the assumption that the k for such reactions will be close to k_c even at ISC temperatures. Such has often been tacitly assumed in models predicting interstellar molecular abundances and the general success of the models gives some credence to this assumption. Extensive lists of the k for fast ion-molecule reactions have been published [8-10]. However, much less attention has been given to product distributions and to the variation of these distributions with temperature for reactions of interstellar significance in which multiple products are evident. Fortunately, there is little experimental evidence to suggest that large changes occur in the product distributions for exoergic binary reactions when the temperature is varied in the thermal range, and in any case small changes would not greatly influence the results of the relatively crude interstellar chemical models.

Exoergic binary reactions which are slow at room temperature. Experiments have shown unequivocally that the k for these reactions invariably change, sometimes dramatically, with temperature and so the k appropriate to ISC temperatures are not readily predictable. Many such reactions have been identified and some are included in useful lists [8-11]. Data have also been presented in graphical form to clearly show how unpredictable the k are with temperature [11], and they forcibly demonstrate the dangers in adopting room temperature k values for such reactions in interstellar models. The classic example - from the interstellar viewpoint - is the reaction:

$$NH_3^+ + H_2 \longrightarrow NH_4^+ + H \tag{1}$$

k(1) at 300K is very small (5×10^{-13} $cm^3 s^{-1}$) and FA studies [20] showed that k(1) increases approximately exponentially with increasing temperature above 300K. For this reason, it was concluded that k(1) would be extremely small at ISC temperatures and that reaction (1) could not be involved in the synthesis of interstellar NH_3. This has presented a major difficulty in explaining the observed NH_3 abundances in ISC. Subsequent to the FA experiments, SIFT experiments at 80K [21] showed that k(1) did not decrease as predicted at temperatures below 300K and recent ion trap experiments [22] performed at very low temperatures have shown that k(1) actually increases at temperatures below about 80K, reaching a value of 2×10^{-12} $cm^3 s^{-1}$ at 10K. These new data gave a strong warning to modellers and alleviate, but do not solve, the problem of gas phase NH_3 synthesis in ISC [28]. It is important to note that the ion trap values for k(1) may be somewhat low since H_2 with a non-equilibrium ortho/para ratio was used in the experiments. Similar problems also arise in the ion trap determinations of k(2) and k(21). Further discussion relating to the NH_3 problem is presented below.

Other striking examples of reactions for which the k vary with temperature include the $O_2^+ + CH_4$ [23] and the $O^+ + N_2$ [24] reactions. Many other slow reactions remain to be further investigated experimentally (e.g. see the list in ref. 11). The study of these reactions at different temperatures will

surely provide further insight into the nature of relatively
inefficient ion-molecule reactions (this is discussed in several
recent papers e.g. refs. 23, 25 and 26).

Slightly endoergic binary reactions. Clearly, ion-molecule
reactions need only be slightly endoergic to severely inhibit
them for reactants which are thermalised at the very low temperatures of ISC. Yet such reactions may have k values which are
appreciable fractions of k_c at room temperature. These reactions
therefore represent another potential pitfall in ISC ion chemical
models. Several such reactions which are directly relevant to
interstellar chemistry, have been identified. The reaction

$$N^+ + H_2 \longrightarrow NH^+ + H \tag{2}$$

is often considered to be the first stage in the synthesis of NH_3
in ISC. At 300K, k(2) is measured to be 4×10^{-10} cm^{3}s^{-1} [8,27]
and so reaction (2) was thought to be fast enough to initiate
the chain:

$$N^+ \xrightarrow{H_2} NH^+ \xrightarrow{H_2} NH_2^+ \xrightarrow{H_2} NH_3^+ \xrightarrow{H_2} NH_4^+ \xrightarrow{e} NH_3 \tag{3}$$

but recent measurements of k(2) at very low temperatures in an
ion trap experiment (Dunn, priv. comm.) have shown that it
decreases rapidly with decreasing temperature, consistent with
an endoergicity for reaction (2) of only ~8 meV. Thus in the
ion trap experiment k(2) is only 5×10^{-12} cm^{3}s^{-1} at 20K and it
might be thought that k(2) would be similarly small at ISC
temperatures. However, it is important to realise that N^+ ions
are most probably formed in ISC via the reaction

$$He^+ + N_2 \longrightarrow (N_2^+)^* \longrightarrow N^+ + N \tag{4}$$

and thus possess a small but significant amount of kinetic energy
(~140 meV) which in collision with an H_2 molecule at 10K is
equivalent to a KE_{cm} of about 40 meV. This is quite sufficient to
overcome the endoergicity of reaction (2) and thus to ensure
that, under these conditions, k(2) will be an appreciable
fraction of k_c. That an elevated KE_{cm} above thermal does indeed
promote reaction (2) has been proved using our VT-SIFDT [28].
From these VT-SIFDT measurements, we have derived an endoergicity
of ~13 meV for reaction (2) which is somewhat greater than that
derived from the low temperature ion trap experiment. Very
recently, this reaction has also been studied in the CRESU
experiment (Rowe, priv. comm.) at temperatures down to 8K and
these measurements indicate an endoergicity of ~4 meV. However,
it must be appreciated that normal H_2 was used in both of the
very low temperature experiments and so the derived endoergicities
refer largely to ortho-H_2 as the reactant neutral. In our
VT-SIFDT experiment, the rotational states of the reactant H_2
were equilibriated at 300K and so a larger value for the

endoergicity is expected from this experiment, as is indeed obtained.

So it appears that a relatively large value for k(2) is appropriate even under the low temperature conditions of cool ISC, and that excess kinetic energy in reactant ions can be important in determining the rates of slightly endoergic reactions of ions with H_2. Kinetically excited ions are also present in shocked regions of ISC and this must, of course, be taken in account when considering the ion chemistry [28] (see below).

The reactions of $C_2H_2^+$ and C_3H^+ with H_2 are considered to be important in interstellar chemistry and in these cases the ion kinetic energies are critical in determining the nature of the reactions. Laboratory experiments [29,30] at 300K have shown that in both reactions, binary and ternary channels occur:

$$C_2H_2^+ + H_2 \longrightarrow C_2H_3^+ + H \tag{5a}$$

$$\xrightarrow{+He} C_2H_4^+ + He \tag{5b}$$

$$C_3H^+ + H_2 \longrightarrow C_3H_2^+ + H \tag{6a}$$

$$\xrightarrow{+He} C_3H_3^+ + He \tag{6b}$$

As the temperature is decreased, the H atom abstraction channels (5a) and (6a) become less important while the association channels (5b) and (6b) become more important. We have shown that this is because the binary channels are only slightly endoergic (~50 meV, ref. 30). So for thermalised ions under cool ISC conditions, the atom abstraction reactions could not proceed at significant rates (radiation association can then occur - see below). Again, however, the ion kinetic energies, and hence KE_{cm}, will be important in controlling k(5) and k(6) since we have shown, again using our VT-SIFDT [30], that elevated KE_{cm} does promote channels (5a) and (6a) and diminishes channels (5b) and (6b).

Exoergic proton transfer reactions are an essential element of interstellar ion chemistry. Protonated species such as HCO^+, N_2H^+, etc. originate from proton transfer reactions between H_3^+ and CO, N_2, etc. The ergicities of these reactions are given by the differences in proton affinities between H_2 and the acceptor molecules (CO, N_2 etc.). Since such exoergic reactions are generally rapid [31], it had been thought that proton transfer from H_3^+ to O_2 could occur in cool ISC because studies at 300K of the reactions

$$O_2H^+ + H_2 \underset{k_r}{\overset{k_f}{\rightleftharpoons}} H_3^+ + O_2 \tag{7}$$

had shown that k_r (7) > k_f (7) [32] and this had been taken to indicate that the proton affinity of O_2 was greater than H_2. However, our recent studies of these reactions [33] at 300K and 80K show that quite the opposite is the case and that protons cannot be transferred to O_2 from H_3^+ under cool ISC conditions. That k_r (7) > k_f (7) at 300K is due to the relatively large entropy change in these reactions and again forcibly demonstrates the dangers of predicting kinetic behaviour at low temperatures from the evidence of room temperature data alone.

<u>Appreciably endoergic binary reactions</u>. Obviously, reactions which are appreciably endoergic cannot occur under low temperature thermalised conditions of ISC. However, such reactions are not precluded under shocked conditions and may explain the presence of unpredictably large concentrations of CH^+ in some diffuse ISC. Elitzur and Watson [34] proposed that the endoergic reaction

$$C^+ + H_2 \longrightarrow CH^+ + H - 0.4 \text{ eV} \tag{8}$$

was being promoted in shocked regions of the clouds by kinetically excited C^+ (much earlier Stecher and Williams [35] had made the interesting suggestion that this reaction might be promoted by vibrational excitation of the H_2). We have studied reaction (8) in our VT-SIFDT apparatus and demonstrated that elevated KE_{cm} does indeed promote the reaction [28]. For example, we found at a KE_{cm} = 0.4 eV (i.e. the threshold energy for reaction (8)), that $k(8) = 3 \times 10^{-11}$ $cm^3 s^{-1}$. Similar results have been obtained in Lindinger's laboratory where the rate coefficient of the reverse of reaction (8) as a function of KE_{cm} has also been obtained (Lindinger, priv. comm.). Clearly, therefore, reaction (8) could occur in shocked regions of ISC, the requirement being that the C^+/H_2 relative velocity approaches about 5 km s^{-1} for KE_{cm} to be near to the threshold value of 0.4 eV for the reaction. Utilizing the MHD shock model due to Draine et al [17], we have shown that the observed column densities of CH^+ could indeed be produced in shocked ISC [28].

No satisfactory explanation of the H_2S abundances in ISC has been obtained using gas phase ion chemistry. Under low temperature thermalised conditions the reactions:

$$S^+ + H_2 \longrightarrow SH^+ + H - 0.87 \text{ eV} \tag{9}$$

$$SH^+ + H_2 \longrightarrow H_2S^+ + H - 0.6 \text{ eV} \tag{10}$$

$$H_2S^+ + H_2 \longrightarrow H_3S^+ + H - 0.35 \text{ eV} \tag{11}$$

cannot occur because they are all appreciably endoergic. However, it is clear that even for moderate increases in the ion velocities relative to H_2 (as would occur in MHD shocks) then reactions (9), (10) and (11) could be initiated and neutral H_2S then generated either via charge transfer reactions of H_2S^+ (say with metal atoms) or via dissociative recombination of H_3S^+. As yet there are no laboratory data on these reactions. It remains a real possibility that the failure of gas phase ion-chemical models to correctly predict the relative abundances of some interstellar molecules, including H_2S, is due to the widespread occurrence of non-thermal conditions in ISC.

2.2. Isotope Exchange in Binary Ion-Molecule Interactions

During the last four years, since the development of the variable-temperature SIFT, we have made a thorough study of isotope exchange reactions. The work was stimulated by the astronomical observations of enhanced abundances (relative to solar system abundances) of the rare (heavy) isotopes of several elements (e.g. ^{13}C, ^2H) in some interstellar molecules [36]. The results of these wide-ranging studies have been presented in recent reviews [37,38]. They are summarised here and the more interesting results from an interstellar viewpoint, are briefly described.

If isotope exchange in ion-molecule reactions is to be properly understood, it is essential to be able to determine both the forward (k_f) and reverse (k_r) rate coefficients for such reactions as a function of temperature, in order to obtain the equilibrium constants for the reactions at various temperatures. Then, the enthalpy and entropy changes, ΔH^o and ΔS^o, for the reactions can be determined to assist in predicting rate coefficients at the very low temperatures of ISC. However, these thermodynamic data are not in themselves sufficient and an appreciation of the kinetics of isotope exchange reactions is also required. Using the VT-SIFT, we have been able to measure k_f and k_r as a function of temperature for many such reactions and to formulate a simple kinetic rule which allows k_f and k_r for isotope exchange reactions to be confidently predicted at low temperatures. Amongst the many reactions studied [39-43] are the important interstellar reactions:

$$D^+ + H_2 \underset{k_r}{\overset{k_f}{\rightleftharpoons}} H^+ + HD \tag{12}$$

$$CH_3^+ + HD \rightleftharpoons CH_2D^+ + H_2 \tag{13}$$

$$H_3^+ + HD \rightleftharpoons H_2D^+ + H_2 \tag{14}$$

$$^{13}C^+ + {}^{12}CO \rightleftharpoons {}^{12}C^+ + {}^{13}CO \tag{15}$$

$$H^{12}CO^+ + {}^{13}CO \rightleftharpoons H^{13}CO^+ + {}^{12}CO \tag{16}$$

All of these reactions are exoergic to the right (represented by k_f, as indicated) and they are listed in order of decreasing ΔH^o which ranges from -40 meV (reaction (12)) to -1 meV (reaction (16)). The ΔH^o and ΔS^o for these reactions have been determined from van't Hoff plots (i.e. $\ln k_f/k_r$ vs $1/T$) according to the well-known thermodynamic relationship $\ln k_f/k_r = -\Delta H^o/RT + \Delta S^o/R$. In all cases, the experimentally derived ΔH^o are in good agreement with values calculated from the zero-point-energies of the reactants and products (however, it should be noted that the ΔH^o for reaction (14) is very temperature dependent below 300K [42]). Reaction (12) is considered to be responsible for HD production in ISC [44]. The ΔH^o for this reaction is large and thus at ISC temperatures, k_r (12) as insignificant and the reaction proceeds only to the right. Reactions (13) and (14) produce the deuterated species CH_2D^+ and H_2D^+ which, via subsequent ionic reactions, generate some of the observed deuterated interstellar molecules. Reactions (15) and (16) lead to the observed enrichment of ^{13}C in interstellar CO and HCO^+. This 'isotope fractionation' is an essential part of interstellar ion chemistry and must be understood if overall isotopic abundances in cool ISC material are to be estimated [36].

Fortunately, as it turns out, the kinetic behaviour for these isotope exchange reactions is simple. All our studies show that exoergic isotope exchange reactions become increasingly efficient with decreasing temperature and that k_f approaches the collisional rate coefficient, k_c, at sufficiently low temperature ($T \lesssim |\Delta H^o/R|$). Thus at the low temperatures of cool ISC, it can be assumed that $k_f = k_c$, except for those reactions with very small ΔH^o (e.g. reaction (16)) when the relationship $k_f + k_r = k_c$ is apparently more appropriate [41,43]. The application of these simple kinetic rules together with the ΔH^o and ΔS^o values allows both k_f and k_r to be obtained at any (low) temperature and thus greatly facilitates interstellar modelling of isotope fractionation.

We have also studied the reactions

$$HCO^+ + C^{18}O \rightleftharpoons HC^{18}O^+ + CO \tag{17}$$

$$HC^{18}O^+ + {}^{13}CO \rightleftharpoons H^{13}CO^+ + C^{18}O \tag{18}$$

$$^{14}N_2H^+ + {}^{14}N^{15}N \rightleftharpoons {}^{14}N^{15}NH^+ + {}^{14}N_2 . \tag{19}$$

The ΔH^o for these reactions are very small (similar to that for reaction (16)) but they could result in fractionation of ^{18}O and ^{15}N into interstellar HCO^+ and N_2H^+ [43,45].

It is most important to appreciate that even when isotope exchange between an ion and a molecule can be shown to be exoergic, the exchange will not necessarily occur at a measurable rate because of the existence for some interactions of potential barriers to the exchange, very short complex lifetimes, steric effects, etc. Many such systems have been recognized, including some of significance to interstellar chemistry. For example, H/D exchange between HD molecules and HCO^+, N_2H^+, NH_4^+ and CH_5^+ does not occur for ground state ions and obviously such reactions should not be included in models describing isotope fractionation in ISC. Recently, however, following a suggestion by A. Dalgarno, we have studied the reactions of HCO^+ and N_2H^+ with D atoms and found that exchange occurs in these reactions and becomes increasingly efficient at low temperatures. The reverse reactions of DCO^+ and N_2D^+ with H atoms are predictably inhibited at low temperatures. We have made preliminary estimates of ΔH^o for these reactions and these are, within error, in accordance with vibrational frequencies determined in recent spectroscopic experiments (Saykally, priv. comm.). A detailed paper on these reaction rate data will be written in due course. However, it is already clear that the kinetic rules which apply to isotope exchange in ion-molecule reactions also apply to ion-atom reactions and these should be seriously considered in ISC models of isotope exchange. We have discussed the details of the laboratory studies of isotope exchange in ion-molecule reactions in several papers and reviews [37-40] and considered the interstellar implications in other papers [41-43,45].

Our studies of isotope exchange to date mostly involve relatively simple interstellar species in which isotope fractionation is seen to have occurred. Since many of the isotopically enriched species in ISC are complex, it is now necessary to direct our attention to reactions which can lead to the incorporation of isotopes into the larger species.

2.3. Ternary Association and Radiative Association Reactions

Although ternary reactions cannot occur at the low pressures pertaining to ISC, studies of these reactions in the laboratory can provide essential data relating to binary radiative association reactions which are now considered to be involved in interstellar molecular synthesis [46,59]. Specifically, from the measurements of ternary association rate coefficients, k_3, the lifetimes against unimolecular decomposition, τ_d, of the excited intermediate complexes formed in the reactions can be estimated. Then, if an estimate of the radiative lifetime of the excited complexes, τ_R, can be made, the radiative association rate coefficient, k_R, can be obtained. A good deal of theoretical work has been carried out to establish the relationship between ternary and radiative association, and laboratory measurements of k_3 as a function of temperature for many association reactions have been made, including several reactions of interstellar interest [47-53]. Most of the data have been obtained down to only 80K, much of it in our laboratory using the VT - SIFT, and so it has been necessary to estimate the values of τ_d at ISC temperatures by extrapolation of these data.

Detailed studies of the association reactions of CH_3^+ ions with several interstellar molecular species (including H_2, N_2, O_2, CO, CO_2, H_2O, NH_3, and HCN) have been carried out [48]. The k_3 for these reactions differ by orders-of-magnitude at a given temperature. Some ternary reactions are very slow implying that the τ_d are too short to allow radiative association to proceed at a significant rate, even at the very low temperatures of ISC. However, others are rapid (e.g. the CH_3^+ reactions with H_2, CO, CO_2) and some are very rapid (e.g. the CH_3^+ reactions with H_2O, HCN) and their radiative association analogues could well be the most important routes to some observed interstellar molecules. The ternary reaction

$$CH_3^+ + H_2 + He \longrightarrow CH_5^+ + He \quad (20)$$

generates CH_5^+ and CH_4 could be produced following dissociative recombination of this ion with electrons. At 80K, k(20) = 1.5×10^{-27} cm^{6}s^{-1} and varies $\sim T^{-2.5}$ in accordance with recent theoretical predictions for the association of a polyatomic ion and a diatomic molecule [50-53]. An extrapolation of the k(20) data to 13K to estimate $\tau_d(20)$ at this temperature and assuming a τ_R of 10^{-3} secs (as is reasonable for excited polyatomic ionic species) provides an estimate of $k_R(21)$ for the radiative association reaction:

$$CH_3^+ + H_2 \longrightarrow CH_5^+ + h\nu \quad (21)$$

Thus $k_R(21) \approx 10^{-13}$ cm^{3}s^{-1}. Very recently, the binary reaction (21) has been measured directly to be 1.1×10^{-13} cm^{3}s^{-1} at 13K in an ion trap experiment [54]. This is in good agreement with the estimate made from the ternary rate data. This gives credence to the kinetic model which predicts τ_d, to the experimental ternary association data and its extrapolation and, not least, to the choice of the value for τ_R. Therefore this model can be used with some confidence to estimate k_R at ISC temperatures.

As stated above, the ternary association reactions

$$CH_3^+ + H_2O + He \longrightarrow CH_3H_2O^+ + He \quad (22)$$

$$CH_3^+ + HCN + He \longrightarrow CH_3HCN^+ + He \quad (23)$$

are very rapid. Reaction (23) is especially rapid at 300K, k(23) = 5×10^{-25} cm^{6}s^{-1} which is an enormous value [55]. The radiative association analogues:

$$CH_3^+ + H_2O \longrightarrow CH_3H_2O^+ + h\nu \qquad (24)$$

$$CH_3^+ + HCN \longrightarrow CH_3HCN^+ + h\nu \qquad (25)$$

are probably the synthetic routes to interstellar CH_3OH and CH_3CN. Reaction (25) has apparently been observed directly in an ion cyclotron resonance cell [56,57] and this is indicative of a large τ_d or a very short τ_R, of the $(CH_3 HCN^+)^*$ excited complex even at interaction energies of ~0.1 eV (typical cool ISC interaction energies are ~1 meV!).

Of course, reactions of ions with H_2, including association reactions, are of importance in ISC even when the rate coefficients are relatively small because of the dominance of H_2 in ISC. However, normal binary (bond rearrangement) reactions of ions with H_2 are often not energetically possible (unless the ions are internally or kinetically excited as we have shown in Section 2.1), but then radiative association reactions may be important. However, efficient association of most ions with H_2 does not occur, but a few interesting reactions have been observed including reaction (20) above and, in particular, reactions (5b) and (6b) for which k(5b) and k(6b) approach 10^{-26} cm^6s^{-1} at 80K. The very efficient association in these latter two reactions has been explained in terms of the prolongation of τ_d due to endoergic H-atom transfer within the $(C_2H_4^+)^*$ and $(C_3H_3^+)^*$ intermediate complexes [25]. Thus the analogous radiative association reactions:

$$C_2H_2^+ + H_2 \longrightarrow C_2H_4^+ + h\nu \qquad (26)$$

$$C_3H^+ + H_2 \longrightarrow C_3H_3^+ + h\nu \qquad (27)$$

will have appreciable k_R and apparently represent important paths to C_2H_2, C_2H_3, C_3H and C_3H_2 [58,59].

A similar phenomenon had previously been invoked to explain why ternary association of CD_3^+ with H_2 is much more rapid than that of CH_3^+ with H_2. Since this is considered to be due to endoergic exchange of H and D within the $(CD_3H_2^+)^*$ complexes, which again effectively prolongs τ_d as does lowering the temperature, the phenomenon was referred to as "isotopic refrigeration" [40]. Thus the radiative association reaction:

$$CH_2D^+ + H_2 \longrightarrow CH_4D^+ + h\nu \qquad (28)$$

is expected to be faster than the analogous reaction (21) in ISC. (Note that the CH_2D^+ will be generated in reaction (13)). Thus reactions (13) and (28) provide an efficient route to CH_4D^+ and thus to partly deuterated methane in ISC.

It must be noted, however, that the slower association reactions with H_2 should not be ignored completely, again because of the large H_2 abundance. Thus the slow radiative association reaction

$$SH^+ + H_2 \longrightarrow H_3S^+ + h\nu \tag{29}$$

could be important in the synthesis of H_2S in ISC [28] and the slow reaction

$$Na^+ + H_2 \longrightarrow Na^+\!\cdot H_2 + h\nu \tag{30}$$

followed by the switching reaction

$$Na^+\!\cdot H_2 + H_2O \longrightarrow Na^+\!\cdot H_2O + H_2 \tag{31}$$

could be a route to the synthesis of NaOH in ISC [60].

3. CONCLUDING REMARKS

Laboratory studies of ion-molecule reactions at low temperatures are providing new insights into the fundamentals of such reactions and an appreciation of new routes to the formation of molecules under ICS conditions. These studies have forcibly demonstrated the dangers inherent in estimating rate coefficients at ISC temperatures by extrapolating data obtained at room temperature. Ion-chemical models have satisfactorily explained the abundances of many interstellar species; however they have failed to correctly predict the relative abundances of some important interstellar molecular species. Further laboratory studies at low temperatures may explain the apparent inadequacies of the models. In this regard, there is a pressing need for more laboratory data on molecular ion/atom reactions (especially for H, C, N, O, S and some metal atoms). Very recent laboratory studies have demonstrated that radiative association is a viable mechanism for interstellar molecular synthesis in accordance with theoretical predictions and inferences from laboratory studies of ternary association reactions. Direct studies of radiative association are, of course, desirable wherever possible. Studies of isotope exchange in ion-molecule reactions at low temperatures have been particularly rewarding. General rules describing the kinetics of this type of reaction have emerged which facilitate estimates of the rates of fractionation of heavy isotopes into molecules under ISC conditions.

Although it is considered that neutral-neutral chemistry is dominant in the shocked regions of ISC [66], ion chemistry also occurs in these regions. Therefore it is desirable to have ion-chemical data to hand appropriate to these conditions. It is also important to appreciate that the product ions of some reactions possess kinetic energy, which can significantly influence their subsequent reactivity with H_2 in ISC. Thus the VT - SIFDT has been developed and is now providing data on the reaction rates of ions at elevated kinetic energies. The combination of such laboratory data and observational data on molecular concentrations in shocked regions of ISC may provide a better understanding of the physics and chemistry of these disturbed regions.

A final note relates to dissociative recombination reactions which are usually considered as the final step in the reaction chains producing the observed neutral interstellar molecules. Again it has been necessary to rely on laboratory recombination coefficients, α, obtained at or near room temperature. Generally, the α at 300K exceed 10^{-7} cm^{3}s^{-1} and show inverse temperature variations [61,62]. Thus the α for all reactions included in models of cool ISC are assumed to exceed 10^{-6} cm^{3}s^{-1}. However, it should be appreciated that the α may be considerably smaller in the higher temperature shocked regions of ISC. It is also worthy of note that the α for the recombination of the very important interstellar ion H_3^+ has recently been shown in our laboratory to be very small ($\lesssim 10^{-8}$ cm^{3}s^{-1}) in accordance with a recent theoretical prediction (see refs. 63 and 64 and the paper by Adams and Smith in these proceedings). This result has important implications to interstellar physics and chemistry and again illustrates the requirement that, ideally, kinetic data for all of the reactions included in interstellar models should be obtained under conditions (particularly of temperature) close to those pertaining in the interstellar regions of interest.

REFERENCES

[1]   Iglesias, E. 1977, Ap. J. 218, p. 697.
[2]   Hartquist, T.W., Black, J.H. and Dalgarno, A. 1978, Mon. Not. R. astr. Soc. 185, p. 643.
[3]   Prasad, S.S. and Huntress, W.T. Jr. 1980, Ap. J. Suppl. Series 43, 1; Ap. J. 239, p. 151.
[4]   Herbst, E. 1983, Ap. J. Suppl. Series 53, p.41.
[5]   Millar, T.J. and Freeman, A. 1984, Mon. Not. R. Astr. Soc., in press.
[6]   Solomon, P.M. and Klemperer, W. 1972, Ap. J. 178, p. 389.
[7]   Herbst, E. and Klemperer, W. 1973, Ap. J. 185, p. 505.
[8]   Albritton, D.L. 1978, Atom Data Nucl. Data Tables 22, p. 1.
[9]   Huntress, W.T. Jr. 1977, Ap. J. Suppl. Series 33, p. 495.
[10]   Anicich, V.G. and Huntress, W.T. Jr. 1984, Ap. J. Suppl.

Series, in press.
[11] Adams, N.G. and Smith, D. 1983, "Reactions of Small Transient Species: Kinetics and Energetics" eds. A. Fontijn and M.A.A. Clyne, Academic Press, London, pp. 311-385.
[12] Smith, D. and Adams, N.G. 1979, "Gas Phase Ion Chemistry", Vol. 1, ed. M.T. Bowers, Academic Press, New York, pp. 1-44.
[13] Adams, N.G. and Smith, D. 1984, "Swarms of Ions and Electrons in Gases" eds. W. Lindinger, T.D. März and F. Howorka, Springer Verlag, Wien, pp. 194-217.
[14] Rowe, B.R., Dupeyrat, G., Marquette, J.B. and Gaucherel, P. 1984, J. Chem. Phys. 80, p. 4915.
[15] Luine, J. and Dunn, G.H. 1984, Phys. Rev. Lett, in press.
[16] Dalgarno, A. 1981, Phil. Trans. R. Soc. Lond. A303, p. 513.
[17] Draine, B.T., Roberge, W.G. and Dalgarno, A. 1983, Ap. J. 265, p. 485.
[18] Lindinger, W. and Smith, D. 1983, "Reactions of Small Transient Species: Kinetics and Energetics" eds. A. Fontijn and M.A.A. Clyne, Academic Press, London, pp. 387-455.
[19] Smith, D., Adams, N.G. and Alge, E. 1984, Chem. Phys. Letts. 105, p. 317.
[20] Fehsenfeld, F.C., Lindinger, W., Schmeltekopf, A.L., Albritton, D.L. and Ferguson, E.E. 1975, J. Chem. Phys. 62, 2001.
[21] Smith, D. and Adams, N.G. 1981, Mon. Not. R. astr. Soc. 197, p. 377.
[22] Luine, J. and Dunn, G.H. 1981, 12th Int. Conf. on Physics of Electronic and Atomic Collisions Gatlinburg, Tennessee, July 1981.
[23] Rowe, B.R., Dupeyrat, G., Marquette, J.B., Smith, D., Adams, N.G. and Ferguson, E.E. 1984, J. Chem. Phys. 80, p. 241.
[24] Lindinger, W., Fehsenfeld, F.C., Schmeltekopf, A.L. and Ferguson, E.E. 1974, J. Geophys. Res. 79, p. 4753.
[25] Ferguson, E.E., Smith, D. and Adams, N.G. 1984, J. Chem. Phys., in press.
[26] Adams, N.G. and Smith, D. 1984, Int. J. Mass Spectrom. Ion Processes, in press.
[27] Smith, D., Adams, N.G. and Miller, T.M. 1978, J. Chem. Phys. 69, p. 308.
[28] Adams, N.G., Smith, D. and Millar, T.J. 1984, Mon. Not. R. astr. Soc., in press.
[29] Adams, N.G. and Smith, D. 1977, Chem. Phys. Letts. 47, p. 383.
[30] Smith, D., Adams, N.G. and Ferguson, E.E. 1984, Int. J. Mass Spectrom Ion Processes, in press.
[31] Bohme, D.K. 1975, "Interaction between Ions and Molecules" ed. P. Ausloos, Plenum, New York, pp. 489-504.
[32] Bohme, D.K., Mackay, G.I. and Schiff, H.I. 1980, J. Chem.

Phys. 73, p. 4976.
[33] Adams, N.G. and Smith, D. 1984, Chem. Phys. Letts. 105, p. 604.
[34] Elitzur, M. and Watson, W.D. 1978, Ap. J. (Letters) 222, L141.
[35] Stecher, T.P. and Williams, D.A. 1972, Ap. J. (Letters) 177, L141.
[36] Watson, W.D. 1980 "Interstellar Molecules" ed. B.H. Andrew, Reidel, Dordrecht, Holland, pp. 341-353.
[37] Smith, D. 1981, Phil. Trans. R. Soc. Lond. A303, p. 535.
[38] Smith, D. and Adams, N.G. 1984, "Ionic Processes in the Gas Phase" ed. M.A. Almoster-Ferreira, Reidel, New York, pp. 41-66.
[39] Henchman, M.J., Adams, N.G. and Smith, D. 1981, J. Chem. Phys. 75, p. 1201.
[40] Smith, D., Adams, N.G. and Alge, E. 1982, J. Chem. Phys. 77, p. 1261.
[41] Adams, N.G. and Smith, D. 1981, Ap. J. 248, p. 373.
[42] Smith, D., Adams, N.G. and Alge, E. 1982, Ap. J. 263, p. 123.
[43] Smith, D. and Adams, N.G. 1980, Ap. J. 242, p. 424.
[44] Dalgarno, A. 1975, "Interaction between Ions and Molecules" ed. P. Ausloos, Plenum, New York, pp. 341-352.
[45] Adams, N.G. and Smith, D. 1981, Ap. J. (Letters), 247, L123.
[46] Herbst, E. "Interstellar Molecules" ed. B.H. Andrew, Reidel, Dordrecht, Holland, pp. 317-321.
[47] Smith, D. and Adams, N.G. 1978, Ap. J. (Letters) 220, L87.
[48] Adams, N.G. and Smith, D. 1981, Chem. Phys. Letts. 79, p. 563.
[49] Böhringer, H., Arnold, F., Smith, D. and Adams, N.G. 1983, Int. J. Mass Spectrom. Ion Phys. 52, p. 25.
[50] Bates, D.R. 1979, J. Phys. B 12, p. 4135.
[51] Herbst, E. 1979, J. Chem. Phys. 70, p. 2201.
[52] Herbst, E. 1980. J. Chem. Phys. 72, p. 5284.
[53] Bates, D.R. 1980, J. Chem. Phys. 73, 1000.
[54] Barlow, S.E., Dunn, G.H. and Schauer, M. 1984, Phys. Rev. Letts. 52, p. 902.
[55] Smith, D. and Adams, N.G. 1984, unpublished VT - SIFT data.
[56] McEwan, M.J., Anicich, V.G., Huntress, W.T. Jr., Kemper, P.R. and Bowers, M.T. 1980, Chem. Phys. Letts. 75, p. 278.
[57] Bass, L., Kemper, P.R., Anicich, V.G. and Bowers, M.T. 1981, J. Amer. Chem. Soc. 103, p. 5283.
[58] Herbst, E., Adams, N.G. and Smith, D. 1983, Ap. J. 269, p. 329.
[59] Herbst, E., Adams, N.G. and Smith, D. 1984, Ap. J., in press.
[60] Smith, D., Adams, N.G., Alge, E. and Herbst, E. 1983, Ap. J. 272, p. 365.
[61] Bardsley, J.N. and Biondi, M.A. 1970, Adv. Atom. Mol. Phys.

6, pp. 1-57.
[62] Alge, E., Adams, N.G. and Smith, D. 1983, J. Phys. B, 16, p. 1433.
[63] Adams, N.G., Smith, D. and Alge, E. 1984, J. Chem. Phys., in press.
[64] Smith, D. and Adams, N.G. 1984, Ap. J. (Letters), in press.
[65] Michels, H.H. and Hobbs, R.H. 1983, Proc. 3rd Int. Symp. on the Production and Neutralization of Negative Ions and Beams, Brookhaven National Research Laboratory, New York.
[66] Hartquist, T.W., Oppenheimer, M. and Dalgarno, A. 1980, Ap. J. 236, p. 182.

ION MOLECULE REACTIONS IN INTERSTELLAR MOLECULAR CLOUDS

Eldon E. Ferguson

Aeronomy Laboratory, NOAA, Boulder, CO 80303 USA

INTRODUCTION

The most important recent development in laboratory astrophysical ion chemistry is the achievement of reaction temperatures below the previous 80 K (liquid nitrogen) limit. At JILA (1,2) several important reactions have been measured down to ∿ 10 K, including some critical radiative association reactions, using a low pressure ion trap. At Meudon (3) a new expanding jet apparatus is being utilized for measurements down to 20 K and at Heidelberg (4) measurements as low as 20 K have been carried out in a helium cooled drift tube. However, only a relatively small fraction of the important reactions relating to molecular production and loss in interstellar clouds have been measured or are likely to be measured at sufficiently low temperatures in the near future. Even fewer of these reactions are susceptible to quantal calculation. It therefore behooves us to develop as much understanding and predictive capability about ion-molecule reactions as we can from interpretations of available data at higher temperature with tractable theoretical models. Several theoreticians are engaged in that endeavor, motivated by their interest in interstellar molecule synthesis. What I will do

G. H. F. Diercksen et al. (eds.), Molecular Astrophysics, 471–490.

here is discuss some recent data and propose some rather empirical guidelines that may be useful, based on interpretation of experimental data in a very simple theoretical framework.

In very many cases, perhaps most of those of interest in interstellar molecular synthesis, ion-molecule reactions occur via an intermediate complex. This has been the explanation of the general increase in rate constant with decreasing temperature below 300 K observed for most slow, positive-ion reactions(5), the increase being attributed to the increase of the complex lifetime with decreased internal energy.

There are several ways to probe the properties of intermediate complexes experimentally. The most common is the measurement of three-body association rate constants,

$$A^+ + B + M \xrightarrow{k_3} AB^+ + M \qquad (1)$$

In the simple, conventional energy transfer mechanism, three-body association is a succession of two-body processes. First, intermediate complex formation with a rate constant k_c and unimolecular decomposition with a rate constant k_u (equal to the reciprocal complex lifetime τ),

$$A^+ + B \underset{k_u = \tau^{-1}}{\overset{k_c}{\rightleftarrows}} [AB^+]^* \qquad (2)$$

This is followed by complex stabilization,

$$[AB^+]^* + M \xrightarrow{k_s} AB^+ + M \qquad (3)$$

This leads to an overall three-body rate constant, $k^3 = k_c k_s \tau$.

The measurement of radiative association,

$$A^+ + B \xrightarrow{\gamma} AB^+ + h\nu \qquad (4)$$

is the process of interest astrophysically. In the usual model, $\gamma = k_c A \tau$, where A is the Einstein A coefficient. Since γ has only been measured in two astrophysically interesting cases, $C^+ + H_2 \rightarrow CH_2^+ + h\nu$ (2) and $CH_3^+ + H_2 \rightarrow CH_5^+ + h\nu$ (1), it is usually inferred from experimental values of k^3, i.e. $\gamma = k^3 A/k_s$. A is not known in any important astrophysical case and probably lies between 10 and 10^3 s^{-1} for most cases. Recently Botschwina (6) has calculated A's for the H stretching mode of the ions HN_2^+, HCS^+, $HOSi^+$, HOC^+ and HCO^+ for the lowest three $\Delta v = 1$ transitions. The values range from 203 s^{-1} for $HCS^+(1 \rightarrow 0)$ to 1352 s^{-1} for $HOC^+(1 \rightarrow 0)$. The $2 \rightarrow 1$ and $3 \rightarrow 2$ values are substantially larger and the radiative lifetimes of the highly vibrationally excited complexes may differ substantially from normal ground state molecules so that the limits on A are not well defined. The value of k_s is approximately equal to the collision rate constant $k_L = 2\pi e \sqrt{\alpha/\mu}$ for cases where M is a heavy third body and $\sim 1/4$ k_L for M = He. Thus, $\gamma \sim 10^{11} k^3$ is a conventional estimate for radiative association. Of course, k^3 is usually measured at 80 K or higher and γ is desired at 50 K or lower so that an extrapolation in temperature is required.

Theorists are attempting to determine the temperature dependence of the complex lifetime and the collisional stabilization efficiency as a function of temperature. If this is successful the laboratory three-body rate constants measured at 80 K and above can be used to estimate complex lifetimes and hence radiative association rate constants at the lower IMC temperatures. Bates (7) and Herbst (8) have developed statistical models which yield $\tau \sim T^{-\ell/2}$, where ℓ is the total number of rotational degrees of freedom of the reactants. Herbst (9) has deduced that the temperature dependence of the collisional stabilization step is $T^{-\delta}$, where $0 < \delta < 1$. Bates (10) has drawn on earlier work of Troe (11) to suggest values of δ also between 0 and 1, depending on the average energy transfer in a stabilizing collision. For large values of (ΔE), greater than a tenth of an eV or so, the so-called strong collision domain, the stabilization efficiency becomes unity and hence $\delta \to 0$. A comparison of the rather sparse data on the dependence of three-body rate constants on temperature suggests that the strong collision assumption is probably a reasonable assumption for heavy third bodies (i.e. with masses comparable to the reactants), while He (which is the most common third body at low temperatures) is less efficient, typically by a factor of three or four. There are clear cut cases where n exceeds $\ell/2 + 1$, i.e. where the statistical model as developed breaks down. This is in the case of weakly bound (e.g. non-polar electrostatic bonds) where the neutral and the ion are effectively free rotors. When the bonding is stronger, e.g. a chemical or polar electrostatic bond, vibrational excitation can become significant and this leads to still larger values of n. When the interaction is such as to lead to values of $k^3 \geq 10^{-27}$ cm^6 s^{-1} it is likely that vibrational excitation plays a role.

An important question is whether n is a constant with temperature, i.e. whether complex lifetimes do indeed follow an exponential law, $\tau \sim T^{-n}$ at all temperatures. Several maxima in three body rate constants have been reported. It appears that these are experimental artifacts and maxima are extremely unlikely from elementary chemical kinetic considerations (12). Bates (10) has shown a physical basis for curvature in $\ln k^3$ vs T plots in the specific case of O_4^+ formation. Rowe, et al. (13) have found the $N_2^+ + 2N_2 \to N_4^+ + N_2$ association to fit $k^3 \sim T^{-1.8}$ down to 20 K. It seems clear that deviations from the power law are going to be quite small generally.

The only quantum mechanically calculated three body association, for $H + H + M \to H_2 + M(14)$, predicted a maximum in k^3 around 100 K but subsequent experiments have not found deviations from a power law (15).

A new method of probing the properties of intermediate ion-molecule complexes has recently been developed, namely vibrational relaxation of molecular ions by neutrals (or of molecular neutrals by ions)(16). The vibrational relaxation rate constant for

$$AB^+(v) + X \xrightarrow{k_q} AB^+(v'<v) + X \tag{5}$$

is given by $k_q = k_c k_{vp} \tau$ (when $k_q < k_c$) where k_{vp} is the vibrational predissociation rate constant, i.e., the rate constant for intramolecular vibrational transfer from the strong ion bond to the weak electrostatic bond,

$$[AB^+(v)\cdot X]^* \xrightarrow{k_{vp}} AB^+(v'<v) + X \tag{6}$$

It has been found that for a range of X's, $O_2^+(v)$ and $NO^+(v)$ have values of $k_{vp} \sim 10^9 - 10^{10}$ s^{-1}. This is the rate for intramolecular energy transfer from the AB^+ bond to the weak electrostatic cluster bond. In most cases k^3 and k_q give the same information, namely an estimate of the complex lifetime, τ, since both k^3 and k_q are proportional to τ. However, in some cases k_q may be experimentally measureable when k^3 is not. In some cases k^3 and k_q may be determined by different effective lifetimes. We will discuss this for the case of O_2^+ complexes with H_2 where the effective lifetime for vibrational relaxation has been suggested to be greater than for three-body formation. This concept is of potential significance since it suggests that the effective lifetime for radiative association may differ from that for three-body association. This would lead to errors in deducing values of γ from values from k^3.

Since three-body rate constants $10^{-27}-10^{-28}$ cm^{6}s^{-1} imply complex lifetimes $\sim 10^{-9}-10^{-10}$ s, it is conceivable that the energy randomization efficiency will change as the rate constant varies through this range (due to temperature change) leading to a larger temperature dependence, i.e. $k^3 \sim T^{-n}$ where $n > \ell/2$. Bates (10) has suggested this as a possible explanation of $n = 2.9 > \ell/2 = 2$ for $O_2^+ + 2O_2 \to O_4^+ + O_2$. However, this cannot be a general solution for large observed n's. For example, the reaction $NO^+ + 2N_2 \to NO^+\cdot N_2 + N_2$ has $n = 4.3$ ($\gg \ell/2 = 2$) and in this case the electrostatic bond energy, 0.19 eV, is less than the vibrational energies of NO^+ or N_2, 0.23 eV, so that vibrational excitation is not energetically possible.

The study of reactions themselves may give information on the reaction mechanism. For example the occurrence of an endothermic binary reaction channel at high temperature (or slightly enhanced kinetic energy) implies a fast three-body reaction at low temperature, or a fast radiative association reaction at low temperature and pressure (17). We will illustrate this in some detail for several cases of hydrocarbon ion reactions with H_2.

METHANE ION CHEMISTRY

The most important results relating to methane chemistry are the JILA results

$$C^+ + H_2 \rightarrow CH_2^+ + h\nu + 4.15 \text{ eV} \tag{7}$$

and

$$CH_3^+ + H_2 \rightarrow CH_5^+ + h\nu + 1.95 \text{ eV} \tag{8}$$

for which the measured rate constants are $\gamma_7(10K) < 2(-15)$ cm^3 s^{-1} (2) and $\gamma_8(13K) = 1.8 \pm 0.3(-13)$ cm^3 s^{-1} (1). The JILA data are incorrectly normalized and should give $\gamma_8(13K) = 1.1(-13)$ cm^6 s^{-1}. An estimate of the value of $\gamma_8(13K)$ from the measured value of the three-body association rate constant, assuming $A = 10^2 s^{-1}$, leads to $\gamma_8(13K) \sim 4(-14)$ cm^{3}s^{-1}, a factor three below the experimental value. Reaction (8) will lead to CH_4 if followed by (9a),

$$CH_5^+ + e \rightarrow CH_4 + H + 7.9 \text{ eV} \tag{9a}$$

$$\rightarrow CH_3 + H_2 + 7.9 \text{ eV} \qquad (9b)$$

The recombination path (9a) rather than (9b) is by no means certain since CH_5^+ has a structure in which two H atoms are bonded differently than the other three and they might well split off together. The determination of the products of dissociative recombination is one of the pressing laboratory experimental fields needing serious attention.

The reaction

$$CH_5^+ + CO \rightarrow HCO^+ + CH_4 + 0.5 \text{ eV} \qquad (10)$$

$k_{10} \sim 1(-9) \text{ cm}^3\text{s}^{-1}$, also produces CH_4 from CH_5^+.

The very low rate constant for (7) makes it difficult to produce the CH_3^+ necessary for reaction (8). The alternative path that has been proposed is

$$H_3^+ + C \rightarrow CH^+ + H_2 + 2 \text{ eV} \qquad (11)$$

So far no C atom reactions have been measured in the laboratory. We are exploring the possibility of doing C atoms reactions in Boulder. Atom and unstable neutral radical reactions in general with ions pose a difficult experimental problem. Most of the measurements of such reactions to date have been carried out by Fehsenfeld and his colleagues in Boulder (18) and only at temperatures of 300 K or higher. The experimental arrangement required to carry out unstable radical reactions with ions are incompatible with the experimental arrangements required to operate at low temperatures. Recently however, Smith and Adams have measured H atom reactions with ions down to 90 K at Birmingham. Hopefully this problem will be solved in the future but in the meantime this situation strengthens our need to have as much understanding of reaction mechanisms as possible to facilitate the required extrapolations to IMC conditions.

If CH_4 exists in IMC's in high abundance, a reaction which becomes potentially significant is

$$O_2^+ + CH_4 \rightarrow CH_3O_2^+ + H + 4.8 \text{ eV} \tag{12}$$

which has been found (3) to be very fast at low temperatures, $k_{12} = 5(-10) (\frac{T}{20})^{-1.8} \text{cm}^3\text{s}^{-1}$ at temperatures below 160 K. The product ion has been established to be protonated formic acid (19, 20) by comparing the properties of this $CH_3O_2^+$ ion with the properties of known protonated formic acid ions. One of those properties is the fast isotope exchange of the two OH hydrogens with D_2O, but not with the CH hydrogen.

The products of dissociative recombination of protonated formic acid are formic acid, formaldehyde and formyl radicals, all observed in IMC's,

$$CH_3O_2^+ + e \rightarrow HCOOH + H + 5.9 \text{ eV} \tag{13a}$$

$$\rightarrow H_2CO + OH + 5.0 \tag{13b}$$

$$\rightarrow HCO + H_2O + 8.1 \tag{13c}$$

The ion $CH_3O_2^+$ produced by three-body (or radiative) association of CH_3^+ with O_2 is not protonated formic acid but a weakly bound (D <1.9 eV) cluster ion. This is because CH_3^+ is a singlet and O_2 a triplet while the ground state of protonated formic acid is a singlet.

Reaction (12) has been so extensively studied that a rather detailed picture of the reaction mechanism has been developed(21). A long-lived complex is formed by hydride abstraction

$$O_2^+ + CH_4 \rightleftarrows [CH_3^+ \ldots HO_2]^* \tag{14}$$

with a lifetime at 300 K $\sim 10^{-9}$ s, increasing to $\sim 10^{-7}$ s at 20 K. The hydride ion abstraction is 0.24 eV endothermic. This endothermicity is supplied on collision by the electrostatic attractive potential, ~ 0.6 eV. This "endothermic trapping" process leads to a long lifetime by absorbing enough of the relative kinetic energy to prevent immediate separation of the reactants. If this 0.24 eV endothermicity is supplied to the O_2^+ prior to collision the CH_3^+ + HO_2 products occur. This has been done both with O_2^+ kinetic energy and O_2^+ vibrational energy $(v \geq 2)$ (21). On the other hand if O_2^+ is attached to CH_4 without providing the 0.24 eV necessary for hydride ion abstraction, reaction between O_2^+ and CH_4 does not occur. This has been done by means of the following reaction,

$$O_2^+ \cdot O_2 + CH_4 \rightarrow O_2^+ \cdot CH_4 + O_2 \tag{15}$$

which is almost thermoneutral so that CH_4 is attached to O_2^+ without falling through the ~ 0.6 eV electrostatic attractive potential.

An additional related reaction has been measured at Birmingham,

$$CH_3^+ + O_2 \rightarrow CH_3O^+ + O + 1.35 \text{ eV} \tag{16}$$

The value of k_{16} is found to be extremely low, $k_{16} \sim 5(-14)$ cm^3 s^{-1}. The reaction listed by Anicich and Huntress(22),

$$CH_3^+ + O_2 \rightarrow HCO^+ + H_2O \tag{17}$$

was not observed. The reason that reaction (17) does not occur is found by noting that it violates spin conservation. The reason reaction (16) is so slow is also evident. While overall spin is conserved in reaction (16), there is a hidden spin problem. The ground state dissociation of $O_2(^3\Sigma)$ yields two $O(^3P)$ atoms. Adding an $O(^3P)$ atom to the singlet CH_3^+ does not lead to the ground state CH_3O^+ ion. The necessity to add an $O(^1D)$ atom to CH_3^+ to produce singlet CH_3O^+ can be expected to lead to a barrier to reaction.

The importance of CH_3O^+ production of course is its potential for producing formaldehyde by dissociative recombination,

$$CH_3O^+ + e \rightarrow CH_2O + H \qquad (18)$$

Proton transfer from CH_3O^+ to CO (or O_2, H_2) is quite endothermic.

Spin conservation deserves a brief discussion. Reactions do occur in which spin is not conserved, the most familiar one to atmospheric physicists being the rapid quenching of $O(^1D)$ in the atmosphere by N_2,

$$O(^1D) + N_2(^1\Sigma) \rightarrow O(^3P) + N_2(^1\Sigma) \qquad (19)$$

which occurs with a large rate constant. A familiar ion-molecule reaction in this class which controls ion chemistry in the atmospheres of Mars and Venus is

$$O^+(^4S) + CO_2(^1\Sigma) \rightarrow O_2^+(^2\pi) + CO(^1\Sigma) \qquad (20)$$

which occurs at the collision rate, $1(-9)$ cm^{3}s^{-1}(23).

Recently we have shown(24) that several charge-transfer reactions occur with rate constants exceeding the collision rate constant multiplied by the statistical weight fraction for spin conservation. The most striking example is the reaction

$$H_2O^+(^2B) + NO_2(^2A) \rightarrow H_2O(^1A) + NO_2^+(^1\Sigma) + 2.8 \text{ eV} \qquad (21)$$

for which $k = 1.2(-9)$ cm^3 s^{-1}, essentially the collision rate constant in spite of the fact that the only exothermic product channel is a singlet channel with only 1/4 of the total doublet plus doublet statistical weight. Presumably such spin non-conservation requires a singlet-triplet curve crossing. This is easily identified for reaction (19) but for reaction (21), and in general, one does not have a sufficient knowledge of the potential curves to make this identification.

An astrophysically interesting reaction for which some spin violation occurs is

$$CO^+(^2\Sigma) + H(^2S) \rightarrow H^+(^1S) + CO(^1\Sigma) \qquad (22)$$

for which k exceeds $1/4\ k_L(25)$, $k_{22} = 7.5(-10)\ cm^3s^{-1}$. The key factor in (22) is the deep potential well accessible to the $CO^+ + H$ in forming the stable HCO^+ which allows a possibility for singlet-triplet curve crossings. Reaction (22) may be of some significance for cometary ion chemistry.

ENDOTHERMIC TRAPPING MODEL

The increased complex lifetime and hence increased reactivity due to increased lifetime has been called "endothermic trapping" and has been used to explain a number of observations(17). Of most astrophysical interest are the three-body association reactions of hydrocarbon ions with H_2. These measurements were used in the conventional way to estimate radiative association rate coefficients. The problem related to a scheme for long chain hydrocarbon synthesis in clouds involving C^+ insertion followed by dissociative recombination to increase the carbon chain length by one. Both these steps remove hydrogen atoms so that rehydrogenation by radiative association is a critical process for the mechanism(26).

It turned out that the three-body association rates with H_2 are generally slow, below the detection limit of $10^{-30}\ cm^6s^{-1}$

with He third body at 80 K. This in fact was to be expected since it had previously been found that the reactions

$$O_2^+ + H_2 + He \rightarrow O_2^+ \cdot H_2 + He \qquad (23)$$

$k(80K) = 7.4(-31)$ cm^{6}s^{-1}(27) and

$$HCO^+ + H_2 + He \rightarrow HCO^+ \cdot H_2 + He \qquad (24)$$

$k(90K) = 8.3(-31)$ cm^{6}s^{-1}(28)

are slow. This is a consequence of the low polarizability of H_2 ($\alpha = 0.8$ Å^3) and the resulting low electrostatic interaction between ions and H_2. The ions $C_2H_3^+$, $C_2H_4^+$, $C_2H_5^+$, $C_3H_2^+$, $C_3H_3^+$, $C_3H_4^+$, $C_3H_5^+$ and $C_3H_7^+$ all have values of $k^3 < 10^{-30}$ cm^{6}s^{-1}(26). On the other hand, the reaction

$$C_2H_2^+ + H_2 + He \rightarrow C_2H_4^+ + He \qquad (25)$$

has $k(80K) = 7(-27)$ cm^6 s^{-1} and the reaction

$$C_3H^+ + H_2 + He \rightarrow C_3H_3^+ + He \qquad (26)$$

has $k(300K) > 2.6(-27)$ cm^{6}s^{-1}. In these cases the endothermic trapping mechanism is operative,

$$C_2H_2^+ + H_2 \underset{k_u = \tau^{-1}}{\overset{k_c}{\rightleftarrows}} [C_2H_3^+\text{---}H]^*, \quad \Delta E = -0.07 \text{ eV} \qquad (27)$$

and $C_3H^+ + H_2 \underset{k_u = \tau^{-1}}{\overset{k_c}{\rightleftarrows}} [C_3H_2^+\text{---}H]^*$, $\Delta E = -0.04$ eV (28)

The endothermicities are less than the estimated electrostatic energy, ~ 0.15 eV, and greater than kT so that the reactants are effectively trapped. In addition to the enhanced lifetime due to trapping, the endothermic binary reaction separates the H atoms so that the chemical well corresponding to the overall reaction exothermicity is accessible. For example, in reaction (27) it is necessary that the two hydrogen atoms go on different carbon atoms in the acetylene ion.

Therefore the possibility of a fast three-body (and therefore radiative association) reaction with H_2 can be estimated from the thermochemistry of the intermediate complex ions and an estimate of the electrostatic interaction potential, which depends on the neutral polarizability and dipole moment. There may also be a "chemical" attractive well in addition to the "electrostatic" attractive well to overcome an endothermic trapping threshold. However, this will only be known generally in the simplest cases, and of course where a reaction potential surface is known one would not have recourse to such naive models as this.

Using the formula $\gamma = \dfrac{k^3 A}{k_s}$ to estimate radiative association rate constants, one estimates for

$$C_2H_2^+ + H_2 \rightarrow C_2H_4^+ + h\nu \qquad (29)$$

a value $\gamma(80K) \cong 5(-15)$ cm^{3}s^{-1}, estimating $A = 10^2$ s^{-1} and the collisional stabilization efficiency of He to be 0.25. Similarly we estimate

$$C_3H^+ + H_2 \rightarrow C_3H_3^+ + h\nu \qquad (30)$$

to have $\gamma(300K) > 2(-15)$ cm^3 s^{-1}.

If there is an exothermic binary channel then generally one should not consider the possibility of a radiative association process in interstellar clouds, according to an argument made by Bates(29). This is certainly generally true but there are undoubtedly exceptions in which exothermic binary reactions do not occur on a radiative lifetime scale, e.g.

$$O_2^+ + 2N_2 \rightarrow O_2^+ \cdot N_2 + N_2, \quad k = 1.0(-31)(\frac{300}{T})^{3.2} \quad (31)$$

produces a very stable ion, $O_2^+ \cdot N_2$, in spite of the fact that the reaction

$$O_2^+ + N_2 \rightarrow NO^+ + NO \quad (32)$$

is exothermic.

Bates(30) has given an empirical formula for γ at 30 K,

$$\gamma(30K) = 10^{-21} \frac{A[6\Delta E(eV) + n-2]^{3n-7}}{(3n-7)!} \quad (33)$$

where ΔE is the overall reaction exothermicity and n is the total number of reactant atoms. For the above reactions this formula gives $\gamma_{29}(30) = 4(-13)$ cm^{3}s^{-1} and $\gamma_{30}(30) = 3(-10)$ cm^{3}s^{-1}, assuming $A = 10^2$s^{-1}.

In the first case this corresponds to n = 4.5 for the temperature dependence in order to force agreement. This is clearly too large. Using the Bates and Herbst values of n = $\ell/2$ = 2 would give $\gamma_{29}(30) = 4(-14)$ cm^3 s^{-1} from the measured k^3, about an order of magnitude below the Bates formula.

The real problem comes for those cases where k^3 is very small. k$^3 < 10^{-30}$ cm^6 s^{-1} implies $\gamma < 10^{-18}$ cm^3 s^{-1}. For radiative association of $C_2H_3^+$ with H_2, $\gamma(30) = 2(-13)$ cm^{3}s^{-1}, for $C_3H_2^+$ with

H_2, $\gamma(30) = 5(-12) \text{cm}^3\text{s}^{-1}$ and for $C_3H_3^+ + H_2$, $\gamma(30) = 2(-14) \text{cm}^3\text{s}^{-1}$ from equation (33). These values are grossly too large. The difficulty is that the expression for $\gamma(30)$, Eq. (33), involves only the overall reaction exothermicity and implicitly assumes that this potential well is accessible, which in general it is not for the hydrocarbon ion reactions with H_2.

Another example of the "endothermic trapping" model is the relatively fast three-body association of Si^+ ions with O_2,

$$Si^+ + O_2 + He \rightarrow SiO_2^+ + He \qquad (34)$$

for which $k = 1(-29) \text{cm}^6\text{s}^{-1}$ at 300 K(31). This is much faster than similar ion associations with O_2, e.g. Fe^+, Mg^+, Na^+, Ca^+, Li^+, or K^+. The difference is that for (34) the slightly endothermic trapping mechanism

$$Si^+ + O_2 \rightleftarrows [S_iO^+ \text{---} O]^* \qquad (35)$$

occurs(24).

The value of $\gamma(30K)$ deduced with Eq. (33) for $Si^+ + O_2 \rightarrow SiO_2^+ + h\nu$ is $4(-17) \text{cm}^3\text{s}^{-1}$, assuming $A = 10^2 \text{s}^{-1}$. The value of $\gamma(300K)$ deduced from k^3 is $\sim 1(-17) \text{cm}^3\text{s}^{-1}$, which would extrapolate to $\gamma(30K) \sim 10^{-16} \text{cm}^3\text{s}^{-1}$ using $\tau \sim T^{-1} (n=\ell/2)$.

Another application of the "endothermic trapping" model is the so-called "isotope refrigerator" effect of Smith et al. (32), in which an isotopic selection effect on hydrocarbon ion association with H_2 occurs. There is a systematic effect in a sequence of six reactions which can be illustrated by two members of the sequence,

$$CD_3^+ + H_2 + He \rightarrow CD_3H_2^+ + He, \quad k(80K) = 8.4(-27) \text{cm}^3\text{s}^{-1} \qquad (36)$$

$$CH_3^+ + H_2 + He \rightarrow CH_5^+ + He, \quad k(80K) = 1.5(-27) \text{cm}^3\text{s}^{-1} \qquad (37)$$

The greater efficiency of (36) is attributed to the endothermic trapping,

$$CD_3^+ + H_2 \rightleftarrows CD_2H^+ + HD \qquad (38)$$

which is 0.7 eV endothermic due to zero point vibrational energy differences of H and D bonds.

Another mechanism for energy "absorption" is rotational excitation of either of the reactant species, converting relative kinetic energy into rotational energy of one or both reactants, which may appear as a libration if the complex is sufficiently tight although this is not generally expected. This of course is not a new idea, it is explicit in the statistical models of Bates (7) and Herbst(8) for the complex lifetime that both ion and neutral reactants are rotationally excited. It is particularly easy to visualize the rotational excitation of the neutral by the ion, since the large anisotropy of the polarizability ($\alpha_{11}/\alpha_\perp = 1.5$ for H_2) as well as the neutral quadrupole moment lead to an anisotropic potential so that the ion exerts a torque on the neutral via the long range interaction. In the case of H_2, with its large rotational constant, a rotational excitation ($\Delta J = 2$) from $J = 0 \rightarrow J = 2$ involves an energy of 1 kcal mol^{-1} and from $J = 1$ to $J = 3$ an energy of 1.7 kcal mol^{-1}, which is sufficient to cause trapping, i.e. exceeds kT at 300K (0.6 kcal mol^{-1}) and below but is less than the electrostatic energy $\sim$ 4 kcal mol^{-1}. A substantial H_2/D_2 isotope effect of a factor of 4 in $O_2^+(v=1)$ vibrational quenching (33), H_2 being more efficient, is consistent with this model since the H_2 rotational constant is twice that of D_2, yielding a larger energy transfer for a given ΔJ transition. Rotational capture alone cannot lead to an enhanced three-body association rate constant since a true potential well must exist for three-body stabilization to occur. We have attributed an anomalous vibrational relaxation rate constant for $O_2^+(v)$ relaxation by H_2 to this effect

(16).

Recently Schelling and Castleman(34) have carried out ion-neutral trajectory calculations which show that the anisotropy of the neutral polarizability does lead to enhanced complex lifetimes. The complex lifetimes are not randomly distributed and the addition of anisotropy to the interaction potential leads to a small fraction of greatly enhanced lifetimes, enhanced well beyond the computed mean lifetimes.

RECENT RESULTS

Some recently measured binary reactions of astrophysical interest are the following. Federer et al.(25) have measured reactions of several ions at 300 K with atomic H. One of these, the thermo-neutral charge-transfer

$$O^+(^4S) + H(^2S) \rightarrow H^+(^1S) + O(^3P) \qquad (39)$$

has a measured rate constant $k_{39} = 6.0(-10)$ cm^{3}s^{-1} over the experimental range of mean average kinetic energies from 0.05 to 0.15 eV. This measurement confirms the theoretical value of k_{39} as well as the value of k_{39} deduced from measurements of the reverse reaction. The measured reverse reaction yields a value of $k_{39} = 6.8(-10)$ cm^{3}s^{-1} ± 50% by detailed balance (35). Theoretical quantal calculations (36) yield $k_{39} = 5.2(-10)$ cm^{3}s^{-1} ± 15% at 300 K, increasing to 7(-10) at 1000 K. The calculations show that the charge-transfer is due to strong radial coupling between the $B^3\Sigma^-$ O^+ + H and the $X^3\Sigma^-$ O + H$^+$ potential curves at large internuclear distances, 8-12 a_o. The calculated value of k_{39} is then close to the spin weighted fraction (3/8) of the Langevin collision rate constant which yields 7.5(-10) cm^{3}s^{-1}.

Another recent measurement (25) is

$$CH^+ + H \rightarrow C^+ + H_2 \tag{40}$$

The value of $k_{40} = 8(-10)$ cm^{3}s^{-1} at 0.06 eV average CM K.E., decreasing to about $5(-10)$ cm^{3}s^{-1} at 0.1 eV. The energy dependence suggests that (40) will be near Langevin, $k_L = 2(-9)$ cm^{3}s^{-1} at low temperatures but substantially less at the higher temperatures in shocked regions. The theoretical value of Solomon and Klemperer (37) which has been in use, $k_{40} = 9.4(-12)(\frac{T}{300})^{1.25}$, has an incorrect sign of the temperature dependence, leading to values far too low at interstellar molecular cloud temperatures. Recently Chesnavich, et al. (38) have deduced theoretical values for k_{40} from phase space calculations that agree quite well with the measured values, both in magnitude and temperature dependence.

The cross-section for the endothermic reverse reaction,

$$C^+ + H_2 \rightarrow CH^+ + H \tag{41}$$

has been measured (39) as has the rate constant (40,41). This has also been calculated (38). All results appear to be in reasonable harmony although a detailed comparative analysis remains to be done. Reaction (41) is presumably responsible for CH^+ production in shocked regions of the interstellar medium (42).

REFERENCES

1. Barlow, S.E., Dunn, G.H. and Schauer, M. 1984, Phys. Rev. Letters 52, p. 902.
2. Luine, J. and Dunn, G.H. 1984, Phys. Rev. Letters (in press).
3. Rowe, B.R., Dupeyrat, G., Marquette, J.B., Smith, D., Adams N.G. and Ferguson, E.E. 1984, J. Chem. Phys. 80, p. 241.
4. Böhringer, H. and Arnold, F. 1982, J. Chem. Phys. 77, p. 5534.
5. Ferguson, E.E., Bohme, D.K., Fehsenfeld, F.C. and Dunkin, D.B. 1969, J. Chem. Phys. 50, p. 5039.
6. Botschwina, P. 1984, Chem. Phys. Lett. 107, p. 535; J. Mol. Spectrosc. (in press).
7. Bates, D.R. 1979 J. Phys. B 12, p. 4135; 1979 J. Chem. Phys. 71, p. 2318.
8. Herbst, E. 1979, J. Chem. Phys. 70, p. 2201; 1980, 72, p. 5284; 1981, 75, p. 4413.
9. Herbst, E. 1982, Chem. Phys. 68, p. 323.
10. Bates, D.R. 1984, J. Chem. Phys. 81, p. 298.
11. Troe, J. 1977, J. Chem. Phys. 66, p. 4745, p. 4758; Heyman, Hipler, and Troe, 1984, ibid 80, p. 1853.
12. Ferguson, E.E. 1983, Chem. Phys. Lett. 101, p. 141.
13. Rowe, B., Dupeyrat, G., Marquette, J.B., and Gaucherel, P. 1984, J. Chem. Phys. 80, p. 4915.
14. Roberts, R.E., Bernstein, R.B. and Curtiss, C.F. 1968, Chem. Phys. Lett. 2, p. 366; 1969 J. Chem. Phys. 50, p. 5163.
15. Ham, D.O., Trainer, D.W. and Kaufman, F. 1970, J. Chem. Phys. 53, p. 4395.
16. Ferguson, E.E. 1984, Vibrational Excitation and De-excitation and Charge-Transfer of Molecular Ions in Drift Tubes, Swarms of Ions and Electrons in Gases, Lindinger, Howorka, and Märk, Eds. Springer-Verlag, Vienna.
17. Ferguson, E.E., Smith, D. and Adams, N.G. 1984, J. Chem. Phys. 81, p. 742.
18. Fehsenfeld, F.C. 1976, Ap. J. 209, p. 638; Viggiano, A.A., Howorka, F., Albritton, D.L., Fehsenfeld, F.C., Adams, N.G. and Smith, D. 1980, Ap. J. 236, p. 492.
19. Villinger, H., Richter, R. and Lindinger, W. 1983, Int. J. Mass Spectrom. Ion Phys. 51, p. 25.
20. Villinger, H., Saxer, A., Richter, R. and Lindinger, W. 1983, Chem. Phys. Lett. 96, p. 513.
21. Durup-Ferguson, M., Böhringer, H., Fahey, D.W., Fehsenfeld, F.C. and Ferguson, E.E. 1984 J. Chem. Phys. (in press - Sept. 1).
22. Anicich, V.G. and Huntress, W.T. 1984, Jan. 12, A Survey of Bimolecular Ion-Molecule Reactions, J.P.L.
23. Fehsenfeld, F.C., Schmeltekopf, A.L. and Ferguson, E.E. 1966, J. Chem. Phys. 44, p. 3022.
24. Ferguson, E.E. 1983, Chem. Phys. Lett. 99, p. 89.

25. Federer, W., Villinger, H., Howorka, F., Lindinger, W., Tosi, P., Bassi, D., and Ferguson, E.E. 1984, Phys. Rev. Lett. 52, p. 2084.
26. Herbst, E., Adams, N.G. and Smith, D. 1983, Ap. J. 209, p. 329.
27. Adams, N.G., Bohme, D.K., Dunkin, D.B., Fehsenfeld, F.C. and Ferguson, E.E. 1970, J. Chem. Phys. 52, p. 3133.
28. Fehsenfeld, F.C., Dunkin, D.B. and Ferguson, E.E. 1974, Ap. J. 188, p. 43.
29. Bates, D.R. 1983, Ap. J. 267, p. L121.
30. Bates, D.R. 1983, Ap. J. 269, p. 329.
31. Fahey, D.W., Fehsenfeld, F.C., Ferguson, E.E. and Viehland, L.A. 1981, J. Chem. Phys. 75, p. 609.
32. Smith, D., Adams, N.G. and Alge, E. 1982, J. Chem. Phys. 77, p. 1261.
33. Böhringer, H., Durup-Ferguson, M., Fahey, D.W., Fehsenfeld, F.C. and Ferguson, E.E. 1983, J. Chem. Phys. 79, p. 4201.
34. Schelling, F.J. and Castleman, A.W. 1984, Chem. Phys. Lett., (in press).
35. Fehsenfeld, F.C. and Ferguson, E.E. 1972, J. Chem. Phys. 56, p. 3066.
36. Chambaud, G., Launay, J.M., Levy, B., Millie, P., Roueff, E. and Tran Minh, F.; 1980, J. Phys. B. 13, p. 4205.
37. Soloman, P.M. and Klemperer, W. 1982, Ap. J. 178, p. 389.
38. Chesnavich, W.J., Akin, V.E. and Webb, D.A. 1984, Ap. J. (submitted).
39. Erwin, K.M. and Armentrout, P.B. 1984, J. Chem. Phys. 80, p. 2978.
40. Lindinger, W. 1984, private communication.
41. Smith, D. and Adams, N.G. 1984, private communication.
42. Elitzur, M. and Watson, W.D. 1978, Ap. J. 222, p. L141.

THEORETICAL PREDICTIONS OF THE INFRARED AND MICROWAVE SPECTRA OF SMALL MOLECULES

P. R. Bunker

Herzberg Institute of Astrophysics, National Research Council of Canada, Ottawa, Ontario, Canada K1A OR6.

The identification of molecules in astrophysical sources depends on their spectra having been obtained in the laboratory, but laboratory spectra are not always available. For ions and free radicals this is because of their reactivity, which makes it difficult to produce them in concentrations large enough for easy laboratory detection. The successful search for such spectra in the laboratory is greatly assisted by theoretical predictions. In this talk the accuracy and utility of such predictions of microwave and infrared spectra will be discussed with emphasis on quasilinear molecules, such as CH_2 and C_3, and quasiplanar molecules, such as H_3O^+ and SiH_3.

1. INTRODUCTION

The molecular constituents of interstellar clouds, of comets, and of stellar and planetary atmospheres are identified by their spectra. Such identification depends on appropriate laboratory spectra being available for comparison. However, some of the molecules of interest are ions or free radicals which, because of their high reactivity, are hard to produce in the laboratory in concentrations large enough for easy detection, and for which, therefore, laboratory spectra are not always available. This is particularly a problem for microwave and infrared spectroscopy in which a high resolution search for an unknown ion or free radical at high sensitivity can be agonizingly laborious. In these circumstances an accurate theoretical prediction of the spectrum (including its intensity) can save a lot of laboratory search time, and can also stimulate the laboratory search for an otherwise unknown spectrum.

G. H. F. Diercksen et al. (eds.), Molecular Astrophysics, 491–496.

In this talk I will discuss what is involved in making such predictions for infrared and microwave spectra, and discuss some of the results obtained. I will particularly concentrate on the special problems that arise in predicting the microwave and infrared spectra of quasilinear molecules, such as CH_2 and C_3, and quasiplanar molecules, such as H_3O^+ and SiH_3.

2. THEORETICAL CALCULATIONS

To calculate the appearance of the microwave or infrared spectrum of a molecule it is necessary to calculate the quantum mechanically allowed energy levels of the molecule as it rotates and vibrates, and to calculate the dipole moment of the molecule and the change in dipole moment with vibrational distortion. The first step in such a calculation is the ab initio calculation of the potential energy surface of the molecule in its ground electronic state. The second step is the calculation of the rotation-vibration energies occurring within the potential, and the final step is the determination of the intensities of the allowed transitions.

2.1. The Ab Initio Calculation

This part of the work involves using a computer programme to calculate the potential energy holding the molecule together; such a calculation must be done at many different nuclear geometries. Each computation requires the nuclear geometry, the nuclear charges, the number of electrons and an appropriate electronic wavefunction basis set as input. The choice of basis set and the choice of technique used to allow for the electron correlation are the places where the various methods can differ and where user skill is required.

For a triatomic molecule or a four-atomic molecule calculations at as many as 50 to 100 different nuclear geometries may be required to give a good coverage of the potential surface, with energies up to about 10000 cm^{-1} above equilibrium. The accuracy of a given method deteriorates with increasing number of electrons in the molecule, but calculations at spectroscopic accuracy are possible for molecules such as H_3O^+ (1-3), HCO^+ (4), and C_3 (5).

2.2. The Rotation-Vibration Problem

Having calculated the ab initio internuclear potential function the next step is to calculate the rotation and vibration energy levels of the molecule and of any of its isotopes that are of interest. There are several different ways of doing this depending on the nature of the molecule

under investigation and on which energy levels are required.
All methods require that the potential be interpolated between
the points obtained in the ab initio calculation. This is
usually accomplished by using an analytic function for the
potential and adjusting the parameters in it in order to fit
the ab initio points. Given the interpolated internuclear
potential function there are three commonly used techniques for
solving the rotation-vibration problem and these involve using
either perturbation theory, variation theory, or numerical
integration.

The perturbation theory technique for solving the rotation-
vibration problem is explained in the book by Papousek and
Aliev (6), and in the paper by Hoy, Mills and Strey (7).
Fundamental to the applicability of this technique is that the
potential surface have a deep minimum that is well approximated
by a quadratic (i.e. harmonic) function in all the vibrational
displacement coordinates. In this approach harmonic
oscillator vibrational wavefunctions, and rigid molecule
rotational wavefunctions, are used as a starting approximation.
The effects of anharmonicity in the potential (i.e. non-
quadratic terms), the effects of centrifugal distortion, and
the effects of Coriolis forces on the vibrations as the
molecule rotates, are all treated by perturbation theory. These
effects must be small for this approach to work. Using this
technique the rotation-vibration energy levels (neglecting
degeneracies) become expanded as a power series in the rotational
and vibrational quantum numbers with coefficients (called
rotation-vibration constants) that depend on the nuclear
masses, the equilibrium geometry, and the parameters of the
potential function. Using a computer programme developed
from the work of ref. (7) the rotation-vibration constants are
calculated from the ab initio potential, the equilibrium
geometry, and the nuclear masses. The most important constants
are the harmonic vibrational frequencies ω_i, which largely
govern the vibrational energies, and the rotational constants
A_e, B_e and C_e which largely govern the rotational energies.

An alternative and related technique involves again using
harmonic oscillator and rigid rotor wavefunctions as starting
functions, but this time the variational method is used to
solve the rotation-vibration wave equation using the ab initio
potential surface (8,9).

The above two methods require some care when calculating
high rotational and/or vibrational energies, and for situations
with wide rather flat potential minima (i.e. large amplitude
vibrations). For molecules with multiple potential minima, and
significant vibrational tunneling between these minima, these
methods are not usually appropriate.

Numerical methods have been developed for integrating the rotation-vibration wave equation for the special cases of quasilinear and quasiplanar molecules (2, 10-13). These are molecules for which the harmonic oscillator and rigid rotor do not provide a good starting approximation. Good examples of quasilinear molecules are provided by CH_2, C_3 and HCNO in their ground electronic states, and NH_3 and H_3O^+ are good examples of quasiplanar molecules in their ground electronic states. Work in this area is reviewed in ref. (14). For quasilinear molecules the nonrigid bender Hamiltonian (10-12) is used to calculate the rotation-vibration energy levels, and for a quasiplanar molecule the nonrigid invertor Hamiltonian (2,13) is used.

2.3. The Dipole Moment Calculation

The intensity of a rotation-vibration transition can be calculated from the rotation-vibration wavefunctions, obtained as discussed above, if we know the dipole moment function. The dipole moment function is the expression for the dipole moment of the molecule as a function of the nuclear geometry. Thus, at the ab initio stage of the calculation we calculate the dipole moment of the molecule at its equilibrium configuration and for several appropriately distorted configurations. It is very important before beginning the search for a spectrum to have an estimate of the intensities. For example theoretical calculations (15) predict the ν_1 and ν_3 stretching fundamentals of CH_2 in its ground electronic state have vanishingly small intensities, whereas the ν_3 stretching fundamental of C_3 is predicted to be very intense (5). None of these bands has been seen in the gas phase.

3. APPLICATIONS

The successful search in the laboratory for the microwave spectrum of HOC^+ (16) was greatly helped by having the ab initio plus perturbation theory prediction (4). The successful laboratory search for the infrared spectrum of H_3^+ (17) and D_3^+ (18) depended crucially on the ab initio plus variational predictions of theory (9). The successful search for the infrared and microwave spectrum of CH_2 (19) depended on ab initio (20) plus nonrigid bender predictions, and finally the successful search for the infrared spectrum of H_3O^+ (21,22) depended on the predictions of ab initio plus nonrigid invertor calculations (1,2). Of these four molecular species none has yet been reliably detected in an astrophysical source; but now it is known exactly where to look.

4. CAPABILITIES AND LIMITATIONS OF THIS WORK

For the low lying rotation-vibration energy levels of three- and four-atomic molecules the rotation-vibration part of the calculation, from a known potential curve, proceeds with little difficulty in all but a few special cases (when the molecule is very flexible). Also given observed rotation-vibration energies covering a reasonable energy range it is possible to determine the shape of the potential curve and thus to calculate unobserved rotation-vibration energies (23-25). The limitation in this work is with the ab initio calculation of the potential surface. Modern large scale ab initio calculations lead to rotation-vibration energies that have an accuracy of better than 5%, and for small molecules involving only first row atoms accuracies of better than 1% can be achieved.

The accuracy of these predictions is not high enough to be useful for unambiguously assigning an interstellar spectral feature. However these predictions have proved themselves to be enormously helpful in improving the efficiency of laboratory searches for microwave and infrared spectra. This is clearly the aim of this type of work in the immediate future.

REFERENCES

1. Colvin, M.E., Raine, G.P., Schaefer, H.F., and Dupuis, M. 1983, J. Chem. Phys. 79, pp. 1551-1552.

2. Bunker, P.R., Kraemer, W.P. and Spirko, V. 1983, J. Mol. Spectrosc. 101, pp. 180-185.

3. Botschwina, P., Rosmus, P. and Reinsch, E.-A. 1983, Chem. Phys. Letters 102, pp. 299-306.

4. Kraemer, W.P. and Diercksen, G.H.F. 1976, Astrophys. J. 205, pp. L97-L100.

5. Kraemer, W.P., Bunker, P.R. and Yoshimine, M. 1984, J. Mol. Spectrosc. in press.

6. Papousek, D. and Aliev, M.R. 1982, "Molecular Vibrational-Rotational Spectra", Academia, Prague.

7. Hoy, A.R., Mills, I.M. and Strey, G. 1972, Mol. Phys. 24, pp. 1265-1290.

8. Whitehead, R.J. and Handy, N.C. 1975, J. Mol. Spectrosc. 55, pp. 356-373.

9. Carney, G.D. and Porter, R.N. 1976, J. Chem. Phys. 65, pp. 3547-3565.

10. Hougen, J.T., Bunker, P.R. and Johns, J.W.C. 1970, J. Mol. Spectrosc. 34, pp. 136-172.

11. Hoy, A.R. and Bunker, P.R. 1974, J. Mol. Spectrosc. 52, pp. 439-456.

12. Jensen, P., 1983, Comp. Phys. Reports 1, pp. 1-55.

13. Spirko, V. 1983, J. Mol. Spectrosc. 101, pp. 30-47.

14. Bunker, P.R., 1983, Ann. Rev. Phys. Chem. 34, pp. 59-75.

15. Bunker, P.R. and Langhoff, S.R. 1983, J. Mol. Spectrosc. 102, pp. 204-211.

16. Gudeman, C.S. and Woods, R.C. 1982, Phys. Rev. Lett. 48, pp. 1344-1348 and p. 1768.

17. Oka, T. 1980, Phys. Rev. Lett. 45, pp. 531-534.

18. Shy, J.T., Farley, J.W., Lamb, W.E. Jr. and Wing, W.H. 1980, Phys. Rev. Lett. 45, pp. 535-537.

19. Sears, T.J., Bunker, P.R. and McKellar, A.R.W. 1981, J. Chem. Phys. 75, pp. 4731-4732.

20. Harding, L.B. and Goddard, W.A. 1978, Chem. Phys. Lett. 55, pp. 217-220.

21. Begemann, M.H., Gudeman, C.S., Pfaff, J. and Saykally, R.J. 1983, Phys. Rev. Lett. 51, pp. 554-557.

22. Haese, N.N. and Oka, T. 1984, J. Chem. Phys., 80, pp. 572-573.

23. Bunker, P.R., Sears, T.J., McKellar, A.R.W., Evenson, K.M. and Lovas, F.J. 1983, J. Chem. Phys. 79, pp. 1211-1219.

24. Bunker, P.R. and Jensen, P., 1983, J. Chem. Phys. 79, pp. 1224-1228.

25. Bunker, P.R., Amano, T. and Spirko, V. 1984, J. Mol. Spectrosc. in press.

Recent Theoretical Results of Rotational-Translational Energy Transfer

Joachim Schaefer

Max-Planck-Institut für Physik und Astrophysik
Institut für Astrophysik, Karl-Schwarzschild-Str. 1
8046 Garching, FRG

Abstract

Theoretical work is reviewed on scattering calculations and tests of some ab initio interaction potentials representing simple systems by subsequent calculations of differential, integral, and effective cross sections. The interaction potentials of $He-H_2$ published by Meyer, Hariharan, and Kutzelnigg have been composed in one vibrotor potential grid which has been used to calculate effective cross sections describing relaxation phenomena. Good agreement with recently measured data has been achieved. Deviations from results previously derived from the well-known Shafer-Gordon potential fit are discussed. The present quality of the H_2-H_2 interaction potential is shown: Some comparisons with experiments give evidence for sufficient accuracy of the rotationally inelastic cross sections of H_2-H_2 and isotopic systems, in the low temperature regime. The remaining disagreement with measurements of the undulatory structure of the total differential cross sections and of low temperature second virial coefficients has been removed by determining a distance dependent scaling function applicable to the full vibrotor potential, under the condition that the measured second virial coefficients can be obtained quantitatively, and the relative anisotropies of the system are conserved. Dimer bound states and pair correlation functions have been calculated in this fit procedure. A first step of improvement is reported for an ab initio potential of He-CO published by Thomas, Kraemer, and Diercksen. Measurements of some time of flight profiles and one total differential cross section have been reproduced theoretically after applying scaling procedures to the potential.

G. H. F. Diercksen et al. (eds.), Molecular Astrophysics, 497–516.

This paper presents a review of recent theoretical work on rotational-translational energy transfer of relatively simple systems, H_2-H_2 and H_2-He, which are most abundant in cold interstellar and gaseous planetary matter. The two interaction potentials governing the dynamics of these systems and of their isotopic species have been basically determined in ab initio calculations of W. Meyer and extended by including ab initio calculations of several other authors.

The weak anisotropy of both potentials and the rotational level spacing of H_2, HD, and D_2 allowed to obtain the scattering matrices accurately which means: the finite basis expansions retained in the close coupled representations of the scattering problems and the numerical errors due to the computational procedures gave generally theoretical errors of only a few percent for the scattering matrices. This has been tested by extending the basis sets and by increasing the accuracy of the scattering program codes. As far as the experimental error bars are of the order of 10% or less, there is a practical one-to-one relation between the measured quantities and the interaction potentials used for reproducing them theoretically. The measured quantitites can be differential cross sections (time of flight measurements) or any effective cross section, "effective" meaning energy averaged and temperature dependent. I thought it should be interesting enough, especially under the general aimes of this workshop, to show how well these ab initio potentials can do.

The first set of examples to be discussed probes the interaction potential of H_2-He. This potential has been composed from two ab initio potentials published by Meyer, Hariharan, and Kutzelnigg[1] in 1980. It is represented by a 3-term Legendre expansion and, most important for the Raman line shape cross sections, the representation of the potential in the intramolecular distance ranges from 0.9 to 2.0 a.u. which properly covers the range of the radial rotational eigenfunctions of the H_2 molecule involved in the dynamics of the system up to about 500 K. We can therefore assume that the contrifugal stretching effects of the H_2 molecule are properly handled.

This potential has been used recently, transformed for the HD-He system, in a very successful application[2] of the Waldmann-Snider kinetic theory describing magnetic field effects in transport, relaxation, and reorientation phenomena. It has been shown apparently in this paper that at least the spherical part of the H_2-He potential is very reliable because quantitative agreement could be achieved with all measurements available.

The critical test of the anisotropy of this potential must be done in comparisons with measurements that are especially sensitive to variations of the anisotropy. Very recently new experimental data became available which exactly fulfill this requirement. In particular, the pressure broadening of depolarized Rayleigh and rotational Raman light scattering has been thoroughly studied for a large temperature range by P. Hermans in his thesis work[3], and also a significantly improved measurement of NMR relaxation times in ortho-H_2-He by Lemaire and Armstrong[4] came to our knowledge most recently.

I will show now several examples of the theoretical work[5] which I have done together with W. Köhler (Erlangen). It will be an attractive aspect of this presentation to include the corresponding results derived from the well-known Shafer-Gordon potential[6] which for more than ten years has been considered the most reliable potential of this system. The same formalism as described by Shafer and Gordon has been used for determining the relevant cross sections.

1) The rotational relaxation cross sections of para-hydrogen and ortho-hydrogen infinitely dilute in a He bath have been calculated from scattering matrix elements obtained for a set of 85 energy points, the lower half of which is important for very low temperatures up to about 100 K, while the upper half is important at room temperature and above. The kinetic theory expression for the relaxation cross section $\sigma(0001)$ valid for molecules with arbitrary excited rotational levels is

$$\sigma(0001) = \frac{\pi k_B}{C_{rot} Z_{rot}} \frac{\hbar^2}{4\mu(k_B T)^4} \sum_{jj'} \int_0^\infty dE_{rel}\, e^{-(E_{rel}+E_{rot}(j))/k_B T}$$

$$\times (E_{rot}(j)-E_{rot}(j'))^2 \sum_{J\ell\ell'} (2J+1)\, |S^J_{j\ell,j'\ell'}|^2 \,, \quad (1)$$

where Z_{rot} and C_{rot} are the rotational partition function and the heat capacity per molecule, respectively, μ is the reduced mass, and the integration is over the final relative kinetic energy. The temperature dependence has been studied separately for j_{max} = 2, 4, 6 and for j_{max} = 3, 5 in order to demonstrate the influence of the higher rotational states in p-H_2 and o-H_2, respectively. The results are shown in Fig. 1. Two experimental points published by Jonkman et al.[7] are available for the p-H_2-He system, at 90 K and 170 K. The results obtained from the Shafer-Gordon potential at these temperatures agree accurately with the measurements which is not surprising because the $V_2(R)P_2(\cos\gamma)$ term of this potential has been determined to do so. Since the disagreement of the ab initio results is significant, I should try to explain it a little bit further. Let us have a look at

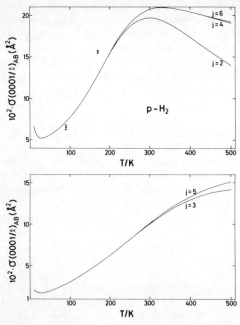

Fig. 1: Rotational relaxation cross sections for para-H_2 and ortho-H_2 infinitely dilute in a He bath

the potentials. In Fig. 2 the Shafer-Gordon fit and the ab initio potential taken at the vibrational averaged r distance of the v, j = 0 state are shown. We can clearly see the large relative and absolute difference of both potential terms in the small R range. It should be emphasized as well that the relative anisotropy, V_2/V_0, is also larger for the Shafer-Gordon fit in this R range. This indeed explains qualitatively why the ab initio result at 170 K is significantly smaller than experiment. However, it is wellknown that ab initio potential points are most reliable especially in the small R range. There is simply no way of reproducing the fit potential by improved ab initio calculations. Since the disagreement is rather singular as I will show in the following examples a repeated measurement should be very helpful. Let me proceed to the next sensitive test of the anisotropy.

2) The rotational Raman broadening and shift of an isolated Raman line is described, according to Shafer and Gordon[6] by the microscopic cross section

$$\sigma^{RR}(j_\alpha j_\beta; j_\alpha j_\beta) = \frac{\pi}{k_{j_a}^2} \sum_{J_\alpha J_\beta \ell \ell'} (2J_\alpha+1)(2J_\beta+1) \begin{Bmatrix} j_\alpha & 2 & j_\beta \\ J_\beta & \ell & J_\alpha \end{Bmatrix}$$

$$\begin{Bmatrix} j_\alpha & 2 & j_\beta \\ J_\beta & \ell' & J_\alpha \end{Bmatrix} \left[\delta_{\ell \ell'} - S^{*J_\beta}(j_\beta \ell, j_\beta \ell', E_\beta) S^{J_\alpha}(j_\alpha \ell, j_\alpha \ell', E_\alpha) \right] \quad (2)$$

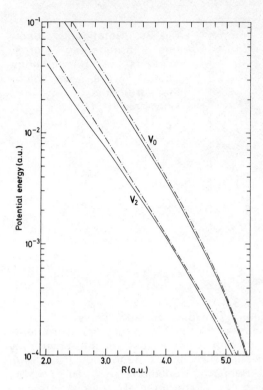

Fig. 2: Legendre expansion of the He-H_2 interaction potential. —·—: Shafer-Gordon rigid rotor fit; ———: ab initio potential of the para-H_2 ground state.

where the $\{\ldots\}$ are 6-j Wigner symbols. The S-matrix elements in (2) must be calculated at different total energies but the same relative kinetic energies for the rotational states j_α and j_β which makes the cross sections complex.

$$E^{rel}_\alpha = \hbar^2 k^2_{j_\alpha}/2\mu = E_\alpha - E_{rot}(j_\alpha) = E_\beta - E_{rot}(j_\beta)$$
$$= E^{rel}_\beta = \hbar^2 k^2_{j_\beta}/2\mu.$$

The energy averaging is obtained as

$$<\sigma^{RR}(j_\alpha j_\beta; j_\alpha j_\beta)> = \frac{1}{(k_B T)^2} \int_0^\infty dE^{rel} E^{rel} \, e^{-E^{rel}/k_B T} \sigma^{RR}(j_\alpha j_\beta, j_\alpha j_\beta).$$

The Raman line width is determined by the real part of the cross section, while the shift is determined by the imaginary part:

$$(\Delta\omega^{RR})_{width} = 2n<v>Re<\sigma^{RR}>,$$
$$(\Delta\omega^{RR})_{shift} = -n<v>Im<\sigma^{RR}>,$$

where n is the number density of the molecules and $<v>$ is the average thermal velocity $(8k_B T/\pi\mu)^{1/2}$.

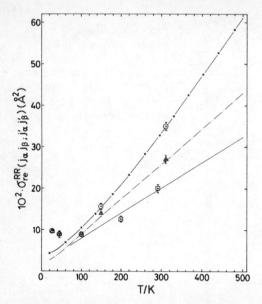

Fig. 3: Rotational Raman line broadening cross sections of H_2 infinitely dilute in a He bath. Measurements of P. Hermans: $S_o(0)$: ⊙; $S_o(1)$: △; $S_o(2)$: □.

The Raman line broadening cross section vanishes for a spherical potential. The comparison with the recent measurements of P. Hermans is therefore another critical test of the anisotropy of the ab initio potential. Results for line broadening are shown in Fig. 3, for three Raman transitions, $S_o(0)$ (j=0→2), $S_o(1)$ (j=1→3), and $S_o(2)$ (j=2→4). Experimental results are included in the figure. Especially at the higher temperatures the agreement is completely satisfying. Significant deviations can be seen at temperatures below 100 K. A possible explanation of this could be the abundance of dimer hydrogen at these low temperatures. The bump of the $S_o(0)$ transitions occurring in the measurements has not been obtained theoretically.

Different results for $S_o(0)$ and $S_o(1)$ have been obtained at room temperature by Shafer and Gordon (0.339 A^2, 0.263 A^2) and by Hermans and McCourt[4] (0.312 A^2, 0.270 A^2) although the same potential had been used. This discrepancy could possibly be due to the energy averaging. Our results for room temperature are 0.324 A^2 and 0.251 A^2.

The Raman shift cross sections can hardly be obtained as accurate as the broadening cross sections with the same amount of computational effort, and the same seems to hold for the experiment. The $S_o(1)$ shift cross sections are shown in Fig. 4, where again experimental results of P. Hermans are included, showing increasing errors bars for decreasing temperature. An estimate of the theoretical error should be similar because the energy averaging includes positive contributions at low energies and negative contributions at larger energies which cancel out at

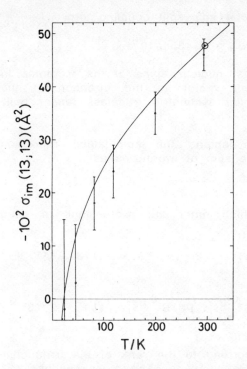

Fig. 4: Rotational Raman line shift cross section of H_2 infinitely dilute in a He bath. Evaluated points of P. Hermans ($\mathrm{\bar{I}}$) for the $S_0(1)$ transition and one theoretical point of Gordon and Shafer are included.

about 25 K. It is noted as well that the results are very sensitive vs. potential variations. The rigid rotor potential of Shafer and Gordon gave a cross section less than 50% in magnitude at room temperature[6]. They therefore extended the potential slightly: the best agreement with our calculations could be reached by changing the equilibrium distance of the potential from 3.3815 A to 3.3836 A, only for the upper (j=3) state which gave -0.477 A^2 (our result: -0.474 A^2). Less agreement than shown in Fig. 4 must be reported for the $S_0(0)$ and $S_0(2)$ shift cross sections[5]. For the reasons just mentioned we cannot decide whether theory or experiment is wrong (cf. a discussion in ref. 5).

A final example should be shown where the cross sections entering are also vanishing for a pure spherical potential.

3) The NMR (or spin lattice) relaxation time T_1 for ortho-H_2-He has been calculated following again the formalism used by Gordon and Shafer[6]. The details of this formalism have been described elsewhere[6,5]. I should briefly mention that two Hamiltonians contribute to the relaxation of the nuclear spin orientation of the hydrogen molecules in collisions with the He atoms of the bath (lattice). There is a spin-rotation interaction term

$$H_{SR} = \omega_{SR} \underline{I} \cdot \hbar \underline{J} \quad , \quad \omega_{SR} = 7.1567 \cdot 10^5 \text{ sec}^{-1},$$

where $\underline{J} = \underline{j}_1 + \underline{j}_2$, and a dipolar spin-spin coupling term

$$H_d = \hbar\omega_d \; \underline{j}_1\underline{j}_2 : \overbrace{\hat{r}\,\hat{r}}, \quad \omega_d = 5.4354\cdot 10^6 \text{sec}^{-1},$$

where $\underline{j}_1$, $\underline{j}_2$ are the individual nuclear spins of the H atoms in the molecule, $\hat{r}$ is the unit vector in the direction of the molecular axis, and $\hat{r}\,\hat{r}$ is the symmetric traceless tensor built up from the components of $\hat{r}$.

According to the Hamiltonians the spin-lattice relaxation frequency is determined by the sum of two terms as

$$T_1^{-1} = T_{1SR}^{-1} + T_{1d}^{-1}.$$

The basic cross sections which enter the expressions of each inverse relaxation time are

$$\sigma^{(k)}(jj;j'j') = \frac{\pi}{k_j^2} \sum_{J_\alpha J_\beta \ell \ell'} [(2j+1)/(2j'+1)]^{1/2} (2J_\alpha+1)(2J_\beta+1)$$

$$\times \begin{Bmatrix} j & k & j \\ J_\beta & \ell & J_\alpha \end{Bmatrix} \begin{Bmatrix} j' & k & j' \\ J_\beta & \ell' & J_\alpha \end{Bmatrix} (\delta_{jj'}\delta_{\ell\ell'} - S^{*J_\beta}(j\ell,j'\ell')S^{J_\alpha}(j\ell,j'\ell')). \quad (3)$$

where k is now 1 or 2, according to the rank of the irreducible tensor in j appearing in the Hamiltonians. The cross sections (3) are real because S^{J_α} and S^{J_β} have to be evaluated at the same total energies. The final relaxation times have been obtained in μsec/amagat, where 1 amagat = $2.693\cdot 10^{19}$ particles per cm^3.

The ab initio results, the previous Shafer-Gordon results[6], two sets of measurements of Riehl and co-workers[8,9], and a very recent set of measurements of very high accuracy provided to us by Lemaire and Armstrong[4] are shown in Fig. 5. The theoretical results have been displayed so that the contributions for j_{max} = 1,3,5 are shown separately. The surprising agreement of the ab initio results with the very accurate measurements of Lemaire and Armstrong raises the question whether or not the approximate assumptions made by Fano[10], Ben-Reuven[11], and Baranger[12] in the development of the formalism used could have caused larger deviations from the measured results than the differences obtained by using the ab initio potential in their formalism. In contrast to this successful test the five calculated points reported by Shafer and Gordon[6] are significantly above the measurements.

The agreement achieved in the last test is so much surprising that I should add a remark: the computations have been completed before the set of accurate measurements came to

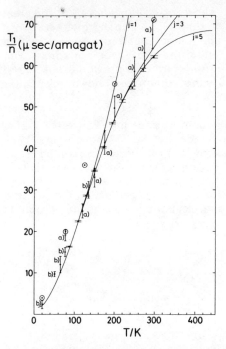

Fig. 5: NMR relaxation time of ortho-H_2 infinitely dilute in a He bath vs. temperature. The solid lines display j_{max} = 1, 3, 5 contributions. Included are theoretical results of Shafer and Gordon: ⊙; two measured sets (a and b), of Riehl et al., and one measured set of Lemaire and Armstrong (∗).

our knowledge. I am grateful to Prof. Armstrong for allowing me to present their results prior to publication.

In summarizing this part of the paper I should say, the reliability of the ab initio potential of H_2-He has been tested successfully. Several other tests not mentioned here[3, 5] gave also quantitative agreement. The only exception to the general findings was the test of the rotational relaxation. Maybe we can find a clean solution of this problem after a new measurement. The quantum chemists should be encouraged by the results of the potential tests to extend the ab initio grid as much as necessary for vibrationally inelastic collision calculations.

In general our present experience with ab initio or empirical potentials of other atom-diatom or diatom-diatom systems is different. The normal procedure of improvement goes step by step in a long run, although I should say, the starting position provided by reliable ab initio potential points in the small center of mass distance range and the knowledge of van der Waals expansion coefficients determining the asymptotic properties of the interaction potential is extremely valuable.

The only interaction potential of atoms and molecules which comes close to the perfection of the H_2-He potential and which has also been entirely determined in ab initio calculations is the

H_2-H_2 potential. I am going to show now the effect of the last steps of improvement, and by doing this, I can give a more detailed description of the very last step of improvement which unavoidably became necessary after several comparisons with measurements mentioned by Prof. Buck in his previous talk, and after a significant shortcoming of the potential shown by R.O. Watts[13].

The crucial set of ab initio points and the van der Waals expansion terms have been provided by W. Meyer in 1979 (M79). These data represent kind of a first order corrected rigid rotor potential because the variation over the intramolecular distance r is done only for two distances, 1.28 a.u. and 1.449 a.u., the latter being the vibrationally averaged distance of the para-H_2 ground state. The potential expansion in triple products of spherical harmonics,

$$V(\underline{r}_1,\underline{r}_2,\underline{R}) = \sum_{\lambda_1 \lambda_2 \lambda} V_{\lambda_1 \lambda_2 \lambda}(r_1,r_2,R)$$

$$\times \sum_{\mu_1 \mu_2} (\lambda_1 \mu_1, \lambda_2 \mu_2 / \lambda \mu) Y_{\lambda_1 \mu_1}(\hat{\underline{r}}_1) Y_{\lambda_2 \mu_2}(\hat{\underline{r}}_2) Y^*_{\lambda \mu}(\hat{\underline{R}}) \quad (4)$$

has been obtained approximately by the ab initio calculations which determined 6 terms: the isotropic V_{000}, the two leading anisotropic terms V_{202} and V_{022} responsible for direct $|\Delta j|=2$ transitions as well as reorientation effects of one molecule with $j \neq 0$, the long-range quadrupole-quadrupole interaction term V_{224}, and two other terms, V_{220}, V_{222}, which are negligible. This set of potential terms represents all direct $|\Delta j|=2$ transitions of both molecules. Since it has been derived from six geometric configurations preferred for reasons of symmetry, the representation of the anisotropic terms was partly incomplete. Improvement could be reached by extending the number of geometries and the expansion to 19 terms, including now all terms responsible for direct $|\Delta j|=4$ transitions of both molecules. This was necessary only at small R distances (3.0 and 4.0 a.u.). The ab initio calculations for this correction have been done in collaboration with B. Liu[14]. Fig. 6 shows the corrections formed for the 4 leading potential terms (M80). There is only little change in the spherical potential, but significant change (given in percentages in Fig. 6) in the main anisotropy. This provides a change of 30-40% in sensitive cross sections.

To give an example of the improvement reached, I show a comparison of measured rotational relaxation cross sections of Jonkman et al.[15] for para-H_2 gas and the theoretical results derived from both potentials. The kinetic theory gives the formula

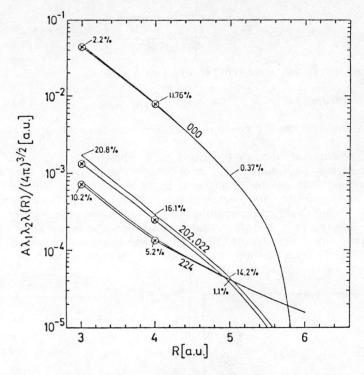

Fig. 6: Leading ab initio interaction potential terms of H_2-H_2 (W. Meyer). The upper curves (M79) are derived from 6 geometries, the lower curves (M80) show the corrected potential derived from 19 geometries, at R=3.0 and 4.0 a.u.

$$\sigma(0001) = \frac{\pi k_B}{C_{rot}} \frac{1}{z_{rot}^2} \sum_{j_1 j_2 j_1' j_2'} \iint \exp(-\epsilon_{j_1} - \epsilon_{j_2} - \gamma^2) \gamma^2 \gamma'$$

$$\times (\Delta\epsilon)^2 \sum_{all\ m} |a_{m_1 m_2, m_1' m_2'}^{j_1 j_2, j_1' j_2'}|^2 \sin\phi\, d\phi\, d\gamma, \qquad (5)$$

where $\epsilon_i = E_{rot}(j_i)/k_B T$, $\Delta\epsilon = \epsilon_{j_1} + \epsilon_{j_2} - \epsilon_{j_1'} - \epsilon_{j_2'}$, γ denotes a final relative velocity in units of $\sqrt{2k_B T/\mu}$, and the scattering amplitudes used in this formula are

$$a_{m_1m_2,m'_1m'_2}^{j_1j_2,j'_1j'_2}(E,\hat{e}) = (\pi/kk')^{\frac{1}{2}} \sum_{J\ell\ell'} \sum_{jj'mm'} i^{\ell'-\ell+1}(2\ell'+1)^{\frac{1}{2}}$$

$$\times (j'm', \ell'o/Jm')(j'_1m'_1, j'_2m'_2/j'm')(j_1m_1, j_2m_2/jm)$$

$$\times (jm, \ell m'-m/Jm')[\delta_{\alpha,\alpha'}-S^J(\alpha,\alpha',E)]Y_{\ell m'-m}(\hat{\hat{e}}). \quad (5a)$$

where α is the full set of channel quantum numbers (j_1, j_2, j, ℓ).

The results of the calculations are shown in Fig. 7, as derived from the potentials M79 and M80. The correction of the potential turned out to be sufficient by giving results within the assumed 10% experimental error bars. This potential M80 has also been tested very successfully[16] in an extensive study of Senftleben-Beenakker effects and relaxation phenomena for para-H_2 gas, in comparisons with experimental data provided by the Molecular Physics Group in Leiden. I should note that the disagreement never exceeded 10%.

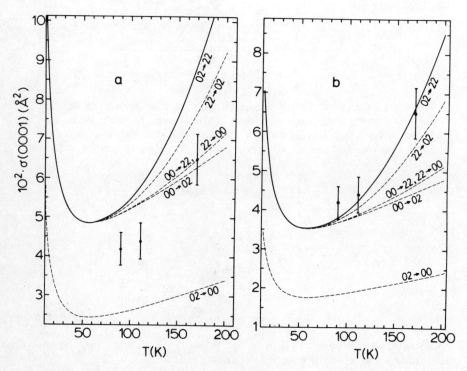

Fig. 7: Rotational relaxation cross sections for para-H_2. Results are derived from the M79 potential (a) and from the M80 potential (b). The dashed curves show the contributions from inelastic transitions.

Another very nice test of the M80 potential has been done in collaboration with U. Buck et al.[17c] who measured differential elastic and $\Delta j=2$ cross sections of D_2-H_2. The comparison of theory and experiment at $E_{c.m.}$ = 84meV is shown in Fig. 8. Please note that the $\Delta j=2$ excitation cross section for D_2 differs strongly depending upon whether para-H_2 or ortho-H_2 is colliding with D_2. The difference of the two cross sections is due to the long range quadrupole-quadrupole interaction term (V_{224}) of the potential which only contributes when both molecules rotate in the final state and/or in the initial state. Some significant disagreement found in this comparison can hardly be seen in this figure: the undulatory structure of the elastic differential cross section is a little bit out of phase. This has to be discussed now when I proceed to the last step of improvement to be applied to the H_2-H_2 interaction potential.

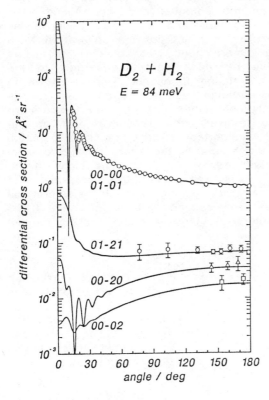

Fig. 8: Differential cross sections of D_2-H_2 determined from time of flight measurements (Buck et al.) and theoretical curves obtained from the ab initio M80 potential, at $E_{c.m.}$ = 84 meV.

The remaining shortcoming of the M80 potential has been revealed experimentally by U. Buck and coworkers[17] in beam experiments for isotopic systems and theoretically by R. Watts[13] who noticed a significant disagreement of the thermodynamical properties and especially of the second virial coefficient above the

Boyle temperature (being too large). Several other potential fits determined from different kinds of experiments indicate a need for the same kind of correction, as e.g. McConville's[18] multiple property fit of the isotropic potential which includes the measured second virial coefficients. The required change of the potential applies to the spherical potential term. It is roughly described by a small shift of the attractive potential towards smaller R, and possibly by a relatively small increase of the well depth.

We have chosen the measured second virial coefficients of para-H_2 between 16 K and 40 K[19,20] to appropriately control the scaling and the shift of V_{000} at r = 1.449 a.u. Quantum mechanical calculations of the second virial coefficients have been performed after obtaining numerical scattering wave functions for a sufficiently dense energy grid, and subsequently pair correlation functions for temperatures up to 40 K. (A possible alternative quantity intermediate between the potential and the second virial coefficient is the set of eigenphases and their energy derivatives determined on the same energy grid and for the full partial wave expansions.) There is one dimer bound state of p-H_2/p-H_2 the eigenfunction and eigenvalue of which contribute largely to the very low temperature pair correlation functions. This is shown in Fig. 9. As a result of this particular feature the attractive part of the spherical potential mainly determines the very low temperature second virial coefficients. However, the repulsive part of the potential is by no means unimportant. Three sensitive potential parameters have been noticed in the fit procedure: the zero point of the potential (R_o), the well depth (ϵ_o), and the slope in the repulsive range below R_o. It is well known from classical and semiclassical calculations that the potential cannot be determined in a unique way from measured second virial coefficients, i.e., the three mentioned parameters of the well region can be chosen

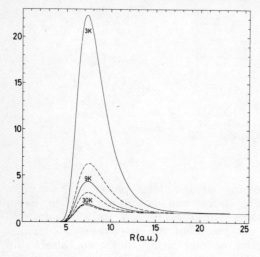

Fig. 9: Pair correlation functions of para-H_2 gas at 3 K, 9 K, and 30 K. Dashed curves: scattering wave function contributions; solid lines: bound state included.

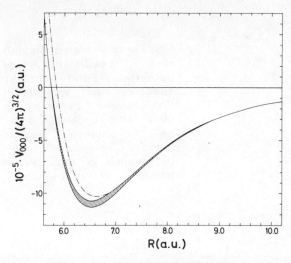

Fig. 10: Attractive range of the V_{ooo}. The two solid curves represent potentials both providing agreement with measured second virial coefficients.
—.—: ab initio M80.

such that the slight variation of one can be compensated by changes of the two others, e.g. R_o can be reduced slightly when the well depth is reduced and/or the magnitude of the slope is enlarged appropriately or, when R_o is kept fixed, a larger well depth can be compensated by a steeper increase in the repulsive range. In Fig. 10 is shown how the spherical ab initio potential of two para-H_2 ground state molecules has been scaled and shifted in two different ways when R_o is kept fixed. Both resulting potentials give quantitative agreement with the measured second virial coefficients below 40 K, and the same should hold for potentials in between them. The more attractive version increases steeper in the repulsive range and therefore comes closer to the original ab initio potential at small R distances. From the quantum chemistry point of view it is the more reasonable version due to the minor change in the very small R range and due to the more continuous corrections over the whole R range. One could possibly think of an extra dispersion attraction effect missed in the previous ab initio calculations. Maybe improved calculations will show.

The calculation of second virial coefficients by using approximate – classical and quantum corrected classical[21] – methods has failed below 40 K. The semiclassical approximation did very well for the new potential fit in the vicinity of the Boyle temperature, as shown in Fig. 11. Quantum mechanical calculations for testing the new potential fit at these temperatures are in progress.

The best possible procedure of conserving the high quality of the inelastic cross sections in the new fit potential as obtained from the ab initio M80 potential is quite simple. The same shift and scaling functions determined for the spherical potential have

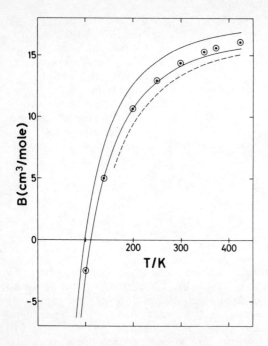

Fig. 11: Calculated second virial coefficients of para-H_2. Solid curves: semiclassical approximation, upper curve: ab initio M80; lower curve: new fit potential; dashed curve: classical approximation of the new fit potential; ⊙, experiment.

been applied to all potential terms and, since no additional information is available, also to the potential representation which is responsible for the centrifugal stretching effects. Thus the relative anisotropies of the potential M80 have been conserved in the shift. Results of this kind of potential variation have been shown in previous papers, for HD-D_2[17b], and for H_2-D_2[17c], where the ratios of the differential cross sections at fixed scattering angle have been found conserved.

There was still another shortcoming of the ab initio M80 potential found in comparisons with beam measurements for HD-D_2[17a,b]. The theoretical j = 0 → 1 cross sections for HD came out too large by about 10% at 45.4 meV and by about 15-20% at 70 meV relative kinetic energy. Since the leading anisotropic term of this system, V_{101}, is mainly determined by V_{000} of the H_2-H_2 potential, it was near at hand to account for this effect as well when determining a V_{000} fit which properly reproduces the phase of the undulatory structure of the elastic cross section. However, this strategy turned out to be wrong for two reasons: 1) the V_{000} fit of Buck et al.[17a,c] does not reproduce the low temperature second virial coefficients, 2) test calculations with the new fit potential transformed the same way as before to the HD-D_2 system gave approximately the same amount of error as derived from the M80 potential. Since the transformation to the 10-term HD-D_2 potential gave a converged V_{101} term as checked previously, a possible source of the disagreement could be another anisotropic term of the HD-D_2 potential which has been obtained unconverged in the potential transformation. And this indeed has been found. After expanding the HD-D_2 potential in 43 terms (by accounting for all direct HD transitions up to $|\Delta j| = 5$, and all D_2 transitions up to $|\Delta j| = 4$) it was found that the V_{121} term became much smaller. This term only contributes to the dynamics when the D_2 molecule rotates initially or finally as one can see easily by looking at the expression for the reduced potential matrix elements.

After 12 leading terms have been taken out of 43 of the HD-D_2 potential expansion a new calculation at 45 meV relative kinetic energy gave satisfying agreement with experiment. A preliminary display of this finding is shown in Fig. 12. The procedure of carefully comparing theory and experiment is described in a previous paper[17a]. Since the theoretical results have been completed very recently, there was no time for properly adapting them to experiment by obtaining averages over calculated distribution functions of the experimental arrangement. Because of this both theoretical cross sections will be replaced finally by slightly damped curves, but the phase of the undulations and the ratio σ_1/σ_{el}[17a] of the cross sections at larger scattering angles will not change significantly. The experimental ratios σ_1/σ_{el} have been used for testing the inelastic cross section assumed the theoretical elastic cross sections are correct. There is quantitative agreement. The phase of undulation of the

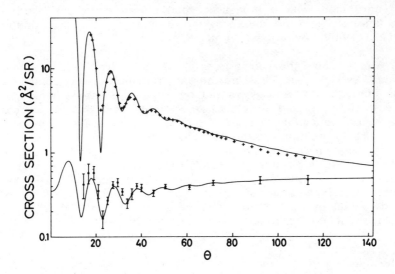

Fig. 12: Comparison of the measured HD-D_2 elastic and 0→1 transition cross sections with close coupled results derived from the new H_2-H_2 potential fit.

elastic cross section is also completely satisfying, although the shift of the new potential fit obtained from the ab initio M80 potential appears to be only half as much as suggested in the previous papers[17a,c].

I should mention that production has been started recently for calculating accurate rotational transition cross sections of HD colliding with H_2 and subsequently rate coefficients and line broadening parameters of this system. These data will be needed in the near future when satellite-borne measurements of the HD spectrum observed from interstellar clouds become available.

Compared to the two interacting systems and their isotopic species discussed by now the number of tests of accuracy of the interaction potentials and the volume of theoretical as well as experimental data obained for any other atom-molecule or molecule-molecule system of astrophysical interest is rather poor. From this point of view it is interesting to discuss a successful improvement of the He-CO interaction potential achieved recently be W. Dilling in his thesis work. The He-CO system can be mentioned in a workshop like this because it can be considered a forerunner of the astrophysically more interesting H_2-CO system for several reasons. Dilling has started with an ab initio CI potential published by Thomas, Krämer, and Diercksen[22]. His aim was to change this potential as much as necessary for reaching agreement with a measured total differential cross section and two time of flight profiles provided at just one relative kinetic energy (27.7 meV) by Faubel et al.[23].

Problems on the experimental side have been found complicated by the fact that no single transition peak could be resolved although the energy resolution of this measurement was extremely good compared to other experiments. On the theoretical side the rather complex dynamics of the system caused by two large anisotropic potential terms (V_1, V_2) did not allow to find a one-to-one relation of a single transition cross section and a single potential term. Actually 12 different inelastic and 3 elastic differential cross sections contributed significantly to the theoretically obtained time of flight diagram, and almost all of them had undulatory structure. Therefore the finding of a track leading to satisfying agreement with experiment was quite a hard job.

The leading three terms of the Legendre expansion of the potential have been varied independently. It is noted that the economy of this variation is essential for reaching success in reasonable computer time. The independent scaling of the center of mass distance R by just one constant parameter for each potential term worked out nicely. As I have already explained above this procedure is appropriate for changing an ab initio potential. Finally three scaling parameters have been found which reproduce the undulatory structure of the measured total differential cross section in phase with the measured points and both measured time of flight profiles. Certainly no general validity can be claimed for this potential fit because of the lack of experimental information for this system, but, without any doubt, this fit is now the best potential available.

I should mention some technical data concerning the computational efforts of this project. The close coupled scattering calculations have been done with a fully optimized collision code on a CRAY1. Generally the number of coupled channels for the full basis was between 55 and 45, up to 40 partial waves, and a smaller number of channels for the higher partial waves. Any full run for one energy and three different initial rotational states ($j=0,1,2$) took about 3 CPU hours which is still reasonable computational effort.

References

1) W. Meyer, P.C. Hariharan, W. Kutzelnigg, J. Chem. Phys. 73, 1880 (1980).
2) W.E. Köhler, J. Schaefer, Physica 120A, 185 (1983).
3) P.W. Hermans, Thesis, Leiden 1982. See also P.W. Hermans, L.J.F. Hermans, J.J.M. Beenakker, Physica 122A, 173 (1983).
4) C. Lemaire, R.L. Armstrong (private communication).
5) J. Schaefer, W.E. Köhler, to be published in Physica.
6) R. Shafer, R.G. Gordon, J. Chem. Phys. 58, 5422 (1973)
7) R.M. Jonkman, G. Prangsma, R.A. Keijzer, H.F.P. Knaap, J.J.M. Beenakker, Physica 38, 456 (1968).
8) J.W. Riehl, J.L. Kinsey, J.S. Wangh, J.H. Rugheimer, J. Chem. Phys. 49, 5276 (1968).
9) J.W. Riehl, C.J. Fisher, J.D. Baloga, J.L. Kinsey, J. Chem. Phys. 58, 4571 (1973).
10) U. Fano, Phys. Rev. 131, 259 (1963).
11) A. Ben-Reuven, Phys. Rev. 141, 34 (1966); 145, 7 (1966); A4, 2155 (1977).
12) M. Baranger, Phys. Rev. 111, 481 (1958); 112, 855 (1958).
13) J. Schaefer and R.O. Watts, Molecular Physics 47, 933 (1982).
14) B. Liu and J. Schaefer, not published.
15) C.G. Sluijter, H.F.P. Knaap, and J.J.M. Beenakker, Physica 31, 915 (1965); R.M. Jonkman, G.J. Prangsma, I. Ertas, H.F.P. Knaap, and J.J.M. Beenakker ibid. 38, 441 (1968); L.M. Valley and R.C. Amme, J. Chem. Phys. 50, 3190 (1969).

16) W. E. Köhler and J. Schaefer, J. Chem. Phys. 78, 4862 (1983); ibid. 78, 6602 (1983).
17 a) U. Buck, F. Huisken, J. Schleusener, J. Schaefer, J. Chem. Phys. 74, 535 (1981).
  b) U. Buck, F. Huisken, G. Maneke, J. Schaefer, J. Chem. Phys. 78, 4430 (1983).
  c) U. Buck, F. Huisken, A. Kohlhase, O. Otten, J. Schaefer, J. Chem. Phys 78, 4439 (1983).
18) G. T. McConville, J. Chem. Phys. 74, 2201 (1981).
19) E. A. Long and O. L. I. Brown, J. Am. Chem. Soc. 59, 1922 (1937).
20) R. D. Goodwin, D. E. Diller, H. M. Roder, and I. A. Weber, J. Res. Nat. Bur. Stand. 68A, 121 (1964).
21) J. O. Hirschfelder, Ch. F. Curtiss, and R. B. Bird, Molecular Theory of Gases and Liquids, 1967, p. 420.
22) L. D. Thomas, W. P. Krämer, and G. H. F. Diercksen, Chem. Phys. 51, 131 (1980).
23) M. Faubel, K. H. Kohl, nad J. P. Toennies, J. Chem. Phys. 73, 2506 (1980), see also K. H. Kohl, thesis work 1982.

ROTATIONAL AND VIBRATIONAL ENERGY TRANSFER IN DIATOMIC AND
POLYATOMIC MOLECULES

Gert Due Billing
Department of Chemistry, Panum Institute, University of
Copenhagen, DK-2200 Copenhagen N, Denmark

Exact quantum mechanical calculations on vibrational/rotational
transitions in molecules are only possible or numerically feasible
for at few - mainly - hydrogen containing (low mass) systems. It
has, however, turned out that semiclassical (classical path)
methods are sufficiently accurate to provide quantitative results
for both light and heavy molecules. For the latter the semiclassical approach is furthermore about the only possibility for
obtaining vib/rot cross sections. We have developed a hierarchy
of methods which are able to give accurate cross sections and
rate coefficients for diatomic and small polyatomic molecules
and qualitative information on larger polyatomic systems.
    A brief outline of the semiclassical method which involves
a partial quantization of the various degrees of freedom, is given
and comparison with quantum calculations is made for a few systems.
Finally results on a number of systems ranging from atom-diatom,
diatom-diatom to polyatomic systems are discussed.

1. INTRODUCTION

    In order to describe energy transfer, in even moderate sized
systems, it is necessary to introduce some approximations. The
goal of these approximations is to reduce the set of coupled
second order differential equations one obtains when using a full
quantum mechanical formulation. If for instance the rotational
motion of the molecules can be described classically the quantum
m-state degeneracy is avoided and the number of coupled equations
is reduced drastically. Other ways of decoupling the rotational
projection states within the quantum framework is to introduce
the "coupled states" approximation or the "large j-limit" of the

coupling elements. These socalled decoupling methods may also be introduced in the semiclassical description where only the relative translational motion is treated classically. For polyatomic systems it may also be necessary to "decouple" the rotational states in order to reduce the problem to a solvable one (less than 100-200 coupled differential equations). Within the quantum mechanical formulation the IOS (infinite order sudden) approximation, which treats the rotational levels as degenerate, has been popular. However, recent investigations [1] have shown that the IOS approximation is questionable at least for small vib/rot transitions and a pure classical mechanical description of the rotational motion is therefore to be preferred. For polyatomic molecules even the number of energetically allowed vibrational states is often too large for the usual state expansion methods to be feasible. Here one may introduce also a classical mechanical description of some of the vibrational modes [2] or introduce an operator approach which is capable of treating the problem to infinite order within the harmonic oscillator (HO) approximation [3]. This zero'th order solution may then be used when expanding the final solution including anharmonic terms by perturbation theory [4].

2. THE A+BC SYSTEM

As an example we consider for simplicity the atom-diatom collision treating only the relative translational motion classically. The hamiltonian may be expressed as:

$$H = (2\mu)^{-1}(P_R^2 + \hbar^2 \ell^2/R^2) + V(R,r,\gamma) + H_o \qquad (1)$$

where

$$H_o = p_r^2/2m + v(r) + \hbar^2 j^2/2I \qquad (2)$$

R is the distance from the atom A to the center of mass of BC, r the BC distance, $\mu = m_A(m_B+m_C)/(m_A+m_B+m_C)$ the reduced mass, $m = m_B m_C/(m_B+m_C)$, I the moment of inertia $I = mr^2$, P_r the momentum of the relative motion, p_r the momentum of the vibrational motion of BC, ℓ the orbital and j the rotational angular momentum. The intramolecular potential is $v(r)$ and the intermolecular potential is $V(R,r,\gamma)$ where γ is the angle between R and r.

If we neglect the vib/rot coupling i.e. introduce the rigid rotor approximation with $I = m\bar{r}^2$ where $\bar{r}$ is the equilibrium distance we obtain as target states

$$\varphi_\alpha^o = \varphi_n^o(r) Y_{jm}(\theta,\varphi) \qquad (3)$$

where φ_n^o is the vibrational wavefunction, $Y_{jm}(\theta,\varphi)$ the spherical harmonics with the angles θ,φ specifying the direction of the rotor in a spacefixed coordinate system. Thus we have

$$\hat{H}_o \varphi_\alpha^o = E_\alpha \varphi_\alpha \tag{4}$$

where

$$E_\alpha = E_n^{vib} + E_j^{rot} \tag{5}$$

and $\hat{H}_o$ is obtained by introducing the quantum operators for p_r and j in eq. (2). The rot/vib coupling may easily be included in the formulation using perturbation theory and that:

$$r^{-2} = \bar{r}^{-2} - 2(r-\bar{r})\bar{r}^{-3} + \ldots \tag{6}$$

for further details see ref. [5].

The translational motion, which is treated classically, induces a timedependent force upon the diatom and by using time-dependent perturbation theory we obtain a set of coupled equations in the expansion coefficients for the wavefunction ψ (see below). However, instead of expanding ψ in the states $\varphi_\alpha^o(\theta,\varphi)$ it is convenient [6] to introduce rotated wavefunctions $Y_{jm}(\gamma,\xi)$ where γ and ξ are the polar angles in a bodyfixed coordinate system having its z-axis along the intermolecular R-axis. This new coordinate system may be obtained by rotating the spacefixed by angles Θ and Φ which specify the orientation of $\vec{R}$ in the space-fixed system, i.e.

$$X = R \sin \Theta \cos \Phi \tag{7a}$$
$$Y = R \sin \Theta \sin \Phi \tag{7b}$$
$$Z = R \cos \Theta \tag{7c}$$

Thus we have:

$$Y_{jm}(\gamma,\xi) = \sum_{m'} D_{mm'}^j(0,\Theta,\Phi) \, Y_{jm'}(\theta,\varphi) \tag{8}$$

where $D_{mm'}^j$ are matrix elements of the rotation operator. The wavefunction ψ is now expanded as:

$$\psi = \sum_{\alpha=(njm)} a_\alpha(t) \exp(iE_\alpha t/\hbar) \varphi_n^o(r) Y_{jm}(\gamma,\xi) \tag{9}$$

From the timedependent Schrödinger equation we obtain a set of coupled equations in the expansion coefficients a_α:

$$i\hbar \dot{a}_\alpha = \sum_\beta a_\beta \{V_{\alpha\beta} \exp(i(E_\alpha - E_\beta)t) - i\hbar U_{\alpha\beta} \dot{\Theta}\} \tag{10}$$

where we have used that the relative motion occur in a plane, i.e. Φ = constant. The potential coupling elements is:

$$V_{\alpha\beta} = <\varphi_n^o, Y_{j'm'} | V(R,\gamma,r) | \varphi_n^o Y_{jm}> \qquad (11)$$

The coupling term $U_{\alpha\beta}$ in eq. (10) arises from the timedependence of the wavefunction $Y_{jm}(\gamma,\xi)$ through $O(t)$ (eq. (18)). We have [6]:

$$U_{\alpha\beta} = \tfrac{1}{2}\delta_{nn'}\delta_{jj'}\{[(j'-m'+1)(j'+m')]^{\tfrac{1}{2}}\delta_{m,m'-1}$$
$$-[(j'+m'+1)(j'-m')]^{\tfrac{1}{2}}\delta_{m,m'+1}\} \qquad (12)$$

we note that $\underline{\underline{U}}$ is diagonal in the vibration and rotation quantum numbers but couple the projection states. The potential coupling elements in the basis $Y_{jm}(\gamma,\xi)$ are [7]:

$$V_{\alpha\beta} = [(2j'+1)/(2j+1)]^{\tfrac{1}{2}} \sum_{ik} V_{ki}(R) M_{nn'}^{(i)} <kj'00|j0><j'km'0|jm> \qquad (13)$$

where we have used the following expansion of the intermolecular potential:

$$V(R,r,\gamma) = \sum_{ik} V_{ki}(R) u^i P_k(\cos\gamma) \qquad (14)$$

where $u = (r-\bar{r})/\bar{r}$. The matrix elements $M_{nn'}^{(i)}$ are defined by:

$$M_{nn'}^{(i)} = <\varphi_{n'}|u^i|\varphi_n> \qquad (15)$$

and $<....|..>$ are Clebsch-Gordon coefficients [8]. The potential matrix elements (13) are diagonal in the m-quantum number due to the factor $<j'km'o|jm>$. The semiclassical coupled states approximation is obtained by neglecting the $\underline{\underline{U}}$-matrix, i.e. the m-states decouple and the coupled equations (10) only depend parametrically upon m.

Including now the inter multiplet coupling due to the $\underline{\underline{U}}$-matrix to infinite order we may introduce the expansion coefficients b_α such that:

$$i\hbar\dot{b}_\alpha = \sum_\beta W_{\alpha\beta} \exp\{i(E_\alpha - E_\beta)t/\hbar\} b_\beta \qquad (16)$$

where

$$\underline{\underline{W}} = \exp(\Theta\underline{\underline{U}})\underline{\underline{V}}\exp(-\Theta\underline{\underline{U}}) \qquad (17)$$

and

$$\underline{b} = \exp(\Theta\underline{\underline{U}})\underline{a} \qquad (18)$$

The matrix elements $W_{\alpha\beta}$ are [7]:

$$W_{\alpha\beta} = \sum_{ik} V_{ki}(R) M_{nn'}^{(i)} <kj'00|j0> \sum_{q=-k}^{k} d_{qo}^k(\Theta) <kjqm|j'm'> \qquad (19)$$

where the m-states within a given rotational state are coupled. Introducing however the large j-limit of the Clebsch-Gordon coefficient [8]:

$$\langle kjqm\, j'm'\rangle \sim (-1)^{k+j'-j} d^{k}_{j'-j,q}(X) \tag{20}$$

where $\cos X = m'/j'$. Using the closure relation [8] for the rotational matrix elements we get:

$$W_{\alpha\beta} = \sum_{ik} V_{ki}(R) M^{(i)}_{nn'} \langle kj'00|j0\rangle d^{k}_{j'-j,0}(X+\Theta) \tag{21}$$

i.e. also in this limit do we get a parametrical dependence upon the m-quantum number. Finally we may decouple the m-states by treating the rotational motion classically.

The equations of motion are obtained by using hamiltons principle on the effective hamiltonian defined by:

$$H_{eff} = (2\mu)^{-1}(P_R^2 + \hbar^2 \ell^2/R^2) + \langle\psi|V|\psi\rangle \tag{22}$$

Thus from $\dot{H}_{eff} = 0$ we obtain [9]

$$\dot{R} = P_R/\mu \tag{23}$$

$$\dot{P}_R = -\partial V_{av}/\partial R + \hbar^2 \ell^2/\mu R^3 \tag{24}$$

$$\dot{q}_\ell = \hbar^2 \ell/\mu R^2 \tag{25}$$

$$\hbar \frac{dl}{dt} = -\sum_{\alpha\beta} a^*_\alpha a_\beta [\underline{V},\underline{U}]_{\alpha\beta} \exp(i(E_\alpha - E_\beta)t/\hbar) \tag{26}$$

where $V_{av} = \langle\psi|V|\psi\rangle$ and $q_\ell = \Theta$. The equations (10) and (23-26) are solved simultaneously using the initial conditions:

$$a_\alpha(-\infty) = \delta_{\alpha I} \quad (I = \text{initial state}) \tag{27}$$

$$R \text{ (large)} \tag{28}$$

$$\ell \in [J-j; J+j] \quad (J \text{ total angular momentum}) \tag{29}$$

$$q_\ell \in [0; 2\pi] \tag{30}$$

$$P_R^I = -(2\mu(E-E_I))^{\frac{1}{2}} \tag{31}$$

However, in order to obtain detailed balance fulfilled an initial average momentum should be used, i.e. for the transition $I \to n$ one uses $\bar{P} = \frac{1}{2}(P_R^I + P_R^n)$. For some systems (low mass) and high accuracy it is furthermore necessary to calculate both forward and reverse probability, i.e. $P_{I \to n}$ and $P_{n \to I}$ and define $P_{In} = P_{nI} = (P_{I \to n} P_{n \to I})^{\frac{1}{2}}$. For details concerning this symmetrization procedure the reader is referred to refs. [10].

From the probabilities we obtain the cross sections as:

$$\sigma_{njm \to n'j'm'}(E,J) = \frac{\pi}{k_{nj}^2} (2j+1) \sum_{\ell=|J-j|}^{J+j} P_{\alpha\beta} \qquad (32)$$

where k_{nj} is the wavenumber. Rate constants for transitions between vib/rot levels are obtained by averaging over an initial Boltzmann distribution of kinetic energy and m-states, i.e.

$$k_{nj \to n'j'}(T) = (8kT/\pi\mu)^{\frac{1}{2}} \int_0^\infty d(\beta\varepsilon)\, \beta\varepsilon \exp(-\beta\varepsilon)\, \sigma_{nj \to n'j'} \qquad (33)$$

where

$$\sigma_{nj \to n'j'} = (2j+1)^{-1} \sum_J (2J+1) \sum_{mm'} \sigma_{njm \to n'j'm'} \qquad (34)$$

The semiclassical model formulated here for A+BC systems can be extended to much larger systems if also the rotational motion is treated classically (see refs. [11]).

Results obtained within the semiclassical framework have been compared with exact or approximate (coupled states) quantum calculations both for rotational, vibrational and combined vib/rot transitions. Here only a few results will be presented. For further documentation and discussion the reader is referred to the litterature [9,12-16]. Tables 1 and 2 show semiclassical and quantum results for rotational transitions in H_2 and NH_3 colliding with ^{4}He. In the He-H_2 case the same analytical potential surface was used in the two calculations whereas two different fits to the same ab initio electron gas data were used for He+NH_3. This may cause differences of at most 5-10% upon the large cross sections. Table 3 shows vibrational excitation probabilities for some collinear atom-diatom systems. Finally table 4 compares quantum and semiclassical cross sections to combined rot/vib transitions in He+H_2.

3. A REVIEW OVER RECENT RESULTS OBTAINED USING A SEMICLASSICAL APPROACH TO ENERGY TRANSFER IN MOLECULAR SYSTEMS

The energy transfer between colliding molecules may be characterized by rate constants, cross sections or probabilities for the involved processes. Table 5 gives an overview of the order of magnitude relation between the above quantities and of the processes involved from the fast rotational relaxation to the slow vibration/translation energy transfer. The rate of the processes is determined by the intermolecular forces, energy spacing, translational energy (temperature) and molecular constants as masses, moments of inertia etc. For vibrational transitions the rate constant normally decreases with temperature. However, if the long range multipole interaction is sufficiently strong the rate constant may increase with decreasing temperature. As an example we may consider $HF(v=1) + HF(v=0) \rightarrow 2HF(v=0) + 3958$ cm^{-1}. The rate constant for this process has a minimum of $8 \cdot 10^{-13}$ cm^3/sec at about 700 K and increases to a value of 1.5-2 10^{-12} cm^3/sec at 300 K. The reason for this effect of long range forces upon a rate constant with a large energy mismatch is the possibility of forming a hydrogen bond and thereby a long lived collision complex (lifetimes $\sim 10^{-11}$ sec). Thus the molecules are momentarily trapped and "collide" several times leading to an increase in VT relaxation. The same collisional trapping which is observed below 700 K for HF+HF is found also for systems as N_2+N_2 or CO+CO but at much lower energies. At 100 cm^{-1} total energy about 60-80% of the contribution to the vibration-vibration (VV) rates comes from collision complexes [25]. Normally however the long range forces are too weak to have any influence upon transitions with energy mismatch larger than 200-300 cm^{-1}. Such processes are typically VV processes as e.g. $CO_2(00^o1) + {}^{14}N_2(0) \rightarrow CO_2(00^o0) + {}^{14}N_2(1) + 18.8$ cm^{-1} which at low temperatures (below 500 K) is induced by the transition-dipole/quadrupole interaction. For ${}^{15}N_2$ the energy mismatch increases to 97.8 cm^{-1} and we find that the long range forces are too weak to support the transition at low temperatures hence the rate constant decreases with temperature [26]. Another example is $CO(0) + N_2(1) \rightarrow CO(1) + N_2(0) + \Delta\omega$ where $\Delta\omega = 187$ cm^{-1}, 149 cm^{-1} and 108 cm^{-1} for ${}^{14}N_2$, ${}^{14}N{}^{15}N$ and ${}^{15}N_2$ respectively. The transition may be induced by the dipole-quadrupole interaction term. The VV rates for the three isotopes decrease with temperature indicating that long range forces cannot be dominating. But the rate constant for CO+${}^{15}N_2$ only decreases very little from $3.7 \cdot 10^{-14}$ cm^3/sec at 300 K to $2.8 \cdot 10^{-14}$ cm^3/sec at 100 K showing that the long range forces contribute significantly to the transition with the smallest energy mismatch [27]. If the rate constant is small even weak long range forces may introduce a near resonant transition. For instance is the transition $CO(v=1) + H_2(j=2) \rightarrow CO(v=0) + H_2(j=6) + 83.3$ cm^{-1} almost entirely induced by the dipole hexadecapole term and has $k \sim 5 \cdot 10^{-16}$ cm^3/sec at 300 K whereas the off resonant transition $CO(v=1) + H_2(j=2) \rightarrow CO(v=0) + H_2(j=4) + 1329$ cm^{-1} has $k \sim 8.5 \cdot 10^{-17}$ cm^3/sec [28]. Another example is the transition $CO_2(02^o0) + X \rightarrow CO_2(10^o0) + X + 102.7$ cm^{-1} which is induced by the cubic derivative of the potential $\partial^3 v/\partial Q_1 \partial Q_2^2$. An interesting influence of long range interaction upon vibrational relaxation is observed by the temperature dependence of transitions among levels which are Coriolis coupled as for instance the $00^o1 \rightarrow 01^o0$ transition in CO_2. In the timedependent semiclassical theory where the CO_2 rotation is treated classically this transition is induced by a term [29] which depend on $\frac{dj}{dt}$. Since the anisotropy of the long range interaction potential may be large enough to induce rotational transitions, i.e. $\frac{dj}{dt} \neq 0$ we get a long range induction of transition among Coriolis coupled levels. This is visualized in an anomalous temperature dependence of the process $00^o1 \rightarrow 01^10 + 1682$ cm^{-1}. The rate constant decreases only slowly with temperature for Ne and Ar+CO_2 and for N_2+CO_2 it even increases slightly with temperature below 500 K [11]. Due to the large energy mismatch the rate constant is however small 10^{-17} - 10^{-16} cm^3/sec at 300 K. Since VT conversion is a slow process there will at low temperatures be an increasing tendency to find near resonant path ways, i.e. to conserve translational energy. However, if the transition has to be induced by successive rotational excitation (by a P_2 term e.q.) there is a competition from "conservation of angular momentum", i.e. the transition does not automatically go through the resonant or near resonant pathway. As an example we may consider the process $H_2(v=1,j)$ + ^{4}He → $H_2(v=0,j')$ + ^{4}He + ΔE. The intermolecular potential has a large P_2 term (~ 20% of the P_0 term) whereas the P_4 term is extremely small. The leading transitions at low energies are 10-04 and 10-06 with ΔE = 2992 cm^{-1} and ΔE = 1749 cm^{-1} but not 10-08 with ΔE = 108 cm^{-1} (see also table 4). The reason is that transition has to be supported by the potential by successive use of the P_2 term during the collision, i.e. through 10-02-04-(06) process.

For polyatomic molecules the energy spacing between the vibrational levels decreases significantly. Thus the mode with the smallest frequency in SF_6 has ω_6 = 350 cm^{-1}. These soft vibrational modes absorbes as much energy as does the rotational degrees of freedom in 1 eV collisions with Ar [30]. For larger polyatomic molecules or polyatomic molecules colliding with atoms or molecules with smaller masses, as He there will be an increasing tendency towards vibrational excitation as compared to rotational excitation. This effect is already seen in CO_2 where the rate constant for deactivation of the bending mode $CO_2(01^10)$ + X → $CO_2(00^00)$ + X is about 10^{-13} cm^3/sec if X = He but about 10^{-15}cm^3/sec if X = Ne at 300 K. The change in rotational energy of CO_2 induced by He collisions is however a factor of 2-3 smaller than that induced by Ne [31]. In order to understand the energy transfer in small and moderate sized polyatomic molecules it is therefore necessary to include an accurate description of both the rotational, the vibrational degrees of freedom and the coupling between them.

TABLE 1

Comparison of quantum mechanical and semiclassical cross sections for rotational excitation of H_2 colliding with ^4He.

	Transition	Cross sections in a_o^2		
E (eV)	j → j'	SCS[a]	CC[b]	SC[c]
0.1 eV	0 → 2	0.71	0.73	0.69
0.9 eV	0 → 2	12.35	11.34	12.00
0.9 eV	0 → 4	0.83	0.81	0.80
0.9 eV	2 → 4	3.75	3.74	3.60

a) Semiclassical couples states results [13].
b) Quantum mechanical (close coupling) [17].
c) Quantum mechanical (coupled states) [17].

TABLE 2

Comparison of semiclassical and quantum mechanical calculations of ^{4}He+para NH$_3$ at E = 250 cm^{-1}.

Transition		Cross sections (Å^2)			
jkε →	j'k'ε'	CC[a]	CS[b]	SCC[c]	SCS[d]
11+ →	11−	0.81	0.89		0.64
	21+	1.00	1.00		1.01
	21−	0.21	0.15		0.27
	22+	0.0006	0.0005		0.002
	22−	1.56	1.47		1.38
	31+	0.069	0.14		0.13
	31−	0.36	0.36		0.45
	32+	0.41	0.42		0.40
	32−	0.22	0.49		0.44
	41+	0.12	0.15		0.18
	41−	0.0076	0.0050		0.0085
	42+	0.053	0.053		0.12
	42−	0.033	0.048		0.041
	44+	0.044	0.045		0.101
	44−	1.24	1.15		1.45
21+ →	21−	0.62	0.65	0.61	0.60
	22+	0.87	0.89	0.81	0.82
	22−	0.16	0.23	0.28	0.31
	31+	0.85	0.94	0.90	0.91
	31−	0.070	0.053	0.10	0.075
	32+	0.018	0.019	0.029	0.031
	32−	0.81	0.85	0.80	0.79
	41+	0.019	0.055	0.023	0.023
	41−	0.085	0.089	0.11	0.11
	42+	0.20	0.20	0.22	0.20
	42−	0.075	0.13	0.13	0.14
	44+	0.91	0.94	0.93	0.92
	44−	0.14	0.13	0.23	0.25
31+ →	31−	0.30	0.34		0.33
	32+	0.77	0.82		0.75
	32−	0.041	0.047		0.071
	41+	0.72	0.73		0.73
	41−	0.023	0.030		0.04
	42+	0.018	0.022		0.03
	42−	0.48	0.51		0.48
	44+	0.17	0.19		0.27
	44−	0.51	0.52		0.51

a) Quantum mechanical (close coupling) [18].
b) Quantum mechanical (coupled states) [18].
c) Semiclassical (close coupling) [14].
d) Semiclassical (coupled states) [14].

TABLE 3

Comparison of semiclassical and quantum mechanical vibrational transition probabilities for collinear atom-diatom collisions.

System[a]	Energy (E)[b]	Transition	Semiclassical[c]	Quantum[a]
1	7.6	0 - 1	4.29(-5)	4.30(-5)
		0 - 2	1.38(-11)	1.28(-11)
	8.8	0 - 1	2.02(-4)	2.03(-4)
		0 - 2	1.13(-9)	1.13(-9)
		1 - 2	2.21(-5)	2.23(-5)
	10.0	0 - 1	6.57(-4)	6.58(-4)
		0 - 2	2.51(-8)	2.51(-8)
		1 - 2	1.51(-4)	1.52(-4)
	12.0	0 - 1	2.84(-3)	2.85(-3)
		0 - 2	9.40(-7)	9.43(-7)
		1 - 2	1.31(-3)	1.32(-3)
		2 - 3	2.25(-4)	2.28(-4)
	16.0	0 - 1	1.90(-2)	1.92(-2)
		0 - 2	8.04(-5)	8.12(-5)
		1 - 2	1.64(-2)	1.66(-2)
		2 - 3	8.38(-3)	8.49(-3)
	20.0	0 - 1	6.21(-2)	6.39(-2)
		0 - 2	1.20(-3)	1.24(-3)
		1 - 2	7.00(-2)	7.14(-2)
		2 - 3	5.42(-2)	5.5(-2)
6	12.8365	0 - 1	0.276	0.317
		0 - 2	4.05(-2)	4.86(-2)
		0 - 3	2.10(-3)	2.50(-3)
		1 - 2	0.305	0.335
		1 - 3	4.04(-2)	4.47(-2)
	14.8365	0 - 1	0.328	0.378
		0 - 2	7.65(-2)	9.49(-2)
		0 - 3	7.64(-3)	9.68(-3)
		1 - 2	0.343	0.376
		1 - 3	9.46(-2)	0.110
	16.8365	0 - 1	0.357	0.409
		0 - 2	0.119	0.150
		0 - 3	1.89(-2)	2.50(-2)
		1 - 2	0.323	0.343
		2 - 3	0.308	0.321
		1 - 3	0.157	0.188
		1 - 4	2.42(-2)	2.97(-2)
		2 - 4	0.147	0.167
		2 - 5	0.016	0.017
8	6.0	0 - 1	2.14(-2)	2.21(-2)
		1 - 2	8.65(-4)	8.98(-4)

TABLE 3 (continued)

System[a]	Energy (E)[b]	Transition	Semiclassical[c]	Quantum[a]
	8.0	0 - 1	9.76(-2)	0.108
		1 - 2	4.06(-2)	4.18(-2)
		2 - 3	1.26(-3)	1.33(-3)
		0 - 2	1.16(-3)	1.22(-3)
		1 - 3	1.41(-5)	1.46(-5)
	10.0	0 - 1	0.211	0.252
		1 - 2	0.171	0.182
		2 - 3	5.81(-2)	5.93(-2)
		0 - 2	1.31(-2)	1.52(-2)
		1 - 3	3.19(-3)	3.31(-3)
	12.0	0 - 1	0.310	0.394
		1 - 2	0.308	0.345
		2 - 3	0.226	0.233
		0 - 2	5.17(-2)	6.78(-2)
		1 - 3	3.34(-2)	3.70(-2)
		2 - 4	5.85(-3)	6.00(-3)
	16.0	0 - 1	0.354	0.434
		1 - 2	0.242	0.220
		2 - 3	0.258	0.250
		0 - 2	0.188	0.291
		0 - 3	4.26(-2)	7.12(-2)
		1 - 3	0.204	0.261
		1 - 4	3.90(-2)	5.12(-2)
		2 - 4	0.165	0.189
		2 - 5	1.50(-2)	1.64(-2)
	20.0	0 - 1	0.250	0.218
		1 - 2	2.10(-2)	8.54(-3)
		2 - 3	2.55(-3)	1.76(-2)
		0 - 2	0.264	0.366
		0 - 3	0.143	0.267
		0 - 4	4.13(-2)	8.91(-2)
		1 - 3	0.199	0.170
		1 - 4	0.167	0.240
		1 - 5	4.75(-2)	7.69(-2)
		2 - 4	0.189	0.169
		2 - 5	0.165	0.194
		2 - 6	2.90(-2)	3.71(-2)

a) Refers to the 8 different systems investigated in [19].
b) In units of $\frac{1}{2}\hbar\omega$.
c) From ref. [15] (4.29(-5) = $4.29 \cdot 10^{-5}$).
d) From ref. [19].

TABLE 4

Comparison of semiclassical and quantum mechanical cross sections for vib/rot transitions in H_2 colliding with ^4He.

Energy (eV)	$\sigma_{00 \to 1}(E)$[a] Å^2	
	Semiclassical[b]	Quantum mechanical[c]
1.0	2.5(-7)	
1.1	2.1(-6)	
1.2	1.0(-5)	9.7(-6)
1.3	3.2(-5)	
1.6	2.8(-4)	3.3(-4)
1.8	7.7(-4)	1.0(-3)

a) $\sigma_{00 \to 1} = \sum_{j'} \sigma_{00 \to 1j'}$

b) From ref. [20] using a classical description of the rotational motion of H_2, a Morse vibrator and the Alexander-Berard [21] surface.

c) From ref. [22], coupled states calculation using a Morse oscillator/rigid rotator and the AB surface.

Energy (eV)[a]	$\sigma_{00 \to 0j}$ (Å^2)					$\sigma_{10 \to 0}$ (Å^2)
	j=0	2	4	6	8	
0.7908	0.8(-10)	1.6(-10)	4.6(-10)	1.1(-10)	1.7(-14)	8.0(-10)[b]
	1.5(-10)	2.5(-10)	1.5(-9)	3.3(-10)		2.2(-9)[c]
0.7968	2.0(-10)	4.1(-10)	1.2(-10)	2.8(-10)	7.1(-14)	2.0(-9)[b]
	2.7(-10)	4.4(-10)	2.5(-9)	7.3(-10)		3.9(-9)[c]
0.8098	8.1(-10)	1.6(-9)	4.8(-9)	1.1(-9)	7.1(-13)	8.3(-9)[b]
	9.5(-10)	1.3(-9)	6.6(-9)	2.9(-9)		1.2(-8)[c]
0.8398	7.4(-9)	1.5(-8)	4.2(-8)	1.1(-8)	3.2(-11)	7.5(-8)[b]
	5.9(-9)	8.2(-8)	4.4(-8)	2.0(-8)		7.8(-8)[c]
0.8998	1.0(-7)	1.7(-7)	5.8(-7)	1.7(-7)	2.5(-9)	1.0(-6)[b]
	5.8(-8)	8.0(-8)	4.2(-7)	2.3(-7)	2.1(-9)	0.8(-6)[c]

a) Threshold energy E_{th} = 0.78584 eV.

b) Quantum mechanical coupled states calculation [23] using a Morse oscillator and the Gordon-Secrest [24] surface.

c) Semiclassical coupled states calculation [5] using a rigid rotator-Morse vibrator and the GS-surface.

TABLE 5

$\log_{10}(k\ cm^3/sec)$	Cross section Å^2	$\log_{10}$(probability)				
-10	20	0	⎤ Rotational Relaxation	CO_2		⎤
-12	0.2	-2	⎦	$H_2(2\to 0)$ ⎤	HF+HF	
-14	2(-3)	-4				
-16	2(-5)	-6		Vibration-Translation/Rotation rates		Vibration-Vibration rates
-18	2(-7)	-8				
-20	2(-9)	-10				
-22	2(-11)	-12		⎦	N_2+N_2	⎦

REFERENCES

[1] Price, R.J., Clary, D.C. and Billing, G.D. 1983, Chem.Phys. Lett. 101, pp. 269.
Jolicard, G. and Billing, G.D. 1984, Chem.Phys. 85, pp. 253.

[2] Billing, G.D. 1978, Chem.Phys. 33, pp. 227; 1979, 41, pp. 11; Chem.Phys.Lett. 1982, 89, pp. 337.

[3] Billing, G.D. 1980, Chem.Phys. 51, pp. 417.

[4] Billing, G.D. 1980, Chem.Phys. 49, pp. 255.

[5] Billing, G.D. 1978, Chem.Phys. 30, pp. 387.

[6] Lawley, K.P. and Ross, J. 1965, J.Chem.Phys. 43, pp. 2943.

[7] Billing, G.D. 1976, J.Chem.Phys. 65, pp. 1.

[8] Brink, D.M. and Satchler, G.R. 1968, "Angular Momemtum", (Clarendon Press, Oxford).

[9] Billing, G.D. 1983, Chem.Phys.Lett. 100, pp. 535.

[10] Billing, G.D. 1975, Chem.Phys.Lett. 30, pp. 391; 1976, J.Chem.Phys. 64, pp. 908; Muckermann, J.T., Rusinek, I., Roberts, R.E. and Alexander, M. 1976, J.Chem.Phys. 65, pp. 2416; Billing, G.D. and Jolicard, G. 1982, Chem.Phys. 65, pp. 323.

[11] Billing, G.D. 1982, Chem.Phys. 64, pp. 35; 1983, 79, pp. 179.

[12] McCann, K.J. and Flannery, M.R. 1975, Chem.Phys.Lett. 35, pp. 124; DePristo, A.E. 1983, J.Chem.Phys. 78, pp. 1237.

[13] Billing, G.D. 1977, Chem.Phys.Lett. 50, pp. 320.

[14] Billing, G.D. and Poulsen, L.L. (in press), J.Chem.Phys.

[15] Billing, G.D. 1978, "Introduction to the theory of inelastic collisions in chemical kinetics", Copenhagen.

[16] Billing, G.D. 1975, Chem.Phys. 9, pp. 359.

[17] Shimoni, Y. and Kouri, D.J. 1977, J.Chem.Phys. 66, pp. 675.

[18] Green, S. 1976, J.Chem.Phys. 64, pp. 3463.

[19] Secrest, D. and Johnson, B.R. 1966, J.Chem.Phys. 45, pp. 4556.

[20] Billing, G.D. 1975, Chem.Phys. 9, pp. 359.

[21] Alexander, M.M and Berard, E.V. 1974, J.Chem.Phys. 60, pp. 3950.

[22] Alexander, M.M. 1974, J.Chem.Phys. 61, pp. 5167.

[23] Alexander, M,M. and Mc Guire, P. 1976, J.Chem.Phys. 64, pp. 452.

[24] Gordon, M.D. and Secrest, D. 1970, J.Chem.Phys. 52, pp. 127.

[25] Cacciatore, M and Billing, G.D. 1981, Chem.Phys. 58, pp. 395.

[26] Billing, G.D. 1979, Chem.Phys. 41, pp. 11.

[27] Billing, G.D. 1980, Chem.Phys. 50, pp. 165.

[28] Poulsen, L.L. and Billing, G.D. 1982, Chem.Phys. 73, pp. 313.

[29] Billing, G.D. 1981, Chem.Phys. 61, pp. 415.

[30] Billing, G.D. 1983, Chem.Phys. 79, pp. 179.

[31] Billing, G.D. and Clary, D.C. 1982, Chem.Phys.Lett. 90, pp. 27.

CALCULATION OF RADIATIVE LIFETIMES OF ALLOWED AND FORBIDDEN TRANSITIONS

Jens Oddershede

Department of Chemistry
Odense University
DK-5230 Odense M
Denmark

ABSTRACT The calculation of radiative lifetimes of electronically excited states of molecules is reviewed. Both dipole allowed and forbidden electronic transitions will be treated. A literature survey shows a strongly increasing activity in this area of molecular physics within the last few years.

1. INTRODUCTION

Optical spectroscopy of interstellar matter gave the first proof of the existence of molecules in interstellar space [1]. Molecules like CH, CH^+, and CN were first discovered by optical spectroscopy and later identified by comparison with laboratory experiments. In recent years microwave spectroscopy has taken over as the branch of spectroscopy which provides most discoveries of new molecules in interstellar space. Still, however, optical spectroscopy continues to give important new information about molecular line spectra. In particular, estimates of molecular abundances are often obtained from measurements of equivalent widths of optical absorbtion lines.

In order to extract information about abundances from equivalent widths it is necessary to know the oscillator strength of the transition in question. Laboratory measurements and theoretical calculations of band absorption oscillator strengths or radiative lifetimes of electronically excited states are thus an essential link in the series of information that is needed to calculate molecular abundances and consequently to provide the input for the theories of the chemical development and composition of interstellar matter. In addition

G. H. F. Diercksen et al. (eds.), Molecular Astrophysics, 533–549.

to standard molecular spectroscopy which is used to identify spectra seen in interstellar environments also time-resolved spectroscopy is needed to measure the intensities of the transitions. This branch of spectroscopy, even though rather new, has contributed substantially to our knowledge of line intensities. A recent review is given by Dumont and Remy [2].

The theoretical calculations of oscillator strength have been less abundant. At the time of my first review [3] of this area in 1979 only 11 calculations on 8 diatomic molecules had been reported. Lately the theoretical activity has been increasing as will be demonstrated in sec. 2. The reason for the rather low number of calculations of intensities is probably that in general it is "easier" to compute energies than wavefunction to any given degree of accuracy. Furthermore, most of the methods used in theoretical molecular physics are based on the variational principle, i.e. an energy principle which does not yield wavefunction of the same overall quality as the energy. In fact, from perturbation theory we know that to obtain an energy to order 2p-1 we only require knowledge of the wavefunction to order p [4]. Also, the basis set problem, i.e. the truncation error introduced by the use of finite basis sets, seems to effect wavefunctions and thus transition probabilities much more seriously than total energies. The calculation of Larsson et al. [5] on CH is a good illustration of the kind of basis set problems one may encounter. In the configuration interaction (CI) method, which is the most commonly used quantum chemical method in electronic structure calculations, it is standard usage to perform separate energy optimization on individual states often using different basis sets for different states. The states thus obtained are non-orthogonal and the calculation of transition matrix elements between them is not a trivial problem. Until recently [6] this problem was a stumbling block in many CI calculations of transition probabilities.

The polarization propagator method [7-9] is an alternative to the state approaches like CI and multiconfiguration SCF (MCSCF). In the propagator method the output of the calculation are energy differencies and the transition matrix element between the states, rather than the states themselves. A propagator calculation thus gives as output the quantities needed to compute radiative lifetimes. Relative to CI and MCSCF approaches the polarization propagator method is rather new in molecular physics, and only a few calculations using propagator methods have been reported (see sec. 3). Also polarization propagator calculations apply finite basis set expansions, in this case of the excitation operators. This means that like in CI and MCSCF type calculations the choice of basis set is often crucial to the outcome of the calculation, again, in particular with respect to the quality of the electronic transition moments. Energy differences are less affected by the choice of basis set.

In the next section we will briefly review the definition of the radiative lifetime of an excited state. The discussion will apply to both dipole allowed and dipole forbidden transitions.

2. RADIATIVE LIFETIMES

Spontaneous emission from state $|i\rangle$ to state $|f\rangle$ is determined by the Einstein coefficient

$$A_{if} = \frac{64\pi^4 \tilde{\nu}_{if}^3}{3h} \{ |\langle i|\vec{M}_{E1}|f\rangle|^2 + |\langle i|\vec{M}_{M1}|f\rangle|^2 +$$

$$\frac{3}{10}\tilde{\nu}_{if}^2 \pi^2 |\langle i|\vec{M}_{E2}|f\rangle|^2 + \ldots \} \qquad (1)$$

where

$$\vec{M}_{E1} = \sum_i e\vec{r}_i \qquad (2)$$

$$\vec{M}_{M1} = \frac{e\hbar}{2m} \sum_i (\vec{l}_i + 2\vec{s}_i) \qquad (3)$$

$$\vec{M}_{E2} = \sum_i e\, \vec{q}_i \qquad (4)$$

and the quadrupole operator $\vec{q}$ has the components

$$q(\pm 2) = \tfrac{1}{2}[x^2 - y^2 \pm i(xy + yx)] \qquad (5)$$

$$q(\pm 1) = \tfrac{1}{2}[x^2 + zx \pm i(yz + zy)] \qquad (6)$$

$$q(0) = (\tfrac{2}{3})^{1/2}[z^2 - \tfrac{1}{2}(x^2 + y^2)] \qquad (7)$$

In Eq.(1) the multipole expansion is terminated after the quadrupole (E2) terms and $\tilde{\nu}_{if} = (E_f - E_i)/hc$. For dipole allowed transitions the first term dominates. Almost all calculations so far have been on allowed electronic transitions, but recently [10] we have computed the last two terms in Eq.(1) for a dipole forbidden transition, the Lyman-Birge-Hopfield (a $^1\Pi_g \leftarrow X\,^1\Sigma_g^+$) band system of N_2. For electric dipole forbidden transitions both the second and third term in Eq.(1) may become important unless, of course, one of them is identically zero for symmetry reasons. The a $^1\Pi_g \leftarrow X\,^1\Sigma_g^+$ transition in N_2 represents a system for which both the M1 and E2 transition moments contribute to

A_{if} even though we do find that the magnetic dipole term gives more than 90% of the Einstein coefficient in that particular case.

Spin-forbidden transitions represent another class of forbidden transition for which only very few calculations are available [11-12]. The calculations of Einstein coefficients for these transitions are based on the concept of "intensity borrowing": the spin-forbidden transition borrows intensity from a close-lying spin-allowed transition. Formally, this is accomplished by computing the perturbation correction to the excited state wavefunction $|i>$ originating for the spin-orbit perturbation operator H_{SO}

$$|i> = |i^o> + \sum_{k \neq o} \frac{<i^k|H_{SO}|i^o>}{E^o - E^k} |i^k> \qquad (8)$$

Here $\{|i^k>\}$ must be a set of states with different spin multiplicity than $|i^o>$ in order for the last turn to be non-vanishing. Typically, they are singlet excited states if the final state has triplet spin symmetry, that is we consider an excitation from a singlet initial state to a final triplet state. If we use the perturbed state $|i>$ of Eq.(8) to calculate the first term in Eq.(1), we find

$$A_{if} = \frac{64\pi^4 \tilde{\nu}_{if}^3}{3h} \left| \sum_{k \neq o} \frac{<i^o|H_{SO}|i^k><i^k|\vec{M}_{E1}|f>}{E^o - E^k} \right|^2 \qquad (9)$$

In certain cases [11] it may become necessary to insert Eq.(8) in the second or third term in Eq.(1) to obtain a non-zero contribution to A_{if}.

For allowed as well as for forbidden transitions the radiative lifetime τ_i of an excited state $|i>$ is obtained by summing the Einstein coefficients for all lower lying states $|f>$

$$\tau_i = g_i \left[\sum_{f<i} A_{if} \right]^{-1} \qquad (10)$$

where g_i is the degeneracy of the initial state. Using the Born-Oppenheimer approximation we express the states as

$$|i> = |n'v'J'> \qquad (11)$$

$$|f> = |n''v''J''> \qquad (12)$$

where n, v, and J are electronic, vibrational, and rotational quantum numbers, respectively. We thus compute radiative lifetimes of rovibronic levels of electronically excited states. The computational expressions that we obtain when Eqs.(11) and (12) are introduced in Eqs.(1) or (9) are given many places [3,13,14] and will not be repeated here.

Equations (1), (9), and (10) show that in order to compute the radiative lifetime of an excited state we need to know energy differences and transition moments. In state formulations like CI and/or MCSCF procedures these quantities are calculated from knowledge of the wavefunctions and energies of the initial and final states. This is the most widely used approach at present as we will see in sec. 3. However, as mentioned earlier, it is also possible to compute energy differences and transition moments directly without knowing the wavefunctions of the individual states, namely as poles and residues, respectively, of the polarization propagator [7-9].

3. LITERATURE SURVEY

As already discussed in sec. 1 it was not until recently that we have seen a rapid growth in number of papers dealing with calculation of radiative lifetimes. This is evident from Table 1. Almost all the papers that are included have appeared since 1980.

A few explanatory remarks about Table 1 are probably in place. Only calculations which have been published since my last review [3] have been included. As mentioned previously, the bottleneck in lifetime calculations is the evaluation of the electronic transition moment

$$\vec{M}_e(R) = <n'|\vec{M}|n''> \qquad (13)$$

where $|n'>$ and $|n''>$ are defined in Eqs.(11) and (12) and $\vec{M}$ is the appropriate transition moment (see Eqs.(2)-(4)). Thus, also entries that computes only $M_e(R)$ but not lifetimes have been included in Table 1.

The more or less standard abbreviations are used to describe the theoretical methods in Table 1. Methods like configuration interaction (CI), self-consistent field (SCF), and multi-configurational SCF (MCSCF) probably need not be explained here. The self-consistent electron pair (SCEP) method and its application to calculation of electronic transition moments have been described by Werner [53]. The multi-reference double CI (MRD-CI) method which has been used for several molecules is the subject of ref. [42]. The complete active space SCF (CASSCF) method is a variant of the MCSCF method in which the configurations used in the MCSCF procedure are all configurations originating from a specified set of (active) orbitals. The generalized valence bond (GVB) method also resembles the MCSCF methods. Details can be found in ref. [54]. The polarization propagator method (in Table 1 referred to as PROP) has been discussed previously and a description of the method and its applications is given elsewhere [7-9,45]. The calculation reported by us [3,10,12] is performed with the so-called second order polarization propagator method while those of Weiner et al. [44-45] and Sangfelt et al. [38] employ an antisymmetrized geminal product (AGP) representation of the reference state.

There are at present only a few calculations of electronic transition moments for polyatomic molecules. There is a calculation of band oscillator strenghts of the $^1A_2 \leftarrow {}^1A_1$ band system of formaldehyde [55]. Separate SCF calculations are performed for the two states. We have also applied the polarization propagator method to two polyatomic molecules, namely $C_3H_3^+$ [56] and SiC_2 [57]. In the latter case we find a sizeable difference between the radiative lifetime of the lowest excited state for SiC_2 in its linear ($C_{\infty v}$) and in its triangular (C_{2v}) conformation. The lifetime of the triangular conformation agrees well with laboratory measurements [58] of τ of SiC_2 thus confirming recent experimental [59] and theoretical [60] findings indicating that SiC_2 is triangular in its most stable conformation. We do find, however, that there is only a very small barrier towards linearity (~1 kcal/ mole).

There may be other calculations of τ or $f_{v'v''}$ of polyatomic molecules, but it is evident that calculations on polyatomic molecules is still a field awaiting the expansion seen in similar calculations on diatomic molecules in the last few years.

ACKNOWLEDGEMENTS

I would like to thank the organizers of this workshop for the invitation to contribute to the scientific program. I would also like to acknowledge useful discussions with P. Erman and M. Larsson on the subject covered in this review.

Table 1. Survey of calculated radiative lifetimes, τ, of diatomic molecules.

Molecule	band system[a] excited state	$\tau(0)$[b]	v'_{max}[c]	$\tau(v'_{max})$[b]	$f^d_{v'v''}$	$\vec{M}_e(R)$[d]	method[e]	basis set[e]	ref/year
AlH$^+$	$A^2\Pi$	257	0	257	–	+	MCSCF	77G	[15] 82
	$B^2\Sigma^+$	–	–	–	–	+	MCSCF	77G	[15] 82
BH$^+$	$A^2\Pi$	1.065μs	0	1.065μs	–	+	MCSCF	60G	[15] 82
	$B^2\Sigma^+$	–	–	–	–	+	MCSCF	60G	[15] 82
C$_2$	$d^3\Pi_g \to a^3\Pi_u$	–	–	–	–	+	SCF+CI	46S	[16] 78
	$d^3\Pi_g \to a^3\Pi_u$	–	–	–	–	+	MRD–CI	60G	[17] 81
	$e^3\Pi_g \to a^3\Pi_u$	–	–	–	–	+	MRD–CI	60G	[17] 81
	$b^3\Pi_g \to a^3\Pi_u$	–	–	–	–	+	SCF+CI	46S	[18] 81
	$d^3\Pi_g \to a^3\Pi_u$	–	–	–	–	+	SCF+CI	46S	[18] 81
	$e^3\Pi_g \to a^3\Pi_u$	–	–	–	–	+	SCF+CI	46S	[18] 81
	$A^1\Pi_u$	11.1μs	9	4.0μs	+	+	MRD–CI	68G	[19] 83
	$A^1\Pi_u \to X^1\Sigma_g^+$	–	–	–	+	+	SCF+CI	70S	[20] 83
	$2^1\Pi_u \to X^1\Sigma_g^+$	–	–	–	+	+	SCF+CI	70S	[20] 83

cont...

Molecule	band system[a] / excited state	$\tau(0)$[b]	v'_{max}[c]	$\tau(v'_{max})$[b]	$f^d_{v'v''}$	$\vec{M}_e(R)$[d]	method[e]	basis set	ref/year
C_2	$b^3\Sigma^-_g$	16µs	9	7.3µs	+	+	MRD-CI	60G	[21] 83
	$A^1\Pi_u$	10.7µs	9	3.8µs	+	+	MRD-CI	60G	[21] 83
	$C^1\Pi_g$	27	3	33	+	+	MRD-CI	60G	[21] 83
	$D^1\Sigma^+_u \leftarrow X^1\Sigma^+_g$	14.0	5	16.2	+	+	MRD-CI	60G	[21] 83
	$B^2\Sigma^+_u \leftarrow X^2\Sigma^+_g$	–	–	–	–	+	MRD-CI	54G	[22] 78
	$A^2\Pi_u$	49.9µs	10	18.8µs	+	+	SCEP	–	[23] 84
	$B^2\Sigma^+_u$	76.5	10	90.0	+	+	SCEP	–	[23] 84
CH	$A^2\Delta$	525	2	660	+	+	CASSCF	52G	[5] 83
CH^+	$A^1\Pi$	850	2	1.15µs	+	+	CASSCF	44G	[24] 83
CN	$A^2\Pi$	11.2µs	19	4.5µs	+	+	GVB-CI	38G	[25] 82
	$B^2\Sigma^+$	62.1	16	126	+	+	GVB-CI	38G	[25] 82
	$A^2\Pi$	8.1µs	10	3.9µs	–	+	CASSCF	44G	[26] 83
	$B^2\Sigma^+$	66.6	5	70.7	–	+	CASSCF	44G	[26] 83

cont...

Molecule	band system[a] / excited state	$\tau(0)$[b]	v'_{max}[c]	$\tau(v'_{max})$[b]	$f^d_{v',v''}$	$\vec{M}_e(R)$[d]	method[e]	basis set[e]	ref/year
CO	$A^2\Pi$	11.3µs	9	5.2µs	+	+	SCF+CI	40S	[27] 84
	$B^2\Sigma^+$	72	9	88	+	+	SCF+CI	40S	[27] 84
	$A^1\Pi$	9.9	8	10.9	+	+	SCF+CI	70S	[28] 81
	$B^1\Sigma^+$	11.2	1	8.7	+	+	SCF+CI	70S	[28] 81
	$C^1\Sigma^+$	2.1	1	2.0	+	+	SCF+CI	70S	[28] 81
	$a^3\Pi$	2.5ms	5	0.7ms	+	−	PROP	58G	[12] 84
CO$^+$	$A^2\Pi$	4.70µs	9	2.06µs	−	+	MCSCF	−	[29] 82
CsH	$A^1\Sigma^+$	72.3	35	31.4	−	−	ECP[f]	−	[30] 84
F$_2$	$^1\Pi_u \leftarrow X^1\Sigma_g^+$	−	−	−	+	+	GVB-CI	38G	[32] 79
H$_2^g$	$\{EF, GK, \overline{HH}\}^1\Sigma_g^+, I^1\Pi_g$	203	−	20−1200[h]	−	−	−[i]	−	[33] 84
HCl	$B^1\Sigma^+$	3	0	3	+	+	MRD-CI	47G	[35] 82
	$C^1\Pi - X^1\Sigma^+$	−	−	−	+	+	MRD-CI	47G	[35] 82
HCl$^+$	$A^2\Sigma^+$	2.51µs	6	1.68µs	+	+	SCEP	61G	[36] 84

cont...

Molecule	band system / excited state	$\tau(0)$ [b]	v'_{max} [c]	$\tau(v'_{max})$ [b]	$f^d_{v'v''}$	$\vec{M}_e(R)$ [d]	method [e]	basis set	ref/year
HF^+	$A^2\Sigma^+$	26.2μs	3	32.6μs	+		SCEP	61G	[36] 84
KCl	$A^1\Pi$	–	–	–	–	+	Semiemp.	–	[37] 83
	$B^1\Sigma^+$	–	–	–	–	+	Semiemp.	–	[37] 83
	$C^1\Pi$	–	–	–	–	+	Semiemp.	–	[37] 83
	$D^1\Pi$	–	–	–	–	+	Semiemp.	–	[37] 83
	$E^1\Sigma^+$	–	–	–	–	+	Semiemp.	–	[37] 83
KF	(like KCl)								
Li_2	$A^1\Sigma_u^+$	18.59	9	19.28	–	+	PROP	32G	[38] 84
	$B^1\Pi_u$	6.81	7	7.44	–	+	PROP	32G	[38] 84
LiCl	(like KCl)								
LiF	(like KCl)								
LiH	$A^1\Sigma^+$	29.2^j	8	32.6	+	+	SCF+CI	60S	[39] 81
	$B^1\Pi$	11.3	2	23.5	+	+	SCF+CI	60S	[39] 81

cont...

RADIATIVE LIFETIMES OF ALLOWED AND FORBIDDEN TRANSITIONS

Molecule	band system[a] excited state	$\tau(0)$[b]	v'_{max}[c]	$\tau(v'_{max})$[b]	$f^d_{v'v''}$	$\vec{M}_e(R)$[d]	method[e]	basis set	ref/year
MgH	$A^2\Pi$	–	–	–	+	+	MCSCF+CI	66S	[40] 79
	$B'^2\Sigma^+$	–	–	–	+	+	MCSCF+CI	66S	[40] 79
	$B'^2\Sigma^+$	88	0	88	+	+	SCF+CI	35S	[41] 79
N_2	$C^3\Pi_u$	32[k]	–	–	–	+	MRD–CI	–	[42] 83
	$B^3\Pi_g$	13.4μs	12	4.75μs	+	+	MCSCF+CI	84G	[43] 84
	$W^3\Delta_u$	31.6s	12	79.8μs	+	+	MCSCF+CI	84G	[43] 84
	$B^3\Sigma_u^-$	46.0μs	12	13.6μs	+	+	MCSCF+CI	84G	[43] 84
	$C^3\Pi_u$	36.7	4	39.4	+	+	MCSCF+CI	84G	[43] 84
	$B^3\Pi_g$	12.5μs	14	3.8μs	–	+	PROP	66G	[44] 84
	$b^1\Pi_u$	82	5	101	–	+	PROP	66G	[45] 84
	$c^1\Pi_u$	9	5	12	–	+	PROP	66G	[45] 84
	$o^1\Pi_u$	120	5	70	–	+	PROP	66G	[45] 84
	$c'^1\Sigma_u^+$	88	5	31	–	+	PROP	66G	[45] 84

cont...

Molecule	band system[a] excited state	$\tau(0)$[b]	v'_{max}[c]	$\tau(v'_{max})$[b]	$f^d_{v'v''}$	$\vec{M}_e(R)$[d]	method[e]	basis set[e]	ref/year
N_2^+	$a^1\Pi_g$	51.1μs	9	100.3μs	+	+	PROP	58G	[10] 85
	$A^2\Pi_u$	17.45μs	5	9.69μs	–	+	GVB-CI	38G	[46] 80
	$B^2\Sigma_u^+$	55.15	5	53.28	–	+	GVB-CI	38G	[46] 80
	$D^2\Pi_g$	287	5	356	–	+	GVB-CI	38G	[46] 80
	$C^2\Sigma_u^+ \leftarrow X^2\Sigma_g^+$	60.5	5	49.4	–	+	GVB-CI	38G	[46] 80
	$^4\Pi_g$	247μs	5	153μs	–	+	GVB-CI	38G	[46] 80
NaCl	(see KCl)								
NaF	(see KCl)								
O_2	$a^1\Delta_g$	5270s	0	5270s	–	–	MRD-CI	55G	[11] 84
	$b^1\Sigma_g^+$	11.65s	0	11.65s	–	–	MRD-CI	55G	[11] 84
OD	$A^2\Sigma^+$	–	–	–	+	+	CI	63S	[47] 83
OH	$A^2\Sigma^+$	580–910	2	690–750	–	+	CI/MR	42G[1]	[48] 82
	$B^2\Sigma^+$	235–350	1	410–590	–	+	CI/MR	42G[1]	[48] 82

cont....

Molecule	band system[a] excited state	$\tau(0)$[b]	v'_{max}[c]	$\tau(v'_{max})$[b]	$f^d_{v'v''}$	$\vec{M}_e(R)$[d]	method[e]	basis set[e]	ref/year
	$C^2\Sigma^+$	6.7–8.1	1	6.5–7.9	–	+	CI/MR	42G[1]	[48] 82
	$1^2\Sigma^-\to X^2\Pi$	–	–	–	–	+	CI/MR	42G[1]	[49] 83
	$D^2\Sigma^-\to X^2\Pi$	–	–	–	+	+	CI/MR	42G[1]	[49] 83
	$A^2\Sigma^+\to X^2\Pi$	–m	–	–	+	+	CI/MR	42G[1]	[50] 83
	$B^2\Sigma^+\to X^2\Pi$	–m	–	–	+	+	CI/MR	42G[1]	[50] 83
	$C^2\Sigma^+\to X^2\Pi$	–m	–	–	+	+	CI/MR	42G[1]	[50] 83
	$1-3\,^2\Delta\to X^2\Pi$	–m	–	–	+	+	CI/MR	42G[1]	[50] 83
	$2-5\,^2\Pi\to X^2\Pi$	–m	–	–	+	+	CI/MR	42G[1]	[50] 83
	$1-5\,^2\Sigma^-\to X^2\Pi$	–m	–	–	+	+	CI/MR	42G[1]	[50] 83
PH$^+$	$A^2\Sigma^+$	590	0	590	+	+	SCEP	76G	[36] 84
	$A^2\Delta$	1.40μs	2	1.69μs	–	+	CASSCF	58G	[51] 85
S$_2$	$a^1\Delta_g$	350s	0	350s	–	–	MRD–CI	59G	[11] 84
	$b^1\Sigma_g^+$	3.4s	0	3.4s	–	–	MRD–CI	59G	[11] 84
									cont...

Mole-cule	band system[a] excited state	$\tau(0)$[b]	v'_{max}[c]	$\tau(v'_{max})$[b]	$f^d_{v'v''}$	$\vec{M}_e(R)$[d]	method[e]	basis set[e]	ref/year
SiO$^+$	A$^2\Pi$	336ms	10	140µs	+	+	MCSCF	88G	[32] 82
SO	a$^1\Delta$	450ms	0	450ms	−	−	MRD-CI	69G	[11] 84
	b$^1\Sigma^+$	13.8ms	0	13.8ms	−	−	MRD-CI	69G	[11] 84

[a] If $M_e(R)$ is computed only for a single band system, this system is given. Otherwise the excited state for which the radiative lifetime is computed is given.
[b] In ns(10^{-9}s) unless otherwise stated.
[c] The maximum vibrational quantum number of the upper state for which τ is computed.
[d] If band oscillator strenghts and/or transition moments are computed, this is indicated by a plus sign.
[e] See text for discussion. (G = # contracted Gaussian basis functions and S = # Slater-type-orbitals).
[f] The transition moment is computed using an effective core potential [31].
[g] Non-adiabatic representation.
[h] The range of the J' and K' quantum number are $0 \leq J' \leq 1$ and $1 \leq K' \leq 48$.
[i] The wavefunction includes explicit dependence of the interelectronic coordinates [34].
[j] $\tau(2)$.
[k] τ_e (the vibrational structure has not been resolved).
[l] In addition to the Gaussian basis set they also used STO basis sets consisting of 54 and 48 basis functions.
[m] Photodissociation of OH.

REFERENCES

[1] See e.g. W.W. Duley, and D.A. Williams "Interstellar Chemistry" (Academic Press, London, 1984).

[2] Dumont, M.N., and Remy, F. 1982, Spectrosc. Lett. 15, 699.

[3] Oddershede, J. 1979, Physica Scripta 20, 587.

[4] See e.g. F.L. Pilar "Elementary Quantum Chemistry" (McGraw-Hill, New York, 1968), sec. 10-4.

[5] Larsson, M., and Siegbahn, P.E.M. 1983, J. Chem. Phys. 79, 2270.

[6] Lengsfield, B.H., Jafri, J.A., Phillips, D.H., and Bauschlicher, C.W. 1981, J. Chem. Phys. 74, 6849.

[7] Oddershede, J. 1978, Adv. Quantum Chem. 11, 275.

[8] Oddershede, J. 1983, in "Methods in Computational Molecular Physics", eds. G.H.F. Diercksen and S. Wilson, p. 249 (Reidel, Dordrecht).

[9] Oddershede, J., Jørgensen, P., and Yeager, D.L. 1985, Compt. Phys. Rep., in press.

[10] Dahl, F., and Oddershede, J. 1985, to be published.

[11] Klotz, R., Marian, C.M., Peyerimhoff, S.D., Hess, B.A., and Buenker, R.J. 1984, Chem. Phys. 89, 223.

[12] Dahl, F. Thesis, Odense University 1984.

[13] Hinze, J., Lie, G.C., and Liu, B. 1975, Astrophys. J. 196, 621.

[14] Larsson, M. 1983, Astron. Astrophys. 128, 291.

[15] Klein, R., Rosmus, P., and Werner, H.-J. 1982, J. Chem. Phys. 77, 3559.

[16] Arnold, J.O., and Langhoff, S.R. 1978, J. Quant. Spectrosc. Radiat. Transfer 19, 461.

[17] Chabalowski, C.F., Buenker, R.J., and Peyerimhoff, S.D. 1981, Chem. Phys. Lett. 83, 441.

[18] Cooper, D.M. 1981, J. Quant. Spectrosc. Radiat. Transfer 26, 113.

[19] van Dishoeck, E.F. 1983, Chem. Phys. 77, 277.

[20] Douilly, B., Robbe, J.M., Schamps, J., and Roueff, E. 1983, J. Phys. B. 16, 437.

[21] Chabalowski, C.F. Peyerimhoff, S.D., and Buenker, R.J. 1983, Chem. Phys. 81, 57.

[22] Zeitz, M., Peyerimhoff, S.D., and Buenker, R.J. 1979, Chem. Phys. Lett. 64, 243.

[23] Rosmus, P., and Werner, H.-J. 1984, J. Chem. Phys. 80, 5085.

[24] Larsson, M., and Siegbahn, P.E.M. 1983, Chem. Phys. 76, 175.

[25] Cartwright, D.C., and Hay, P.J. 1982, Astrophys. J. 257, 383.

[26] Larsson, M., Siegbahn, P.E.M., and Ågren, H. 1983, Astrophys. J. 272, 369.

[27] Lavendy, H., Gandara, G., and Robbe, J.M., J. Mol. Spectrosc. 106, 395.

[28] Cooper, D.M., and Langhoff, S.R. 1981, J. Chem. Phys. 74, 1200.

[29] Rosmus, P., and Werner, H.-J. 1982, Mol. Phys. 47, 661.

[30] Telle, H.H. 1984, J. Chem. Phys. 81, 195.

[31] Laskowski, B., and Stallcop, J. 1981, J. Chem. Phys. 74, 4883.

[32] Cartwright, D.C., and Hay, P.J. 1979, J. Chem. Phys. 70, 3191.

[33] Glass-Maujean, M., Quadrelli, P., Dressler, K., and Wolniewicz, L. 1983, Phys. Rev. A 28, 2868.

[34] Wolniewicz, L., and Dressler, K. 1982, J. Mol. Spectrosc. 96, 295.

[35] van Dishoeck, E.F., van Hemert, M.C., and Dalgarno, A. 1982, J. Chem. Phys. 77, 3693.

[36] Werner, H.-J., Rosmus, P., Schätzl, W., and Meyer, W. 1984, J. Chem. Phys. 80, 831.

[37] Zeiri, Y., and Balint-Kurti, G.G. 1983, J. Mol. Spectrosc. 99, 1.

[38] Sangfelt, E., Kurtz, H.A., Elander, N., and Goscinski, O. 1984, J. Chem. Phys., 81, 3976.

[39] Partridge, H., and Langhoff, S.R. 1981, J. Chem. Phys. 74, 2361.

[40] Saxon, R.P., Kirby, K., and Liu, B. 1978, J. Chem. Phys. 69, 5301; Kirby, K., Saxon, R.P., and Liu, B. 1979, Astrophys. J. 231, 637.

[41] Sink, M.L., and Bandrauk, A.D. 1979, Can. J. Phys. 57, 1178.

[42] Buenker, R.J., and Peyerimhoff, S.D. 1983, in "New Horizons in Quantum Chemistry", Eds. P.-O. Löwdin and B. Pullman (Reidel Dordrecht), p. 183.

[43] Werner, H.-J., Kalcher, J., and Reinsch, E.-A. 1984, J. Chem. Phys. 81, 2420.

[44] Weiner, B., and Öhrn, Y. 1984, J. Chem. Phys. 80, 5866.

[45] Weiner, B., Jensen, H.J.Aa., and Öhrn, Y. 1984, J. Chem. Phys. 81, 587.

[46] Collins, L.A. Cartwright, D.C., and Wadt, W.R. 1980, J. Phys. B 13, L613.

[47] Singh, P.D., van Dishoeck, E.F., and Dalgarno, A. 1983, Icarus 56, 184.

[48] Langhoff, S.R., van Dishoeck, E.F., Wetmore, R., and Dalgarno, A. 1982, J. Chem. Phys. 77, 1379.

[49] van Dishoeck, E.F., Langhoff, S.R., and Dalgarno, A. 1983, J. Chem. Phys. 78, 4552.

[50] van Dishoeck, E.F., and Dalgarno, A. 1983, J. Chem. Phys. 79, 873.

[51] Elander, N., Erman, P., Gustafson, O., Larsson, M., Rittby, M., and Rurarz, E. 1985, Physica Scripta, to be published.

[52] Werner, H.-J., Rosmus, P., and Grimm, M. 1982, Chem. Phys. 73, 169.

[53] Werner, H.-J. 1984, J. Chem. Phys. 80, 5080.

[54] Hunt, W.J., Hay, P.J., and Goddard, W.A. 1972, J. Chem. Phys. 57, 738.

[55] Pauzat, F., Levy, B., and Millie, P. 1980, Mol. Phys. 39, 375.

[56] Eyler, J.R., Oddershede, J., Sabin, J.R., Diercksen, G.H.F., and Grüner, N.E. 1984, J. Phys. Chem. 88, 3121.

[57] Oddershede, J., Sabin, J.R., Diercksen, G.H.F., and Grüner, N.E. 1985, J. Chem. Phys., to be published.

[58] Bondybey, V.E. 1982, J. Phys. Chem. 86, 3396.

[59] Michalopoulos, D.L., Geusic, M.E., Langridge-Smith, P.R.R., and Smalley, R.E. 1984, J. Chem. Phys. 80, 3556.

[60] Grev, R.S., and Schaeffer, H.S. 1984, J. Chem. Phys. 80, 3552.

MOLECULAR PHOTOIONIZATION PROCESSES OF ASTROPHYSICAL AND AERONOMICAL INTEREST

P. W. Langhoff

Department of Chemistry, Indiana University, Bloomington, IN 47405 USA
and
Max-Planck-Institut für Astrophysik, Karl-Schwarzschild-Str. 1, 8046 Garching bei München, Federal Republic of Germany

INTRODUCTION

Ionization processes dominate photon attenuation cross sections over wide spectral ranges and, consequently, are relevant in a significant number of astrophysical connections [1-4]. Although heating and the ionization thought to initiate ion-molecule chemistry in interstellar clouds is largely attributed to cosmic rays [5], uv and x-ray photoionization processes can also contribute in diffuse clouds and in dense clouds in the immediate neighborhoods of uv-luminous O- and B-type stars, planetary nebulae, and supernovae [6-8]. Precursor radiation from recombining gas in shock regions can also ionize molecules under appropriate conditions [9]. Studies of stellar [10] and planetary atmospheres [11], and of cometary comae [12], furthermore, involve a variety of photoionization processes. Refined theoretical and computational models in such astrophysical and aeronomical situations require reliable values of relevant photoionization cross sections [13].

In the present report, an account is given of aspects of photoionization processes in molecules, with particular reference to recent theoretical and experimental studies of partial cross sections for production of specific final electronic states and of parent and fragment ions. Such cross sections help provide a basis for specifying the state of excitation of the ionized medium, are useful for estimating the kinetic energy distributions of photoejected electrons and fragment ions, provide parent- and fragment-ion yields, and clarify the possible origins of neutral fragments in highly excited rovibronic states. Section II provides a descriptive account of photoionization phenomena, including tabulation of valence- and inner-shell ionization potentials for some molecules of astrophysical and aeronomical interest. Cross sectional expressions are given in Section III, various approximations currently employed in computational studies are described briefly, threshold laws and high-energy limits are indicated, and distinction is drawn between resonant and direct photoionization phenomena. In Section IV, recent experimental and theoretical studies of partial photoionization cross sections in selected compounds of astrophysical and aeronomical relevance are described and discussed. Concluding remarks are made in Section V.

II. PHOTOIONIZATION PHENOMENA

One-electron photoionization processes

$$M(i) + h\nu(\vec{k}_p, \vec{\varepsilon}_p) \rightarrow M^+(j) + e^-(\vec{k}_e) \quad (1a)$$

$$M^+(j) \rightarrow M^+(f), \text{ parent ions or}$$

$$\rightarrow F^+(f) + N(f), \text{ fragment-neutral pairs} \quad (1b)$$

dominate the interactions between radiation and molecules for sufficiently low photon energy [14]. In Eq. (1a), $M(i)$ refers to a molecule in an initial rovibronic state "i", $M^+(j)$ is a suddenly ($\sim 10^{-16}$ sec) produced doorway-state ion, the incident photon has momentum $\hbar\vec{k}_p$ and polarization direction $\vec{\varepsilon}_p$, and the ejected electron has momentum $\hbar\vec{k}_e$. Equation (1b) refers to subsequent ($\sim 10^{-14}$ sec) relaxation of $M^+(j)$ into final parent-ion states $M^+(f)$ or fragment ion-neutral pairs $F^+(f) - N(f)$. In the energy region of general astrophysical and aeronomical interest here ($h\nu \leq 50$ keV), photon momentum, magnetic interactions, Compton scattering, and pair production are generally small [14], conditions under which the dipole approximation to radiation-matter interaction is valid [15]. These circumstances are illustrated in Figure 1, which depicts the total photon attenuation cross section for elemental copper, as well as individual contributions due to photoionization (τ), coherent (σ_{coh}) and incoherent (σ_{incoh}) Compton scattering, pair production (κ), and nuclear photoeffect ($\sigma_{ph.n.}$) [16]. Because of a strong dependence of scattering and nuclear photoeffect cross sections on atomic number [14], photoionization contributions to total attenuation cross sections for the lighter atoms of which molecules of astrophysical and aeronomical interest are comprised are even more dominant than shown in Figure 1 for copper.

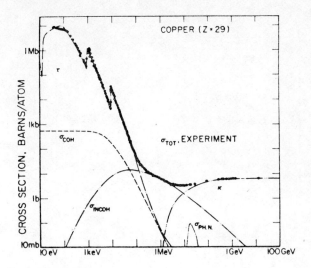

Fig. 1: Contributions to photon attenuation cross section of elemental copper; (τ) photoionization, (σ_{coh}) coherent scattering, (σ_{incoh}) incoherent scattering, (κ) pair production, ($\sigma_{ph.n.}$) nuclear photoeffect. 1 Barn = $10^{-24} cm^2$. Adopted from reference [16].

Temperature and other conditions appropriate to the physical circumstances of interest determine the distribution of initial molecular states $M(i)$ of Eq. (1a). By contrast, the excited final ionic states $M^+(f)$ and fragment-neutral pairs $F^+(f)-N(f)$ are produced in accordance with the spectral distribution of incident radiation and the relevant molecular ionization cross sections. In Figure 2 is given a plausible approximate rendering of the interstellar radiation field initiating the ionization processes of Eqs. (1a) and (1b) under general ambiant conditions [17]. Also shown in the figure is an approximation to the solar spectrum in the extreme uv region (~ 1000 to 100 Å) at a distance of 1 AU from the sun [18, 19]. As indicated, the latter data refer to averaged values smoothed over emission line features. Of course, electromagnetic energy densities in the uv and x-ray spectral intervals will be significantly greater than depicted in Figure 2 in the immediate vicinity of hot objects [7-12]. By contrast, extinction due to dust grains and shielding due to strong absorption will reduce the photon flux in specific spectral intervals, effects which can be estimated from radiative-transfer considerations [7].

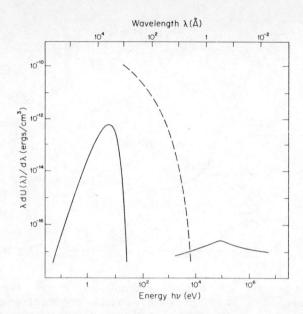

Fig. 2: Interstellar (———) and extreme ultraviolet solar (---) electromagnetic radiation spectra, adopted from references [17-19]. The solar spectrum shown is appropriate for a distance of 1 astronomical unit (AU) from the sun, and refers to smooth average values over quiescent emission line features.

Considerable experimental and theoretical ingenuity is generally required in determining partial cross sections that refer to the individual physical processes or final ionic channels of Eqs. (1), since many such contributions are usually degenerate at a given photon energy [20]. In distinguishing among these various partial cross sections it is helpful to take cognizance of the different time scales relevant to electron ejection ($\sim 10^{-16}$ secs) and subsequent vibrational motion ($\sim 10^{-14}$ secs.), as intimated by Eqs. (1a) and (1b). On this basis one can partition the total photoionization cross section in accordance with Eq. (1a) as a sum over partial-channel cross sections to electronic resolution in the form

$$\sigma_{ion}(h\nu) = \sum_f \sigma_f^{(e\ell)}(h\nu). \qquad (2a)$$

Here, the electronic cross sections $\sigma_f^{(e\ell)}(h\nu)$ correspond to sums over contributing vibrational partials, including dissociative contributions, with these in turn referring to sums over rotational components. Branching ratios at a given photon energy for production of specific electronic states are obtained from areas under photoelectron kinetic energy spectra, provided these can be uniquely assigned and separated [20]. This gives an operational definition of the cross sections $\sigma_f^{(e\ell)}(h\nu)$ as sums over contributions from ions $M^+(j)$ corresponding to suddenly produced rovibronic states. Cross sectional values for vibronic states are obtained by multiplying branching ratios into the total one-electron ionization cross section [20]. In addition to providing the distribution of ionic states in photoionization, the cross sections $\sigma_f^{(e\ell)}(h\nu)$ also give some indication of the kinetic energy spectrum of photoejected electrons.

It is convenient to refer also to partial cross sections for parent and fragment-ion production, which provide a partioning of the total ionization cross section in the form

$$\sigma_{ion}(h\nu) = \sum_f \sigma_f^{(frag)}(h\nu) \qquad (2b)$$

Here, the parent- or fragment-ion cross sections $\sigma_f^{(frag)}(h\nu)$ correspond to sums over all contributing vibronic states in accordance with Eq. (1b). Photoionization mass-spectrometric or electron-ion coincidence techniques are generally required in measurements of such fragments or parent-ion cross sections [20], which provide yields as well as approximate estimates of kinetic-energy distributions.

It is convenient to write the one-electron ionization cross section in the form

$$\sigma_{ion}(h\nu) = \gamma(h\nu)\,\sigma_{abs}(h\nu) \qquad (3)$$

where $\gamma(h\nu)$ is the one-electron ionization efficiency and $\sigma_{abs}(h\nu)$ is the total photoabsorption cross section. It should be noted that the total one-electron ionization cross section differs from the total absorption cross section above threshold by multiple-electron, ion-pair, fluorescence, and possibly Auger contributions [20], so that generally $\gamma(h\nu) < 1$.

On basis of cosmic or interstellar abundances [1], cross sections of H, He, and H_2 are of central astrophysical importance, whereas that of H^- is relevant to the solar atmosphere and is prototypical of negative ions. In Figure 3 are shown calculated cross sections that refer specifically to production of H^+, of ground-state $(1s)^2 S_{\frac{1}{2}} He^+$ ions, of $(1\sigma_g)^2 \Sigma_g^+ H_2^+$ ions, and of $(1s)^2 S_{\frac{1}{2}}$ hydrogen atoms. The values shown for H are exact [14], whereas those for He and H^- [21] and for H_2 [22] should provide highly reliable approximations to the indicated one-electron cross sections in regions free from

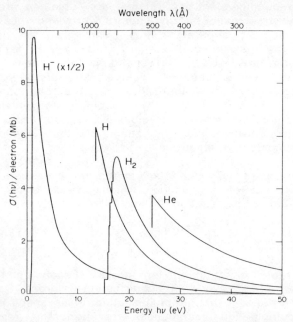

Fig. 3: Ground state to ground state photoionization cross sections of H, He, H_2, and H^-, shown on a per electron basis, to electronic resolution, adopted from references [20-22]. Note that the H^- cross section is multiplied by a factor of ½.

autoionization and Feshbach resonances, and away from threshold structures in H_2 [20]. Evidently, the cross sections for ionization of H, He, H_2, shown on a per electron basis, are generally similar in nature. The major significant differences are the numerical values of the three ionization potentials and the elimination of a pronounced threshold edge in H_2 due to vibrational structures, the indicated cross section referring to a sum over final vibrational states [22]. By contrast, the cross section for H^- ionization has a significantly different spectral shape than the other three cross sections, as a consequence of the very different initial and final scattering or continuum ionization states in this case, discussed further below.

Ionization of ubiquitous elemental hydrogen [Figure 3] is responsible for a cut off of uv radiation [Figure 2] in the Galaxy above $h\nu$=13.6 eV, a nominal number density of one H atom per cm^3 providing a threshold photon mean free path of only ~0.05 pc (~ 0.2 light years). Consequently, molecular photoionization in the interstellar medium is relevant largely to molecules having ionization potentials ≤ 13.6 eV, at frequencies sufficiently above the strong H atom absorption interval ($h\nu$ ≳ 100 eV), or in

Table 1: Ionization Potentials for Diatomic and Triatomic Molecules of Astrophysical and Aeronautical Interest[a]

Molecule	Ionization Potential[b] valence/L, K edges	Molecule	Ionization Potential[b] valence/L, K edges
H_2	16.0 / -	CH_2	10.5/(290)
HD	16.2 / -	H_2O	12.8/540
C_2	(12.0)/(290)	D_2O	12.8/540
N_2	15.7/410	NH_2	12.0/(410)
O_2	12.3/544	H_2S	10.5/187, 2470
S_2	9.4/(187, 2470)	HCN	13.6/294, 407
CH	10.6/(290)	HNC	12.6/(290, 410)
NH	13.5/(410)	HC_2	(11.0)/(290)
OH	13.0/(535)	HCO	9.3/(290, 535)
NaH	6.9/(30, 1075)	HNO	10.3/(410, 535)
MgH	6.9/(55, 1310)	HO_2	11.5/(535)
AℓH	8.4/(77, 1560)	CO_2	13.8/298, 541
SH	10.4/(187, 2470)	OCS	11.2/187, 295, 540, (2470)
SiH	8.0/(110, 1850)		
HCℓ	12.7/208, (2830)	CS_2	10.1/201, (290, 2470)
CaH	5.9/(365, 4044)		
CN	14.1/(290, 410)	C_3	(12.8)/(290)
CO	14.0/296, 543	N_2O	12.9/441, 541
CS	11.4/(187, 2470)	SO_2	12.5/(187, 535, 2470)
NO	9.3/411, 544		
MgO	7.5/(55, 535, 1310)	O_3	12.8/543
NS	8.9/(187, 410, 2470)	NO_2	11.0/412, 541
SO	10.3/(187, 535, 2470)	SiC_2	10.2/(110, 290, 1850)
SiO	11.6/(110, 535, 2470)		
SiS	9.8/(110, 187, 1850, 2470)	NaOH	- /(30, 535, 1075)

a Vertical ionization potentials in eV from electron spectroscopy measurements [28-31]; values in parenthesis refer to theoretical estimates for valence shells in the absence of measured data [26], to experimental elemental values for the L and K edges in the absence of molecular data [27].

b Outer-most valence-shell values reported only; L- and K-edge values refer to ns/np, multiplet, or chemical-shift splitting averages, when appropriate [27, 28]. Hartree-Fock Koopmans estimates are employed for ionization potentials not reported in the literature cited [26-31].

spatial regions that are in close proximity to strong uv emitters [7-9]. The existence of significant numbers of so-called uv stars in the Galaxy, postulated to account for the relatively high temperature of the intercloud medium [7], would also give rise to molecular photoionization on the surface boundaries of interstellar clouds.

In Table I is given a list of molecules of astrophysical and aeronomical interest [23-26], and of their valence- and inner-shell ionization potentials [27-31]. A significant number of these compounds are seen to have ionization potentials below the 13.6 eV H-atom cutoff, and all have L- and/or K-edge thresholds well above the strong H-atom ionization region of Figure 3. As a consequence, studies of ionization processes in these compounds are of relevance in varying degrees in astrophysical and aeronomical connections. Considerable progress has been made recently in experimental [32-36] and theoretical [37-39] studies of molecular partial-channel photoionization cross sections in uv and x-ray spectral regions, to which attention is now directed.

III. THEORY OF PHOTOIONIZATION

A descriptive account is provided in this Section of the energy range and nature of valence and inner-shell molecular ionization processes. Cross sectional expressions are given, computational approximations currently employed are described briefly, aspects of threshold laws and high-energy limits are indicated, and types of possible resonance phenomena are classified.

In order to help characterize the nature of molecular photoionization and the intermediate doorway-state ions $M^+(j)$ of Eqs. (1), it is helpful to cite the Koopmans approximation [40], which refers to these ions in terms of so-called electronic hole states, wherein a single electron is removed from any one of the ground-state canonical Hartree-Fock orbitals [41]. In this approach, Hartree-Fock orbital energies consequently provide useful first approximations to the spectrum of ionization protentials associated with molecular ions. Numerical values range from $\sim$ 5 to 20 eV for the so-called outer-valence orbitals of typical second and third row light-atom molecules, from $\sim$ 20 to 50 eV for so-called inner-valence orbitals, and to a few 100 eV to a few 1,000 eV for L- and K-shell orbitals [27-31]. The electronic distributions associated with Koopmans hole-state ions relax on a very short time scale ($\sim 10^{-16}$ sec) to eigenstates of the correct ionic Born-Oppenheimer Hamiltonian, with subsequent nuclear motion occuring on a longer time scale ($\sim 10^{-14}$ sec) over the ionic-state potential surface [42]. In the case of outer-valence orbitals, electronic relaxation effects are generally small, and the

Koopmans hole states are in one-to-one correspondence with close lying relaxed parent-ionic states at the vertical ground-state nuclear configuration. By contrast, inner-valence Koopmans hole states generally provide poor approximations to the adiabatic electronic states of the ion, since there are infinitely many of the latter converging on the vertical electronic two-electron ionization threshold [43]. Consequently, ionic relaxation or reorganization effects are generally significant for inner-valence ionization. Similarly, significant ionic core relaxation occurs for L- and K-shell hole states, although there is generally only one strongly allowed ionic state in these cases. Dissociative ionization generally occurs through inner-valence ionic states, whereas L- and K-shell ionic states can relax by fluorescence, Auger effect, and subsequent fragmentation [20].

a) <u>Cross Sectional Expressions</u>

Partial-channel photoionization cross sections are conveniently written in the general form [44]

$$d\sigma_{i \to j}(h\nu, \vec{k}_e, \vec{\epsilon}_p) = (\pi e^2 h/mc)(df_{i \to j}(h\nu, \vec{k}_e, \vec{\epsilon}_p)/d\epsilon)d\Omega \quad , (4)$$

where i, j, $h\nu$, $\vec{k}_e$ and $\vec{\epsilon}_p$ are as in Eqs. (1), the collection of atomic constants $(\pi e^2 h/mc) = 1.098 \times 10^{-16} cm^2$ eV and $df_{i \to j}(h\nu, \vec{k}_e, \vec{\epsilon}_p)/d\epsilon$ in eV^{-1} is the so-called differential dipole oscillator-strength density for ionization into the energy interval $d\epsilon$ centered at $h\nu$ and for electron ejection with asymptotic momentum $\hbar \vec{k}_e$ into the solid angle $d\Omega$. Note that in this dipole expression the cross section of Eq. (4) does not depend on the photon propagation direction ($\vec{k}_p$) but only on its polarization direction $\vec{\epsilon}_p$. The asymptotic photoejected electron momentum and photon energy are related by [cf. Eq. (1a)]

$$E_i + h\nu = E_j^+ + |\hbar \vec{k}_e|^2/2m \quad (5a)$$

or

$$|\hbar \vec{k}_e|^2/2m = h\nu - IP_{i \to j} \quad (5b)$$

where $IP_{i \to j}$ is the ionization potential for formation of the ionic state "j" from initial molecular state "i". Ionization potentials shown in Table I refer to ground vibronic state molecules M(i) and vertically excited ions $M^+(j)$, corresponding to unresolved rotational envelopes in each case. Small changes in translational energy during ionization are neglected in Eqs. (5). The differential oscillator strength density of Eq. (4) is given by the quantum-mechanical expression [44]

$$df_{i \to j}(h\nu, \vec{k}_e, \vec{\epsilon}_p)/d\epsilon = (2m/\hbar^2)h\nu |<\vec{k}_e \Phi_j^{(-)}|\vec{\epsilon}_p \cdot \vec{\mu}|\Phi_i>|^2, \quad (6)$$

where Φ_i is the initial-state molecular wave function, $k_e\Phi_j^{(-)}$ is the corresponding wave function for the final-state channel "j", and μ is the total dipole moment operator of the molecule. The function Φ_i describes the electrons and nuclei in the initial rovibronic state "i", and can be approximated employing the four standard Hund's classifications of coupling sequences [45]. The function $k_e\Phi_j^{(-)}$ describes a photoejected electron having asymptotic momentum $\hbar k_e$, as well as an (N-1)-electron ion in the rovibronic state "j", and satisfies incoming wave boundary conditions [46]. The form of Eq. (6) assumes that the continuum portion of the final-state function $k_e\Phi_j^{(-)}$ is normalized in the Dirac delta-function sense

$$<k_e'\Phi_{j'}^{(-)}|k_e\Phi_j^{(-)}> = \delta_{j',j}\delta(\epsilon'-\epsilon)\delta(\Omega'-\Omega) \qquad (7)$$

where $\epsilon = |\hbar k_e|^2/2m$ is the asymptotic electron kinetic energy, and Ω specifies the corresponding photoejection direction. It is appropriate to note that the cross section of Eq. (4) satisfies the general symmetry rule [47, 48]

$$(2J_i+1)^{-1}\sum_{M_i}\sum_{M_j} d\sigma_{i\to j}(h\nu, k_e, \epsilon_p) =$$

$$(\sigma_{i\to j}(h\nu)/4\pi)(1+\beta_{i\to j}(h\nu)P_2(\cos\Theta))d\Omega \qquad (8)$$

Here, Θ is the angle between the vectors k_e and ϵ_p, J_i is the total angular momentum quantum number of the initial state, and the sums are over the associated degenerate magnetic sublevels M_i, M_j, the former assumed to be equally populated and, therefore, corresponding to an isotropic initial state. The partial cross section $\sigma_{i\to j}(h\nu)$ and so-called anisotropy factor $\beta_{i\to j}(h\nu)$ depend on $h\nu$ but not on the directions of k_e and ϵ_p, as indicated. These fundamental quantities provide the point of comparison between measured values and theoretical calculations, to which attention is now directed.

b) **Computational Approximations**

Because electrons photoejected from molecules generally feel an anisotropic potential field, partial-wave expansion of the scattering functions results in a coupled-channel problem in ℓ-waves, as well as in energetically distinct channel labels associated with different final ionic states. As a consequence, explicit expressions for the two fundamental quantities $\sigma_{i\to j}(h\nu)$ and $\beta_{i\to j}(h\nu)$ in terms of one-electron matrix elements are more complicated for molecules than they are for atoms [14, 49-51].

Nevertheless, these have been written out employing ℓ-wave representations of the scattering functions [52-55], and calculations have been carried out largely in separated energetic-channel approximations. Most calculations, but not all, reported to date employ Hartree-Fock descriptions of the electronic portions of initial and final ionic states, continuum functions calculated for fixed molecular frames essentially in Hund's cases a or b, and refer largely to cross sections to vibrational resolution [37-39]. In the simplest approximation employed, partial cross sections involve products of London-Hönl and Franck-Condon factors and discrete-to-continuum electronic matrix elements calculated at equilibrium internuclear molecular configurations. The approximations employed in construction of body-frame continuum functions for photoejected electrons include plane waves [56], one- and two-center Coulomb functions [57, 58], local-potential, muffin-tin x_α calculations [59], and static-exchange potentials involving correct treatment of the noncentral and nonlocal molecular potential [60]. Plane-wave results are generally unsatisfactory for ionization of neutral molecules, whereas Coulomb wave functions can reproduce the correct high-energy limits of cross sections. However, important shape-resonance features, described further below, are not included in Coulomb results. Although local potentials can reproduce qualitatively shape resonance features, and the asymptotic Coulombic behavior can be correctly incorporated, ambiguities in choice of potential parameters preclude use of the approximation as a quantitatively reliable ab initio procedure. By contrast, single-channel static-exchange calculations can provide quantitatively reliable partial cross sections and anisotropy factors in the absence of significant channel coupling or relaxation effects. Coupled-channel studies in time-dependent Hartree-Fock [61,62], static-exchange [63], or optical potential [64] approximations are employed in the latter cases.

c) **Threshold Laws, High-Energy Limits, and Sum Rules**

Rules governing the behaviors of photoionization cross sections immediately above thresholds are familiar from atomic and nulear considerations [65]. Certain aspects of these results carry over to molecules with little modification [66]. Of particular interest is the finite step law that obtains at threshold for ionization of neutral systems, as illustrated in Figure 3 for H, He and H_2. Total molecular cross sections can be expected to exhibit Franck-Condon steps at the opening of each new vibrational channel, and could possibly reveal London-Hönl steps if sufficient resolution is achieved.

Closely related to the generally finite cross sectional value at threshold for ionization of neutral targets is the continuity rule for discrete f numbers at the first Coulombic threshold in the form [67]

$$\operatorname{Lim}(n\to\infty) n^3 f_n \to \sigma(h\nu_t) \quad , \tag{9}$$

where $h\nu_t$ is the threshold energy for ionization, and the total cross section $\sigma(h\nu_t)$ is in Hartree atomic units ($1a_0^2 = 0.2800 \times 10^{-16} cm^2$). In separated-channel approximations theoretical expressions similar to Eq. (9) are also appropriate for higher thresholds, where f_n refers to f numbers for the series converging on the threshold of interest, and $\sigma(h\nu_t)$ is the appropriate partial cross section. For the higher thresholds such series generally autoionize, complicating measurements of f numbers, although computed values in single-channel approximation will extrapolate in accordance with Eq. (9).

As is evident from Figure 3, the threshold behavior of the photoionization cross section of the H^- ion starts at a zero value and consequently is significantly different than those for the neutral species. The so-called Wigner threshold law gives in this case the behavior $\sim \epsilon^{\ell+\frac{1}{2}}$, where ϵ is the detached electron kinetic energy and ℓ ($= 1$ here) is the dominant outgoing partial-wave contribution. These considerations are generally valid for molecular ions [68,69], although more than one partial wave can contribute in these cases, and separation of the various rotational thresholds requires high resolution in experimental determinations [70].

The high photon energy behaviors of ionization cross sections are inferred from Coulomb wave functions for ionization of neutrals, and from plane waves for electron photodetachment of negative ions. In the former case, the behavior is $\sim \epsilon^{-(\ell+5/2)}$, whereas in the latter case it is $\sim \epsilon^{-(\ell+\frac{1}{2})}$, with ℓ the dominant angular momentum of the ejected electron [65]. The behavior in molecules is complicated by contributions to the scattering functions from more than one value of angular momentum, although it is clear for neutrals that the ultimate limiting behavior is $\sim \epsilon^{-7/2}$. By contrast, the behaviors of partial cross sections will generally differ, and will be controlled by the dominant orbital angular momentum of the occupied molecular orbital that is ionizated.

Cross sectional sums or power moments of the form [71]

$$S_i(k) = \int_{-\infty}^{\infty} (h\nu)^k \sigma_i(h\nu) d(h\nu) \tag{10}$$

provide useful constraints on calculated or measured values. Here

$$\sigma_i(h\nu) = \sum_j \sigma_{i \to j}(h\nu) \qquad (11)$$

is the total cross section for initial state "i", and $\sigma_{i \to j}(h\nu)$ are the partial cross sections of Eq. (8). Note that the integral of Eq. (10) includes discrete state contributions as well as continuum values. In the cases k=2,1,0 and -1 closure can be employed to reduce Eq. (10) to sum rules the evaluation of which requires knowledge of only the initial state function Φ_i [71]. The case k=0 corresponds to the so-called Thomas-Reiche-Kuhn or f sum rule, the value of which is proportional to the total number of molecular electrons N. Failure of the Born-Oppenheimer approximation results in deviations from N that are of the order of the ratio of electron to proton mass. It is a well-known empirical observation that the discrete state contributions to the f sum rule are generally ≤1, and that individual shell cross sections to electronic resolution integrate approximately, but not precisely, to the number of shell electrons present. Local potential computational approximations satisfy these conditions, but nonlocal ones generally do not.

d) Resonances

The variation with photon energy $h\nu$ of an individual molecular partial-channel photoionization cross section [Eqs. (4) and (8)], corresponding to production of a final ionic state "j" from the initial molecular state "i", is determined largely by the change in spatial characteristics of the final-state wave function with kinetic energy, in accordance with Eq. (6). Thus, if the electronic and nuclear portions of the wave function are slowly changing with kinetic energy, the relevant transition moment is similarly slowly varying, and the corresponding channel cross section is a smooth function of photon energy. Various circumstances can lead, however, to relatively rapid changes in the electronic or nuclear spatial characteristics of a wave function as the kinetic energy is changed. Although there are general theoretical (Feshbach-Fano) procedures for dealing with such resonance phenomena [72,73], it is helpful to clarify the origins of such behavior at first in terms of separated channel models. Indeed, Feshbach-Fano procedures generally require the a priori selection of plausible zeroth-order descriptions of "resonance states" the spatial characteristics of which differ significantly from a "background" continuum which varies slowly with energy through the resonances. Moreover, although, strictly speaking, resonances are characterized formally in scattering theory by a change in π radians of a phase shift or eigenphase sum, it is helpful to adopt a somewhat more general terminology which uses this designation for any pronounced cross sectional feature.

Pronounced features in measured partial photoionization cross sections can be attributed to (i) shape resonances, (ii) autoionization resonances, (iii) Feshbach resonances, and (iv) threshold effects. Shape resonances in photoionization are attributed to electronically compact ionization functions which can increase the cross section significantly above "background", generally over a ∼1 to 10 eV specral range. In this case only a single zeroth-order ionization channel or electronic configuration is involved, the spatially compact continuum function arising from the shape of the potential, generally a barrier [37], or from a virtual valence molecular orbital [38]. Such features are widespread in molecules, appearing above L- and K-shell as well as valence thresholds [74]. Shape resonances that are relatively broad can provide backgrounds for more narrow features, complicating their detection [75].

Electronic autoionizing resonances correspond to singly or multiply excited electronic configurational states that appear above one or more channel thresholds. Such states generally precede higher-lying ionic thresholds, and would form Rydberg series in the absence of configuration mixing with the underlying ionization continua. Rydberg states can also autoionize into their own continua through vibrational and/or rotational states that are above the adiabatic ionization threshold [37]. Predissociation of vibrational states can be regarded as autoionization in the nuclear degrees of freedom, and so comes under this general heading, as well. As a consequence of such electronic and vibrational autoionization, photoionization cross sections in the spectral ranges between ionization thresholds are generally highly structured, requiring careful experimental and theoretical study [76].

Feshbach resonances involve the trapping of scattering or ionized electrons in potentials produced by excited targets or cores, respectively [73], whereas thresholds effects involve the formation of structures in partial cross sections due to the opening of other energetic channels [64]. Such effects are more common in electron scattering situations than in the context of photoionization of neutrals, although negative-ion cross sections can exhibit Feshbach resonances [77].

IV. RECENT PHOTOIONIZATION STUDIES

Experimentally and theoretically determined molecular photoionization cross sections are conveniently tabulated and reviewed in a recent report [78]. In this section, representative measured and calculated valence-shell and K-edge values are presented for N_2 and CO molecules.

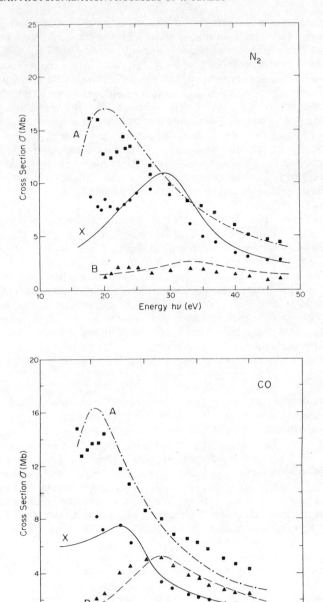

Fig. 4: Valence-shell partial-channel photoionization cross sections of N_2 and CO as functions of photon energy, obtained from recent experimental and theoretical studies [78].

In Figure 4 are shown measured and calculated partial cross sections for the valence-shell channels of N_2 and CO molecules as functions of photon energy, reported to electronic resolution. The experimental values shown are from so-called dipole (e, 2e) measurements, whereas the theoretical values are from vertical-electronic static-exchange calculations [78]. Figure 4 illustrates similaritiles and differences in the photoionization cross sections of homo- and heteronuclear diatomic molecules. The π-shell ($1\pi_u$ and 1π) cross sections are seen to be highly similar functions of photon energy. These correspond largely to 2p→kd atomic-like cross sections, the formation of N-N and C-O π bonds resulting primarily in 2p orbitals stretched along the internuclear line perpendicular to their directions.

By contrast, the four σ-orbital cross secitons of Figure 4 exhibit significant differences. The $3\sigma_g$ channel of N_2 includes a $k\sigma_u$ (σ^*) shape resonance approximately 10 eV above threshold, whereas the $2\sigma_u$ channel does not include this feature due to the parity selection rule. Much of the f number for the $2\sigma_u$ channel is included in the discrete $2\sigma_u \to 1\pi_g(\pi^*)$ transition that appears below threshold, and is not shown in Figure 4. Because of symmetry breaking, both 5σ and 4σ channels in CO include a $k\sigma(\sigma^*)$ resonance approximately 10 eV above threshold. Such shape resonance features are widespread in diatomic and polyatomic molecules, particularly in σ channels across short bonds [79].

Small differences between calculated and measured values in Figure 4 can be attributed to neglect of electronic relaxation, reorganization effects, and vibrational averaging in the theoretical development, and to resolution considerations in the experiments [78]. That is, highly structured autoionization features present in some of the channels of Figure 4 are not resolved by the measurements, the reported cross sections corresponding to spectral averages. Indeed, higher resolution studies exhibit significant cross sectional variation through autoionizing regions between the first and higher ionization thresholds of molecules [20].

The outer-valence electronic cross secitons of Figure 4 refer to production of parent ions in X, A, or B states, the latter two generally decaying by fluorescence under appropriate conditions. By contrast, deeper-lying inner-valence ionic states in N_2 and CO, associated with removal of $2\sigma_g$ and 3σ orbital electrons, respectively, result largely in dissociative neutral-ion pairs. In Figure 5 are shown cross sections for production of $N-N^+$, $C-O^+$, and $O-C^+$ pairs in N_2 and CO. Various features observed in the figure correspond to the opening of specific ionic states. Theoretical calculations of these inner-valence ionic states, in which there is a general failure of the Koopmans

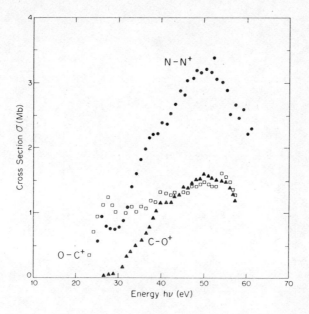

Fig. 5: Dissociative photoionization cross sections of N_2 and CO for production of N-N$^+$ and C$^+$-O, O$^+$-C fragments as functions of photon energy, obtained from recent experimental studies [78].

approximation, are in excellent accord with the measured values, and with related photoelectron spectroscopy measurements [41]. Vibrationally resolved studies in these spectral regions can show highly structure photoionization cross sections as a consequence of the many individual vibronic channels accessible and of their expected interactions.

In Figure 6 are shown as representative examples the carbon and oxygen K-edge cross sections of CO in the near threshold regions. Both cross sections are seen to include strong pre-edge features, which can be attributed to $1\sigma \rightarrow \pi^*$ and $2\sigma \rightarrow \pi^*$ orbital excitations. Higher resolution studies reveal the presence of vibrational peaks in these strong features, indicating these metastable states have lifetimes of a few vibrational periods. Subsequent decay is by Auger effect, autoionization, fluorescence, and ultimate fragmentation. Both cross sections also exhibit post-edge shape resonance features approximately 10 eV above threshold. As in Figure 4, these resonances can be attributed to a valence σ^* orbital positioned at positive energy in the continuous spectrum [79], resulting in enhanced cross-sectional values.

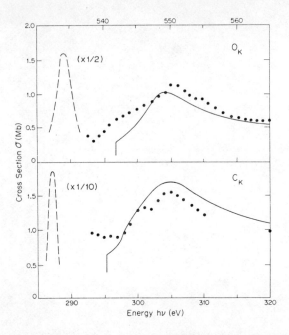

Fig. 6: K-edge photoabsorption cross sections of CO for production of $1s_C^{-1}$ and $1s_O^{-1}$ hole states as functions of photon energy, obtained from recent experimental and theoretical study [60].

The foregoing results for N_2 and CO are prototypical of the $\sigma-$ and $\pi-$electron photoionization cross sections of light-atom diatomic molecules, and illustrate the general variation with incident photon energy in these cases. Partial-channel, dissociative, and L/K-edge photoionization cross sections in polyatomic molecules generally exhibit more complex variations with photon energy, although in many cases $\sigma-$ and $\pi-$like behaviors are also apparent, with $\sigma \rightarrow \sigma^*$ shape resonances and pre-edge structures particularly in evidence [78]. Heavier-atom diatomic and polyatomic molecules generally exhibit larger photoionization cross sections at lower photon energies than do light-atom molecules. This is a consequence of the lower ionization potentials and of the greater spatial extents of the orbitals of heavier-atom molecules. Referring to Table 1, comparisons of entries for CO/CS, NO/NS, OH/SH, CO_2/CS_2, and H_2O/H_2S illustrate the differences in ionization potentials for O- and S-atom containing molecules. Experimental and theoretical studies indicate significantly larger threshold cross sectional values for the S-atom molecules in each case [78], suggesting the possible relevance of photoionization of heavier-atom compounds in astrophysical and aeronomical connections.

V. CONCLUDING REMARKS

Ultraviolet and x-ray photons can initiate a great variety of molecular photoionization processes, resulting in the production of parent ions and fragment-neutral pairs in various rovibronic states, and in subsequent fluorescence, Auger, and dissociative decay. Such processes are relevant in a number of astrophysical and aeronomical connections, particularly in the immediate vicinity of uv luminous O- and B-type stars and soft x-ray sources. Although molecular photoprocesses have long been subjects of experimental study, it is only comparatively recently that synchrotron-radiation and equivalent electron-impact sources have provided the energetic photons necessary for measurements of partial and fragment cross sections over wide spectral ranges. Correspondingly, while the theory of radiation-matter interaction is well-established, viable computational approximations have been developed for studies of partial-channel and fragmentation processes in complex molecules only recently.

In the present report, an overview of molecular ionization processes is given with particular attention focused on recent experimental and theoretical developments. The account is intended for general readers in astrophysics and aeronomy, and cites references that tabulate photo cross sections and related data for relevant compounds. This information should provide a basis for refined studies of the role of photoionization processes in various astrophysical and aeronomical connections. Of particular interest would be examination of heating of diffuse clouds by uv ionization of trace molecules, in which connection the associated radiative recombination coefficients would be useful, and of the effects of soft x-ray ionization of molecules on dense cloud temperatures and abundances, taking cognizance of the higher ionization thresholds and channels. In light of the possible critical role of fractional ionization in the collapse of clouds and in the consequent formation of stars, through coupling of the charged medium with the galactic magnetic field, refined studies of the degree of ionization of the interstellar medium would seem to be of some importance.

ACKNOWLEDGMENTS

It is a pleasure to acknowledge the financial support of the Chemistry and International Programs Divisions of the National Science Foundation, of the National Aeronautics and Space Administration, and of the American Chemical Society during the early stages of this work. The cooperation and assistance of coworkers, particularly G.H.F. Diercksen, S.R. Langhoff, C.E. Brion, M.R. Hermann, and B.W. Fatyga at various stages is also gratefully acknowledged.

REFERENCES

1. L. Spitzer, Jr., *Diffuse Matter in Space* (Wiley, New York, 1968).
2. L. Spitzer, Jr., *Physical Processes in the Interstellar Medium* (Wiley, New York, 1978).
3. J.E. Dyson and D.A. Williams, *Physics of the Interstellar Medium* (Manchester University Press, Manchester, 1980).
4. W.W. Duley and D.A. Williams, *Interstellar Chemistry* (Academic, London, 1984).
5. T.C. Cravens and A. Dalgarno, Astrophys. J. $\underline{219}$, 750 (1978).
6. T. de Jong, A. Dalgarno, and W. Boland, Astron. Astrophys. $\underline{91}$, 68 (1980).
7. G.B. Field, in *Atomic and Molecular Physics and the Interstellar Matter*, R. Balian, P. Encrenaz, and J. Lequeux, Editors (North Holland, Amsterdam, 1975), pp. 467-531.
8. J. Danzinger and P. Gorenstein, Editors, *Supernova Remnants and Their X-ray Emission*, IAU Symposium 101 (Reidel, Dordrecht, 1983).
9. D.J. Hollenbach and C.F. McKee, Astrophys. J., Supp. Ser. $\underline{41}$, 555 (1979).
10. D. Mihalas, *Stellar Atmospheres* (Freeman, San Francisco, 1978).
11. J.W. Chamberlain, *Theory of Planetary Atmospheres* (Academic, New York, 1978).
12. W.F. Huebner and P.T. Giguere, Astrophys. J. $\underline{238}$, 753 (1980).
13. A. Dalgarno, in *Applied Atomic Collision Physics*, H.S.W. Massey and D.R. Bates, Editors (Academic, New York, 1982), pp. 427-467.
14. H.A. Bethe and E.E. Salpeter, *Quantum Mechanics of One- and Two- Electron Atoms* (Academic, New York, 1957).
15. U. Fano and J. Cooper, Rev. Mod. Phys. $\underline{40}$, 441 (1968).
16. J.H. Hubbell and W.J. Veigele, *Comparison of Theoretical and Experimental Photoeffect Data, 0.1 keV to 1.5 Mev*, NBS Technical Note 901 (Office of Standard Reference Data, Washington, D.C., 1976).
17. D.W. Watson, Rev. Mod. Phys. $\underline{48}$, 513 (1976).
18. H.E. Hinteregger, J. Atmos. Terr. Phys. $\underline{38}$, 791 (1976).
19. D.G. Torr, M.R. Torr, H.C. Brinton, L.H. Brace, N.W. Spencer, A.C. Hedin, W.B. Hanson, J.H. Hoffmann, A.O. Nier, J.C.G. Walker, and D.W. Rusch, J. Geophys. Res. $\underline{84}$, 3360 (1979).
20. J. Berkowitz, *Photoabsorption, Photoionization and Photoelectron Spectroscopy* (Academic, New York, 1979).
21. P.W. Langhoff, C.T. Corcoran, J.S. Sims, F. Weinhold, and R.M. Glover, Phys. Rev. $\underline{A14}$, 1042 (1976).

22. S.V. ONeil and W.P. Reinhardt, J. Chem. Phys. $\underline{69}$, 2126 (1978).
23. A.P.C. Mann and D.A. Williams, Nature (London) $\underline{283}$, 721 (1980).
24. S. Green, Ann. Rev. Phys. Chem. $\underline{32}$, 103 (1981).
25. Rh. Lüst, Topics in Current Chem. $\underline{99}$, 37 (1981).
26. W. von Niessen, L.S. Cederbaum, J. Schirmer, G.H.F. Diercksen, and W.P. Kraemer, J. Electron Spectrosc. $\underline{28}$, 45 (1982).
27. S. Hagström, C. Nordling, and K. Siegbahn, in Alpha-, Beta-, and Gamma-Ray Spectroscopy, K. Siegbahn, Editor (North Holland, Amsterdam, 1965) Vol. 1, pp. 845-862.
28. K. Siegbahn, C. Nordling, G. Johansson, J. Hedman, P.F. Heden, K. Hamrin, U. Gelius, T. Bergmark, L.O. Werme, R. Manne, and Y. Baer, ESCA Applied to Free Molecules (North-Holland, Amsterdam, 1969).
29. D.W. Turner, C. Baker, A.D. Baker, and C.R. Brundle, Molecular Photoelectron Spectroscopy (Wiley, London, 1970).
30. H.M. Rosenstock, K. Draxl, B.W. Steiner, and J.T. Herron, J. Phys. Chem. Ref. Data $\underline{6}$, S1 (1977).
31. A.P. Hitchcock, J. Electron Spectrosc. $\underline{25}$, 245 (1982).
32. E.E. Koch and B.F. Sontag, in Synchrotron Radiation, C. Kunz, Editor (Springer, Berlin, 1979), pp. 269-415.
33. M.O. Krause, in Synchrotron Radiation Research, H. Winich and S. Doniack, Editors (Plenum, New York, 1980), pp. 101-157.
34. J.A.R. Samson, in Electron Spectroscopy: Theory, Techniques, and Applications, C.R. Brundle and A.D. Baker, Editors (Academic, New York, 1981), Vol. 4, pp. 361-396.
35. C.E. Brion and A. Hamnett, Advan. Chem. Phys. $\underline{45}$, 2 (1981).
36. C.E. Brion, in Physics of Electronic and Atomic Collisions, S. Datz, Editor (North Holland, Amsterdam, 1982), pp. 579-593.
37. J.L. Dehmer, D. Dill, and A.C. Parr, in Photophysics and Photochemistry in the Vacuum Ultraviolet, S. McGlynn, G. Findley, and R. Huebner, Editors (Reidel, Dordrecht, Holland, 1983).
38. P.W. Langhoff, in Methods in Computational Molecular Physics, G.H.F. Diercksen and S. Wilson, Editors (Reidel, Dordrecht, Holland, 1983), pp. 299-334.
39. R.R. Lucchese, K. Takatsuka, D.K. Watson, and V. McKoy, in Electron-Atom and Electorn-Molecule Collisions, J. Hinze, Editor (Plenum, New York, 1983), pp. 29-49.
40. T.A. Koopmans, Physica $\underline{1}$, 104 (1933).
41. P.W. Langhoff, J. Mol. Sci. $\underline{2}$, 15 (1982).
42. H.W. Meldner and J.D. Perez, Phys. Rev. $\underline{A4}$, 1388 (1971).

43. P. W. Langhoff, S. R. Langhoff, T. N. Rescigno, J. Schirmer, L. S. Cederbaum, W. Domcke, and W. von Niessen, Chem. Phys. 58, 71 (1981).
44. H. A. Bethe, in Handbuch der Physik, S. Flugge, Editor (Springer, Berlin, 1933) Vol. 24, pp. 457-487.
45. L. D. Landau and E. M. Lifschitz, Quantum Mechanics, NonRelativistic Theory (Pergamon, Oxford, 1962).
46. G. Breit and H. A. Bethe, Phys. Rev. 93, 888 (1954).
47. C. N. Yang, Phys. Rev. 74, 764 (1948).
48. U. Fano and D. Dill, Phys. Rev. A6, 185 (1972).
49. J. Cooper and R. N. Zare, J. Chem. Phys. 48, 942 (1968).
50. D. Dill, A. F. Starace, and S. T. Manson, Phys. Rev. A11, 1596 (1975).
51. M. White, Phys. Rev. A26, 1907 (1982).
52. J. C. Tully, R. S. Berry, and B. J. Dalton, Phys. Rev. 176, 95 (1968).
53. A. D. Buckingham, B. J. Orr, and J. M. Sichel, Philos. Trans. Roy. Soc. London, A268, 147 (1970).
54. D. Dill and J. L. Dehmer, J. Chem. Phys. 61, 692 (1974).
55. R. A. Bonham and M. L. Lively, Phys. Rev. A29, 1224 (1984).
56. W. Rabalais, Principles of Ultraviolet Photoelectron Spectroscopy (Wiley, New York, 1977).
57. S. Iwata and S. Nagukura, Mol. Phys. 27, 425 (1974).
58. F. Hirota, J. Electron Spectrosc. Relat. Pheom. 9, 149 (1976).
59. J. L. Dehmer and D. Dill, in Electron-Molecule and Photon-Molecule Collisions, T. N. Rescigno, V. McKoy, and B. Schneider, Editors (Plenum, New York, 1979), pp. 225-263.
60. P. W. Langhoff, N. Padial, G. Csanak, T. N. Rescigno, and B. V. McKoy, Int. J. Quantum Chem. S14, 285 (1980).
61. G. R. J. Williams and P. W. Langhoff, Chem. Phys. Lett. 78, 21 (1981).
62. Z. Levine and P. Sovin, Phys. Rev. Lett. 50, 2074 (1983).
63. P. W. Langhoff, in Electron-Atom and Electron-Molecule Collisions, J. Hinze, Editor (Plenum, New York, 1983), pp. 297-314.
64. L. A. Collins and B. Schnieder, Phys. Rev. A 29, 1695 (1984).
65. E. P. Wigner, Phys. Rev. 73, 1002 (1948).
66. T. E. H. Walker, P. M. Dehmer, and J. Berkowitz, Phys. Rev. A 59, 4292 (1973).
67. J. Hargreaves, Proc. Camb. Philos. Soc. 25, 91 (1928).
68. B. K. Janousek and J. I. Brauman, Phys. Rev. A 23, 1673 (1981).

69. P.A. Schulz, R.D. Mead, W.C. Lineberger, Phys. Rev. A 27, 2229 (1983).
70. R.D. Mead, K.R. Lykke, and W.C. Lineburger, in Electronic and Atomic Collisions, J. Eichler, I.V. Hertel, and N. Stolterfoht, Editors (Elsevier, Amsterdam, 1984), pp. 721-730.
71. A. Dalgarno, Rev. Mod. Phys. 35, 522 (1963).
72. U. Fano, Phys. Rev. 124, 1866 (1981).
73. H. Feshbach, Ann. Phys. (NY) 43, 410 (1967).
74. F.C. Brown, in Synchrotron Radiation Research, W. Winick and S. Doniach, Editors (Plenum, New York, 1980), pp. 61-100.
75. I. Nenner, Laser Chem. 3, 339 (1983).
76. V. Schmidt, Applied Optics 19, 4080 (1980).
77. J.T. Broad and W.P. Reinhardt, Phys. Rev. A14, 2159 (1976).
78. J.W. Gallagher, C.E. Brion, P.W. Langhoff, J.A.R. Samson, J. Phys. Chem. Ref. Data (to be published).
79. P.W. Langhoff, in Resonances in Electron-Molecule Scattering, van der Waals Complexes, and Reactive Chemical Dynamics, D.G. Truhlar, Editor (ACS, Washington, D.C., 1984), pp. 113-138.

DISSOCIATIVE RECOMBINATION OF INTERSTELLAR IONS:
ELECTRONIC STRUCTURE CALCULATIONS FOR HCO$^+$

W.P. Kraemer
Max-Planck-Institut für Physik und Astrophysik, Institut für
Astrophysik, Karl-Schwarzschild-Str. 1, 8046 Garching, FRG

and

A.U. Hazi
Lawrence Livermore National Laboratory
University of California
Livermore, CA 94550, USA

I. INTRODUCTION

In the gas-phase theory of the chemistry in dense interstellar clouds most polyatomic neutral molecules are assumed to be formed by dissociative recombination reactions in which a positive molecular ion captures an electron and dissociates into neutral fragments. The physical process of dissociative recombination was introduced by Bates and Massey [1] to account for the rapid rate of recombination of various molecular ions in the upper atmosphere of the Earth. Since then this reaction has also come to be recognized as an important process in plasmas, lasers, and in interstellar molecular clouds. The subject has been reviewed by several authors [2-8].

The capture of an electron by a positive molecular ion requires that the electron gives up energy when moving from a free to a bound state. This energy may be transferred to a third body or it may be radiated away as a result of a free-bound transition. Three-body encounters become probable only at fairly high concentrations of neutral molecules or electrons, while the radiative process occurs only with small probability and usually leads to very small recombination rates. Especially under the very low density conditions in interstellar molecular clouds the three-body reactions can be rouled out because they are extremely improbable. However, according to the theory of Bates and Massey, the dissociative recombination reaction can be thought of as a process in which the positive molecular ion carries along its own third body to act as an energy removing agent. The process can be described schematically as:

$$AB^+ + e \rightleftharpoons (AB)^{**} \longrightarrow A + B \qquad (1)$$

where A and B represent atoms or molecular fragments. In the collision of an electron with a molecular ion AB^+ a temporary neutral molecular state AB^{**} is formed. In this initial reaction step the incident electron excites a target electron and itself falls into an unoccupied orbital of the molecule, i.e. the capture of the electron occurs through exchange of energy between the incident and a target electron. The resulting intermediate state AB^{**} is unstable because an electron can be reemitted through autoionization. After electron capture the molecular potential controlling the motion of the nuclei is that of the neutral AB^{**} state rather than that of the initial ion. If this potential happens to be repulsive with respect to one or several dissociation channels, the captured electron can be trapped via dissociation of the intermediate AB^{**} into neutral fragments.

In this picture the dissociative recombination reaction is described as a two-step process consisting of the formation and the dissociation of the intermediate state AB^{**}. There are two features of this process that are not found in atoms. First, the electron capture can take place for electrons of almost any thermal energy. In order that electron capture does occur without transfer of energy between electronic and nuclear motion it is necessary only that at some point on the potential energy surfaces of the ion AB^+ and of the AB^{**} state the vertical difference between the two surfaces be equal to the energy brought in by the incident electron. This point on the potential energy surfaces is called the capture point R_c. The second unique feature of the molecular process is that the dissociation of the intermediate AB^{**} provides an efficient mechanism for trapping the captured electron. The lifetimes of autoionizing states against electron emission are typically of the order of 10^{-14} sec. This time must be compared with the stabilization time, i.e. the time taken by the nuclei to separate along the repulsive AB^{**} potential from the capture point R_c past the so-called stabilization point R_s, the crossover point of the two potentials. From any point on the AB^{**} potential beyond R_s ionization becomes highly improbable assuming the Born-Oppenheimer approximation to be valid. If the potential of AB^{**} intersects the Franck-Condon region of AB^+ the stabilization time is much shorter than the autoionization time and the intermediate state AB^{**} will therefore dissociate rather than reionize by emitting an electron. For electron recombination with atomic ions at moderate densities there is no stabilization mechanism of comparable efficiency. For this reason the recombination rates of molecular ions are normally several orders of magnitude larger than those for atomic ions. The reaction mechanism described above is called the direct process. At medium electron temperatures the dependence of the recombination rate of this direct process on the electron temperature can be derived to be approximately $T^{-1/2}$.

Several earlier experimental studies of the temperature dependence of the recombination rates for a number of reactions indicated however a stronger temperature dependence of about $T^{-3/2}$. This led Bardsley [4] to introduce an alternative three-step reaction mechanism, called the indirect process:

$$AB^+ + e^- \rightleftharpoons (AB)^* \rightleftharpoons (AB)^{**} \longrightarrow A + B. \tag{2}$$

In the first step, a Rydberg state $(AB)^*$ is formed in an excited vibrational level. This involves energy transfer from the incident electron directly to vibrational motions of the nuclei. This initial capture step is followed by predissociation of the Rydberg state. This predissociation can be considered as a two-step process. The step $(AB)^* \to (AB)^{**}$ arises from configuration interaction between the Rydberg state and the valence state $(AB)^{**}$. If the potential of $(AB)^{**}$ is repulsive (in the region where configuration mixing with $(AB)^*$ is large), recombination will be completed by dissociation.

II. EXPERIMENTAL AND THEORETICAL STUDIES

Dissociative recombination reactions of molecular ions are studied in the laboratory by different experimental techniques such as plasma decay measurements, intersecting beam experiments, and more recently by using the trapped-ion technique. An excellent review of recent experimental work that has been done in this field is given by Mitchell and McGowan [8]. So far experimental studies have focused mainly on the determination of total cross sections or thermal rate coefficients. Experiments are usually performed at medium electron energies and at ion temperatures that are much higher than the very low temperatures characteristic for molecular cloud conditions. Extrapolation of these results to molecular cloud temperatures is not possible. More detailed information (intermediate product state, state of dissociation products) is mostly not available from experiment. In this situation theoretical studies can be rather helpful to supplement experimental information and to gain a better understanding of the reaction mechanism of dissociative recombination reactions.

Experimental studies of dissociative recombination reactions cover a large number of diatomic and polyatomic ions. Many of the molecular ions studied in the laboratory are those that are known from processes in the upper atmosphere of the Earth. With only a few exceptions the rate constants at 300K for diatomic ions are typically of the order $\alpha(300K) \approx (2-4) \times 10^{-7} cm^3 sec^{-1}$. For polyatomic ions the α-values at 300K are usually obtained in the range 2×10^{-7} to $1 \times 10^{-6} cm^3 sec^{-1}$. Laboratory measurements are also done for cluster ions such as $H_3^+ \cdot H_2$, $H_3O^+(H_2O)_n$, $NH_4^+ \cdot NH_3$, $N_2^+ \cdot N_2$, $O_2^+ \cdot O_2$, $CO^+ \cdot CO$. As an interesting result of these measurements it turns out that the $\alpha(300K)$-values of these clusters are about an order of magnitude larger than those of the unclustered ions. This may have also some implications on the chemistry in molecular clouds.

Recently it has been pointed out by Herbst [9] that no successful experiments have yet been performed to determine the relative amounts of various possible dissociation products. Polyatomic electron-ion dissociative recombination reactions are in fact often quite exoergic for a variety of dissociation channels. In order to obtain branching ratios for various sets of neutral dissociation products Herbst applied the statistical phase-space theory. Calculations were done for a number of interstellar molecular ions such as $HCNH^+$, CH_3^+, NH_4^+, OH_3^+. The general conclusion reached from these calculations is that exoergic polyatomic electron-ion reactions usually do give a variety of different dissociation products. However, the minimal disruption that can occur is found to be usually the major reaction channel.

Although dissociative recombination has been studied in the laboratory since many years, only a few rigorous theoretical studies appeared in the literature so far. This is mainly due to the fact that dissociative recombination cross sections are extremely sensitive to the shape and the relative position of the electronic potentials for the ion AB^+ and the neutral species AB^{**} to each other and that accurate potentials are usually not available for most systems. Existing theoretical work is therefore limited to diatomic ions where the necessary potentials can often be generated from spectroscopic data. Another major problem in theoretical studies arises from the uncertainty about the role of vibrationally excited Rydberg states in the recomination process and from the computational problems in calculating these Rydberg states.

In view of these difficulties, the reaction of H_2^+ ions with electrons is of particular interest because this system offers the best possibility for accurate theoretical calculations to which experimental results can be compared. A problem in comparing experimental results for H_2^+ with theory comes however from the fact that excited vibrational states of the H_2^+ ion are very long lived and that the cross section for H_2^+ recombination depends strongly on the vibrational state of the ion. Despite of the difficulty to produce a plasma with H_2^+ as the dominant ion, several laboratory studies of H_2^+ dissociative recombination were performed recently using the intersecting beams technique [10-14]. Most accurate cross section measurements at energies below 3eV [13] show narrow structures superimposed upon a monotonically decreasing background with an energy dependence close to $E^{-0.9}$. According to McGowan [15] this resonance structure could be a manifestation of the indirect mechanism for dissociative recombination of H_2^+. An alternative explanation of the resonance structure was recently presented by O'Malley [16]. Considering the $e-H_2^+$ reaction as a direct process, O'Malley's analysis shows that curve crossings of the repulsive H_2^{**} resonance state with various Rydberg states could also give rise to the resonant structure in the dissociative recombination cross section of H_2^+.

The first fully theoretical treatment of the $e-H_2^+$ collision process was done by Bottcher [17,18] using the projection operator formalism of resonance scattering processes in molecules developed originally by Feshbach [19]. There remained however significant discrepancy between theory and experiment particularly regarding the cross sections at low vibrational energies of the ion and the branching ratios for the formation of different dissociation products. Several theoretical studies were performed since then assuming that at low energies the recombination process proceeds primarily through the lowest doubly excited $(1\sigma_u)^2\ {}^1\Sigma_g^+$

state of H_2 and investigating the importance of Rydberg states for the dissociation step [20-22]. A comprehensive discussion of the large degree of disagreement among various theoretical calculations of the lowest doubly excited $^1\Sigma_g^+$ state of H_2 is presented in the recent paper by Hazi et al. [23]. In this study the energy and width of the $(1\sigma_u)^{2}\,^1\Sigma_g^+$ state of H_2 were calculated using Feshbach projection operators and the Stieltjes moment method. The results of these calculations were used by Giusti et al. [24] to re-investigate the $e-H_2^+$ collision process. Calculations were performed for electron energies below 0.5 eV and the lowest three vibrational levels of H_2^+. It was found that in agreement with experiment [13] Rydberg states lead to narrow structures in the cross sections and that the cross section is very small for the v = 0 level of the ion becoming larger however for v = 1 and 2. While the overall agreement between the results of these calculations and experiment is reasonable, it appears on the other hand that a quantitative comparison between theory and experiment will only be possible if measurements can be made for single vibrational levels of the H_2^+ ion.

The actual need for a detailed comparison between experiment and theory becomes evident also in view of the large disagreement between experimental and theoretical results for more complicated systems such as $e-NO^+$, $e-N_2^+$, $e-O_2^+$. These diatomic atmospheric ions were studied in the laboratory by various experimental techniques [25-33]. Generally there is good agreement between experimental results obtained with different techniques even though the vibrational populations encountered in the different methods are probably quite different. This indicates that there is only little vibrational state dependence of the total recombination rate coefficients for these diatomic ions which is in apparent contradiction to the vibrational dependence found for rare gas recombination rates [34]. Theoretical studies on NO^+ and N_2^+ were performed using potential curves that were generated from spectroscopic data. In the case of N_2^+, the experimental measurements are in total disagreement with the theoretical calculations by Michels [35] who found that the rate coefficients for the v = 1 and 2 vibrational states are much smaller than the v = 0 value. Subsequent laboratory measurements by Zipf [33] however showed that in fact the v = 0, 1, 2 rate coefficients for N_2^+ are all approximately equal. In the theoretical treatment of the NO^+ ion Lee [36] used experimental data to impose boundary conditions for the various dissociation channels of the resonance complex. The uncertainty of the theoretical results however is still estimated to be as large as 50%. To study the recombination reaction of O_2^+ Guberman [37] performed fairly extensive electronic structure calculations of the diabatic potential curves of several excited states of O_2 which are important for electron recombination of O_2^+ in the lowest five vibrational states. From these calculations Guberman was able to give qualitative explanations for example for the experimental finding by Zipf [33] that the percentage of $O(^1S)$ formation in the dissociative recombination of O_2^+ is critically dependent on the initial vibrational state of the ion.

III. ELECTRONIC STRUCTURE CALCULATIONS FOR HCO^+

The present study of the interstellar formyl ion HCO^+ is the first attempt to investigate dissociative recombination for a triatomic molecular ion using an entirely theoretical approach. In the following we will describe a number of fairly extensive electronic structure calculations that were performed to determine the reaction mechanism of the $e-HCO^+$ process. Similar calculations for the isoelectronic ions HOC^+ and HN_2^+ are in progress.

The ions HCO^+ and HN_2^+ are known to be rather abundant interstellar molecular species playing an important role in the interstellar cloud chemistry. Searches for the isoformyl ion HOC^+ remained unsuccessful for a long time. Only recently HOC^+ was observed in the laboratory [38] and a tentative identification of the ion was made in the source Saggitarius B2, a dense interstellar cloud near the Galactic center [39]. However, the existence of HOC^+ in interstellar clouds still remains to be doubtful and several attempts were made recently to give an explanation for the difficulties of observation of HOC^+ in dense interstellar clouds [39-43].

Under molecular cloud conditions the predominant formation reaction of HCO^+ and HN_2^+ occurs through proton transfer from abundant H_3^+ ions to CO and N_2, respectively. Dissociative recombination with thermal electrons is generally assumed to be the main loss reaction for both ions. Electron recombination rates for HCO^+ were measured in a microwave afterglow/mass spectrometer experiment [44] as $\alpha(HCO^+) = (3.3 \pm 0.5)$ and $(2.0 \pm 0.3) \times 10^{-7}$ cm^3/sec at 205 and 300 K, respectively, leading to a strong temperature dependence ($\sim T^{-1}$) for α. With this temperature dependence a linear extrapolation to the low temperatures in molecular clouds gives α-values in the range of $(3.0-0.6) \times 10^{-6}$ cm^3/sec for cloud temperatures between 20 and 100 K. Current models describing interstellar cloud chemistry use instead a smaller α-value of 2.0×10^{-7} cm^3/sec which is probably a more realistic estimate for the recombination rate of HCO^+ in the lowest vibrational state. Recombination cross sections for HN_2^+ were measured over a wide range of electron energies using a merged electron-ion beam technique [45]. Assuming a Maxwellian distribution of the electron velocities, the rate coefficient for HN_2^+ at 100 K is derived to be as large as $1.5 \; 10^{-6}$ cm^3/sec.

There are two major problems with the above experiments. First, the vibrational populations of the initial target ions are not known. Since at the very low temperatures in dense interstellar clouds only the lowest vibrational level of the ions is populated, the interstellar rate coefficients may be quite different from the values extrapolated from measurements at much higher temperatures. Secondly, no information about the electronic states of the dissociation products is available from the experiments. For an understanding of the dissociative recombination reaction mechanism however it is important to know in which electronic states the dissociation products are obtained because this gives information about the resonance state which is formed in the electron capture step.

In the electronic structure calculations the molecular orbitals were approximated by linear combinations of Gaussian-type basis functions located at the nuclear centers. Single-configuration Hartree-Fock self-consistent field (SCF) calculations were performed to optimize the expansion coefficients. The basis set employed here consisted of two groups of basis functions. The first group contains those basis functions that are required to approximate the molecular orbitals of the ion and the neutral resonance state. Functions of this type are used in any electronic structure calculations of molecular valence states. This valence basis was augmented by a second group containing diffuse Gaussian functions that are able to generate a sequence of Rydberg-type orbitals in the molecular orbital basis. These diffuse Rydberg-type orbitals are essential for describing the background continuum of the non-resonant scattering wave function and the series of Rydberg states converging against the ion limit and interacting with the repulsive resonance state.

The valence basis selected for the present calculations consisted of (11s 7p 1d) functions for carbon and oxygen and of (6s 1p) functions for hydrogen, contracted to (5s 4p 1d) and (3s 1p), respectively. This basis set is equivalent to basis sets of triple-zeta plus polarization quality used in other high accuracy electronic structure calculations. In augmenting this valence basis by adding diffuse functions a compromise had to be made because the complexity of the computations increases rapidly with the size of the basis set used. Additional diffuse functions with exponents decreasing in geometrical sequence were therefore distributed in the following way: (7s 5p) functions at the carbon center, (3s 3p) functions at oxygen, and (3s 2p) functions at hydrogen. Maximum flexibility is thus added at the central carbon atom because functions in this location are expected to be most effective. The diffuse functions, while not contributing to the energy of the molecular valence states, are capable to produce a sufficiently large number of diffuse Rydberg-type orbitals. The valence part of the basis set consists of 52 contracted functions whereas the total augmented basis counts as many as 95 functions.

The reliability of the valence basis set was checked by comparing the results of some basic electronic structure calculations with corresponding experimental data. At the simple SCF level of approximation the dissociation energy of $HCO^+(X^1\Sigma^+)$ was obtained as $D_e = 6.16$ eV, which is fairly close to the value of 6.40 eV deduced from most recent proton affinity measurements [46] of CO. This indicates that single-determinant SCF calculations are able to provide a rather good description of the ground electronic state of the HCO^+ ion. Good agreement was further obtained at the SCF level of approximation for the ionization potential of hydrogen: IP(SCF) = 13.60 eV as compared to the true value of 13.6058 eV. The SCF value was calculated as the energy difference between the dissociation limits of HCO and HCO^+ in their ground electronic states. Apart from these results, correlation energy effects are expected to become more important for other energy properties such as ionization potentials of CO and HCO and excitation energies of CO. Using canonical SCF orbitals, single-reference state full single and double excitations CI calculations were performed for the ground electronic states of HCO,

HCO^+, CO^+ and for CO in the ground and the two lowest triplet excited states. Vertical ionization potentials and excitation energies that are deduced from these calculations, are collected in Table 1 and compared with corresponding experimental results. The rather poor agreement between theoretical and experimental results for the ionization potential of HCO (X^2A') is due to the fact that the HCO equilibrium geometry was not fully optimized in the present calculations. At the end of this discussion about the reliability of the basis set, it should also be mentioned that previous accurate theoretical determinations of the rotational and vibrational energy levels of the series of isoelectronic triatomic molecular species such as HCN, HNC, HCO^+, HOC^+, HN_2^+ [47-49] were based on electronic structure calculations employing basis sets that are equivalent to the one used in the present study. These calculations provide very sensitive tests of the correct shapes of the molecular potentials in the minimum region. The average error of the calculated rotational transition frequencies was found to be about 0.5% and that of the vibrational frequencies of the order of a few percent.

TABLE 1:

Energy values (in eV) determining the relative positions of the potentials relevant for dissociative recombination of HCO^+: ionization potentials for $H(^2S)$, $CO(X^1\Sigma^+)$, $HCO(X^2A^1)$, dissociation energy of $HCO^+(X^1\Sigma^+)$, and excitation energy of $CO(X^1\Sigma^+ - a'\ ^3\Sigma^+)$.

	SCF	CI	Exptl.
$\dfrac{H(^2S)}{IP}$	13.60	13.60	13.6058
$\dfrac{CO(X^1\Sigma^+)}{IP}$	13.56	13.74	14.0139
$T_e(X^1\Sigma^+ - a'\ ^3\Sigma^+)$	–	7.12	6.92
$\dfrac{HCO^+(X\ ^1\Sigma^+)}{D_e}$	6.16	6.46	6.40
$\dfrac{HCO(X^2A^1)}{IP}$	9.0	9.2	9.88

The CI calculations on HCO^+ were extended to calculate the entire dissociation potential of $HCO^+(^1\Sigma^+) \to H^+ + CO(^1\Sigma^+)$. In these calculations the CO distance was held fixed at the equilibrium value in HCO^+, namely at 2.10 a_o, which is close to the equilibrium bond length in carbon monoxide $R_e(CO) = 2.132\ a_o$. The calculations showed that the correlation energy is changing very slowly as a function of the HC separation, which explains the rather good SCF result for the dissociation energy of HCO^+.

Similar CI calculations were also done for the lowest $^2\Sigma^+$ excited state of HCO. The corresponding HC dissociation potential is known to be repulsive with respect to dissociation to $H(^2S) + CO(^1\Sigma^+)$. This HCO state was previously assumed to be possibly responsible for an effective electron recombination reaction of the HCO$^+$ ion at low temperatures. The calculations proved however that there is no crossing between the HCO ($^2\Sigma^+$) and HCO$^+$($^1\Sigma^+$) potentials. Variation of the CO distances in both species did not move the two curves closer to each other. According to the present calculations the lowest $^2\Sigma^+$ valence state of HCO does therefore not contribute to direct dissociative recombination of the HCO$^+$ ion.

By far the most important configuration of HCO$^+$ in the $^1\Sigma^+$ ground state is the Hartree-Fock configuration (core) $3\sigma^2 4\sigma^2 5\sigma^2 1\pi^4$, where "core" represents the doubly-occupied oxygen and carbon 1s-orbitals. Since the bending potential of HCO$^+$ was recently calculated to be rather steep [49], it is safe to assume that electron capture by the HCO$^+$ ion will lead to a linear resonance state, particularly if the process occurs at the low energy conditions in dense interstellar clouds. Existing theoretical studies on HCO [50-53] do not provide sufficiently detailed information on which excited states of HCO could be responsible for the dissociative recombination reaction of the HCO$^+$ ion. Qualitative considerations based on preliminary calculations of several excited HCO states show that there are essentially two possibilities to form HCO resoncance states in low energy e-HCO$^+$ collisions.

In the first case, the scattering electron excites a 5σ target electron to a $n\sigma$ orbital (n > 6) and falls itself into the 6σ unoccupied orbital. Particularly for larger n-values the $^2\Sigma^+$ states thus formed belong to a series of Rydberg states converging against the HCO$^+$ ($^3\Sigma^+$) ion limit. This excited state of HCO$^+$ is unstable with respect to the dissociation products $H(^2S) + CO^+(^2\Sigma^+)$. The energy levels at the dissociation limits of this $^3\Sigma^+$ excited state and of the $^1\Sigma^+$ ground state of HCO$^+$ are separated only by 0.408 eV, i.e. the energy difference between the ionization potentials of carbon monoxide and the hydrogen atom. The location of the Rydberg states relative to the ionization limit were obtained using the simple formula

$$E_n^{Ryd}(R) = E^{Ion}(R) - \frac{0.5}{(n-\delta)^2} \qquad (3)$$

where n is the primary quantum number and δ the quantum defect. It turns out that the lowest repulsive Rydberg states cross the ion potential in an energy region that is much too high to be accessible under interstellar cloud conditions.

The other possibility to form an HCO resonance state in low energy e-HCO$^+$ collisions is that the energy transfer from the scattering electron leads to a $\pi\pi^*$-excitation of a target electron while the captured electron ends up in the 6σ unoccupied orbital. The $^2\Sigma^+$ state thus formed may be described as a combination of a hydrogen with the $\pi\pi^*$ states of CO. The electronic configuration of this state is essentially $(1-4\sigma)^2$

$5\sigma^2 6\sigma 1\sigma^3 2\pi$. For large HC separations the 6σ orbital becomes a hydrogen 1s orbital and the dissociation products of the lowest lying member of these states are $H(^2S) + CO(a'^3\Sigma^+)$. On bending this $^2\Sigma^+$ state goes over into the B $^2A^1$ state that was recently studied by Tanaka and Takeshita [52] and that was assigned as the upper state of one of the hydrocarbon flame bands (A-band). At intermediate HC separations there is a strong configuration mixing of this $^2\Sigma^+$ state with a charge-transfer type configuration $(1-4\sigma)^2 5\sigma 6\sigma^2 1\pi^4$. This is in agreement with the findings of Tanaka and Davidson [53] in their study of the potential surfaces of the lower electronic states of HCO.

Electronic structure calculations of the dissociation potential of this $^2\Sigma^+$ state were performed. These calculations show that the $^2\Sigma^+$ excited state of HCO discussed here is in fact the lowest possible resonance state crossing the potential of the HCO$^+$ ion close to its minimum. Dissociation of this HCO state leads to the a' $^3\Sigma^+$ excited state of CO. It should be noticed that the CO bond length of this a' $^3\Sigma^+$ state (R_{CO}=1.359 Å) is considerably larger than that of the $^1\Sigma^+$ ground state of CO (R_{CO} = 1.128 Å) and the CO bond length in HCO$^+$ (R_{CO} = 1.105 Å). Relaxation of the CO bond length is therefore important in studying the dynamics of the dissociative recombination process of HCO$^+$.

Summarizing the present electronic structure calculations it appears that there is a sparsity of HCO resonance states that are accessible at low energy e-HCO$^+$ collisions. In this respect the situation seems to be similar to the one found for the e-H$_2^+$ reaction.

APPENDIX: METHOD TO CALCULATE ELECTRON CAPTURE WIDTH

Dissociative recombination in a resonant electron scattering process. Computationally the most promising method available at present to deal with such processes is based on the Feshbach formalism [19] and the Stieltjes moment theory [54,55]. The method thus uses projection operator techniques and the golden-rule definition of the resonance width. Square-integrable, L^2, basis functions are employed to describe both the resonant and the non-resonant part of the scattering wave function. Stieltjes moment theory is finally applied to extract a continuous approximation for the width from the discrete representation of the background continuum. For a detailed description of this method covering also various computational applications we refer to recent work by Hazi and co-workers [56-58,23]. In the following we will discuss very briefly a few computational aspects and their implications for the basic electronic structure calculations.

In Feshbach's resonance theory the width of a resonance decaying into an open channel is given by the so-called golden-rule formula

$$\Gamma(E) = 2\pi \left| < \Psi_R (H-E) \Psi_E^+ > \right|^2 \tag{4}$$

where Ψ_R describes the resonance state and where $\Psi_E^\pm$ is an energy-normalized scattering function representing the non-resonant, background continuum at energy E. The operator H is the full many-electron Hamiltonian. The problem in numerical applications is to determine Ψ_R and $\Psi_E^{\pm}$ in such a form that the above matrix element can be easily evaluated. The resonance state wave function Ψ_R is obtained directly by quantum chemical configuration interaction (CI) procedures. The difficulty arises with the calculation of the nonresonant scattering function $\Psi_E^{\pm}$ because of the non-spherical and non-local nature of the electron-molecule interaction potential. In the Stieltjes moment method this difficulty is avoided by expanding $\Psi_E^{\pm}$ in terms of square-integrable, L^2, many-electron basis functions which are constructed to be orthogonal to the resonance wave function Ψ_R through the use of projection operators.

In the computational method used here, a total function space must first be generated from an orthogonal set of many-electron basis functions (configurations) in such a way that it can be partitioned into a resonant and a non-resonant part from which the wave functions Ψ_R and Ψ_E^{+} are obtained. These wave functions are used in a formula analogous to (3) to calculate a discrete spectrum for the width. Apart from the initial generation of the total function space, there are three computational steps in the procedure to determine the resonance width: (i) Construction of projection operators after partitioning of the total function space and calculation of Ψ_R and an approximation to Ψ_E^{+} by diagonalizing the full Hamiltonian in the two respective subspaces, (ii) evaluation of the width matrix elements connecting the two subspaces from the golden-rule formula, and (iii) application of Stieltjes moment theory techniques to extract correctly normalized widths from the discrete representation of the background continuum.

In order to be able to generate a suitable set of orthogonal many-electron basis functions defining the total function space, one first has to select an orthogonal one-electron function basis that can accurately describe both the resonance state as well as the target ion plus the scattering electron. The choice of this molecular orbital basis is thus governed by the electronic structure of the ion, the expected nature of the resonance state, and also by the needs to represent the background continuum. The basis must contain two groups of orbitals: those which are required to describe the resonance state and the target ion, and those which can approximate the wave function of the scattering electron. Since both the resonance and the target states are bound states, the molecular orbitals in the first group are those normally used in any electronic structure calculation of ground and low-lying excited states. The second group of orbitals on the other hand must contain numerous diffuse basis functions providing a dense representation particularly of the lower part of the background continuum.

Using this one-electron orbital basis, the total function space $(P_o + Q_o)$ is generated from all those many-electron configurations that are required for an accurate description of the resonance state. It has to be noticed however, that in this total CI space internal consistency between the levels of approximation used for the resonant (Q_o) and the non-resonant (P_o) subspaces is more important than achieving the best possible accuracy for the resonance state itself. The subspace P_0 is constructed from those configurations which to first order describe the target ion and the scattered electron. All the remaining configurations are included in the Q_0 subspace. The resonance wave function Ψ_R and the corresponding resonance energy ϵ_R are obtained by diagonalizing the transformed Hamiltonian Q_0HQ_0. If P_0 contains all the configurations that are essential for describing the target ion, the lowest eigenstate or the few lowest eigenstates of Q_0HQ_0 will correspond to the resonance: $\Psi_R = \{\Phi_{Ri}\}$.

After the determination of the resonance function Ψ_R a repartitioning of the function space is performed. This repartioning is found to be essential to incorporate many-electron correlation and polarization effects in the description of the non-resonant continuum. In the case of many-electron target ions such effects must be accounted for if accurate widths are to be obtained. Two new projection operators P and Q are defined as:

$$Q = |\Psi_R\rangle\langle\Psi_R| \tag{5a}$$

and

$$P = 1 - Q = P_0 + (Q_0 - Q). \tag{5b}$$

The operator P thus contains not only those configurations which approximate the decay channel (originally in subspace P_0) but also the higher, non-resonant solutions of Q_0HQ_0. The discrete representation of the non-resonant continuum is finally obtained by diagonalizing PHP in the basis of all configurations of the total function space. The resulting eigenfunctions $\{X_n\}$ with non-zero eigenvalues ϵ_n form an orthonormal set and they are orthogonal to Ψ_R by construction.

For sufficiently large basis sets each solution X_n approximates Ψ_E^+ with $E = \epsilon_n$ in a region near the nuclei, except for an overall normalization factor. Since the golden-rule formula (4) contains the L^2 function Ψ_R which is localized near the nuclei, the matrix element has no significant contribution from the asymptotic region where X_n fails to approximate Ψ_E^+ [38,39,40]. It is therefore possible to use the set of functions $\{X_n\}$ to calculate the width matrix elements:

$$\gamma_n = 2\pi|\langle\Psi_R H X_n\rangle|^2. \tag{6}$$

However, X_n cannot be used directly to approximate $\Gamma(\epsilon_n)$ because of the different normalization. By analogy to the work on accurate photoionization cross sections by Langhoff and co-workers [54,55], Stieltjes moment theory can be employed to extract appropriately normalized resonance widths from the discrete spectrum obtained from (6).

ACKNOWLEDGMENT

This study was initiated when one of the authors (WPK) visited the NASA-Ames Research Center as a NRC Senior Research Associate. The NRC grant and the hospitality of the Computational Chemistry group at NASA-Ames are gratefully acknowledged. We are also grateful for a NATO Research Grant supporting this project.

REFERENCES

[1] D.R. Bates and H.S.W. Massey: Proc. Roy. Soc. (London) A192, 1 (1954)
[2] J.N. Bardsley: J. Phys. B. 1, 349, 365 (1968)
[3] J.N. Bardsley and F. Mandl: Rep. Prog. Phys. 31, 471 (1968)
[4] J.N. Bardsley and M.A. Biondi: Adv. Atom. Molec. Phys. 7, 1 (1970)
[5] Ch. Bottcher: Proc. Roy. Soc. (London) A340, 301 (1974)
[6] K.T. Dolder and B. Peart: Rep. Prog. Phys. 39, 693 (1976)
[7] D.R. Bates: Adv. Atom. Molec. Phys. 15, 235 (1979)
[8] B.A. Mitchell and J.W. McGowan, in "Physics of Electron-Ion and Ion-Ion Collisions", ed. Brouillard (Plenum Press, New York, 1982), p.279
[9] E. Herbst: Astrophys. J. 222, 508 (1978)
[10] B. Peart and K.T. Dolder: J. Phys. B 7, 236 (1974)
[11] M. Vogler and G.H. Dunn: Phys. Rev. A 11, 1983 (1975)
[12] R.A. Phaneuf, D.H. Crandall, and G.H. Dunn: Phys. Rev. A 11, 528, (1975)
[13] D. Auerbach, R. Cacak, R. Caudano, T.D. Gaily, C.J. Keyser, J.W. McGowan, J.B.A. Mitchell, and S.F.J. Wilk: J. Phys. B 10, 3797 (1977)
[14] D. Mathur, S.U. Khan, and J.B. Hasted: J. Phys. B 11, 3615 (1978)
[15] J.W. McGowan, R. Caudano, and J. Keyser: Phys. Rev. Lett. 36, 1447 (1976)
[16] T.F. O'Malley: J. Phys. B 14, 1229 (1981)
[17] Ch. Bottcher and K. Docken: J. Phys. B 7, L5 (1974)
[18] Ch. Bottcher: J. Phys. B 9, 2899 (1976)
[19] H. Feshbach: Ann. Phys. 5, 357 (1958)
[20] V.P. Zhadanov and M.I. Chibisov: Sov. Phys. JETP 47, 38 (1978)
[21] C. Derkits, J.N. Bardsley, and J.M. Wadehra: J. Phys. B 12, L529 (1979)
[22] V.P. Zhadanov: J. Phys. B 13, L311 (1980)
[23] A.U. Hazi, C. Derkits, and J.N. Bardsley: Phys. Rev. A 27, 1751 (1983)
[24] A. Giusti-Suzor, J.N. Bardsley, and C. Derkits: Phys. Rev. A 28, 682 (1983)
[25] F.L. Walls and G.H. Dunn: J. Geophys. Res. 79, 1911 (1974)
[26] F.L. Walls and G.H. Dunn: Physics Today 27, 34 (1974)
[27] P.M. Mul and J.W. McGowan: J. Phys. B 12, 1591 (1979)
[28] C.M. Huang, M.A. Biondi, and R. Johnson: Phys. Rev. A 11, 901 (1975)
[29] F.J. Mehr and M.A. Biondi: Phys. Rev. 181, 264 (1969)

[30] A.J. Cunningham and R.M. Hobson: J. Phys. B $\underline{5}$, 2328 (1972)
[31] E.C. Zipf: Geophys. Res. Lett. $\underline{7}$, 645 (1980)
[32] W.H. Kasner and M.A. Biondi: Phys. Rev. $\underline{174}$, 139 (1968)
[33] E.C. Zipf: J. Geophys. Res. $\underline{85}$, 4232 (1980)
[34] A.J. Cunningham, T.F. O'Malley, and R.M. Hobson: J. Phys. B $\underline{14}$, 773 (1981)
[35] H.H. Michels: Proc. 3rd Int. Conf. Atom. Phys. Boulder $\underline{1}$, 73 (1981)
[36] C.M. Lee: Phys. Rev. A $\underline{16}$, 109 (1977)
[37] S.L. Guberman: Int. J. Quantum Chem. $\underline{S13}$, 531 (1979)
[38] C.S. Gudeman and R.C. Woods: Phys. Rev. Lett. $\underline{48}$, 1344 (1982)
[39] R.C. Woods, C.S. Gudeman, R.L. Dickman, P.F. Goldsmith, G.R. Huguenin, A. Hjalmarson, L.-A. Nyman, and H. Olofsson: Astrophys. J. $\underline{270}$, 583 (1983)
[40] A.J. Illies, M.F. Jarrold, and M.T. Bowers: J. Chem. Phys. $\underline{11}$, 5847 (1982)
[41] S. Green: Astrophys. J. $\underline{277}$, 900 (1984)
[42] D.J. DeFrees, A.D. McLean, and E. Herbst: Astrophys. J. $\underline{279}$, 322 (1984)
[43] D.A. Dixon, A. Komornicki, and W.P. Kraemer: J. Chem. Phys. (in press)
[44] M.T. Leu, M.A. Biondi, and R. Johnson: Phys. Rev. A $\underline{8}$, 420 (1973)
[45] P.M. Mul and J.W. McGowan: Astrophys. J. $\underline{227}$, L.157 (1979)
[46] S.G. Lias: private communication
[47] W.P. Kraemer and G.H.F. Diercksen: Astrophys. J. $\underline{205}$, L97 (1976)
[48] P. Hennig, W.P. Kraemer, and G.H.F. Diercksen: Technical Report MPI/PAE Astro 135, Max-Planck-Institute of Physics and Astrophysics, Munich, Nov. 1977
[49] W.P. Kraemer and P.R. Bunker: J. Mol. Spectrosc. $\underline{101}$ 379 (1983)
[50] P.J. Bruna, S.D. Peyerimhoff, and R.J. Buenker: Chem. Phys. $\underline{10}$, 323 (1975)
[51] P.J. Bruna, R.J. Buenker, and S.D. Peyerimhoff: J. Mol. Struct. $\underline{32}$, 217 (1976)
[52] K. Tanaka and K. Takeshita: Chem. Phys. Lett. $\underline{87}$, 373 (1982)
[53] K. Tanaka and E.R. Davidson: J. Chem. Phys. $\underline{70}$, 2904 (1979)
[54] P.W. Langhoff, in "Electron-Molecule and Photon-Molecule Collisions", eds. T. Resigno, V. McKoy, and B. Schneider (Plenum Press, New York, 1979), p.183
[55] P.W. Langhoff: Int. J. Quantum Chem. $\underline{S8}$, 347 (1974)
[56] A.U. Hazi: J. Phys. B $\underline{11}$, L259 (1978)
[57] A.U. Hazi, in "Electron-Molecule and Photon-Molecule Collisions", eds. T. Resigno, V. McKoy, and B. Schneider (Plenum Press, New York, 1979), p.281
[58] A.U. Hazi, in "Electron-Atom and Electron-Molecule Collisions", ed. J. Hinze (Plenum Press, New York, 1983), p.103
[59] W.H. Miller: Chem. Phys. Lett. $\underline{4}$, 627 (1970)
[60] A.P. Hickmann, A.D. Isaacson, and W.H. Miller: J. Chem. Phys. $\underline{66}$, 1483 (1977)

CONTRIBUTED PAPERS

DETECTION OF INTERSTELLAR ROTATIONALLY-EXCITED CH

L.M. Ziurys and B.E. Turner

Dept. of Chemistry and Radio Astronomy Laboratory,
UC Berkeley, CA
National Radio Astronomy Observatory, Charlottesville, VA

ABSTRACT

Rotationally-excited CH has been detected for the first time in the ISM. The first excited level (F_1, N=1, J=3/2) was observed via its Λ-doubling transition at 700 MHz towards W 51--the lowest frequency molecular lines detected to date. The two main hyperfine components were seen in absorption, with approximately optically-thin, LTE intensities. The spectra qualitatively resemble the masering line profiles of the ground state at the same position, suggesting both transitions orginate in the same gas. Column density derived for the excited state is $NL > 1.2-1.4 \times 10^{14}$ cm^{-2}, comparable to that of the ground state in W 51 and most other sources. The large excited state population cannot easily be explained by FARIR radiative dust excitation; collisions with gas densities $n(H_2) > 10^6$ cm^{-3} are probably the dominant source of excitation. CH is therefore present with considerable abundance at very high densitites in W 51, contrary to predictions of chemical models. The F_1, N=1, J=3/2 measurements are consistent with the Gwinn-Townes model for collisional pumping of the ground state maser, but are not conclusive evidence for it.

The CH radical is certainly one of the more important known interstellar molecules. Its observed widespread distribution in giant molecular clouds, dark clouds (Zuckerman and Turner 1975, Rydbeck, et al. 1976, Hjalmarson, et al. 1977) and in diffuse gas (Lang and Willson 1978) is evidence for its important chemical role. In its 3.3 GHz ground-state transitions, CH exhibits weak masering and often non-LTE hyperfine ratios (e.g., Genzel, et al. 1979); resulting theoretical studies of its ex-

G. H. F. Diercksen et al. (eds.), Molecular Astrophysics, 591–601.

citation (e.g., Bertojo, Cheung and Townes 1975, Elitzur 1977) have contributed to the interpretation of interstellar masers in $^2\Pi$ molecules.

Aside from optical observations, radio, mm, and sub-mm measurements of CH have been limited to the 3.3 GHz ground-state Λ-doubling transitions (F_2, N=1, J=1/2). Thus only limited data have existed for defining the molecule's excitation and abundance. Past searches for the excited rotational states of CH in the radio and FARIR have been unsuccessful. Recently, however, high-resolution laboratory spectroscopy by Brazier and Brown (1983) Rogey, Demuycnk, and Destombes (1983) and Brown (1984) has resulted in very accurate values for CH Λ-doubling transitions.

Figure 1 shows the energy level diagram of CH for the lowest four rotational levels. The F_1, N=1, J=3/2 level lies 25.6 K, above ground state. We successfully searched for the F_1, N=1, J=3/2 Λ-doublet of CH at 700 MHz, towards W 51 using the Arecibo antenna and the NRAO 300 ft. telescope—the first detection of the molecule in a rotationally-excited state. The Λ-doublet was also detected towards W 43, W3, and Ori B.

All four hyperfine lines were searched for in W 51 at Arecibo. Only the two main hyperfine components of the F_1, N=1, J=3/2 doublet were detected (F=2-2, F=1-1); the much weaker satellite lines were not seen which is expected if the transition is in LTE. (See Table 1.) Figure 2 shows the spectra of the main lines taken at Arecibo, both in absorption, toward W 51. Figure 3 is the main component (F=2-2) absorption spectrum from the 300 ft. antenna, as well as the line profiles of the three ground-state CH transitions of Genzel, et al. (1979).

Two conclusions can be reached from a qualitative examination of the CH data: first, the F_1, N=1, J=3/2 doublet does not show any of the non-LTE effects that characterize the ground state. Second, the qualitative similarity between the excited and ground state spectra (in terms of LSR velocity and full-width at zero-power) strongly indicates that both transitions originate in the same gas. Additional detections of F=2-2 component with the 300 ft. telescope towards W3, W 43, and Ori B, also are absorption lines that resemble the ground-state emission.

The ratio of the peak antenna temperatures, as well as the integrals $\int T_A dv$, is roughly the LTE, optically-thin value for the two detected lines. However, the signal to noise is not adequate to rule out the possibility of a moderate optical depth. Such an optical depth is seemingly inconsistent with the low values derived for the ground-state doublet in W 51 (τ<0.015; Genzel, et al. 1979), but such low opacities are determined by averaging over a 5' beam. If the beam-filling factor is < 1, the opacity in the ground state would be higher. However, if large opacities are present in the ground state, there should be considerable maser gain and more spectacular maser effects would be expected than are observed.

If one assumes that the CH cloud uniformly covers the background continuum source, the optical depth for the F_1, N=1, J=3/2 doublet can be estimated from the line to contiuum ratio, T_L/T_C. The opacity of the F=2-2 line is $\tau \cong 0.0021$ and $\tau \cong 0.0014$ for the F=1-1 line, from the Arecibo data and the optical depth of the F=2-2 transition, estimated from the 300 ft. measurements, is $\tau \cong 0.0013$. The lower opacity derived from the 300 ft. data indicates that the source of excited CH is < 19 arcmin. in extent, the 300 ft. beam size.

Column Density and Rotational Temperature

Using the data from Table 1 and an excitation temperature of $T_{ex} > 2.7$ K, the column density for the F_1, N=1, J=3/2 Λ-doublet is NL > $1.2-1.4 \times 10^{14}$ cm^{-2}. Even using the lower limit to the column depth of $1.2-1.4 \times 10^{14}$ cm^{-2}, the amount of CH in this level is comparable to what has been measured for the ground-state doublet at this position. With a HPBW of 5', Genzel, et al. found NL (ground state) = 1.8×10^{14} cm^{-2}. Ground state CH column densities toward other sources have typically been found to be NL < 10^{14} cm^{-2} (Genzel, et al. 1979; Zuckerman and Turner 1975; Rydbeck, et al. 1975).

A lower limit to the rotational temperature between the two levels was calculated to be $T_{rot} > 17$ K. For thermalization in both levels, $T_{rot} = T(kinetic) = 50$ K. The true rotational temperature probably lies in between these values. T_{rot} in this range is consistent with upper limits found for the F_2, N=2, J=3/2 level by Matthews, et al. (1983) and the ground-state column density of Genzel, et al.

Excitation of the F_1, N=1, J=3/2 Λ-Doublet

For collisions alone to account for the 700 MHz doublet population, densities of $n(H_2) > 5 \times 10^6$ cm^{-3} are needed, using a kinetic temperature of 50 K and a collisional cross section of 10^{-15} cm^2. The presence of such high-density gas towards the CH W 51 position is evidenced by observations of HCN:J=4-3 (White, et al. 1982) and CS:J=5-4 (White, et al. 1983). However, the calculated rotational temperature suggests that the densities are not large enough to thermalize the rotational transition from the 700 MHz level to ground at 560 μm. Densities greater than 10^7 cm^{-3} are required for thermalization, so even at $n(H_2) = 5 \times 10^6$ cm^{-3}, the excitation is subthermal.

Radiative excitation of the F_1, N=1, J=3/2 doublet could occur via direct pumping by FARIR dust photons at 560 μm. However, there apparently is not much dust radiating at this wavelength. Using the peak dust optical depth in a 42" beam measured at 400 μm towards our W 51 position (Jaffe, Becklin, and Hildebrand 1984), and scaling the opacity by λ^{-2}, the peak dust optical depth at 560 μm is $\tau_d \sim 0.087$. Assuming a perfect filling factor of 1 for CH molecules and dust photons, an upper limit to the radiative excitation rate was calculated and found to be a factor of 4-5 too small to account for the observed lower limit for the excited state column density (T_{dust} from Jaffe, et al.). In the best of circumstances, dust emission at 560 μm could be responsible for the F_1, N=1, J=3/2 population, but the CH excited-state gas would have to be confined to the 400 μm peak. Towards this peak, $n(H_2)$ has been estimated to be $> 10^6$ cm^{-3} (Genzel, et al. 1982); Jaffe, et al. set the gas density to be <u>at least</u> 2×10^5 cm^{-3}. At such densitites, collisions are competitive or nearly so. To get enough photons at 560 μm, large dust densities are required and hence large gas densities. It is thus difficult to explain the F_1, N=1, J=3/2 doublet excitation by the 560 μm dust radiation alone.

An indirect radiative process involving excitation to the F_2, N=2, J=3/2 level from ground state via 149 μm dust photons, followed by spontaneous decay to the F_1, N=1, J=3/2 doublet is also possible. At 149 μm there are more FARIR dust photons; however, the estimated F_2, N=2, J=3/2 population from 149 μm dust excitation is at least an order of magnitude too low to account for the lower limit to the observed 700 MHz level column density under optimal conditions.

<u>Implications for Ground-State Excitation</u>: The inversion thought to occur in the CH ground-state Λ-doublet is believed to result from collisional excitation to F_1, N=1, J=3/2 level, followed by radiative decay back to ground-state--the "Gwinn-Townes" model (Bertojo, et al. 1975; Elitzur 1977).

The F_1, N=1, J=3/2 measurements in W 51 do fit into the Gwinn-Townes scheme. The doublet does appear to be collisionally excited, at least in part, and the excitation is not thermalized such that the main deexcitation process is spontaneous emission. The presence of at least some FARIR radiation at 560 μm could account for the non-LTE ground-state intensities. Furthermore, the excited state lines are not inverted and are in absorption, as might be expected if collisions preferentially populate the lower half of the Λ-doublet. However, more subtle tests are needed to prove the correctness of the model than are afforded by these observations.

<u>Implications for CH Chemistry</u>: Previous work on the ground state (e.g., Rydbeck, et al. 1976; Genzel, et al. 1979) have suggested that CH abundance decreases at higher densities: chemical models have shown this effect also (e.g., Graedel, Langer, and Frerking 1983). The 700 MHz column density in W 51 has been shown to be comparable or greater to that of the ground state in almost any source, though it traces very high densities. It appears at least in W 51 that CH abundance <u>does not</u> decrease with increasing density.

A fractional abundance relative to H_2 can be estimated using the column depth of molecular hydrogen deduced by Jaffe, et al. However, part of this column density could represent material <u>behind</u> G49.5-0.4, where much CH does not appear to be present. The derived CH/H_2 ratio, $f > 3 \times 10^{-10}$, thus represents a lower limit. For cloud models with $n(H_2) = 10^5 - 10^6$ cm^{-3}, CH/H_2 = $10^{-11} - 10^{-12}$ (Leung, Herbst, and Huebner 1984; Graedel, et al. 1983; Prasad and Huntress 1980), when the cloud has reached steady-state. Leung, et al. predicts $CH/H_2 = 10^{-10}$ at an earlier cloud age, more in agreement with the W 51 results. At an earlier cloud age, however, the predicted HCN/H_2 abundance is $10^{-5} - 10^{-6}$ too high than that observed in the source (White, et al. 1982); steady-state abundances must be invoked to explain this ratio. Even the early cloud abundance is barely large enough to satisfactorily explain the high concentration of CH in W 51.

TABLE 1 CH: $^2\Pi$, F_1, N=1, J=3/2 Data Towards W 51 [a]

Hyperfine Transition	Line Strength	Rest Frequency (MHz)	T_A(K)	V_{LSR}(Kms^{-1})	$\Delta V_{\frac{1}{2}}$(Kms^{-1})
$F=2^--2^+$	1.2059	701.677±0.010 [c]	0.75 [e] 0.20 [f]	60.0 60.0	14 16
$F=1^--1^+$	0.6700	724.788±0.010 [c]	0.50 [e]	60.0	14
$F=1^--2^+$	0.1340	722.303±0.020 [d]	$\lesssim$0.3 [e]	60.0	—
$F=2^--1^+$	0.1339	704.175±0.020 [d]	≲0.85 [e]	60.0	—

a) Data taken towards $\alpha=19^h21^m27^s$; $\delta=14°24'36''$ (1950.0).
b) From Brown and Evenson 1983.
c) Astronomically-derived rest frequencies,
d) Predicted rest frequency from Brown 1984.
e) Arecibo measurement; upper limits are 3σ.
f) NRAO 300 ft. measurement.

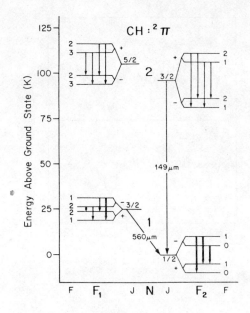

Figure 1. The energy level diagram of the four lowest-lying Λ-doublets of CH. All higher-lying levels are >220 K above ground state. The electric-dipole allowed Λ-doubling transitions are indicated in the figure by arrows; the two allowed rotational transitions to ground state are also shown (at 149 and 650 μm). The darker arrows show the only CH transitions as yet observed; all three hyperfine components of the ground state Λ-doublet, and the two main hyperfine lines (F=2-2 and F=1-1) of the first excited state detect ions presented here. The levels are labeled by Hund's Case (b) notation; total parity for the doublet pairs is given by the plus and minus signs.

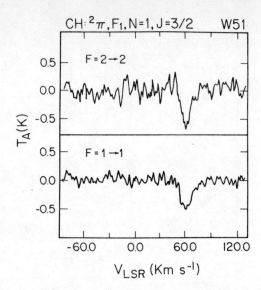

Figure 2. The spectra of the F=2-2 and F=1-1 hyperfine components of the F_1, N=1, J=3/2 excited Λ-doublet detected in this work, seen in absorption towards W 51 ($\alpha=19^h21^m27^s$ $\delta=14°24'36"$; 1950.0) using the Arecibo dish ($\theta_b \cong 7.'7$). The data were Hanning-smoothed once. The much weaker satellite components were not detected.

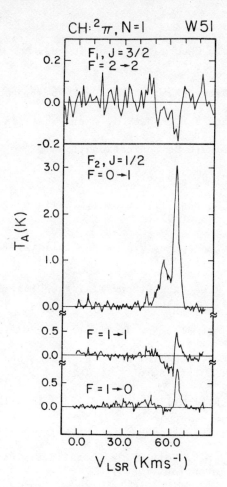

Figure 3. The spectrum of the main hyperfine line (F=2-2) of the F_1, N=1, J=3/2 Λ-doublet towards W 51--the same line as shown in Figure 2, but taken with the NRAO 300 ft. antenna ($\theta_b \cong 19'$). Also shown are the three hyperfine components of the CH ground state (F_2, N=1, J=1/2) Λ-doublet of Genzel, et al. (1979), taken at the same position with the Bonn 100m. ($\theta_b \cong 5'$). The emission features are thought to be masering. All spectra are plotted on the same velocity scale. The ground and excited state spectra show remarkable similarity, suggesting both transitions originate in the same gas.

REFERENCES

1. M. Bertojo, A.C. Cheung, and C.H. Townes, Ap. J. $\underline{208}$, 914 (1976).

2. M. Bogey, C. Demuynck, and J.L. Destombes, preprint (1983).

3. C.R. Brazier and J.M. Brown, J. Chem. Phys. $\underline{78}$, 1608 (1983).

4. J.M. Brown, Private Communication (1984).

5. M. Elitzur, Ap. J. $\underline{218}$, 677 (1977).

6. R. Genzel, D. Downes, T. Pauls, T.L. Wilson, J.H. Bieging, Astro. Ap. $\underline{73}$, 253 (1979).

7. R. Genzel, E.E. Becklin, C.G. Wynn-Williams, J.M. Moran, M.J. Reid, and D.T. Jaffe, Ap. J. $\underline{255}$, 527 (1982).

8. T.E. Graedel, W.D. Langer, and M.A. Frerking, Ap. J. Suppl. $\underline{48}$, 321 (1983).

9. A. Hjalmarson, et al., Ap. J. Suppl. $\underline{35}$, 263 (1977).

10. D.T. Jaffe E.E. Becklin, and R.H. Hildebrand, Ap. J. (Letters) $\underline{279}$, L51 (1984).

11. K.R. Lang and R.F. Willson, Ap. J. $\underline{224}$, 125 (1978).

12. M.L. Leung, E. Herbst, and W.F. Huebner, preprint (1984).

13. H.E. Matthews, M.B. Bell, T.J. Sears, B.E. Turner, and L.J. Rickard, preprint (1984).

14. T.G. Phillips, G.R. Knapp, P.J. Huggins, M.W. Werner, P.G. Wannier, G. Neugebauer, and P. Ennis, Ap. J. $\underline{245}$, 512 (1980).

15. S.S. Prasad and W.T. Huntress, Ap. J. $\underline{260}$, 590 (1980).

16. O.E.H. Rydbeck, E. Kollberg, A. Hjalmarson, A. Sume, J. Ellder, and W.M. Irvine, Ap. J. Suppl. $\underline{31}$, 333 (1976).

17. G.J. White, J.P. Phillips, J.E. Beckman, and N.J. Cronin, M.N.R.A.S. $\underline{199}$, 375 (1982).

18. G.J. White, J.P. Phillips, K.J. Richardson, R.F. Frost, G.D. Watt, J.E. Beckman, and J.H. Davis, M.N.R.A.S. $\underline{204}$, 1117 (1983).

19. B. Zuckerman and B.E. Turner, Ap. J. <u>197</u>, 123 (1975).

INTERSTELLAR CLOUDS: FROM A DYNAMICAL PERSPECTIVE ON THEIR CHEMISTRY

Sheo S. Prasad

Jet Propulsion Laboratory (183-601)
California Institute of Technology
4800 Oak Grove Drive
Pasadena, California 91109, USA

It is possible that the great diversity of physical and chemical properties exhibited by interstellar clouds may be easily comprehended if these clouds are incessantly evolving from initial diffuse to a later dense state and then to star formation which ultimately restructures or disperses the remaining cloud material to re-enact the same evolutionary process. This suggests the need and usefulness of studying the chemistry and dynamics of interstellar clouds in a coupled manner.

1. INTRODUCTION

Two basic problems in molecular astrophysics of interstellar clouds are: (i) the formation and destruction mechanisms of interstellar molecules, and (ii) the great diversity of physical and chemical compositional properties exhibited by interstellar clouds. The first problem has been addressed by other authors in this Proceedings (see, for example, the review by E. Herbst). Our contribution is concerned with the second problem. We consider the possibility that from a hydrodynamical-chemical evolutionary perspective it may be simpler to comprehend the great diversity of physical and molecular compositional properties exhibited by interstellar clouds.

Interstellar clouds span a very broad spectrum of physical and chemical compositional properties. On the one end of the spectrum we have diffuse clouds characterized by low densities on the order of 10-100 cm^{-3}. These clouds are quite transparent (i.e., visual extinction, A_v, <1), warm (i.e., 50 $\lesssim$ T $\lesssim$ a few

G. H. F. Diercksen et al. (eds.), Molecular Astrophysics, 603–612.

100K) and mostly atomic. Even very simple molecules like OH, CO are barely detectable in these clouds. On the other end of the spectrum we have dense clouds characterized by high densities on the order of 10^5-10^6 cm^{-3}. Thse clouds are very dark (i.e., $A_v \gtrsim 16$) and very cold (i.e., $T \sim 10$-$20K$). Chemical processing in these clouds is quite advanced, so that polyatomic molecules with 9, 10, 11 or even 13 atoms are frequently encountered in these clouds. Then we have clouds perturbed by either shocks or intense doses of UV or x-ray radiation. Quite often such clouds have embedded protostellar or young stellar sources. Finally, clouds, such as TMC-1 and L134N, with strikingly similar physical properties exhibit strikingly different chemical composition (1). This adds yet another dimension to the diversity of interstellar clouds.

Hydrodynamical-chemical evolution has the potential to provide a common thread through the diffuse and dense clouds [Tarafdar, Prasad, Huntress, Villere and Black (2), hereinafter referred to as TPHVB]. From dynamical perspective gravitationally unstable, initially diffuse, cloud of interstellar gas would contract, and in course of its contraction, would span the entire range of A_v and core densities that are presented by observed clouds. Thus, diffuse and dense clouds may not be unrelated objects. Instead, as emphasized by TPHVB, they might constitute families. A given family may differ from another by virtue of a few simple initial conditions, such as mass. The idea that diffuse clouds may contract to form dense clouds and ultimately stars is not new [see, for example, reviews by Woolfson (3), Larson (4), McNally (5), Bodenheimer (6)]. However, most previous studies of gravitational collapse were concerned with clouds which were already dense and not far from protostellar state (3). Previous attempts to model collapse of diffuse clouds were limited to clouds which were either unrealistically cold (T $\sim$ 10K) or very massive with mass $\sim 2 \times 10^4$ $M_\odot$ [Villere and Black (7), Gerola and Glassgold (8)]. Thus, the possible existence of a unifying link between diffuse and dark clouds on a general basis has not been demonstrated till now, even though the concept seems very simple at the first sight. Recently, TPHVB were able to make significant progress in demonstrating the feasibility of this simple concept. Their success was due to the use of a realistic temperature distribution in interstellar clouds.

2. TEMPERATURE DISTRIBUTION AND CLOUD COLLAPSE

The heating in the outer region of interstellar clouds by interstellar UV radiation field via grain photoelectrons is quite efficient and cooling due to atomic species is relatively inefficient. In contrast, the cosmic ray heating of the core is weak and cooling due to molecular species such as CO is quite efficient. Consequently, outer layers of interstellar clouds are well known to be hotter than the inner region [Jura (9), Silk (10), Clavel et al. (11)]. Under suitable conditions, this inward decrease of temperature would result in an inward pressure gradient force. If properly modeled, this pressure gradient force would assist gravity in driving the contraction during the early stages of collapse. TPHVB recognized that the observed variation of temperature in interstellar clouds was not being properly represented by either polytropic T-ρ relationship (commonly used in hydrodynamical collapse calculations) or by the temperatures calculated from the first principles as reported in the current literature. TPHVB, therefore, modeled the observed temperatures by a semi-empirically derived temperature formulae.

$$T = 163/\{2.5 + \ln n_H - \ln (1+500\exp(-1.8A_v))\} \qquad (1)$$

Details are discussed by TPHVB. When this semi-empirically derived formula is used in modeling the gravitational collapse, then even low mass (~ 40 $M_\odot$) low density ($n \sim 100$ cm^{-3}) warm diffuse clouds are found to be collapsing without any external triggers because the pressure gradient force is now quite significant.

3. SIGNIFICANT FEATURES OF GRAVITATIONAL COLLAPSE

Figure 1, based on the work of TPHVB, shows the temporal behavior of density and temperature as functions of radial distance from the center during the collapse of an interstellar cloud of 40 $M_\odot$ and initial density 3.3×10^{-22} gm cm^{-3}. These results were obtained on the basis of a simplified one-dimensional model of hydrodynamic collapse which neglects rotation, magnetic field and turbulence. For this purpose we used a computer code developed by Bodenheimer (6) after modifying it to calculate temperature from equation (1) instead of the original polytropic equation. From the point of view of the issue under discussion, a significant feature of the collapse is the rapid increase in the density of the core when the cloud becomes dense. For example, the cloud illustrated in Figure 1 takes only 4.3×10^5 yr to increase its central core density from 10^4 to 10^5 cm^{-3}. The time taken decrease further to 1.4×10^5 yr for an increase in core density from 10^5 to 10^6 cm^{-3}. In contrast, it took 2.51×10^6 yr for visual extinction (A_v) to increase from 0.5 to 2.0. In the course of the adopted model of their dynamical evolution, therefore, interstellar clouds spend most of their lifetime as diffuse clouds, and the **high density** cores are quite short lived.

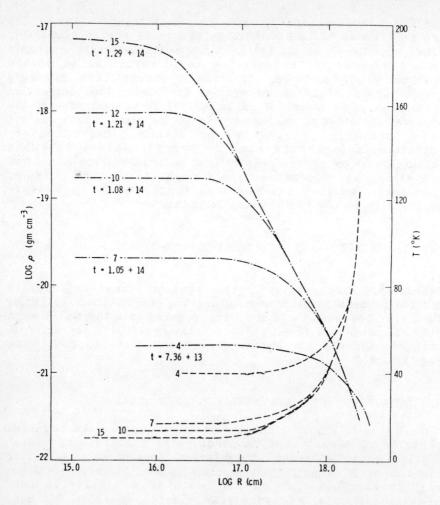

Figure 1. Evolution of density (-·-) and temperature (---) as a function of radial distance for a cloud of 40 M$_\odot$. Numbers such as 1.08 ± 14 = 1.08 × 10^{14} s denotes the evolutionary times. Increasing numbers on the curve indicate the progress of evolution.

4. FEATURES OF CHEMISTRY IN COLLAPSING CLOUDS

Models of chemical evolution in collapsing interstellar clouds are quite in dearth. Previously, Kiguchi et al. (12) and Suzuki et al. (13) have reported chemical models of clouds in free fall. Free fall models are, however, extreme idealization of the dynamical situation. Gerola and Glassgold (8) treat the dynamics much more realistically, but confine their study to a single massive cloud of mass 2×10^4 $M_\odot$. TPHVB studied the chemical evolution utilizing the radial distribution of ρ and T as functions of time derived from independent hydrodynamical calculations. This decoupling of the chemical and hydrodynamical evolution was possible in TPHVB's study due to the use of the semi-empirical temperature formula. Details are given in TPHVB. The decoupling substantially reduces the computational burden. As a result, TPHVB were able to study the chemical evolution of a number of clouds and for our purpose their results are in a very convenient format. We shall therefore use their results. For the present discussion, it is important to bear in mind that TPHVB model assumed that interstellar gas phase species stick onto the grains with an effective probability of 0.5 per collision and neglected desorption of molecules from the grain surface.

Figures 2 and 3 show respectively the variation of N(CO), the column density of CO, with the visual extinction A_V during the diffuse phase of evolution and the variation of the fractional abundance f(CO) with hydrogen density in the core during the dense phase of the evolution of model clouds. Theoretical predictions are for clouds of 40, 100 and 1000 $M_\odot$.

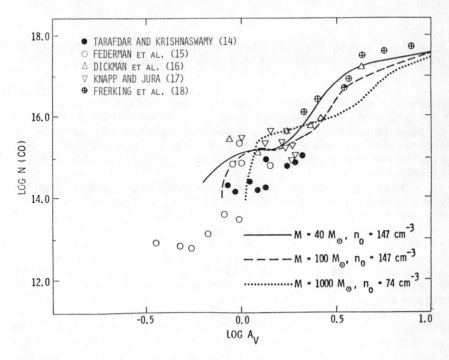

Figure 2. Theoretically predicted and observed variations of N(CO) with A_V.

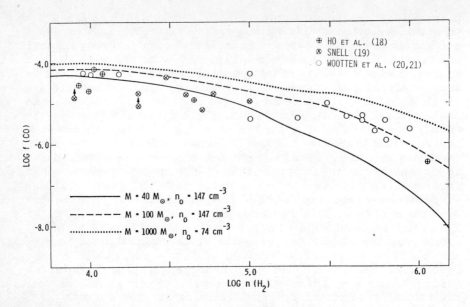

Figure 3. Theoretical and observed variations of the fractional abundance of CO (f(CO)) with the hydrogen density (n_H).

Observational data from diffuse and dense clouds are also shown in the figure for comparison with theoretical results. For both diffuse and dense clouds the theoretical results agree with observations. This agreement is not limited to CO only. Results for other molecules presented by TPHVB also show similar agreement.

Existing static equilibrium models can of course explain the observed variation of N(X) with A_v and of f(X) with n_H. Static models, however, have to assume a different set of physical conditions for each case, and give no feel for any possible relationships between diffuse and dense cloud conditions. In contrast, in dynamically evolving models the observed variations of N(X) with A_v and of f(X) with n_H follow from the first principles of hydrodynamics and chemistry with only a few assumed initial conditions for mass and density. Furthermore, the dynamical perspective reveals a common thread through the various diffuse (i.e., whole range of A_v) and various dense (i.e., whole range of core densities) clouds. The dynamical models do this in the sense that a single mass of interstellar gas appears to exhibit the whole range of physical-chemical properties observed for a variety of diffuse and dense clouds. Also, the spread in the chemical composition for a given A_v or for a given core density appears attributable to differences in the initial conditions, such as the initial mass and density in TPHVB models.

Condensation of gas phase molecules onto the grains is much less of a problem in dynamically evolving models than in static models. From a model calculation by Iglesias (22), it is easy to see that in the static equilibrium model, the gas phase abundances of molecules would tend to vanishingly small values in dense clouds even with a sticking coefficient of 0.5- unless we invoke special mechanisms to bring molecules off the grains. These mechanisms include strong circulation currents suggested by Boland and deJong (23) which bring core material to the surface where photodesorption may restore molecules to the gas phase. The efficacy of this mechanism, however, is not assured. Indeed, Leger, Jura and Omont (24) have argued that this mechanism may not work, and have suggested their own mechanism for desorption from interstellar grains. The efficiency of Leger et al. mechanism also remains to be demonstrated on a firm basis. Given this situation with static models, we note that in dynamical models the fractional abundance of molecules are close to the observed abundances, even without invoking any special desorption mechanisms. This improved tolerance of the dynamical models to condensation of gas phase species onto the grains is a direct consequence of the fact that dynamically evolving clouds spend most of their lifetime as diffuse clouds and the lifetime of the dense cores at any specified density decreases steeply as the density increases.

Recently, the observed large decrease of the fractional abundance of CO (and for that matter of many other species) has become doubtful. This would imply an effective sticking coefficient smaller than 0.5 due to some efficient desorption mechanism such as those suggested by Boland and deJong and by Leger et al. The key point, however, is that dynamical models would put relatively smaller demand on any of these mechanisms (compared to static models) provided the dense cores of interstellar clouds do not last long after attaining a density of 10^6 cm^{-3}. One possible way to satisfy this condition is to assume that the dense cores of gravitationally contracting interstellar clouds do not resist star formation after attaining density of 10^6 cm^{-3}. Detection of infrared sources embedded in a very large fraction of previously known dense clouds by the Infrared Astronomical Satellite (IRAS) mission lends support to this possibility.

Once a star is formed inside the contracting cloud, the stellar wind and expanding HII region associated with the nascent star will drive shocks in the remaining cloud material.

At this stage of evolution the cloud will show quite special chemical compositional features characteristic of high temperature chemistry, such as those discussed by Hartquist et al. (25), Elitzer and Watson (26), Mitchell (27), Hollenbach and McKee (28) and many others. Observationally, there is a widespread occurrence of anomalous chemical compositional features attributable to shocks in interstellar clouds, and quite often these regions are indeed associated with stellar and protostellar sources. Theoretical papers (26-28) cited earlier give useful discussions of the observatonal data. Many young stars are also copious sources of x-rays. Interstellar clouds which produce these stars will, therefore, show significant changes in their chemical composition such as those discussed by Krolick and Kallman (29). Ultimately, through the combined action of winds, expanding HII region, and intense UV/x-ray doses, the nascent star may ultimately restructure or even disperse the placental material which will then re-enact the whole evolutionary process.

Finally, we must emphasize that much studies remain to be done before the possibilities examined in this paper can be said to be on firm ground. We must seek a physical basis for the semi-empirical temperature formula which played a critical role in the collapse of diffuse clouds. The self-shielding effects in the chemistry of CO were neglected in the present study. This drawback of the present chemical model should be removed and the ability of the model to explain the fractional abundances of more complex molecules, such as cyanopolyynes, should be examined.

5. SUMMARY

We have examined the possibility that in the course of its dynamical evolution, a single mass of interstellar gas would exhibit properties of diffuse clouds, dense clouds and finally also of clouds perturbed by shocks or intense UV or x-ray radiation generated by a star of its own creation. This concept then provides a common thread through the bewildering diversity of physical and chemical compositional properties shown by interstellar clouds. From this perspective, instead of being static objects, interstellar clouds are possibly incessantly evolving from initial diffuse to a later dense state and then to star formation which ultimately restructures or disperses the remaining cloud material to begin the whole evolutionary process once again. These ideas are, however, based upon a simplified study of interstellar chemistry from a dynamical perspective. They have been presented here primarily to entice some thought on the future direction of molecular astrophysics, e.g., the need to consider chemical behavior of interstellar clouds in conjunction with, rather than in isolation from, their dynamical behavior.

6. ACKNOWLEDGEMENT

The research reported in this paper was performed at the Jet Propulsion Laboratory, California Institute of Technology, under a contract with the National Aeronautics and Space Administration.

7. REFERENCES

1. Irvine, W. M., Good, J. C. and Schloerb, F. P. 1983, Astron. Astrophys. 127, pp. L10-L13.

2. Tarafdar, S. P., Prasad, S. S., Huntress, Jr., W. T., Villere, K. R. and Black, D. C. 1985, Ap. J. (In Press).

3. Woolfson, M. M. 1979, Phil. Transc. Roy. Soc., A291, pp. 219.

4. Larson, R. B. 1973, Ann. Rev. Astron. Astrophys. 11, pp. 219.

5. McNally, D. 1971, Rep. Prog. Phys. 34, pp. 71.

6. Bodenheimer, P. 1968, Ap. J. 153, pp. 483.

7. Villere, K. R. and Black, D. C. 1982, Ap. J. 252, pp. 524.

8. Gerola, H. and Glassgold, A. E. 1978, Ap. J. Suppl. 37, pp. 1.

9. Jura, M. 1978 in "Protostar and Planets" (Ed. T. Gehrel), University of Arizona Press, Tucson, Arizona, pp. 165-171.

10. Silk, J. 1973, Pub. Astron. Soc. Pacific 85, pp. 704.

11. Clavel, J., Viala, Y. P. and Bel, N. 1978, Astron. Astrophys. 65, p. 435-448.

12. Kiguchi, M., Suzuki, H., Sata, K., Miki, S., Tominatsu, A. and Nakagawa, Y. 1974, Publ. Astron. Soc. Japan 26, pp. 499.

13. Suzuki, H., Miki, S., Sata, K., Kiguchi, M. and Nakagawa, Y. 1976, Prog. Theo. Phys. (Japan) 56, pp. 1111.

14. Tarafdar, S. P. and Krishnaswamy, K. S. 1982, Mon. Not. R.

Astr. Soc. 200, pp. 431.

15. Federman, S. R., Glassgold, A. E., Jenkins, E. B. and Shaya, E. J. 1980, Ap. J. 242, pp. 545.

16. Dickman, R. L., Somerville, W. B., Whittet, D. C. B., McNally, D. and Blades, J. C. 1983, Ap. J. Suppl. 53, pp. 55.

17. Knapp, G. and Jura, M. 1976, Ap. J. 209, pp. 782.

18. Ho, P. T. P., Martin, R. N. and Barrett, A. H. 1981, Ap. J. 246, pp. 761.

19. Snell, R. L. 1981, Ap. J. Suppl. 45, pp. 121.

20. Wootten, A., Evans II, N. J., Snell, R. L. and Vanden Bout, P. 1978, Ap. J. 225, pp. L143.

21. Wootten, A., Bozyan, E. P., Garrett, D. B., Loren, R. B. and Snell, R. L. 1980, Ap. J. 239, pp. 844.

22. Iglesias, E. 1977, Ap. J. 218, pp. 697.

23. Boland, W. and deJong, T. 1982, Ap. J. 261, pp. 110.

24. Leger, A., Jura, M. and Omont, A. 1984, "Desorption from Interstellar Grains." Preprint submitted to Astron. Astrophys.

25. Hartquist, T. W., Oppenheimer, M. and Dalgarno, A. 1980, Ap. J. 236, pp. 182.

26. Elitzer, M. and Watson, W. D. 1980, Ap. J. 236, pp. 172.

27. Mitchell, G. F. 1983, Ap. J. Suppl. 54, pp. 81-101.

28. Krolick, J. H. and Kallman, T. R. 1983, Ap. J. 267, pp. 610.

MODELS OF GALACTIC MOLECULAR SOURCES

T.J. Millar

Mathematics Department, UMIST, Manchester M60 1QD.

A chemical model developed for the purpose of investigating complex molecule formation has been applied to conditions typical of the cold dark clouds TMC-1 and L183. Excellent agreement between theory and observation was found for the time-independent calculations. Time-dependent models have also been studied. Due to the high CI abundance at early times ($\gtrsim 10^5$ yr.) the hydrocarbon species are overproduced with respect to their observed values. Models which include the adsorption of gas-phase species onto grains have also been investigated. In this case essentially everything is removed from the gas on a time-scale of 10^6 yr for a normal gas:dust ratio.

1. INTRODUCTION

The cold dark dust cloud Taurus Molecular Cloud 1 (TMC-1) contains many of the most complex molecules detected in the interstellar medium. To date, 28 different species, not counting isotopes, have been detected there including the cyanopolyyne molecules, HC_nN (n = 3, 5, 7, 9). Complex species have been observed in the dark cloud L183 (L134 N) which has similar physical characteristics to TMC-1 but, in general, has lower abundances of the complex molecules.

We have developed a model incorporating 120 species and around 500 reactions to describe the chemistry of these dark clouds. The species and reactions were specifically chosen to model complex molecule formation while reproducing the observed abundances of the more simple species. There are several advantages in modelling these dark clouds. Since they are quiescent (the line widths are not much greater than the thermal width at

G. H. F. Diercksen et al. (eds.), Molecular Astrophysics, 613–619.

10K), cold (T ~ 10K) and dark ($A_v > 5$ magnitudes), there are no complicating factors such as turbulent motions, collapse, embedded sources of radiation, or high temperature chemistry. Photoprocesses are unimportant although they are explicitly contained in the model. In addition, the great wealth of molecular information available for TMC-1 and L183 put important constraints on many of the free parameters contained in the model, such as the cosmic-ray ionization rate and the elemental depletions. The reaction set was restricted to a (relatively) small number of reactions since we did not wish to go too far beyond the available laboratory data. This is an advantage in that the results may be fairly insensitive to further laboratory studies, but obviously a disadvantage in that many more reactions are playing a role in real interstellar clouds.

The chemistry of the simpler species involving H, C, N and O is adapted from Prasad and Huntress (1980). For the hydrocarbon chemistry five main types of reaction occur:- (i) C^+ insertion reactions which build up the carbon chain backbone, (ii) abstraction reactions involving H_2 which serve to hydrogenate the hydrocarbon ions. Laboratory studies have shown that several of these are slightly endoergic and will not go at cloud temperatures (Herbst et al. 1983), (iii) condensation reactions involving the hydrocarbon neutrals and ions. These reactions increase the length of the carbon backbone as well as hydrogenate ions, (iv) atomic nitrogen reactions with hydrocarbon ions which produce the cyanopolyynes and certain other of the N-bearing ions, and (v) dissociative recombination of the ions with electrons. The latter two categories of reaction have little laboratory data available.

2. STEADY-STATE MODELS

Results which assume chemical steady-state has been reached have been published for both TMC-1 (Millar and Freeman 1984a, hereafter Paper I) and L183 (Millar and Freeman 1984b) and for a full discussion the reader is referred to these papers.

Table 1 shows a comparison between observed and calculated abundances for both TMC-1 and L183. The calculated abundances differ slightly from those published previously because the reaction set has been updated slightly. In particular, I have excluded the reaction (Smith and Adams 1984)

$$H_3^+ + e \rightarrow H_2 + H$$

and included

$$N + H_3^+ \rightarrow NH_2^+ + H \quad \text{- which enhances } NH_3 \text{ formation}$$

$$C_3H^+ + H_2 \rightarrow C_3H_3^+ + h\nu \quad \text{- Raksit and Bohme (1984)}$$

MODELS OF GALACTIC MOLECULAR SOURCES

	TMC-1		L183	
Species	Observed	Calculated	Observed	Calculated
C	-	1.9(-7)	$\geq$1(-6)	1.2(-8)
N	-	3.4(-6)	-	1.3(-6)
O	-	3.8(-5)	-	4.3(-5)
CO	3(-5)[a]	1.9(-4)	2.5(-5)	6.2(-5)
OH	3(-8)	6.4(-8)	2(-8)	4.4(-7)
HCO^+	8(-9)	3.6(-9)	6(-9)	5.9(-9)
H_2CO	6(-9)	1.3(-8)	6(-9)	1.0(-8)
CN	>4(-9)	6.2(-9)	c	2.1(-9)
HCN	6(-9)	3.1(-8)	5(-9)	5.0(-9)
HNC	4(-9)	1.2(-8)	3(-9)	2.5(-9)
N_2	-	1.7(-5)	-	6.8(-6)
N_2H^+	b	7.8(-11)	2(-10)	5.0(-10)
NH_3	5(-8)	9.0(-9)	1(-7)	1.2(-8)
CH_4	-	1.5(-7)	-	4.1(-8)
C_2H	4(-9)	4.4(-9)	5(-10)	1.4(-9)
C_2H_2	-	3.0(-8)	-	3.9(-9)
C_3H	b	4.1(-8)	-	4.2(-9)
C_3N	1(-9)	4.0(-9)	$\lesssim$1(-10)	3.3(-10)
HC_3N	1(-8)	3.9(-8)	2(-10)	2.9(-9)
CH_3CCH	3(-9)	8.9(-11)	<5(-10)	4.3(-12)
C_4H	1(-8)	4.5(-10)	6(-10)	5.4(-12)
CH_3CN	3(-10)	6.8(-11)	c	3.1(-12)
CH_2CHCN	2(-10)	8.4(-11)	c	9.7(-13)
CH_3C_3N	3(-10)	1.2(-10)	c	3.7(-13)

Table 1. A comparison of observed and calculated abundances for the dark clouds TMC-1 and L183. Notes: (a) 3(-5) signifies 3×10^{-5}; (b) species detected but abundance not determined; (c) species searched for but not detected, upper limit available.

and $C_3H_3^+ + C \rightarrow C_4H_2^+ + H$ - Herbst (1983).

These latter two reactions serve to enhance the C_4H formation rate above that found in Paper I, although this is not a significant factor in the equilibrium models since the calculated CI abundance is not large.

Table 1 contains 16 species for which observed abundances are available in TMC-1. One sees that, from the calculated abundances in this table, only CH_3CCH and C_4H are in flagrant disagreement with the observations. The reasons for this disagreement are discussed in Paper I.

The results for L183 also show excellent agreement with the observations, although a glaring exception is CI, for which a high abundance is observed. This high CI abundance led Herbst

(1983) to argue that CI-fixation reactions may be extremely important in the hydrocarbon chemistry. In L183, the abundances of CO, HCO^+, H_2CO, HCN, HNC, N_2H^+ and C_2H are all within a factor of 5 of those observed, while the abundances of C_3N, CH_3CCH, CH_3CN, CH_2CHCN and CH_3C_3N are in agreement with their observed upper limits.

One of the strong conclusions of this modelling was that C_3H should be detectable in TMC-1. This has now been achieved (Rydbeck and Hjalmarson, these proceedings) and shows that the carbon chain backbone grows efficiently through the insertion of C^+ ions; CH, C_2H, C_3H and C_4H have now all been detected in TMC-1. A search for C_5H should also be made. A further consequence of this modelling is that we would expect a line from C_3H in L183 to be a factor of 5-10 weaker than the corresponding line in TMC-1.

3. TIME-DEPENDENT MODELS

The steady-state models described above implicitly assume that the chemical lifetime of a dark cloud is in excess of 10^7 yr and in addition ignore the adsorption of molecules onto grain surfaces for which a time-dependent description must be employed. This chemical lifetime may only just be compatible with that estimated for the Taurus region (Cohen and Kuhi 1979). In addition the high CI abundance observed in L183 may indicate that cold dark clouds are young objects, that is, we may be observing them at a time, $\lesssim 10^5$ yr, before all the carbon has been processed into CO. For these reasons, I have examined time-dependent models of dark clouds, using a chemical set and physical parameters identical to those used in the steady-state calculations for TMC-1.

Figure 1(a) shows a plot of $y(t)/y(s.s.)$ versus time t, where the abundances, $y(t)$, are calculated in the time-dependent model, while the steady-state abundances, $y(s.s.)$, are taken from the results presented in section 2. For most of the species shown in this figure, the steady-state abundances are within a factor of 5 of those observed (see Table 1). The time-dependent calculations however, show significant enhancements over the steady-state values of up to several orders of magnitude. In particular the abundances of C_3N, HC_3N, CH_3CN and H_2CO as well as many other species, conflict severely with the observations for times less than about 10^6 yr. The inclusion of time into the chemistry introduces yet another parameter and it is not yet clear whether such discrepancies indicate that the reaction set is inadequate or that some important physical process has been left out.

One process usually ignored is that of adsorption of molecules onto grains. For a normal gas/dust ratio, the e-folding time for adsorption is around $4 \times 10^9/n$ yr, where $n(cm^{-3})$ is the cloud density. A model identical to that in figure 1(a), but with the inclusion of adsorption, was calculated and the results are shown in figure 1(b). For this calculation it was assumed that all species, with the exception of H_2 and He, stick to the

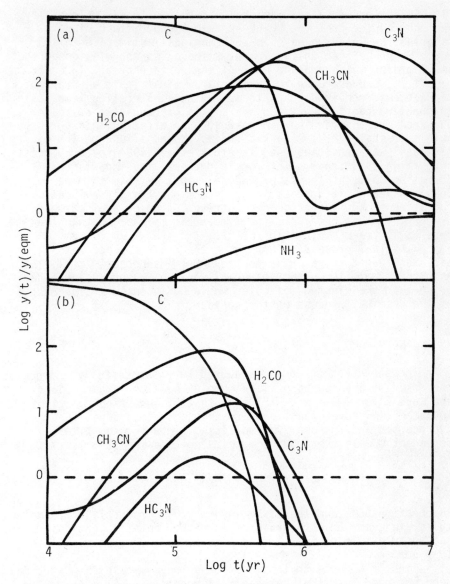

Figure 1. This plot shows the abundances of various species as a function of time when grain adsorption is (a) excluded and (b) included. NH_3 is always off-scale in figure 1(b).

grains upon collision. One sees that all molecules are completely removed from the gas on a time-scale of around 10^6 yr.

Since we do detect molecules in the cold dark clouds, one needs to find a 'way of escape' from the results of figure 1(b).

(1) The surface area of dust per unit volume in dark clouds may be much reduced from the value pertaining to diffuse cloud material for which we have an observational estimate. This could be due either to mantle formation or to coagulation of the small grains which provide most of the surface area. In order to be effective a decrease in surface area of at least an order of magnitude is required so as to increase the accretion time-scale to greater than 10^6 yr. However it is difficult to envisage a physical mechanism which can have such an effect within a reasonable time-scale. (2) A dark cloud may spend a long period of its life at lower densities (see Tarafdar et al. 1984 for arguments supporting this idea). This would decrease the importance of accretion while at the same time one might be able to make complex species if conditions are such that the ultraviolet radiation field is not too destructive. A point against this argument is that the accretion time-scale is shorter than the free-fall time-scale, $\sim 10^8/n^{\frac{1}{2}}$ yr., for $n \gtrsim 2 \times 10^3$ cm^{-3}. (3) A chemical process occurring on the grain surface may efficiently return material to the gas (d'Hendecourt et al. 1982). An unknown here is whether material would be returned unchanged or whether the ejected molecules would be essentially the results of the chemical process on the surface.

4. CONCLUSIONS

The steady-state model discussed here successfully reproduces the observed abundances of many complex species in dark clouds. The models for TMC-1 and L183 show that C^+ insertion reactions and atomic nitrogen reactions are necessary to account for the hydrocarbon and cyanopolyyne molecules. However the time-dependent models overproduce complex species for $t \leq 10^6$ yr. This is due to the reaction

$$C + H_3^+ \rightarrow CH^+ + H_2$$

which drives CH and the higher hydrocarbon production. If molecules are adsorbed onto grains then either an efficient desorption mechanism exists or the surface area of dust must be much less than normal.

Finally we note that some of this ion-molecule chemistry has been included into the photochemical models of IRC+10216 (Nejad et al. 1984) and that this chemistry can account for the abundance and spatial extent of C_3H which has been detected by Johansson et al. (1984).

REFERENCES

Cohen, M. and Kuhi, L.V. 1979, Astrophys. J. Suppl. Ser. 41, 743.
d'Hendecourt, L.B., Allamandola, L.J., Baas, F. and Greenberg,

J.M. 1982, Astron. & Astrophys. 109, L12.
Herbst, E. 1983, Astrophys. J. Suppl. Ser. 53, 41.
Herbst, E., Adams, N.G. and Smith, D. 1983, Astrophys. J. 269, 329.
Johansson, L.E.B. et al. 1984, Astron. & Astrophys. 130, 227.
Leung, C.M., Herbst, E. and Huebner, W.F. 1984, Ap. J. Suppl. Ser. 56, 321
Millar, T.J. and Freeman, A. 1984a, M.N.R.A.S., 207, 405.
Millar, T.J. and Freeman, A. 1984b, M.N.R.A.S., 207, 425.
Nejad, L.A.M., Millar, T.J. and Freeman, A., 1984, Astron. Astrophys. 134, 129.
Prasad, S.S. and Huntress, W.T. 1980, Astrophys. J. Suppl. Ser. 43, 1.
Raksit, A.B. and Bohme, D.K. 1984, Int. J. Mass Spect. Ion Proc. 55, 69.
Smith, D. and Adams, N.G. 1984, Astrophys. J., in press.
Tarafdar, S.P., Prasad, S.S., Huntress, W.T., Villere, K.M. and Black, D.C. 1984, Astrophys. J. Suppl. Ser., in press.

Millimeter wave spectroscopy of transient species in RF and DC discharges.

J.L. DESTOMBES

Laboratoire de Spectroscopie Hertzienne, associé au CNRS
Université de Lille I, 59655 VILLENEUVE D'ASCQ CEDEX.

Millimeter wave spectra of free radicals and molecular ions are observed in a liquid nitrogen cooled RF or DC discharges. Recent results on CO^+, CH, ^{13}CN, H_2D^+ and HCO_2^+ are reviewed.

Since the first detection of a molecular ion by WOODS and coworkers [1], DC discharges are currently used to produce transient species of very short lifetime.
In Lille we have developped an electrodeless discharge excited by a radiofrequency self oscillator. The absorption cell is 1 m long and can be cooled to liquid nitrogen temperature. In situ discharges are easily ignited in a large pressure range, from some mtorr to several hundred mtorr and they are generally very quiet. The maximum power density attainable is 0.15 W cm^{-3}. As expected, the rotational temperature is near the wall temperature. But the vibrational temperature is always very high for example, rotational spectrum of CN has been observed up to v = 11.
A number of interesting species can be efficiently produced in this discharge, including some molecules of astrophysical interest = CO^+, CH, CN, HCS^+, HCCN,...

1) The CO^+ molecular ion is easily produced by discharging pure CO at low temperature. The analysis of 4 isotopic substitutions in ground and vibrational states led us to the determination of the equilibrium structure and to the study of the breakdown of the Born-Oppenheimer approximation [2].

2) CH was known in the IS medium for about 10 years but its microwave spectrum was not known in the laboratory. We have been able to detect several transitions of this radical [3]. In a

parallel work done at the same time, J. BROWN detected other transitions by Microwave Optical Double Resonance [4] . All these results have been used to predict unobserved transitions of astrophysical interest, and some of them have been recently detected [5].

3) In addition to the J = 1 - 2 transition of HCS^+ observed by C. WOODS and coworkers [6], we have measured 5 successive rotational transitions of this ion first detected in the IS medium [7]. This is a definitive confirmation of the identification of this species. This was not obvious since discharge in H_2S + CO give rise to a number of unidentified lines and then to possibilities of fortuitous coïncidences [8].

4) Owing to the presence of two nuclear spins, the hyperfine structure of the ^{13}CN radical is complicated and we have for example measured 41 components for the two first rotational transitions. Using our analysis [9], french radioastronomers have recently detected ^{13}CN is three IS sources [10].

More recently we have developped another type of discharge which is a negative glow discharge extended by a magnetic field. This technique has been applied to millimeter wave spectroscopy by FC de LUCIA and coworkers [11].

Using a predicted frequency kindly provided by T. AMANO, we have detected the $1_{11} - 1_{10}$ submillimeter transition of H_2D^+ at 372421.34 (20) MHz. The best results have been obtained in an Ar discharge with traces of H_2 and D_2 [12].

We have also observed the K = 0 and K = 2 components of two successive rotational transitions of the protonated carbon dioxide HCO_2^+ [13]. These lines fit very well with the IS lines observed by THADDEUS and coworkers some years ago [7]. This is a confirmation that the IS lines are due to HCO_2^+ and not to the HOCN isoelectronic molecule. The observation of HCO_2^+ is then an indirect detection of CO_2 in the IS medium. This is the second example of a non polar molecule detected through its protonated form.

REFERENCES.

[1] T.A. DIXON and R.C. WOODS, 1975, Phys. Rev. Lett. 34, pp 61.

[2] M. BOGEY, C. DEMUYNCK and J.L. DESTOMBES, 1983, J. Chem. Phys. 79, pp 4704.

[3] M. BOGEY, C. DEMUYNCK and J.L. DESTOMBES, 1983, Chem. Phys. Lett. 100, pp 105.

[4] C.R. BRAZIER and J.M. BROWN, 1983, J. Chem. Phys. 78, pp 1608.

[5] L.C. ZIURYS, this workshop.

[6] C.S. GUDEMAN, N.N. HAESE, N.D. PILTCH and R.C. WOODS, 1981, Astrophys. J. (letters), 246, pp 47.

[7] P. THADDEUS, M. GUELIN and R.A. LINKE. 1981, Astrophys. J. (Letters), 246, pp 41.

[8] M. BOGEY, C. DEMUYNCK, J.L. DESTOMBES and B. LEMOINE, 1984, J. Mol. Spectrosc., in press.

[9] M. BOGEY, C. DEMUYNCK and J.L. DESTOMBES, Can. J. Phys. submitted.

[10] M. GERIN, F. COMBES, P. ENCRENAZ, R. LINKE, J.L. DESTOMBES and C. DEMUYNCK, 1984, Astron. Astrophys. in press.

[11] F.C. DE LUCIA, E. HERBST, G.M. PLUMMER and G.A. BLAKE, 1983, J. Chem. Phys. 78, pp 2312.

[12] M. BOGEY, C. DEMUYNCK, M. DENIS, J.L. DESTOMBES and B. LEMOINE 1984, Astron. Astrophys. Letters in press.

[13] M. BOGEY, C. DEMUYNCK and J.L. DESTOMBES, 1984, Astron. Astrophys. Letters, submitted.

PHOTOFRAGMENT SPECTROSCOPY OF INTERSTELLAR MOLECULES

Margaret M. Graff

Harvard-Smithsonian Center for Astrophysics

Abstract. The applications of photofragment spectroscopy to the study of interstellar molecules is discussed. The measurement of photodissociation cross sections is of direct astrophysical interest; in addition, the use of photodissociation as a half-collision event allows an examination of low energy collisions that are characteristic of interstellar clouds. Examples presented here include a recent half-collision study of CH^+ and the extension of photofragment methods to the study of neutral free radicals CH and OH.

Recent experimental developments have increased our ability to study the photodissociation of molecular ions and neutral free radicals. In photofragment spectroscopy, photon absorption by a particular molecular species is observed by direct detection of the photofragments, allowing a detailed study of the dissociation process. In addition to allowing the measurement of absolute photodissociation cross sections, photofragment spectroscopy may be used to study half-collision events: the molecules, excited from a bound state to a predissociated level, resemble a colliding atom pair that have been prepared in a specific quantum state at short internuclear distance.

The fast ion beam photofragment spectrometer at SRI International (Huber et al. 1977), shown in Figure 1, has contributed greatly to our understanding

of ionic molecules. A beam of molecular ions is
extracted from an ion source, accelerated, and
collimated. The molecular ion of interest is mass-
selected using a magnetic sector, bent into the laser
interaction region by an electrostatic quadrupole and
irradiated with a laser beam either parallel or
perpendicular to the ion beam. A second electrostatic
quadrupole is tuned to bend photofragment ions into
the analysis region. The fragment ions are energy
analyzed prior to detection by a channeltron electron
multiplier. The kinetic energy distribution of the
fragment ions provides the kinetic energy above the
dissociation limit and the anisotropy parameter of the
transition. These features have proven very valuable
in the analysis of photofragment spectra.

This apparatus has been used in a half-collision
study of the formation of CH^+ by radiative
association, the reverse of the photodissociation
process. The radiative association reaction

$$C^+(^2P_{1/2}) + H \rightarrow CH^+ + h\nu$$

has been of significant astrophysical interest since
its initial proposal as a formation mechanism for CH^+

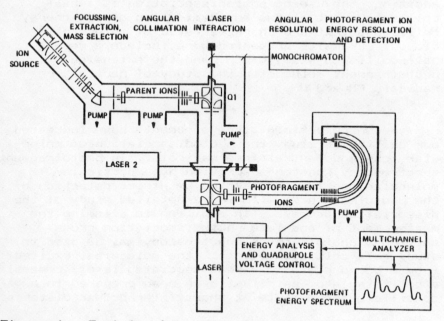

Figure 1. Fast ion beam photofragment spectrometer at
SRI International.

in diffuse interstellar clouds (Bates and Spitzer 1951). Abgrall et al. (1976) showed that the rate coefficient for this reaction is enhanced by shape resonances behind the rotational barrier in the effective potential of the radiating state. Subsequent calculations by Uzer and Dalgarno (1979) indicated that photodissociation by excitation to these shape resonances would be experimentally observable. The experimental observation of these shape resonances (Helm et al. 1982) provided an improved determination of the potential curve of the radiating state. This potential curve was then used (Graff et al. 1983a) to enumerate the resonances that could contribute to the radiative association process. Because the radiating state correlates adiabatically to the upper fine structure dissociation limit, many predissociated levels between the two fine structure limits ($\Delta E/k = 92$ K), in addition to the shape resonance levels, contribute to the radiative association. Because carbon ions in diffuse interstellar clouds exist primarily in the lower fine

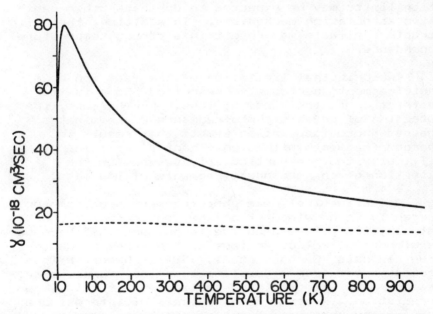

Figure 2. The rate coefficient for the formation of CH^+ by radiative association (full line), including both resonant and nonresonant contributions. The results of the previous quantum mechanical calculation (Abgrall et al., 1977) are shown for comparison (dotted line).

structure level, resonances between the dissociation
limits will have a greater effect than the higher-
energy shape resonances in collisions at typical cloud
temperatures. The radiative association rate coefficient, including the effects of all resonances in the
$A^1\Pi$ radiating state, was shown by Graff et al.
(1983b) to be significantly larger than predicted from
consideration of the enhancement by shape resonances
alone (Figure 2).

This half-collision analysis is a more detailed
study of the radiative association process than would
be possible for larger molecules: even for triatomic
systems, the greater density of states makes enumer-
ation of resonances impractical. As a prototype of
radiative association reactions, the example of CH
illustrates several features that should be considered
in other radiative association studies. In cases
where the splitting between the respective dissociation limits of the separated atoms and the radiating
state is on the order of the relevant temperature, the
effects of predissociated levels between the dissociation limits may be expected to dominate other radiative association mechanisms. In addition, these fine
structure effects may result in a strong temperature
dependence.

The principal limitation of the fast ion beam
photofragment spectrometer described above is its
restriction to the study of ions. Furthermore, the
detection of on-axis photofragments discriminates
against photodissociation events that result in
charged fragments with significant kinetic energy, a
difficulty that is particularly severe for the
detection of charged photofragments of low mass.

A fast neutral beam photofragment spectrometer
currently in development at the Center for
Astrophysics will expand the techniques that have been
developed for molecular ions to the study of neutral
free radicals (Gardner et al. 1984). Negative ions
are extracted from a duoplasmatron ion source, accelerated, mass selected, and collimated prior to laser
irradiation. Neutral free radicals of interest are
prepared by photodetachment of the corresponding
negative molecular ion by an initial laser pulse. A
second laser pulse dissociates the molecules. The
radial and angular distributions of off-axis photofragments, which provide the kinetic energy of dissociation and information on the details of the

electronic transition, are detected by a microchannel plate fitted with a multiple anode array.

Initial experiments will focus on the photodissociation of first-row hydrides CH and OH. The photodetachment method of creating a fast neutral beam results in molecules with relatively little internal excitation, enabling the measurement of absolute photodissociation cross sections. The experimental design has been optimized for the detection of light fragments with substantial separation energy. The apparatus may therefore also be applied to molecular ion photodissociation studies that have been inaccessible with on-axis detection techniques.

REFERENCES

Abgrall, H., Giusti-Suzor, A. and Roueff, E. 1976, Ap. J. (Letters) 207, pp. L69-L72.
Bates, D.R. and Spitzer, L.,Jr. 1951, Ap. J. 113, pp. 441-463.
Gardner, L.D., Graff, M.M. and Kohl, J.L. 1984, Rev. Sci. Instrum. (to be submitted).
Graff, M.M., Moseley, J.T., Durup, J. and Roueff, E. 1983a, J. Chem. Phys. 78, pp. 2355-2362.
Graff, M.M., Moseley, J.T. and Roueff, E. 1983b, Ap. J. 269, pp. 796-802.
Helm, H., Cosby, P.C., Graff, M.M. and Moseley, J.T. 1982, Phys. Rev. A25, pp. 304-321.
Huber, B.A., Miller, T.M., Cosby, P.C., Zeman, H.D., Leon, R.L., Moseley, J.T. and Peterson, J.T. 1977, Rev. Sci. Instrum. 48, pp. 1306-1313.
Uzer, T. and Dalgarno, A. 1979, Chem. Phys. Letters 63, pp. 22-24.

MEASUREMENTS OF ION-MOLECULE REACTION RATE COEFFICIENTS BETWEEN 8 AND 160 K BY THE CRESU TECHNIQUE.

B.R. Rowe, J.B. Marquette and G. Dupeyrat

Laboratoire d'Aérothermique du CNRS, Meudon (France)

Abstract : The CRESU technique uses uniform supersonic jets as flow reactors and allows the determination of ion-molecule reaction rate coefficients at extremely low temperatures (down to 8 K). This technique requires an exceptional pumping capacity and a correct design of contoured axisymmetric Laval nozzles. Several reactions have been studied in oxygen or nitrogen buffer gases ($O_2^+ + CH_4$; $O_2^+ + 2O_2$; $N_2^+ + 2N_2$) as well as in helium buffer gas ($He^+ + N_2$, O_2, CO ; $N^+ + H_2$, CH_4, CO, O_2 ; $N_2^+ + N_2 + He$; $N_2^+ + O_2$; $O^+ + O_2$), some of which being relevant to interstellar cloud chemistry.

1. INTRODUCTION

The discovery of complex molecules in dense interstellar clouds (1) has stimulated the ion-molecule reaction research field since it has been recognized that such reactions play a key role in the formation of these molecules (2,3). Numerous ion-molecule reactions relevant to interstellar cloud chemistry have already been studied between 80 and 300 K, especially by the so-called SIFT technique (4). However, typical cloud temperatures are in the range 10-100 K and these available results need to be extrapolated ; this can obviously lead to serious problems when one deals with temperatures as low as 10 K.
Designing an experiment for such low temperatures is a considerable challenge. Only two ways seem possible : cryogenic cooling or supersonic expansion. Cryogenic cooling by liquid helium has been used in two experiments, an ion trap by Dunn and coworkers (5) and a cooled drift tube by Böhringer and Arnold (6). These techniques have already given results respectively on

a radiative association reaction (7) and on some three-body association reactions (6,8) below 80 K. However, in such experiments the condensation on the wall of the system allows only hydrogen to be used as neutral reactant for the lowest temperatures.

We have developped in our laboratory a technique which uses a uniform supersonic jet generated by a Laval nozzle as a flow reactor. This jet is very different of the so-called "free jet" widely used by physicists as a source of very low temperatures ; downstream the nozzle exit it does not exhibit gradients in temperature, velocity, density on quite long axial distances. At low pressure exceptional pumping capacities are needed to obtain such flows. We have called these technique "CRESU" (Cinétique de Réactions en Ecoulement Supersonique Uniforme, i.e. Reaction Kinetics in Uniform Supersonic Flow) which has already yielded numerous data (9-11). The results obtained by this CRESU experiment are presented below.

2. EXPERIMENTAL

The CRESU technique has been described in details elsewhere (10) and only the main features are summarized here. A schematic drawing of the apparatus is shown in Figure 1. The nozzle, N, generates a supersonic jet whose isentropic core exhibits uniform Mach number, temperature, pressure and velocity. The nozzle reservoir, NR, can be cooled to liquid nitrogen temperature in order to establish very low jet temperatures at moderate Mach numbers. Buffer and reactant gases, as well as the ion parent gas, are mixed together in the nozzle reservoir. The ions are created directly in the flow by a 20 keV, 100 µA electron beam directed across the jet ; product and primary ions are monitored using a quadrupole mass spectrometer/counting system, QM. The electron gun and the quadrupole mass spectrometer are set on moving cartesian carriages which allow complete exploration of the flow.

In the uniform isentropic core with velocity v, A^+ ions reacting with a neutral reactant R in the buffer gas B will have a density decrease along the axis according to the equation

$$v \, d\{A^+\} / dx = - k \, \{A^+\} \, \{R\} \tag{1}$$

k being the binary rate coefficient k_b for a binary reaction or being equal to $k_t\{B\}$ in case of a ternary reaction. The rate coefficient is then obtained by monitoring the primary ion density decrease versus distance x (from the e-beam) or neutral reactant flow rate (10).

The nature of the buffer, the reservoir temperature, the flow pressure and the Mach number determine the geometry of the nozzles. Details of these calculations can be founded in Ref. 11.

Table 1 summarizes the flow conditions obtained with five different nozzles.

Nozzle	Buffer gas	Mach number	Flow conditions T(K)	p(Torr)
1a	He	2.95	20	2.06×10^{-2}
1b	"	3.1	67.5	6.4×10^{-2}
2a	He	5	8	8.0×10^{-3}
2b	"	5	27	3.9×10^{-2}
3a	N_2-O_2	3.9	20	5.0×10^{-3}
3b	"	4.05	68	3.3×10^{-2}
4b	N_2-O_2	4	70	2.0×10^{-2}
5a	N_2-O_2	1.94	45	1.8×10^{-2}
5b	"	2	163	0.102

Reservoir temperature : a = 77 K ; b = 293 K

Table 1 : Flow conditions of the CRESU experiment

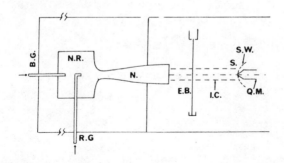

Figure 1. Basic features of the CRESU apparatus
B.G., R.G. buffer and reactant entry port ; N.R. nozzle reservoir ; N nozzle ; E.B. electron beam ; I.C. isentropic core ; S. skimmer ; S.W. shock wave ; Q.M. quadrupole mass spectrometer.

3. RESULTS AND DISCUSSION

3.1. Reactions studied in oxygen or nitrogen buffer gases

Results have been obtained for the reactions $O_2^+ + CH_4 \longrightarrow CH_3O_2^+ + H$, $N_2^+ + 2N_2 \longrightarrow N_4^+ + N_2$ and $O_2^+ + 2O_2 \longrightarrow O_4^+ + O_2$ (9,10) with an excellent consistency, in the overlapping range of temperature, with data obtained by other techniques. The reactant ions were always quickly relaxed on the vibrational and electronic ground level, prior to reaction, by collisions with the parent buffer gas. Since extensive discussions of these results are available (9,10) only the main conclusions are given below.

The reaction $O_2^+ + CH_4$ exhibits a strong temperature dependence between 20 and 300 K. The rate coefficient which has a value of 5.4×10^{-12} cm^3.s^{-1} at room temperature increases dramatically at the lowest temperatures and extrapolation near the absolute zero yields the Langevin rate coefficient. It has been shown by experiments of the Innsbruck group (12) that the product ion is protonated formic acid. A detailed mechanism for this reaction has been proposed by Ferguson (10).

The initial $O_2^+ + CH_4$ encounter leads to hybride ion H$^-$ abstraction from CH_4 by the O_2^+. The $(CH_3^+.HO_2)^*$ intermediate complex yields either the reactants by unimolecular decomposition or the products $CH_3O_2^+ + H$. The probability of the reaction increases with the complex lifetime τ and approaches unity when T --> 0 K because τ --> ∞. The fact that decompositon back to the reactants is favored at all temperatures above 20 K probably reflects the complicated rearrangement of the $CH_3O_2^+$ product.

Concerning the association reaction $N_2^+ + 2N_2$, our results together with those of Böhringer and Arnold (8) are well represented by the power law

$$k_t = 6.5 \times 10^{-29} (300/T)^{1.77} \text{ cm}^6.\text{s}^{-1} \qquad (2)$$

This temperature dependence has been studied in theoritical models by Bates (13) and Herbst (14). In the low pressure limit k_t is given by

$$k_t = (k_a / k_{-a}) k_s \qquad (3)$$

k_s being the complex collisional stabilization rate coefficient. The ratio k_a / k_{-a} is determined by detailed balance as the ratio of the partition functions of the complex and the reactants (14). If a $T^{-\delta}$ dependence for k_s is included (14), the reactants being on their electronic and vibrational ground states, k_t has a $T^{-1/2-\delta}$ behavior where l is the sum of the rotational degrees of freedom of the reactants. In this model the present power law gives $\delta = -0.23$.

For $O_2^+ + 2O_2$ only lower limits of the rate coefficient were determined at 20 and 45 K. However, our results do not indicate any decrease in the rate coefficient below 60 K as suggested by the data of Böhringer and Arnold.

3.2. Reactions studied in helium buffer gas

As in the case of O_2^+ in O_2 or N_2^+ in N_2, He$^+$ ions can be created directly in the flow by the e-beam. Other reactant ions are obtained by mixing a large amount (a few percent of the buffer density) of a suitable parent gas with the buffer ; in this way N^+ and N_2^+ are created by the reaction

$$He^+ + N_2 \longrightarrow N^+ + N + He$$
$$ \longrightarrow N_2^+ + He \tag{4}$$

and O^+ by

$$He^+ + O_2 \longrightarrow O^+ + O + He \tag{5}$$

Here again the parent gas is abundant enough to ensure that the ions are quickly relaxed on their ground state.

Metastable He^M is also created by the e-beam and some ions like N_2^+ or O^+ can be obtained by the Penning reaction

$$He^M + N_2 \longrightarrow N_2^+ + He \tag{6}$$

This reaction could give an additional source of reactant ions downstream the e-beam and therefore induce a lower value of the reaction rate coefficient. This problem does not exist for N^+ since He^M cannot yield this ion from N_2. It was found experimentally that this problem was not very important for temperatures higher than 20 K and for the reactions presented here. These results are summarized in Table 2.

The rate coefficient of fast reactions like He^+ with O_2, N_2, CO or N^+ with O_2, CO, CH_4 is nearly temperature independent. This confirms a conclusion which was induced from results at higher temperatures.

A striking exception is the reaction $N^+ + H_2$; Figure 2 shows the variation of the rate coefficient with temperature which is two orders of magnitude lower at 8 K than at room temperature. Indeed available results show that this reaction is nearly thermoneutral, with a precision which does not allow to decide if it is exo or endothermic. In this last hypothesis, since our experiment is done with a normal mixture of ortho and parahydrogen (3:1), the levels $J = 0$ and $J = 1$ being alone populated, only the reaction of orthohydrogen is probably observed. The reaction of parahydrogen in $J = 0$ would be even much slower and this would have of course implications for interstellar cloud chemistry since this reaction is thought to be important in the process of ammonia synthesis. At the present time, our results only yield from an Arrhénius plot an empirical activation energy $\sim$ 4 meV. Nevertheless, if the rate coefficient of the reverse process $NH^+ + H \longrightarrow N^+ + H_2$ was known, it would be possible from a Van't Hoff plot to deduce the enthalpy and entropy variations for this reaction. In fact, since we probably observe the reaction with $H_2(v = 0, J = 1)$ below 70 K, detailed balance could be applied if the reverse rate coefficient was

Reactions	Behavior 80-300 K	k(300) (a)	k(20)	k(8)	Behavior 8-80 K
$N^+ + H_2$ $NH^+ + H$	nearly constant	4.8(-10)	4.2(-11)	2.9(-12)	$\sim \exp(-43.5/T)$
$NH_3^+ + H_2$ $NH_4^+ + H$	decreases with T	4.5(-13)	<1(-12) (b)	<3.6(-12) (b)	incr. at lowest T's
$He^+ + O_2$ $O^+ + O + He$ $O_2^+ + He$	constant	1.0(-9)	8.5(-10)	1.0(-9)	constant
$He^+ + N_2$ $N^+ + N + He$ $N_2^+ + He$	constant	1.2(-9)	1.3(-9)	1.2(-9)	constant
$He^+ + CO$ $C^+ + O + He$	constant	1.5(-9)	1.4(-9)	1.5(-9)	constant
$N^+ + O_2$ products	nearly constant	6.1(-10)	–	5.5(-10)	nearly constant
$N^+ + CO$ $CO^+ + N$ $NO^+ + C$	nearly constant	5.0(-10)	–	1.1(-9)	nearly constant
$N^+ + CH_4$ products	nearly constant	9.4(-10)	–	8.2(-10)	nearly constant
$N_2^+ + N_2 + He$ $N_4^+ + He$	$\sim T^{-2.3}$	1.9(-29)	4.0(-27)	–	increases (b)
$N_2^+ + O_2$ $O_2^+ + N_2$	increases	4.3(-11)	3.1(-10)	–	increases (b)
$O^+ + O_2$ $O_2^+ + O$	increases	1.9(-11)	1.1(-10)	–	increases (b)

(a) obtained from other authors
(b) preliminary results

Table 2. Results obtained in helium buffer gas.
For each column reaction, the first line gives the reactants and the others the products. Rate coefficients are given in $cm^3.s^{-1}$ or $cm^6.s^{-1}$. $(-n) = 10^{-n}$.

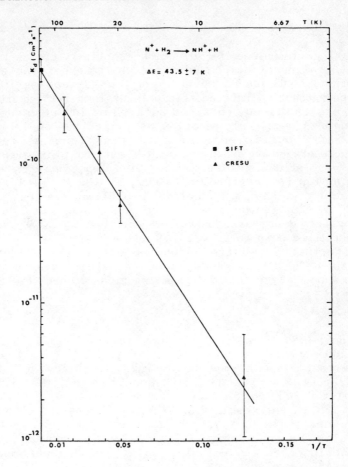

Figure 2. Temperature dependence of the $N^+ + H_2$ reaction rate.

known for the same internal states of NH^+ than those attained by the direct process. Assuming this reverse reaction to be temperature independent, our results would yield the endothermicity of the direct reaction rather than its activation energy.

Others reactions have been studied in the helium buffer. $O^+ + O_2$ and $N_2^+ + O_2$, which are slow at room temperature and are known to be faster at 80 K, have a rate coefficient which increases strongly at 20 K. The association reaction $N_2^+ + N_2 +$ He follows a power law down to 20 K.

4. CONCLUSION

A new technique (CRESU) which completely avoids the problem of neutral reactant condensation on the experimental chamber walls has been developped and allows determination of ion-molecule reaction rate coefficients at temperatures down to 8 K. The results already obtained confirm or indicate some general trends of ion-molecule reaction behaviors at low temperatures. The rate coefficients of fast reactions are temperature independent and an exception like $N^+ + H_2$ could be eventually due to a very slight endothermicity. Slow reactions can become very fast at low temperature ($O_2^+ + CH_4$, $O^+ + O_2$, $N_2^+ + O_2$); however a reaction as $NH_3^+ + H_2$ does not exhibit such a strong increase at 20 K. Concerning association reactions, they do not show important departure from a power law $k \sim T^{-n}$ ($n > 0$) at temperatures down to 20 K.

Further works are of course absolutely needed and the versatile CRESU technique will contribute to provide extensive laboratory data in ion-molecule reaction research field.

References

1. Rank, D.M., Townes, C.H. and Welch, W.J. 1971, Science 174, pp. 1083-1101.
2. Herbst, E. and Klemperer, W. 1973, Ap. J. 185, pp. 505-533.
3. Graedel, T.E., Langer, W.D. and Frerking, M.A. 1982, Ap. J. Suppl. Ser. 48, pp. 321-368.
4. Smith, D. and Adams, N.G. 1979, Gas Phase Ion Chemistry vol. 1, Bowers, M.T. ed., Academic Press, New York, pp. 1-44.
5. Walls, F.L. and Dunn, G.H. 1974, Physics Today 27, pp. 30-35.
6. Böhringer, H. and Arnold, F. 1982, Int. J. Mass Spectrom. Ion Phys. 49, pp. 61-83.
7. Barlow, S.E., Dunn, G.H. and Schauer, M. 1984, Phys. Rev. Lett. 52, pp. 902-905.
8. Böhringer, H. and Arnold, F. 1982, J. Chem. Phys. 77, pp. 5534-5541.
9. Rowe, B.R., Dupeyrat, G., Marquette, J.B., Smith, D. Adams, N.G. and Ferguson, E.E. 1984, J. Chem. Phys. 80, pp. 241-245.
10. Rowe, B.R., Dupeyrat, G., Marquette, J.B. and Gaucherel, P. 1984, J. Chem. Phys. 80, pp. 4915-4921.
11. Dupeyrat, G., Marquette, J.B. and Rowe, B.R. 1984, Phys. Fluids, submitted.
12. Villinger, H., Richter, R. and Lindinger, W. 1983, Int. J. Mass Spectrom. Ion Phys. 51, pp. 25-30.
13. Bates, D.R. 1980, J. Chem. Phys. 73, pp. 1000-1001.
14. Herbst, H. 1982, Chem. Phys. 68, pp. 323-330.

MEASUREMENTS OF ION-MOLECULE REACTION RATE COEFFICIENTS WITH AN ION DRIFT-TUBE METHOD AT TEMPERATURES FROM 18 TO 420 K

H. Böhringer and F. Arnold

Max-Planck-Institut für Kernphysik, Postfach 10 39 80,
6900 Heidelberg, W-Germany

For the understanding of molecule synthesis in interstellar clouds data on the kinetics of ion-molecule reactions at temperatures below 100 K are needed. A helium-cooled ion drift-tube experiment is described which can provide rate coefficients of ion-molecule reactions in the temperature range 18 to 420 K. Recent measurements are presented of three ion-molecule reactions which show a complex kinetics at low temperatures: the binary reactions of He^+ and NH_3^+ with H_2 and the association reaction of N_2H^+ with H_2.

INTRODUCTION

Ion-molecule reactions are considered to be essential for the synthesis of the molecules observed in interstellar clouds (1-5). For the understanding and modeling of the ion chemistry data on the reaction kinetics of ion-molecule reactions are needed for the temperature range from 10 to 100 K. So far only few data exist for rate coefficients at temperatures below 77 K, the temperature of liquid nitrogen. At present there are only three experimental methods that have provided kinetic data for this low temperature regime: the CRESU technique of B.R. Rowe and colleagues (6), the ion-trap experiment of G.H. Dunn and colleagues (7) and the ion drift-tube method used by the authors (8).

Most of the rate coefficients of exothermic binary ion-molecule reactions are in fact not very temperature sensitive and proceed with a reaction rate close to the collision limit (9). This has been fortunate for the construction of ion chemistry models that had to rely on kinetic data measured in the laboratory at higher temperatures.

G. H. F. Diercksen et al. (eds.), Molecular Astrophysics, 639–647.

There are exceptions, however, that deserve further investigation. Some exothermic binary reactions proceed with a low reaction probability per collision and show a strong dependence on temperature. They often show a minimum in the rate coefficient at intermediate collision energies with a strong increase of the coefficient towards lower or higher energy (6,10,11,12). Association reactions are another class of ion-molecule reactions that show a strong increase of the rate coefficient with decreasing temperature (13,14,15). To further refine the models of interstellar cloud ion chemistry laboratory studies of these temperature dependent reactions at the relevant low temperatures are required. For this application a selected ion drift-tube has been developed in our laboratory and has been applied to measurements of ion mobilities in gases and ion-molecule reaction rate coefficients in the temperature range from 18 to 420 K. The studies included a detailed investigation of the temperature dependence of some three-body association reactions (14,16,17). Here the data of recent studies of two binary reactions and an association reaction are presented. These reactions illustrate very well the complex kinetic behavior as a function of temperature described above.

2. EXPERIMENTAL

The selected-ion drift-tube apparatus (8) is shown schematically in Fig. 1. Primary ions produced in an electron impact ion source are mass-selected by a quadrupole mass filter before they are injected into the drift-reaction-cell. The drift cell has a length of 2.9 cm and is usually operated at gas densities from 10^{16} to 3×10^{17} cm^{-3}. Generally reactions are observed by adding small amounts of a reactant gas to an inert buffer gas in the reaction cell. Alternatively, for slow reactions (for example for the reaction He$^+$ + H$_2$, reported below) the measurements are carried out with pure reactant gas in the drift cell. Then either the total pressure or the reaction time is varied to follow the depletion and production of the ions.

The ions cross the drift cell in a drifting motion imposed by an external electric field. The residence time of the ions in the drift cell is measured in each experiment. Mass analysis of the primary and product ions leaving the drift-reaction cell is provided by a second quadrupole mass filter with a channeltron ion detector.

The drift cell can be electrically heated or cooled by either liquid nitrogen or liquid helium to any temperature between 18 and 420 K. The accuracy of the temperature measurement and control is better than ± 1 degree. In general, drifting ions collide with the neutral gas molecules with energies above thermal energy. The present drift tube experiment is mainly applied to measurements at such low E/N values (electric field strength/gas number density) that the deviation from thermal energy conditions can be neglected (14).

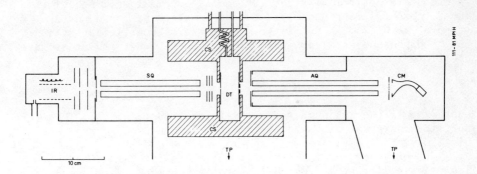

Figure 1: Schematic illustration of the ion drift-tube apparatus. Ions from the source, IR, are mass-selected by the quadrupole mass filter, SQ, and injected into the drift tube, DT. The cryostat system, CS, provides the temperature control of the drift tube. Ions leaving the drift tube are analysed by the quadrupole mass filter, AQ, with ion detector, CM.

3. RESULTS AND DISCUSSION

3.1 $He^+ + H_2$

The reaction (1)

$$He^+ + H_2 \rightarrow H^+ + H + He$$
$$\rightarrow H_2^+ + H \qquad (1)$$
$$\rightarrow HeH^+ + H$$

is an important initial step in the reaction scheme of ion chemistry in interstellar clouds (1,4). It is known from previous measurements (18-20) that the reaction rate coefficient is very small at temperatures from 80 to 300 K.

$$k(1) = 1.1 \times 10^{-13} \text{ cm}^3\text{s}^{-1} \text{ (at 300 K)}$$

If the rate coefficient is equally low at ambient interstellar temperatures He^+ will preferentially react with CO to give C^+, an important reactant ion for the synthesis of larger carbon hydrates. On the other hand an increase of k(1) with decreasing temperature as observed for other binary reactions (6) would imply that reaction (1) competes with the reaction of $He^+ + CO$.

Measurements of rate coefficients for reaction (1) in our laboratory in the temperature range 18 to 408 K showed that the reaction did not follow a correct second order kinetic behavior. The rate coefficient has a pressure dependence as shown for a temperature of 46.5 K in Fig. 2. The measured coefficient k(1) can be expressed as a combination of a second and third order rate coefficient, k_2 and k_3

$$k = k_2 + [N] \cdot k_3 \qquad (2)$$

where $[N]$ is the gas number density in the reaction cell. This result is consistent with the earlier measurements of Johnson et al. (20) for temperatures between 78 and 300 K.

Of relevance for interstellar cloud chemistry is only the low pressure limit, k_2, which is obtained from the measured data by extrapolation to zero pressure. The results obtained for k_2 as a function of temperature are given in Fig. 3 together with the previous results of Johnsen and Biondi (19) and Johnsen et al. (20). At temperatures below 45 K the pressure dependence of reaction (1) was so strong that because of the uncertainty of the extrapolation to zero pressure only upper limits can be given for k_2. The data of Johnsen and Biondi (19) above 330 K are not obtained under thermal energy conditions but by increasing the drift energy of the ions. The data therefore provide only a qualitative picture of the behavior of the rate coefficient at higher energy. Because of follow-up reactions the reaction products could not certainly be identified in our experiments.

The results imply that reaction (1) does not increase enough with temperature to be an important loss process for He^+ ions in interstellar clouds. Several previous theoretical studies deal with the complex kinetics of this reaction. Mahan (21) has shown that there is no adiabatic path on the energy surface of the $HHHe^+$ system that leads from the reactants $He^+ + H_2$ to any of the products given in (1). This explains why the reaction is so slow. Two thermal energy reaction mechanisms have been proposed that lead via tunneling to H^+ (22) and via a radiative transition to H_2^+ (23,24), while at higher energy adiabatic reaction channels are open. Both thermal energy reaction mechanisms depend on the lifetime of the $He^+ - H_2$ collision complex which is reflected by the slight inverse temperature dependence of k_2 between 100 and 300 K. k_2 is roughly inversely proportional to the relative velocity in the reactive collisions. At temperatures below 100 K there is an indication that the rate coefficient decreases again. Recent results by Barlow and Dunn (25) show a rate coefficient even about an order of magnitude lower for 12 K than the upper limits given for the lowest temperatures in the present results. Such a further decrease is difficult to explain because an energy barrier to the formation of the collision complex $He^+ - H_2$ is not expected. Further research on this interesting binary reaction should be conducted.

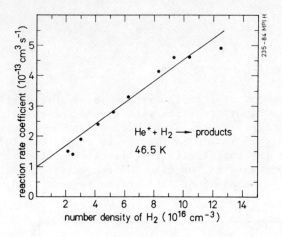

Fig. 2: Measured rate coefficients for the reaction of He$^+$ with H$_2$ in pure H$_2$ as a function of the H$_2$ density at 46.5 K.

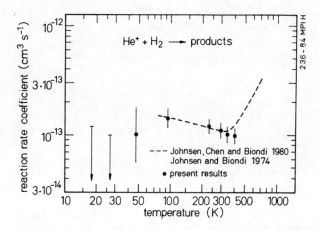

Fig. 3: Low pressure value, k_2, of the rate coefficient for the reaction He$^+$ + H$_2$ = products as a function of the gas temperature. Present data are compared to results from ref. 19 and ref. 20.

3.2 $NH_3^+ + H_2$

The reaction

$$NH_3^+ + H_2 \rightarrow NH_4^+ + H \qquad (3)$$

which is an important step in the synthesis of interstellar NH_3 (3,26) is another binary reaction with a very low reaction probability at room temperature. The reaction rate coefficient further decreases if the temperature is lowered to about 80 K (27). If this decrease of the rate coefficient is extrapolated to interstellar cloud temperatures reaction (3) becomes inefficient. Measurements by Luine and Dunn (28) and subsequent studies in our

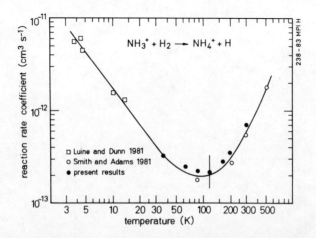

Figure 4: Temperature dependence of the rate coefficient for the reaction $NH_3^+ + H_2 = NH_4^+ + H$. Literature data are from ref. 27 and ref. 28.

laboratory revealed a very surprising result. The rate coefficient shows again a sharp increase at temperatures below 80 K, as shown in Fig. 4.

Similarly to the case of reaction (1) also reaction (3) was found to show a dependence on pressure in a series of measurements conducted around 50 K. This interesting effect is being investigated further. The drift tube results given in Fig. 4 are the low pressure limits of the measured rate constants.

3.3 $N_2H^+ + H_2$

N_2H^+ and HCO^+ are the most abundant molecular ions observed in dense interstellar clouds. Besides binary reactions association reactions with H_2 may be an important loss process for these ionic species.

The three-body association

$$N_2H^+ + 2H_2 \rightarrow N_2H^+ \cdot H_2 + H_2 \qquad (4)$$

was investigated in the ion drift-tube experiment at temperatures from 44 to 192 K. The results are shown in Fig. 5. One observes a strong increase of the reaction rate coefficient with decreasing temperature. The measured coefficients as a function of temperature can be approximated by the expression

$$k(4) = 2.6 \times 10^{-30} \left(\frac{100}{T}\right)^{2.2 (\pm 0.5)} cm^6 s^{-1}$$

In interstellar clouds association reactions can only occur via radiative association because of the low ambient gas density.

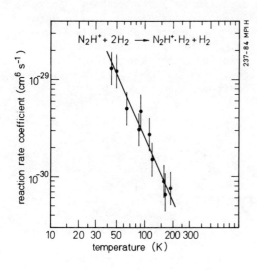

Figure 5:

Temperature dependence of the measured three-body association rate coefficient for the reaction $N_2H^+ + 2H_2 = N_2H^+ \cdot H_2 + H_2$.

Following the discussion in the chapter by E.E. Ferguson in this book one can estimate the rate coefficient for the radiative association

$$N_2H^+ + H_2 \rightarrow N_2H^+ \cdot H_2 + h\nu \qquad (5)$$

from the measured values for reaction (4). One obtains a radiative association rate coefficient of the order of $10^{-17} cm^3 s^{-1}$ for temperatures around 20 to 70 K. This low value implies that the process (5) is not very important in interstellar clouds. A similarly low rate constant has been obtained by Fehsenfeld et al. (29)

for the reaction

$$HCO^+ + 2H_2 \rightarrow HCO^+ \cdot H_2 + H_2 \qquad (6)$$

$$k\,(6) = 8.3 \cdot 10^{-31} \text{ cm}^6\text{s}^{-1} \qquad \text{(at 90 K)}$$

The reason for the low reaction efficiency is the small bond energy of these association products, $D\,(N_2H^+ \ldots H_2) = 7.2$ kcal/Mol and $D\,(HCO^+ \ldots H_2) = 3.9$ kcal/Mol (30). Larger H_2-association rate coefficients are expected for more strongly bound ion complexes as for example for the reaction $CH_3^+ + H_2$ (7).

4. CONCLUSION

The given examples of kinetic studies of ion-molecule reactions show that it is not always possible to extrapolate the temperature dependence of measured rate coefficients down to low temperatures. The need for low-temperature laboratory measurements is obvious. Recently three new experimental methods have become available for such measurements. The good agreement of the data from these methods as presented in the chapter by B.R. Rowe and in this chapter is very promising. Further research with these experiments will not only help in developing a better understanding of interstellar cloud chemistry but should also give new insights in basic problems of reaction dynamics.

REFERENCES

1) Herbst, E., and Klemperer, W. 1973, Astrophys.J. 158, pp. 505-533.
2) Watson, W.D. 1976, Rev.Mod.Phys. 48, pp. 513-552.
3) Dalgarno, A., and Black, J.H. 1976, Rep.Prog.Phys. 39, pp. 573-612.
4) Prasad, S.S., and Huntress, W.T. 1980, Astrophys.J.Suppl.43, pp. 1-35.
5) Graedel, T.E., Langer, W.D., and Frerking, M.A. 1982, Astrophys.J.Suppl. 48, pp. 321-368.
6) Rowe, B.R., Dupeyrat, G., Marquette, J.B., Smith, D., Adams, N.G., and Ferguson, E.E. 1984, J.Chem.Phys. 80, pp. 241-245. See also the chapter by B.R. Rowe in this book.
7) Barlow, S.E., Dunn, G.H., and Schauer, M. 1984, Phys.Rev. Lett. 52, pp. 902-905.
8) Böhringer, H., and Arnold, F. 1983, Int.J. Mass Spectrom. Ion Phys. 49, pp. 61-83.
9) Albritton, D.L. 1978, Atom. Data Nucl. Data Tables 22, pp. 1-101.

10) Albritton, D.L., Dotan, I., Lindinger, W., McFarland, M., Tellinghuisen, J., and Fehsenfeld, F.C. 1977, J.Chem.Phys. 66, pp. 410-421.
11) Lindinger, W., McFarland, M., Fehsenfeld, F.C., Albritton, D.L., Schmeltekopf, A.L., and Ferguson, E.E. 1975, J.Chem.Phys. 63, pp. 2175-81.
12) Durup-Ferguson, M., Böhringer, H., Fahey, D.W., and Ferguson, E.E. 1983, J.Chem.Phys. 79, pp. 265-272.
13) See Chapters by E.E. Ferguson and by D. Smith in this book.
14) Böhringer, H., and Arnold, F. 1982, J.Chem.Phys. 77, pp. 5534-41,
15) Rowe, B.R., Dupeyrat, G., Marquette, J.B., and Gaucherel, P. 1984, J.Chem.Phys. 80, pp. 4915-21.
16) Böhringer, H., Arnold, F., Smith, D., and Adams, N.G. 1983, Int.J. Mass Spectrom. Ion Phys. 52, pp. 25-41.
17) Böhringer, H., Glebe, W., and Arnold, F. 1983, J.Phys.B 16, pp. 2619-26.
18) Fehsenfeld, F.C., Schmeltekopf, A.L., Goldan, P., Schiff, H.I., and E.E. Ferguson 1966, J.Chem.Phys. 44, pp. 4087-94.
19) Johnsen, R., and Biondi, M.A. 1974, J.Chem.Phys. 61, pp. 2112-15.
20) Johnsen, R., Chen, A., and Biondi, M.A. 1980, J.Chem.Phys. 72, pp. 3085-88.
21) Mahan, B.H. 1971, J.Chem.Phys. 55, pp. 1436-46.
22) Preston, R.K., Thompson, D.L., and McLaughlin, D.R. 1979, J.Chem.Phys. 68, pp. 13-21.
23) Hopper, D.G. 1980, J.Chem.Phys. 73, pp. 3289-3293.
24) Jones, G.E., Wu, R.L.C., Hughes, B.M. and Tiernan, T.O. 1980, J.Chem.Phys. 73, pp. 5631-45.
25) Barlow, S.E. 1984, Ph.D. thesis, Colorado University, Boulder, Colorado.
26) Fehsenfeld, F.C., Lindinger, W., Schmeltekopf, A.L., Albritton, D.L., and Ferguson, E.E. 1975, J.Chem.Phys. 62, pp. 2001-2003.
27) Smith, D., and Adams, N.G. 1981, Mon.Not.R.Astr.Soc. 197, pp. 377-384.
28) Luine, J.A., and Dunn, G.H. 1981, Proc. XII Int.Conf.Phys. of Electronic and Atomic Collisions, Gatelinburg 1981.
29) Fehsenfeld, F.C., Dunkin, D.B., and Ferguson, E.E. 1974, Astrophys.J. 188, pp. 43-44.
30) Kebarle, P. 1977, Ann.Rev.Phys.Chem. 28, pp. 445-76.

LABORATORY STUDIES OF ION REACTIONS WITH ATOMIC HYDROGEN

W. Federer, H. Villinger, P. Tosi[a], D. Bassi[a],
E. Ferguson[b] and W. Lindinger
Institut für Experimentalphysik, Univ. Innsbruck,
Karl Schönherrstr. 3, A 6020 Innsbruck, Austria

ABSTRACT

Using a Selected Ion Flow Drift Tube, the following reactions have been investigated in the regime from thermal (300 K) to ~ 0.1 eV center of mass kinetic energy, KE:

$O^+ + H \rightarrow H^+ + O$; $CO^+ + H \rightarrow H^+ + CO$; $CH^+ + H \rightarrow C^+ + H_2$

and $CH_4{}^+ + H_2 \rightleftarrows CH_5{}^+ + H$. For the isotope exchange reaction $DCO^+ + H \rightarrow HCO^+ + D$ a value $k = 1.5 \times 10^{-11}$ cm^3 sec^{-1} has been obtained at KE $\simeq$ 0.06 eV.

INTRODUCTION

Though atomic hydrogen is the most abundant species besides H_2 in diffuse interstellar clouds and shocked regions, few laboratory experiments on ion reactions with H have been investigated which might be of importance in interstellar molecular

Permanent address:
a) Istituto per la Ricerca Scientifica e Technologica
38050, Povo, Trento, Italy
b) Aeronomy Lab., NOAA, Boulder, Colorado, 80303, USA

G. H. F. Diercksen et al. (eds.), Molecular Astrophysics, 649–655.

synthesis. The same holds for low energy ion-H-atom reactions proceeding in the boundary regions of fusion plasmas, especially when the technique of cold gas blankets is applied. One obstacle in attacking H atom reactions in typical swarm experiments, such as Selected Ion Flow Drift Tubes, is the unkown degree of dissociation of hydrogen used as the neutral reactant gas. Even when H_2 is dissociated in very efficient radio frequency sources recombination causes an H_2 density to be present comparable to or greater than that of H so that ions reacting with H are also lost by the usually fast parallel reactions with H_2, thus complicating the analysis of the data. Making use of the fact that CO_2^+ reacting with H only produces COH^+, and CO_2^+ reacting with H_2 results in CO_2H^+ ions but no HCO^+ (1), we have developed a method allowing for the in situ determination of the degree of dissociation, γ, of H_2, in swarm type experiments. Knowing γ, rate coefficients for the reactions of any ions with H are then obtained by observing the declines of the ion signals as a function of hydrogen addition to the reaction region with the radio frequency source on and off respectively, from which data both rate coefficients k_1 and k_2 for the reactions

$$A^+ + H_2 \xrightarrow{k_1} \text{Products} \qquad (1)$$

$$A^+ + H \xrightarrow{k_2} \text{Products}, \qquad (2)$$

are then determined as shown in detail in Ref. (2).

RESULTS AND DISCUSSION

First results on thermal rate coefficients of the reactions of O^+, CO^+ and CH^+ with H as obtained in our Selected Flow Drift Tubes in Innsbruck and Trento have recently been reported (3). In the meantime we have extended these investigations and have obtained the energy dependences of these rate coefficients in the range from 300 K to ~ 0.2 eV mean center of mass KE. These results are shown in Fig. 1 and a detailed discussion of

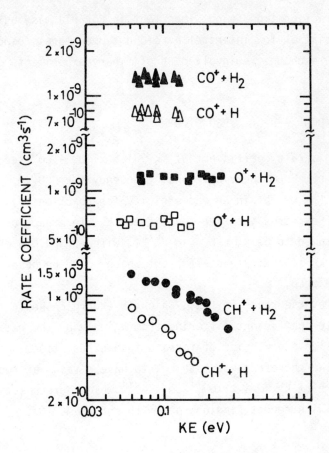

Fig. 1: Rate coefficients for the reactions of O^+, CO^+ and CH^+ with H and H_2 respectively as dependent on KE.

their implications to interstellar molecular synthesis is given in the chapter of Eldon Ferguson in this book.

Isotopic exchange reactions cause an enhancement of heavier isotopes in interstellar molecules as has been shown by Smith and Adams (4). Due to the lower zero point vibrational energy connected with the heavier isotopes bound in molecular systems the replacement of a light isotope by a heavier one is always an exoergic process. As the differences in zero point vibrational energies usually lie in the range of a few to a few tens of °K, isotope fractionation is not important in hot plasmas, but becomes crucial at low interstellar cloud temperatures, especially when H - D exchange is involved. Preliminary experiments on the reaction

$$DCO^+ + H \underset{k_{-3}}{\overset{k_3}{\rightleftarrows}} HCO^+ + D \qquad (3)$$

have shown a rate coefficient of $k_3 \simeq 1.5 \times 10^{-11}$ cm^3 sec^{-1} at KE $\simeq 0.06$ eV. From the vibrational frequency for HCO$^+$, $\nu = 3088$ cm^{-1} (5) we obtain an endoergicity for reaction (3), $\Delta E \simeq 0.056$ eV, thus the rate coefficient k_3, for reaction (3) is assumed to be $k_3 = k_0 \times e^{-\Delta E/kT}$, with $\Delta E = 0.056$ eV, and where k_0 is the rate coefficient which would be observed if ΔE were zero.

The reverse reaction (-3), which is exoergic should thus have a rate coefficient differing from k_0 only by the ratio of the collision frequencies of forward to back reaction, which is $\sim \sqrt{2}$. We therefore estimate k_{-3} to have a value at room temperature of $k_{-3} \simeq 4 \times 10^{-11}$ cm^3 sec^{-1} which agrees satisfactorily with recent findings of Smith and Adams (6). The system

$$CH_4^+ + H_2 \underset{k_{-4}}{\overset{k_4}{\rightleftarrows}} CH_5^+ + H \qquad (4)$$

has been investigated both in the forward (k_4) and reverse direction (k_{-4}) as a function of KE. the respective rate coefficients being shown in Fig. 2. The only rate coefficient known previously for this system, is k_4 (300 K) = 3×10^{-11} cm^3 sec^{-1} (7) in good agreement with the present results. While k_4 decreases with increasing KE, k_{-4} stays constant at $k_{-4} \simeq 1.5 \times 10^{-}$ cm^3 sec^{-1} in the energy regime investigated. Though $k_4 < k_{-4}$ over the whole energy regime from thermal to ~ 0.1 eV KE$_{cm}$, k_4 applies to the exoergic direction of the system.

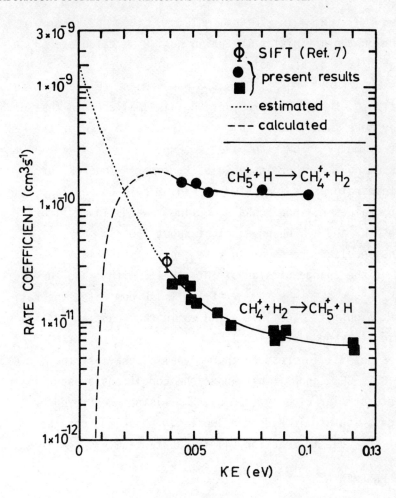

Fig. 2: Rate coefficients k_{-4} and k_4 for the reactions of CH_5^+ with H and of CH_4^+ with H_2 respectively as a function of KE. The values indicated by the dashed curve (- - -) are calculated from the estimated values of k_4 (dotted curve) and the values ΔH_f and ΔS^o_{297} obtained from a Van't Hoff plot of k_4/k_{-4}.

A Van't Hoff plot of the data in Fig. 2 yields an exoergicity of reaction (4) of $\Delta H = 1.2$ kcal/mole and $\Delta S^o_{297} = 7.5$ e.u. This is in accord with reported heats of formation for the species involved: ΔH_f (CH_4^+) is 275 kcal/mole, ΔH_f (H) = 52 kcal/mole and from proton affinities listed in Ref. 8 we calculate ΔH_f (CH_5^+) = (218 ± 3) kcal/mole.

As $S^°_{297}$ (H_2) - $S^°_{297}$ (H) = 3.8 e.u., we calculate from the present data, that the entropy of CH_5^+ is smaller than that of CH_4^+ (by more than 3 e.u.) which is expected in view of the greater symmetry of CH_4^+ compared to CH_5^+.

Assuming a further increase of k_4 towards lower KE or lower temperatures as indicated in Fig. 2 by a dotted line, the computed values k_4 show a sharp decline towards lower T, or KE, (both k_4 and k_{-4} being equal at about 150°K).

The peculiar crossover in rate constants in Fig. 2 is a result of the change of sign of $\Delta G = \Delta H - T\Delta S$ with T and the fact that $k_4/k_{-4} = K_{equ} = e^{-\Delta G/RT}$. At low T, ΔH dominates the ratio k_4/k_{-4}, at high T, $T\Delta S$ dominates and while the reaction (4) is exoergic it is endoentropic.

Besides its interesting thermodynamic behavior, the system CH_4^+-CH_5^+ shows, that under the conditions present in cold interstellar clouds (T<60°K), CH_5^+ is not destroyed by reactions with atomic hydrogen, while in shock regions the loss of CH_5^+ due to reaction (4) is substantial.

ACKNOWLEDGEMENTS

This work was supported in part by Fonds zur Förderung der Wissenschaftlichen Forschung under Project P 5149 and by the Austrian Acedemy of Sciences by Project "Fusionsrelevante Ionen-Neutral-Reaktionen".

REFERENCES

1. Fehsenfeld, F.C., and Ferguson, E.E. 1971, Geophys. Res. 76, 8453
2. Tosi, P., Iannotta, S., Bassi, D., Villinger, H., Dobler, W., and Lindinger, W., 1984, J. Chem. Phys. 80 p. 1905

3. Federer, W., Villinger, H., Howorka, F., Lindinger, W., Tosi, P., Bassi, D., and Ferguson, E., 1984, Phys. Rev. Lett. 52, p. 2084
4. Smith, D., and Adams, N.G., 1981, Int. Rev. Phys. Chem., 1, 271
5. Saykally, R., private communication
6. Smith, D., and Adams, N.G., private communication
7. Smith, D., and Adams, N.G., 1977, Int. J.Mass Spectrom. Ion Phys., 23, 123
8. Walder, R., and Franklin, J.L., 1980, Int. J. Mass Spectrom. Ion Phys., 36, 85.

DISSOCIATIVE RECOMBINATION OF H_3^+, HCO^+, N_2H^+ and CH_5^+

N.G. Adams and D. Smith

Department of Space Research, University of Birmingham,
Birmingham B15 2TT, England

The dissociative recombination coefficients, α, for H_3^+, HCO^+, N_2H^+ and CH_5^+ have been measured in a flowing afterglow/Langmuir probe (FALP) apparatus at 95 and 300K. It was found that $\alpha(H_3^+)$ is very small, in conflict with previous measurements, but that $\alpha(HCO^+)$, $\alpha(N_2H^+)$ and $\alpha(CH_5^+)$ are much larger being in substantial agreement with expectations based on previous laboratory data. The implications of these new data to the chemistry and physics of interstellar gas clouds are mentioned.

1. INTRODUCTION

Dissociative recombination is considered to be the most important neutralization process for interstellar molecular ions and is therefore an important stage in the synthesis of the observed interstellar neutral molecules. It also regulates the electron density in interstellar clouds. Clearly, therefore, it is important to know the dissociative recombination coefficients, α, for all of the abundant interstellar ions at appropriately low temperatures. The laboratory study briefly described here was undertaken to determine the α for several important ions, i.e. H_3^+, HCO^+, N_2H^+ and CH_5^+, at the lowest temperature accessible in our variable-temperature flowing afterglow/Langmuir probe apparatus (the VT-FALP apparatus). This apparatus has been previously described in detail elsewhere [1,2] and has been used successfully to determine ion-ion recombination coefficients [1,3], electron attachment coefficients [4,5], as well as dissociative recombination coefficients for several atmospheric ions [2].

2. RESULTS

Measurements of $\alpha(H_3^+)$, $\alpha(D_3^+)$, $\alpha(HCO^+)$, $\alpha(DCO^+)$, $\alpha(N_2H^+)$, $\alpha(N_2D^+)$ and $\alpha(CH_5^+)$ have been made at 95K and 300K. Additionally, $\alpha(H_3^+)$ has been measured at 550K. The detailed results of these experiments and details of the experimental technique are reported in a paper which is to be published soon [6]. Some of the interstellar implications are given in another short paper [7]. Here, we summarise the major results of the study.

The most surprising and important result, both from the fundamental molecular physics and the interstellar viewpoints, is the very small value indicated for $\alpha(H_3^+)$ (and also for $\alpha(D_3^+)$). An upper limit of $\lesssim 10^{-8}$ cm^{3}s^{-1} was obtained at each of the three temperatures. This low value contrasts with the previous experimental results from Biondi's laboratory [8,9] which have indicated $\alpha(H_3^+)$ to be greater than 10^{-7} cm^{3}s^{-1}, i.e. comparable to most other simple (mostly diatomic) molecular ions. Our result, however, conforms with the recent theoretical prediction of Michels and Hobbs [10] and with much earlier experimental [11] and theoretical [12] studies which also indicated that $\alpha(H_3^+)$ is small. Evidence was obtained from our experiment that vibrationally-excited H_3^+ does however recombine at an appreciable rate (i.e. $\alpha(H_3^+(v)) > 10^{-7}$ cm^{3}s^{-1}) which, again, is in accordance with theoretical expectations [10].

The implications of such a small $\alpha(H_3^+)$ to the physics and chemistry of interstellar clouds may be great. The full significance of the result needs to be ascertained by detailed modelling. It is clear, however, that the number densities of H_3^+ and electrons will be greater than previous estimates had indicated [7] and this has implications to the rates of molecular evolution and gas cloud collapse. In fact the electron density can no longer be determined using the model due to Watson [13] and an alternative approach has to be used [7].

Since recombination of H_3^+ (and also H_2D^+) is no longer important in regulating the H_3^+ and electron densities in cool interstellar clouds, yet H_3^+ is the important initial molecular ion, then proton transfer from H_3^+ to abundant interstellar molecules such as CO, N_2 and perhaps CH_4 controls the loss rate of H_3^+. In such proton transfer reactions HCO^+, N_2H^+ and CH_5^+ ions are formed. Our studies have confirmed that $\alpha(HCO^+)$, $\alpha(N_2H^+)$ and $\alpha(CH_5^+)$ are relatively large, the actual values obtained being in reasonable agreement with previous values [14-16]. Thus at 95K, $\alpha(HCO^+) = 2.9 \times 10^{-7}$, $\alpha(N_2H^+) = 4.9 \times 10^{-7}$ and $\alpha(CH_5^+) = 1.5 \times 10^{-6}$, all in units of cm^{3}s^{-1}. These values together with those at 300K indicate (within the limitations to accuracy imposed by only two data points!) that $\alpha(HCO^+)$ and $\alpha(N_2H^+)$ vary approximately as T^{-1} and that $\alpha(CH_5^+)$ varies approximately as $T^{-0.3}$. No significant differences were obtained for $\alpha(DCO^+)$ and $\alpha(N_2D^+)$ relative to their hydrogenated analogues. Thus in cool molecular clouds $\alpha(HCO^+)$, $\alpha(N_2H^+)$ and $\alpha(CH_5^+)$ should exceed 10^{-6} cm^{3}s^{-1} which is in accordance with the magnitudes usually assumed for these parameters in interstellar chemical models.

These studies of dissociative recombination are just the beginning and we intend to extend the measurements to include more complex ions of interstellar importance. We will also attempt to determine the neutral products of some reactions, which at present can only be guessed.

REFERENCES

[1] Smith, D. and Adams, N.G. 1983, "Physics of Ion-Ion and Electron-Ion Collisions" eds. F. Brouillard and J. Wm, McGowan, Plenum, New York, pp. 501-531.
[2] Alge, E., Adams, N.G. and Smith, D. 1983, J. Phys. B 16, p. 1433.
[3] Smith, D., Church, M.J. and Miller, T.M. 1978, J. Chem. Phys. 68, p. 1224.
[4] Smith, D., Adams, N.G. and Alge, E. 1984, J. Phys. B 17, p. 461.
[5] Alge, E., Adams, N.G. and Smith, D. 1984, J. Phys. B 17, in press.
[6] Adams, N.G., Smith, D. and Alge, E. 1984, J. Chem. Phys. (Aug), in press.
[7] Smith, D. and Adams, N.G. 1984, Ap. J. (Letters), in press.
[8] Leu, M.T., Biondi, M.A. and Johnsen, R. 1973, Phys. Rev. A8, p. 413.
[9] Macdonald, J.A., Biondi, M.A. and Johnsen, R. 1984, Planet Space Sci. 32, p. 651.
[10] Michels, H.H. and Hobbs, R.H. 1983, Proc. 3rd Int. Symp. on the Production and Neutralization of Negative Ions and Beams, Brookhaven National Research Laboratory, New York.
[11] Persson, K.B. and Brown, S.C. 1955, Phys. Rev. 100, p. 729.
[12] Carney, G.D. and Porter, R.V. 1977, J. Chem. Phys. 66, p. 2756.
[13] Watson, W.D. 1977, "CNO Isotopes in Astrophysics" ed. J. Audouze, Reidel, Dordrecht-Holland, pp. 105-114.
[14] Leu, M.T., Biondi, M.A. and Johnsen, R. 1973, Phys. Rev. A8, p. 420.
[15] Mul, P.M. and McGowan, J. Wm. 1979, Ap. J. (Letters) 227, L157.
[16] Mul, P.M., Mitchell, J.B.A., D'Angelo, V.S., Defrance, P., McGowan, J. Wm and Froelich, H.R. 1981, J. Phys. B 14, p. 1353.

MOLECULAR DISSOCIATION FUNCTIONS OBTAINED FROM THERMODYNAMIC AND SPECTROSCOPIC DATA

C. M. Sharp

Max-Planck-Institut für Physik und Astrophysik,
Karl-Schwarzschild-Str. 1, D-8046 Garching b. München,
FRG

ABSTRACT

The dissociation functions of several diatomic and polyatomic molecules obtained from thermodynamic data are compared to those computed from spectroscopic data, where it is found that in many cases the agreement is good.

This work is done for several molecules that can be abundant in the atmospheres of late-type stars, in particular for the highly reactive free radicals for which data is often difficult to obtain.

1. INTRODUCTION

In computing the abundances of atomic, ionic and molecular species in conditions of LTE, a knowledge of the equilibrium constants of these species is necessary over the range of temperatures of interest. In the atmospheres of late-type stars with temperatures in the range of about 2000K to 6000K, the formation of molecules can be very important, and in this work we discuss how molecular equilibrium constants can be obtained from both thermodynamic and spectroscopic data, and compare some results of calculations obtained by modifying a version of the Molecular Equation of State routines, as kindly provided by Huebner (1).

The comparison between the thermodynamic and spectroscopic data is obtained in the form of a deviation of the effective dissociation potential as a function of temperature. So this deviation gives an indication of how consistent are the data, and how good are our partition functions as a function of temperature.

The thermodynamic data is obtained in all cases from the most recent tabulations of the JANAF Thermochemical Tables for the species in question, references (2), (3), (4), and (5), with the dates being respectively 1971, 1974, 1975 and 1982.

In this work we consider a selected sample of atoms and molecules of astrophysical interest as follows:

6 Elements H, C, N, O, S and Ti with 1982 JANAF data, except for Ti which has 1971 data. The energy levels were obtained from Moore (6) and the ionization potentials and electron affinities from Allen (7).

15 Diatomic Molecules as listed in table 1, with the dates of the thermodynamic data indicated. All the spectroscopic data, including the dissociation potentials (from the lowest vibrational level), were obtained from Huber and Herzberg (8). Note that we have included three molecular ions, to show that the method discussed here can be used to check such cases, even though in conditions of LTE in a stellar atmosphere, molecular ions are generally unimportant.

10 Polyatomic Molecules as listed in table 2, with the dates as indicated. The references for the spectroscopic data are given in the last column, and with the exception of HCN, H_2O and C_2H_2, all the data were obtained from the same JANAF reference as the thermodynamic data. All the dissociation potentials (from the lowest vibrational level) to completely dissociate the molecule into its separate atoms, were obtained from the JANAF tables.

Table 1 Diatomic Molecules

Molecule	Date	NDX	Molecule	Date	NDX
H_2(rs)	1982	0	CN	1971	1
CH	1971	1	CO	1971	0
CH^+	1974	1	N_2(rs)	1982	0
NH	1982	0	NO	1971	0
OH	1982	0	O_2(rs)	1982	2
OH^+	1974	0	SO	1982	2
C_2^-	1971	0	TiO	1975	7
C_2	1971	4			

Table 2 Polyatomic Molecules

Molecule	Date	NDX	X_{ij}	Refs.
C_2H(ℓ)	1971	0	NO	JANAF
CH_2	1975	2	NO	JANAF
HCN(ℓ)	1971	0	YES	(9)
HCO	1974	1	NO	JANAF
NH_2	1982	1	NO	JANAF
H_2O	1971	0	YES	(10)
C_3(ℓ)	1971	0	*	JANAF
C_2H_2(ℓ)	1971	0	YES	(11)
CH_3	1971	0	NO	JANAF
CH_4	1971	0	NO	JANAF

where: rs=Reference state; ℓ=linear polyatomic molecule, NDX=Number of excited electronic states used to calculate the partition function, the column headed by X_{ij} indicates whether anharmonic constants are considered. * C_3 is a quasi-linear molecule which is treated specially.

2. COMPUTATION OF PARTITION FUNCTIONS

In order to compute the dissocitation functions from spectroscopic data, partition functions are required as follows:

Atoms:

$$Q = \sum_i g_i e^{-T_i hc/kT} \quad (1)$$

where g_i and T_i are the statistical weight and energy in cm^{-1} respectively of the ith state. The cut-off in the summation is at $20000 cm^{-1}$, which is adequate over the range of temperatures and densities of interest in our context. Although (1) is formally divergent if the sum is carried to infinity, it is found that the contribution of the upper levels can be neglected, except at very low densities, where LTE is not a good approximation anyway.

Molecules:

$$Q = \sum_i Q_{S_i} Q_{E_i} Q_{V_i} Q_{R_i} e^{-T_{O_i} hc/kT} \quad (2)$$

where:

Q_{S_i} = Symmetry partition function of the ith electronic state = $1/\sigma_i$

Q_{E_i} = Electronic partition function of the ith electronic state
$= (2 - \delta_{\Lambda_i, 0})(2S+1)$

Q_{V_i} = Vibrational partition function of the ith electronic state

Q_{R_i} = Rotational partition function of the ith electronic state

T_{O_i} = Energy in cm^{-1} of the ith electronic state after correcting for the zero-point energy.

Note that for polyatomic molecules, Q_S should be inside the summation to allow for different possible symmetries of excited electronic states, and Q_E for those states with Hund's case (a) coupling should be obtained by summing the separate rotational ladders, and applying the appropriate Boltzmann factors. However, it is found in practice, that for all the cases considered here over the temperature range of interest, the expression given for Q_E is a good approximation. Q_V and Q_R are computed from asymptotic expressions derived from Kassel (12) and (13) respectively, and are found to be rapidly convergent, with the exception of C_3, which is discussed later.

3. DISSOCIATION FUNCTIONS

In order to express the formulae of molecular dissociation in a general way, the following notation is adopted:

M = Molecular weight of a molecule Y,
m_i = Atomic weight of an element X_i, which is the ith element in Y,
m_e = Mass of electron,
N = Total number of atoms in Y,
n_i = Number of atoms of the element X_i in Y,
k = Number of different elements in Y,
q = Charge of atom or molecule, $q = -1, 0, 1 \ldots$
N_A = Avogadro's number, so all masses are in amu,
A_o = Standard atmospheric pressure in dyne cm^{-2},
E = Energy of ionization, dissocitation etc, in cm^{-1},

and by definition, we can write

$$M = \sum_{i=1}^{k} n_i m_i - q m_e \quad \text{and} \quad N = \sum_{i=1}^{k} n_i \qquad (3)$$

The general molecular equilibrium reaction for a molecule and its neutral free atoms is written as

$$\sum_{i=1}^{k} n_i X_i \Leftrightarrow Y^q + q e^- \qquad (4)$$

If the molecule is an ion, then depending on the sign of its charge, free electrons appear on the LHS or RHS of (4). The equilibrium constant by number for this reaction is given by

$$K_n(Y^q) = \frac{N(Y^q) N^q(e^-)}{\prod_{i=1}^{k} N^{n_i}(X_i)} \qquad (5)$$

and $\ln(K_n)$ is the definition of the <u>dissociation function</u>. Note that here N has the meaning of number density in cm^{-3} for the species considered.

3.1. Determination of the Thermodynamic Dissociation Function

The thermodynamic dissociation functions can be obtained from the JANAF tables which give K_p^o, which is the equilibrium constant by pressure in atmospheres of the species in question referred to its elements in their <u>standard state</u>. E.g. for CO, K_p^o has the meaning

$$K_p^o (CO) = \frac{P(CO)}{P(C_s) P^{1/2}(O_2)} \qquad (6)$$

where C_s (graphite) and O_2 are the elements in their standard state, and $P = NkT/A_o$, even though in the case of graphite, this "pressure" is fictitious.

From the above discussion and (5), the equilibrium constant by number is seen to be

$$K_n(Y^q) = \frac{K_p^o(Y^q)}{\prod_{i=1}^{k} [K_p^o(X_i)]^{n_i}} \left[\frac{kT}{A_o}\right]^{(N-1-q)} \tag{7}$$

and $K_p^o(e^-) = 1$ by definition. The dissociation function is then

$$\ln[K_n(Y^q)] = \ln[K_p^o(Y^q)] - \sum_{i=1}^{k} n_i \ln[K_p^o(X_i)] + (N-1-q)\ln\left[\frac{kT}{A_o}\right] \tag{8}$$

or using the fact that

$$K_p^o = e^{-\Delta G_f^o/RT} \tag{9}$$

where ΔG_f^o is the Gibbs energy change for the formation of the atomic or molecular species from its elements in their standard state, and R is the gas constant, we can write

$$\ln[K_n(Y^q)] = -\frac{1}{RT}\left[\Delta G_f^o(Y^q) - \sum_{i=1}^{k} n_i \Delta G_f^o(X_i)\right] + (N-1-q)\ln\left[\frac{kT}{A_o}\right] \tag{10}$$

where if, as is usual, ΔG_f^o is in cal mole^{-1}, then R is in cal mole^{-1} deg^{-1}.

3.2. Determination of the Spectroscopic Dissociation Function

The dissociation function can be obtained from spectroscopic data by considering that the equilibrium of a molecule with its N atoms in their dissociated state, can be written as a succession of N-1 equilibria, each of which can be written as a Saha-like equation. If molecular ions are considered, the Saha equation for ionization must be included, and if this is done in a consistent way, we obtain the general expression

$$K_n(Y^q) = \frac{2^q Q(Y^q)}{\prod_{i=1}^{k} Q^{n_i}(X_i)} \left[\frac{2\pi\mu kT}{N_A h^2}\right]^{3(1+q-N)/2} \exp(hc\Sigma E/kT) \tag{11}$$

which is the generalised Saha equation for the equilibrium reaction (4) and where:

$Q(Y^q)$ = Partition function of the molecule Y^q,

$Q(X_i)$ = Partition function of the element X_i.
ΣE = sum of all energies to completely dissociate the molecule into separate atoms, <u>minus</u> the ionization potentials if $q>0$ or <u>plus</u> the electron affinity if $q=-1$.
μ = "multiple" reduced mass, and is given by

$$\mu^{(N-1-q)} = \prod_{i=1}^{k} m_i^{n_i} / m_e^q M \tag{12}$$

The dissociation function is thus obtained by taking the natural logs of (11) and (12)

$$\ln\left[K_n(Y^q)\right] = \ln\left[Q(Y^q)\right] - \sum_{i=1}^{k} n_i \ln\left[Q(X_i)\right] + \frac{hc}{kT} \Sigma E$$

$$+ \frac{3}{2}(1+q-N)\ln T + \frac{3}{2}\left[\ln M - \sum_{i=1}^{k} n_i \ln m_i\right] \tag{13}$$

$$+ q(\ln 2 + \frac{3}{2}\ln m_e) + \frac{3}{2}(1+q-N)\ln\left[\frac{2\pi k}{N_A h^2}\right]$$

where for convenience, we have separated out the T-independent part of (13) into the last three terms. For additonal convenience, numerical values for the constants can be put in (13), and we obtain

$$\ln\left[K_n(Y^q)\right] = \ln\left[Q(Y^q)\right] - \sum_{i=1}^{k} n_i \ln[Q(X_i)] + 1.438830\ \Sigma E/T$$

$$+ \frac{3}{2}(1+q-N)\ln T + \frac{3}{2}\left[\ln M - \sum_{i=1}^{k} n_i \ln m_i\right] \tag{14}$$

$$+ 36.113434\ q + 46.682517(1-N)\ .$$

4. COMPARISON BETWEEN THE THERMODYNAMIC AND SPECTROSCOPIC DATA EXPRESSED AS AN ERROR OF THE EFFECTIVE DISSOCIATION POTENTIAL

If the RHS of (13) or (14) above is instead found from thermodynamic data, and the partition functions are computed from spectroscopic data, then ΣE can be solved as an unknown, and its deviation as a function of temperature from the value obtained in the literature, gives an indication of how consistent are the spectroscopic data with the thermodynamic data, and how good are the partition functions as a function of temperature.

Specifically, if $\Delta\Sigma E$ is the error of the effective dissociation potential in

cm^{-1}, it is seen that it is obtained from

$$\Delta\Sigma E = \frac{kT}{hc} \left\{ \ln\left[K_n(Y^q)\right]_{Th} - \ln\left[K_n(Y^q)\right]_{Sp} \right\} \quad (15)$$

where $\ln[K_n(Y^q)]_{Th}$ is the dissociation function obtained from thermodynamic data in section 3.1 and $\ln[K_n(Y^q)]_{Sp}$ is obtained from spectroscopic data in section 3.2.

If we substitute (10) and (13) into (15), then we can obtain $\Delta\Sigma E$ explicitly in terms of the Gibbs energy changes and other quantities.

$$\Delta\Sigma E = \frac{1}{N_A hc} \left[\sum_{i=1}^{k} n_i \Delta G_f^o(X_i) - \Delta G_f^o(Y^q) \right]$$

$$+ \frac{kT}{hc} \left\{ \frac{5}{2}(N-1-q)\ln T - \ln\left[Q(Y^q)\right] + \sum_{i=1}^{k} n_i \ln\left[Q(X_i)\right] \right.$$

$$- \frac{3}{2}\left[\ln M - \sum_{i=1}^{k} n_i \ln m_i\right] - q\left[\ln 2 + \frac{3}{2}\ln m_e\right] \quad (16)$$

$$\left. + (N-1-q)\ln\left[\left[\frac{2\pi}{N_A}\right]^{3/2} \frac{k^{5/2}}{A_o h^3}\right]\right\} - \Sigma E$$

with all quantities in cgs units except for masses which are in amu, ΣE in cm^{-1} is the total dissociation potential, as defined in section 3.2, that is obtained from spectroscopic or thermodynamic data. If ΔG_f^o is in cal $mole^{-1}$, then the conversion factor of 4.1854×10^7 erg cal^{-1} must be applied to the first term.

In the limit that $T \to 0$, then (16) reduces to

$$\Delta\Sigma E = \frac{1}{N_A hc} \left[\sum_{i=1}^{k} n_i \Delta G_f^o(X_i) - \Delta G_f^o(Y^q) \right] - \Sigma E \quad (17)$$

This equation may be used as a cross-check between the available sources of data.

5. DISCUSSION OF RESULTS

Using the data for the 6 elements and 25 molecules given in section 1, the dissociation function and errors are computed using the theory discussed above, and the results are presented in graphical form as a function of temperature.

Figure 1 gives plots of the dissociation function between 1000 and 6000K of the diatomic molecules computed from the thermodynamic data, with several of the molecules indicated. Note that for the neutral molecules and C_2^-, the dissociation function approaches positive infinity as T→0 as ΣE is positive, but it approaches negative infinity for the two positive molecular ions as ΣE is negative, due to the ionization potential being greater than the dissociation potential of the corresponding neutral molecules.

Figure 2 gives plots of the dissociation function computed from thermodynamic data of the polyatomic molecules over the same temperature range as before. On the scale used in figures 1 and 2, the differences between the dissociation function computed from thermodynamic and spectroscopic data would hardly be visible, so only the former data is used here.

Figure 3 is a plot of the error in the effective dissociation potential in cm^{-1} as defined by (15), for the diatomic molecules down to temperatures for which thermodynamic data is available. The displacement of each curve at low temperatures is an indication of the agreement of the dissociation potentials obtained from Huber and Herzberg (8) and the JANAF data, and the departure of each curve from a horizontal line at higher temperatures indicates how consistent are our partition functions with the tabulated thermodynamic data. In comparison with typical values of dissociation potentials, it is seen that for the sample of molecules considered here, the errors for most of them are small. The dissociation potentials of the molecular ions and the dissociation potentials and ionization potentials or electron affinities of the corresponding neutral molecules, as given by Huber and Herzberg (8) were found to be consistent with the ionization potentials or electron affinities of the corresponding elements given by Allen (7). However, it is certainly seen that in the case of C_2^- or OH^+, this data is not consistent with the thermodynamic data given in the JANAF tables.

Figure 4 is the corresponding plot of errors for the polyatomic molecules on the same scale as before. All the curves converge to $\Delta\Sigma E=0$ at small temperatures as the values of ΣE are obtained from the thermodynamic data. Only at large temperature do the deviations become significant, which in practice is unimportant, as the molecules could then be virtually completely dissociated. Several of the molecules are indicated, but the case of C_3 stands out particularly, and is discussed below.

MOLECULAR DISSOCIATION FUNCTIONS

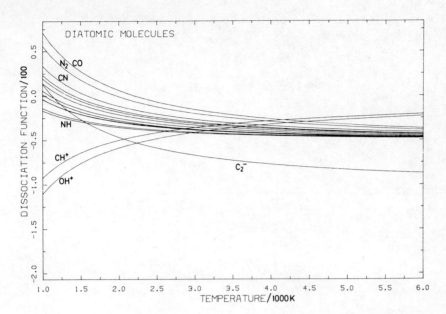

Figure 1. The dissociation function between 1000K and 6000K for diatomic molecules

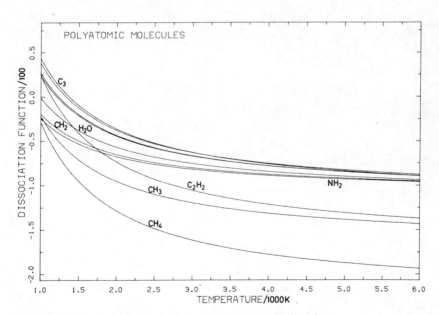

Figure 2. As fig. 1 but for polyatomic molecules

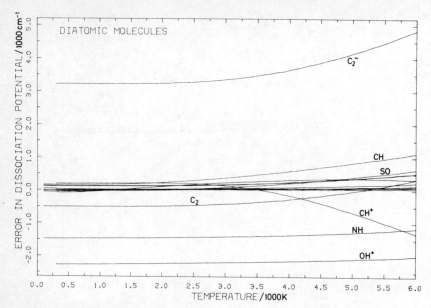

Figure 3. The error of the effective dissociation potential as a function of temperature for diatomic molecules

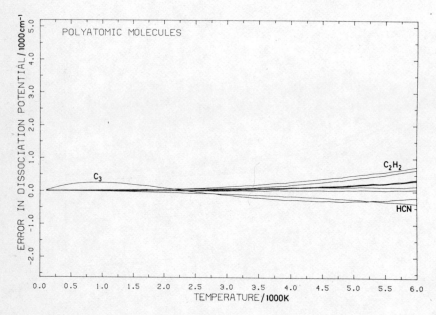

Figure 4. As fig. 3 but for polyatomic molecules

5.1. Special Case of the C_3 Molecule

Using the thermodynamic and spectroscopic data for C_3 from the JANAF 1971 tables and computing the error as a function of temperature, it is found that its error rapidly increases with temperature, and at 6000K the error is about -7.3×10^4 cm^{-1}, which is very large compared to the other molecules.

This large error is due to incorrectly calculating the vibrational partition function by assuming that the three vibrational modes can be approximated to harmonic oscillators, which is certainly wrong for the bending mode which has an anomalously small vibrational constant. Unfortunately, using the vibrational harmonic and anharmonic constants computed from ab initio calculations by Kraemer et al. (14), does not help matters, as these constants are only poorly known and apply to the lowest levels. At temperatures of several thousand degrees, very high bending levels will be populated.

By Hansen and Pearson (15), it appears that some average potential between an harmonic oscillator and a square well potential may be a reasonable approximation. Accordingly, as the dissociation function is already known from thermodynamic data, then the vibrational partition function Q_V can be obtained. If it is assumed that Q_V can be written as the product

$$Q_V = Q_{V_1} Q_{V_2} Q_{V_3} \qquad (18)$$

then Q_{V_2}, the contribution of the bending mode, can be obtained as Q_{V_1} and Q_{V_3}, the contributions of the two stretching modes, are calculated using the data from JANAF. Knowing Q_{V_2} and using the value of the harmonic bending constant ω_2 from JANAF, then the anharmonic constant x_{22} can be found by iteration for several temperatures. It is assumed that the stretching modes behave as harmonic oscillators, and all the other anharmonic constants are zero.

It is indeed found that except at very low temperaturs, x_{22} is only weakly dependent on temperature, and for temperatures relevant to the possible formation of C_3 in stellar atmospheres, an avorage value of about 8 cm^{-1} is obtained, which corresponds to levels of increasing separation with energy and is intermediate between the harmonic and square well potentials. If this value is adopted for all temperatures, the error curve for C_3 is obtained in figure 4.

Unfortunately, due to the large value of x_{22} compared to ω_2, the asymptotic expansions used are not convergent, and the bending contribution to the partition function is obtained directly from

$$Q_{v_2} = \sum_{v=0} (v+1) e^{-[(\omega_2 + 2x_{22})v + x_{22}v^2]hc/kT} \qquad (19)$$

where the summation is cut-off at half the total dissociation energy of C_3, which is not critical as this is at a high enough energy for the uppermost levels to contribute negligibly.

The adoped value of $x_{22} = 8$ cm^{-1} only represents an average value over a large number of levels, and cannot be used for temperatures below about 1500 K.

Acknowledgements

I am grateful to Dr. W. F. Huebner and Dr. W. P. Kraemer for many useful discussions, and to Dr. E. Trefftz for kindly reading the manuscript.

References

(1) Huebner, W. F., Los Alamos National Laboratory, private communication.
(2) JANAF Thermochemical Tables, NSRDS-NBS 37 (1971).
(3) JANAF Thermochemical Tables, J. Phys. & Chem. Ref. Data, $\underline{3}$, 311 (1974).
(4) JANAF Thermochemical Tables, J. Phys. & Chem. Ref. Data, $\underline{4}$, 1 (1975).
(5) JANAF Thermochemical Tables, J. Phys. & Chem. Ref. Data, $\underline{11}$, 695 (1982).
(6) Moore C. E., Selected Tables of Atomic Spectra, NSRDS-NBS3, Sections 6 (1972), 3 (1970), 5 (1975) & 7 (1976) for H, C, N & O respectively; for S & Ti see Atomic Energy Levels I, Circular 467 of the NBS (1949).
(7) Allen C. W., Astrophysical Quantities, 3rd Edition (1973)
(8) Huber K. P. & Herzberg G., Molecular Spectra and Molecular Structure, IV Constants of Diatomic Molecules, Van Nostrand Reinhold Co. (1979)
(9) Strey G. & Mills I. M., Mol. Phys. $\underline{26}$, 129 (1973)
(10) Kraemer W. P., Roos B. O. & Siegbahn P. E. M., ref. (36) of Chem. Phys. $\underline{69}$, 305 (1982)
(11) Strey G. & Mills I. M., J. Mol. Spec. $\underline{59}$, 103 (1976)
(12) Kassel L. S., Phys. Rev. $\underline{43}$, 364 (1933)
(13) Kassel L. S., J. Chem. Phys. $\underline{1}$, 576 (1933)
(14) Kraemer W. P., Bunker P. R. & Yoshimine M., submitted to J. Mol. Spec.
(15) Hansen C. F. & Pearson W. E., Can. J. Phys. $\underline{51}$, 751 (1973)

A THEORETICAL STUDY OF H_2O MASERS

W.H. Kegel, S. Chandra

Institut für Theoretische Physik
Universität Frankfurt/Main, F.R.G.

D.A. Varshalovich

Ioffe Physical-Technical Institute,
USSR Academy of Sciences, Leningrad

ABSTRACT

The influence of different physical effects on the efficiency of H_2O masers has been studied by solving, numerically, the coupled set of statistical and radiative transfer equations in an on-the-spot approximation. The computations were based on a physical model in which the basic excitation is due to collisions. In particular, we studied the influence of the quasi-resonances between the ($J = 2 \rightarrow 0$)-transition in the H_2 molecule and the ($5_{23} \rightarrow 7_{25}$)- and the ($2_{20} \rightarrow 4_{40}$)-transition in the H_2O molecule. The results indicate that the combined effects of resonant collisions and cold dust can lead to luminosities in the 1.35 cm line comparable to those observed. Maser emission at 1.635 mm and 922 μ from para-H_2O is predicted.

INTRODUCTION

Cosmic H_2O masers emitting in the 1.35 cm line are the most powerful masers known. They are characterized by a high brightness temperature ($T_B \approx 10^{13}$ K) and scale sizes of the order of 10^{13}-10^{14} cm if the individual

G. H. F. Diercksen et al. (eds.), Molecular Astrophysics, 673–680.

emission features are considered as single objects, or of the order of 10^{16} cm if the "centers of activity" [1] are taken as single objects. The corresponding luminosities are of the order of 10^{29} and 10^{30} erg/sec, respectively [2].

The various attempts made so far to construct theoretical models have not led to a unique interpretation of the observations. Therefore, it seems worthwhile to study the influence of different physical effects separately. Our calculations are based on a physical model in which the basic excitation is due to collisions with molecular hydrogen and the only external source of radiation is the 2.7 K background.

The numerical problem is to solve the statistical and radiative transfer equations, simultaneously. For the latter an on-the-spot approximation was used in order to keep the computing time within limits.

Calculations were performed for both ortho- and para-H_2O.

ORTHO-H_2O

The 1.35 cm line corresponds to the ($6_{16} \rightarrow 5_{23}$)-transition in the ortho-system of H_2O (Fig. 1). In our calculations the lowest 31 energy levels connected by 92 radiative transitions were taken into account, and three different sets of input data were used.

In the first set of computations, we used collision rates given by Green [3] and disregarded the influence of any continuous absorption or emission. We present the numerical results in a qualitative manner by drawing in a (N_{H_2O}, n_{H_2})-diagram the lines $\tau = 0$, $\tau = -1$, and $\tau = -2$ for the maser line, indicating by this the general region of inversion and the region of larger negative optical depths. Figure 2 gives the results obtained with a kinetic temperature of 500 K, and a line width of $\Delta\nu/\nu = 10^{-5}$. We see that inversion occurs over a wide range of parameters. The region in which $\tau < -1$, however, is restricted to relatively low densities and high column densities.

The data given by Green include collisions with H_2 molecules in the rotational ground state (J=0) only. However, at least at higher temperatures expected to prevail in H_2O maser regions, collisions with rotationally

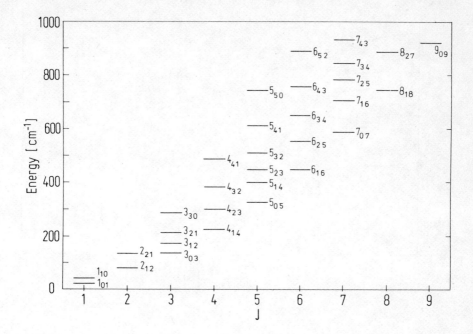

Figure 1. Energy level diagram for ortho-H_2O.

excited H_2 molecules are also important. We note in particular that there exist several quasi-resonances between the H_2O and H_2 molecules which may lead to an exchange of excitation energy [4]. The most important one seems to be between the ($J = 2 \rightarrow 0$)-transition in the H_2-molecule and the ($5_{23} \rightarrow 7_{25}$)-transition in the H_2O-molecule, since there is a direct optical transition from the 7_{25}-state to the upper maser level.

In order to study the influence of this quasi-resonance, we assumed in the second set of computations, for the resonant excitation coefficient at T_{kin} = 500 K the value (in cgs-units)

$$C(7_{25} \leftarrow 5_{23}) = 3 \cdot 10^{-9} \, n_{H_2} \qquad (1)$$

which is about a factor of 400 larger than the value given by Green. All other molecular data were kept as before. Figure 3 shows the results. The region of inversion is wider than that in figure 2, and the degree of inversion is enhanced, but the general structure of the diagram is similar.

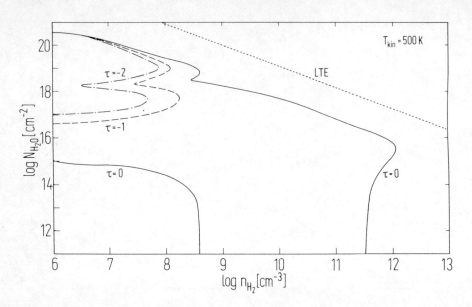

Figure 2. (N_{H_2O}, n_{H_2})-diagram showing the region of inversion for the 1.35 cm line, computed for $T_{kin} = 500$ K with collision rates according to Green [3] and a line width $\Delta\nu/\nu = 10^{-5}$. The curves represent the lines $\tau = 0$, $\tau = -1$, and $\tau = -2$. Above the curve labeled LTE, the occupation numbers of all levels up to the maser levels deviate by less than 0.01% from their LTE values.

In the third set, besides the quasi-resonance, we accounted for the influence of cold dust assumed to be olivine particles. The effect of cold dust is to absorb IR photons out of the spectral lines and to emit the energy at other wave lengths. As can be seen from figure 4, the region of strong maser emission now extends to high densities. This shows that the combined effects of resonant collisions and cold dust constitute a very efficient pumping mechanism leading to models of rather compact and dense H_2O maser sources.

If we choose a model with $N_{H_2O} = 5 \cdot 10^{19}$ cm^{-2}, $n_{H_2} = 3 \cdot 10^9$ cm^{-3}, radius of dust particles $r_d = 0.1\,\mu$, and $n_d/n_{H_2O} = 10^{-6}$, we find $T_B \approx 5 \cdot 10^8$ K. If we assume $n_{H_2O}/n_{H_2} \approx 5 \cdot 10^{-6}$, our parameters imply $N_{H_2} = 10^{25}$ cm^{-2}, a dust to gas ratio of about 2% by mass, and a linear dimension of $3 \cdot 10^{15}$ cm. With this size the total luminosity in the 1.35 cm line is about $3 \cdot 10^{30}$ erg/sec.

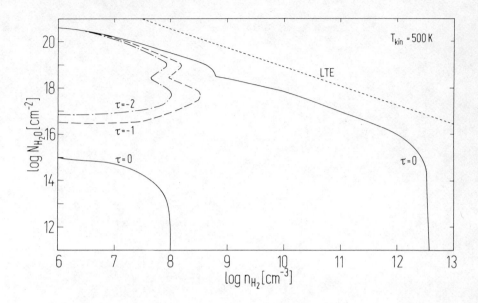

Figure 3. Same as Fig. 2, but with resonant collisions according to equation (1).

The linear dimension and the luminosity of our numerical model are in fair agreement with the observations, if we consider the "centers of activity" as individual objects. The computed brightness temperature is considerably lower than the observed values. This is not surprising because the observed bright spots are much smaller in size than the "centers of activity" and may correspond to inhomogeneities not taken into account in our computations.

PARA-H_2O

In the para-system of H_2O there are two potential maser lines at 1.635 mm and at 922 μ, corresponding to the $(3_{13} \rightarrow 2_{20})$- and the $(5_{15} \rightarrow 4_{22})$-transition, respectively.

In our computations we took into account 26 energy levels connected by 71 radiative transitions. Since the tables of Green [3] include only the ten lowest levels of para-H_2O, we assumed for the non-available deexcitation rate constants

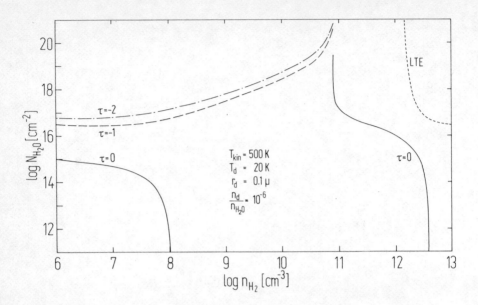

Figure 4. Same as Fig. 3, but taking also into account the influence of cold dust (olivine particles of radius $r_d = 0.1\,\mu$) with $n_d/n_{H_2O} = 10^{-6}$.

$$C(i \leftarrow k) = 10^{-12}\, g_i n_{H_2} \qquad (2)$$

at $T_{kin} = 200$ K. We note that also pumping of the 1.635 mm line may strongly be influenced by a quasi-resonance, the one between the $(2_{20} - 4_{40})$-transition in the H_2O-molecule and the $(J = 2 \rightarrow 0)$-transition in the H_2-molecule. Similar to the case of the 1.35 cm line, we find the largest intensity for the 1.635 mm line in models taking into account the effects of resonant collisions and cold dust. With $T_{kin} = 200$ K, $n_{H_2} = 10^8 \mathrm{cm}^{-3}$, $N_{H_2O} = 3 \cdot 10^{19} \mathrm{cm}^{-2}$, and $n_d/n_{H_2O} = 3 \cdot 10^{-6}$ we find $T_B \cong 1.7 \cdot 10^6$ K for the 1.635 mm line and $T_B = 2.6 \cdot 10^5$ K for the 922 μ line. This shows that maser emission in these two lines is to be expected from regions of star formation under conditions somewhat less extreme than those for the 1.35 cm masers.

The 1.635 mm line has been observed in Orion by Waters et al. [5]. The spatial resolution, however, was not sufficient to determine the linear dimensions of the emission regions and its brightness temperature.

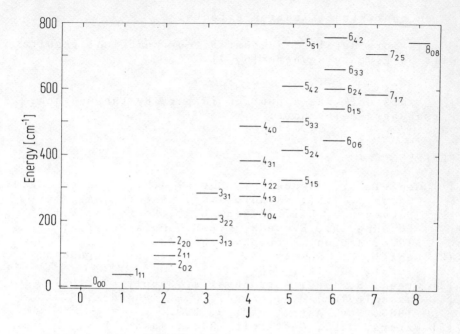

Figure 5. Energy level diagram for para-H_2O

CONCLUSIONS

From our numerical computations it becomes apparent that

a) radiative transfer effects are of importance for the understanding of H_2O masers;

b) taking into account effects of resonant collisions and cold dust leads to models of compact and dense H_2O masers, as required by observation;

c) maser emission is to be expected also in other lines of the H_2O-molecule.

For a more detailed interpretation of the observations progress has to be made into two directions:

a) We need more and better collision rates, including collisions with rotationally excited H_2 molecules.

b) Models have to be constructed in which the radiative transfer is treated more properly taking into account stratification effects.

A more detailed account of our numerical results will be given elsewhere [6, 7].

This work was supported in part by the Deutsche Forschungsgemeinschaft.

REFERENCES

[1] Genzel, R., Downes, D., Moran, J.M., Johnston, K.J., Spencer, J.H., Walker, R.C., Haschick, A., Matveyenko, L.I., Kogan, L.R., Kostenko, V.I., Rönnäng, B., Rydbeck, O.E.H., Moiseev, I.G. 1978, Astron. Astrophys. 66, 13.
[2] Genzel, R., Downes, D. 1977, Astron. Astrophys. Suppl. 30, 145.
[3] Green, S. 1980, Astrophys. J. Suppl. 42, 103.
[4] Varshalovich, D.A., Kegel, W.H., Chandra, S. 1983, Sov. Astr. Lett. 9, 209.
[5] Waters, J.W., Gustincic, J.J., Kakar, R.K., Kuiper, T.B.H., Roscoe, H.K., Swanson, P.N. 1980, Astrophys. J. 235, 57.
[6] Chandra, S., Kegel, W.H., Varshalovich, D.A., Albrecht, M.A. 1984, Astron. Astrophys. in press.
[7] Chandra, S., Kegel, W.H., Varshalovich, D.A. in preparation.

THE EXCITATION OF INTERSTELLAR C_2

Ewine F. van Dishoeck

Sterrewacht Leiden
Postbus 9513
2300 RA Leiden, The Netherlands *

The excitation of the C_2 molecule under interstellar conditions is described and its use as a diagnostic probe of the physical conditions in interstellar clouds is discussed.

1. THEORY

The C_2 molecule is, just as the H_2 molecule, particularly suitable as a diagnostic probe of the physical conditions prevailing in interstellar clouds because it has no permanent dipole moment (1). Its excited rotational levels are therefore long-lived and can be populated significantly through both collisional and radiative processes. It has the advantage over H_2 that some of its absorption lines lie in the red part of the spectrum and are thus accessible from Earth. Also, it can be observed in thicker clouds than can H_2.

A theory has been developed to describe the enhanced rotational populations of interstellar C_2 (2,3). The initial step in the mechanism consists of absorption of photons from the interstellar radiation field into excited electronic states, primarily the $A^1\Pi_u$ state in the 800-1200 nm region. The absorptions are followed by spontaneous emission back into vibration-rotation levels of the ground state, which may subsequently cascade down into rotational levels of the lowest vibrational state through quadrupole- and intercombination transitions. These radiative processes, governed by the strength of the interstellar radiation field in the (infra) red part of the spectrum, compete with collisional (de-) excitation processes, governed by the interstellar temperature and density, in establishing the steady-state population of the rotational levels.

* Present address: Center for Astrophysics, Cambridge Ma 02138.

G. H. F. Diercksen et al. (eds.), Molecular Astrophysics, 681–682.

In order to determine the probabilities of the radiative processes, the quadrupole moment of the ground state of C_2 has been determined (3), and oscillator strengths of the C_2 Phillips system have been calculated (4) using quantum chemical methods. The collisional cross sections are a major uncertainty in the analysis. Calculated rotational populations have been presented for a range of physical conditions (3, see 5 for more extensive tables). The C_2 rotational excitation temperature is found to be generally <u>higher</u> than the kinetic temperature. The use of the C_2 rotational population as a diagnostic probe of the local temperature, density and strength of the radiation field has been illustrated, although all three parameters cannot usually be determined from C_2 data alone (3,5).

2. COMPARISON WITH OBSERVATIONS

Absorption line observations of C_2 in the (2,0) Phillips band have been performed toward a few bright stars in the southern sky, using the Coudé echelle spectrograph on the 1.4 m CAT telescope at ESO (6). Seventeen lines originating from the lowest eight rotational levels have been detected toward χ Oph, and eleven lines originating from the lowest five levels toward HD 154368 and HD 147889. The theory has been used to extract information about the kinetic temperature T and density n in the clouds. Together with a reinterpretation of the observed rotational populations of H_2 (5), the C_2 data indicate that toward χ Oph, the intensity of the radiation field is enhanced in the UV part of the spectrum relative to the IR part. The C_2 data suggest a higher temperature for the material in front of HD 147889 than is inferred from radio observations. The theory has also been used to interpret observations of C_2 toward HD 29647, a highly-reddened B star which lies behind a substantial part of the Taurus molecular cloud complex (7). All observations of C_2 to date, and their interpretation, have been summarized in (5). From the C_2 data, it appears that the physical conditions in six of the nine clouds studied are very similar (T$\approx$30 K, n=n(H)+n(H_2)$\approx$200 cm^{-3}). The results have been compared with those found from the analysis of the rotational populations of H_2 (5). For the ζ Oph and o Per clouds, the H_2 data indicate higher central densities than the C_2 data.

REFERENCES
1. Black, J.H. 1985, this volume.
2. Chaffee, F.H., Lutz, B.L., Black, J.H., Vanden Bout, P.A. and Snell, R.L. 1980, Astrophys. J. 236, pp. 474-480.
3. van Dishoeck, E.F. and Black, J.H. 1982, Astrophys. J. 258, pp. 533-547.
4. van Dishoeck, E.F. 1983, Chem. Phys. 77, pp. 277-286.
5. van Dishoeck, E.F. 1984, Ph.D. Thesis, University of Leiden.
6. van Dishoeck, E.F. and de Zeeuw, P.T. 1984. Mon. Not. R. Astron. Soc. 206, pp. 383-406.
7. Hobbs, L.M., Black, J.H., and van Dishoeck, E.F. 1983, Astrophys. J. 271, pp. L95-L99.

PHOTODISSOCIATION PROCESSES IN THE OH AND HCL MOLECULES

Ewine F. van Dishoeck

Sterrewacht Leiden
Postbus 9513
2300 RA Leiden, The Netherlands *

The photodissociation processes in the OH and HCl molecules are discussed and their importance in various astrophysical environments is demonstrated.

1. PHOTODISSOCIATION CHANNELS OF OH

A systematic investigation of the excited states that may participate in the photodissociation of the OH molecule has been carried out by ab initio self-consistent-field plus configuration-interaction methods (1). Potential energy curves for states of $^2\Sigma^+$, $^2\Sigma^-$, $^2\Pi$ and $^2\Delta$ symmetries have been computed (see Figure 1) and the electric dipole transition moments connecting the states have been obtained. Photodissociation cross sections for absorption from the v"=0 vibrational level of the ground $X^2\Pi$ state into the repulsive $1^2\Sigma^-$, $1^2\Delta$ and $B^2\Sigma^+$ states have been reported.

Photodissociation may take place also following absorption into the repulsive $2^2\Pi$ and bound $3^2\Pi$ states, which are coupled by the action of the nuclear kinetic energy operator (2). The theoretical description of this process has been studied in detail in an adiabatic and a diabatic formulation. The calculated cross section for absorption into the coupled $^2\Pi$ states shows a series of resonances superposed on a broad continuous background. The resonances are located near to the vibrational levels of the bound diabatic potential curve and they have asymmetric Beutler-Fano profiles.

* Present address: Center for Astrophysics, Cambridge Ma 02138.

G. H. F. Diercksen et al. (eds.), Molecular Astrophysics, 683–686.

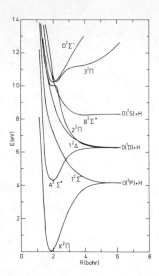 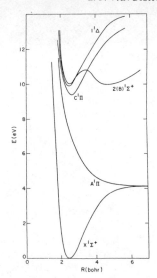

Fig. 1. Potential energy curves for OH (left) and HCl (right).

2. ASTROPHYSICAL APPLICATIONS

The contributions of the different channels to the total photodissociation rate of the molecule in a particular astrophysical environment are determined not only by the magnitudes of the cross sections, but also by the dependence on wavelength of the intensity of the radiation field. Since the wavelength dependence may vary from one region of interstellar space to another, different channels may dominate the photodissociation in different environments.

2.1 Photodissociation rates in interstellar clouds

The intensity of the unattenuated interstellar radiation field decreases with shorter wavelengths, so that lower-lying channels are relatively more important than higher-lying channels. In optically thin clouds, the photodissociation of OH is dominated by absorption into the coupled $3^2\Pi-2^2\Pi$ states and the $1^2\Sigma^-$ and $1^2\Delta$ states. The total rate is about 4×10^{-10} s^{-1} (1). Deeper into the clouds, the rates are diminished due to the attenuation of the radiation field by grains. The depth dependences of the rates have been calculated for three different assumptions about the scattering properties of the grains (3). The effectiveness of the different channels changes with depth into a cloud. Absorption into the $1^2\Sigma^-$ state around 157 nm is usually the main dissociation channel of OH, but with forward scattering grains in clouds of small optical depths, the higher-lying coupled $3^2\Pi-2^2\Pi$ channel around 104 nm is also very effective. For forward scattering grains, photodissociation remains significant into large optical depths, $A_V \approx 10$ mag. Self-shielding is unimportant for OH. Previous calculations of the OH interstellar photodissociation rate have assumed that photodissociation proceeds by absorption into the $C^2\Sigma^+$ state for which a large oscillator strength had been adopted (4). We find that the C state contributes negligibly and that the new rates are larger by a factor which increases from five at small optical depths, to between 10 and 100 at larger depths.

In interstellar regions subjected to fast shocks, the radiation field has a high intensity at the Lyman-alpha wavelength. The calculations demonstrate that OH is dissociated by absorption of Lyman-alpha radiation into the $1^2\Delta$, $B^2\Sigma^+$ and coupled $3^2\Pi-2^2\Pi$ states (3).

2.2 The abundance of OH in diffuse clouds

Using the new rates, column densities of OH have been calculated for several lines of sight. To this end, new models have been constructed of the clouds in front of ζ Per and ζ Oph (5). The new models, which contain an improved treatment of the radiative transfer in the H_2 lines, have central densities of about 325 and 700 cm^{-3}, and central temperatures of about 30 and 25 K, respectively.

The abundance of OH is directly proportional to the cosmic ray ionization rate ζ. The observed OH abundance may thus be used to infer ζ (6). The new calculations of the OH photodissociation rate remove one of the main uncertainties in the analysis. For the ζ Per cloud, $\zeta \approx (1-2) \times 10^{-16}$ s^{-1} has been derived; for the ζ Oph cloud, $\zeta \approx 4 \times 10^{-16}$ s^{-1} (5). Since the analysis assumed that no OH molecules are formed in a possible shocked region of the cloud, the inferred values of ζ are upper limits. Together with the observed HD column density, a deuterium abundance of about 1×10^{-5} has been determined for the ζ Per cloud, and 5×10^{-6} for the ζ Oph cloud.

2.3 The photodissociation of OH and OD in comets

In cometary atmospheres, the OH radical is photodissociated by the solar radiation field. The intensity of the solar radiation field decreases rapidly toward shorter wavelengths and predissociation of the lowest-lying $A^2\Sigma^+$ state, which contributes negligibly to the total rate in interstellar clouds, provides an important pathway in comets. Because the dissociation is stimulated by line absorption in a wavelength region where the solar spectrum is rapidly varying, the corresponding destruction rate of OH depends on the heliocentric radial velocity of the comet (7,8). Continuous absorption into the repulsive $1^2\Sigma^-$ state is also a major photodissociation path (8). The rate for this channel does not vary with velocity, but it changes significantly between solar maximum and minimum. Photodissociation also occurs by absorption of Lyman-alpha radiation into the $1^2\Delta$, $B^2\Sigma^+$ and coupled $3^2\Pi-2^2\Pi$ states. The total lifetimes of OH have been calculated for cometary velocities ranging from -60 to +60 kms^{-1}, both for solar maximum and minimum conditions. The calculations are generally consistent with observations. Photodissociation of OH produces a low velocity component of H atoms and a high velocity component. Their relative production varies through the solar cycle. Photodissociation of OH also leads to metastable $O(^1D)$ and $O(^1S)$

atoms and is an additional source of the red and green line emission of atomic oxygen (8).

The $A^2\Sigma^+$ channel has been shown to be unimportant to the photodissociation of the OD radical leading to a relative enhancement of the OD/OH ratio in comets by a factor between two and three (8).

3. PHOTODISSOCIATION PROCESSES IN THE HCL MOLECULE

A similar study has been performed for the HCl molecule (9). Potential energy curves have been calculated for selected singlet (see Figure 1) and triplet states, and the electric dipole transition moments between the singlet states have been obtained. Direct photodissociation of HCl occurs by absorption from the ground $X^1\Sigma^+$ state into the repulsive $A^1\Pi$ state. Since also experimental cross sections are available for this channel, a bona fide comparison with calculated cross sections can be made for the first time for a neutral diatomic molecule. The agreement is very satisfactory. Photodissociation of HCl takes place also by absorption into the bound $C^1\Pi$ state, followed by predissociation by spin-orbit coupling with 100% efficiency. The theoretical oscillator strengths for the C-X transition are in reasonable agreement with experiment.

The photodissociation rate of HCl in diffuse interstellar clouds has been computed (5,9), and is a factor of three to four larger than the rate which was used in previous studies of interstellar HCl. The enhanced rate diminishes substantially the discrepancy between the theoretical and observed abundances of interstellar HCl, and in conjunction with other modifications to the chemistry may remove it (5).

REFERENCES

1. van Dishoeck, E.F. and Dalgarno, A. 1983, J. Chem. Phys. 79, pp. 873-888; van Dishoeck, E.F., Langhoff, S.R., and Dalgarno, A. 1983, J. Chem. Phys. 78, pp. 4552-4561.
2. van Dishoeck, E.F., van Hemert, M.C., Allison, A.C. and Dalgarno, A. 1984, J. Chem. Phys. in press.
3. van Dishoeck, E.F. and Dalgarno, A. 1984, Astrophys. J. 277, pp. 576-580.
4. Smith, W.H. and Stella, G. 1975, J. Chem. Phys. 63, pp. 2395-2397.
5. van Dishoeck, E.F. 1984, Ph. D. Thesis, University of Leiden.
6. Hartquist, T.W., Black, J.H., and Dalgarno, A. 1978, Mon. Not. R. Astron. Soc. 185, pp. 643-646.
7. Jackson, W.M. 1980, Icarus 41, pp. 147-152.
8. van Dishoeck, E.F. and Dalgarno, A. 1984, Icarus in press; Singh, P.D., van Dishoeck, E.F. and Dalgarno, A. 1983, Icarus 56, pp. 184-189.
9. van Dishoeck, E.F., van Hemert, M.C. and Dalgarno, A. 1982, J. Chem. Phys. 77, pp. 3693-3702.

LOW TEMPERATURE ROSSELAND MEAN OPACITIES FOR ZERO-METAL GAS MIXTURES.

Francesco PALLA

Osservatorio Astrofisico di Arcetri
Florence (Italy)

ABSTRACT

Rosseland mean opacities for a gas of primeval chemical composition are tabulated for a temperature range from 1500K to 7000K. Molecular effects at low temperatures (below 5000K) have been included and their important contribution to the Rosseland mean opacity is discussed. Also, a comparison with previously published tables is presented.

INTRODUCTION

In the recent literature there has been a renewed interest on the topic of the fate of the primeval gas following the recombination time, particularly on the efficiency of molecule formation in these rather peculiar conditions (see the introductory paper by A.Dalgarno, this volume).

A prediction of the standard theory of the Big Bang is that the primordial gas, out of which the first stellar objects have formed (the so-called Population III), did not contain any element heavier than helium, with the only exception of the trace elements Li, Be, and B (see, for example, Audouze 1982). However, due to the very low abundance of these species, the gas mixture can be treated to a very good approximation as consisting of pure hydrogen and helium in their neutral, ionic and molecular components and as lacking of any solid particle (i.e.,

G. H. F. Diercksen et al. (eds.), Molecular Astrophysics, 687–693.

dust grains). This makes the investigation of all possible
absorption and scattering transitions of the gas components
rather simple, and the calculation of the mean opacities does
not require the many complications and uncertainties introduced
by the presence of heavy elements and dust grains, characteristic of the composition of the present-day interstellar medium
(Alexander et al. 1983).

The determination of a correct opacity coefficient at low
temperatures is an obvious input requirement in all calculations
of protostellar collapse of gas clouds, or of the pre-main-sequence of hydrostatic objects in the accretion phase, or of the
envelopes of stars of low effective temperatures. In the framework of the Population III stars, although little observational
evidence to support their existence has been presented so far,
it is quite important to have estimates of the opacity as much
accurate as possible, especially if one attempts to present
theoretical HR diagrams or to evaluate observable properties of
these objects.

Among the many aspects of the problem of the primordial star
formation, the study of the structure and evolution of zero-metal stars has received much attention and a number of model
calculations, which cover the entire life and the mass spectrum,
have been published (Castellani et al. 1983; Guenther and Demarque
1983; Stahler et al. 1984; and Chiosi 1983, for an excellent review). However, in discussing the opacity coefficient most
authors have made extensive use of the data published by Cox and
Tabor (1976, hereafter CT), who, as a part of a large sample of
stellar mixtures, considered also the case of a pure H-He gas.
Although these tables greatly improved and extended previous computations of the Los Alamos group, they did not include the
effects of molecular absorption and scattering, thus limiting
their validity at the low temperatures where molecules form and
eventually dominate the opacity.

It is the aim of this contribution to present data on the
Rosseland mean opacities which take into account the presence
and the contribution of molecular hydrogen, in its neutral and
ionic forms, to the total absorption coefficient and to point
out the difference with previous results. The opacities are
given for a temperature range 1500K to 7000K and for a wide

interval of densities (10^{-15} to 10^{-4} g cm^{-3}).

RESULTS AND DISCUSSION

The Rosseland mean of the opacity coefficient is defined as:

$$K_R = \frac{1}{K} = \frac{\int_0^\infty (\frac{1}{K_\nu})(\frac{\partial B_\nu(T)}{\partial T})\, d\nu}{\int_0^\infty (\frac{\partial B_\nu(T)}{\partial T})\, d\nu} \qquad (1)$$

where, as it is well known, the term $(1/k_\nu)$ in (1) represents the sum of all the possible sources of true absorption and scattering and it depends on the constituents of the mixture considered. A good reference list of the principal processes that must be taken into account when computing the continuous or quasi-continuous opacities can be found in Kurucz (1970). Given the composition of the primordial gas, we specifically included the following processes:

(1) Rayleigh scattering by HI and HeI atoms and H_2 molecules;

(2) electron scattering;

(3) bound-free and free-free absorption by HI atoms and H^- and H_2^+ ions;

(4) free-free absorption by H_2^- and He^- ions;

(5) pressure-induced absorption by H_2 molecules due to collisions between each other and/or H_e atoms, in the form given by Linsky (1969, 1983).

Since the last process turned out to be extremely important, we will discuss it at length later. Electron-conduction opacity does not contribute to the total opacity in the range of temperatures and densities of interest here (Hubbard and Lampe 1969).

The opacity calculations we present here refer to a gas mixture such that the fraction by mass of the components is $X = 0.72$ and $Y = 0.28$ where, as usual, X stands for hydrogen and

Y for helium, respectively. This composition will be referred
to as Standard Mixture. Note that the Standard Mixture differs
from that of CT and called Paczynski III, with X = 0.70, Y = 0.30,
and Z = 0.00. However, the difference in X is only marginal and
still allows for a direct comparison between the two set of data.
In the rest of the discussion we focus the attention only on the
properties of the opacity of the Standard Mixture, while the
complete set of opacity calculations for several chemical compositions is given elsewhere (Stahler et al. 1984).

Since the Rosseland mean is non-additive, it is not straightforward to single out the process responsible for the bulk of the
opacity, unless a comparison between two calculations done with
and without a given source can be made. In our case, since the
main difference with the CT paper is the inclusion of the molecular effects, we can safely attribute to them the possible discrepancies.

Trafton (1964,1966) was the first to point out the potential
role of the pressure- (or collision-) induced absorption of
infrared radiation in the atmospheres of the major planets and to
give analytical expressions for the frequency dependence of the
translational absorption coefficient for a pure H_2 gas at very
low temperatures (T<600K). Linsky (1969) extended Trafton's
calculations over a larger temperature range and considered also
the case of binary collision of H_2 molecules with He atoms. Also,
in addition to the translational transitions he included the
rotation-translational bands, the vibration-rotation-translational
bands, and the overtone and double vibration bands for H_2-H_2 collisions, and all but the last for H_2-He interactions. His functional forms for the various monochromatic absorption coefficients
have been adopted in the present calculations. Their validity
is restricted in the temperature interval 600K to 4000K, with the
exception of the vibration bands excited by H_2-H_2 collisions for
which Patch (1971) gives analytic expressions valid up to 7000K.
In these papers the absorption coefficient is expressed in units
of cm^6 s^{-1}; to have it in cm^2 g^{-1} it must be multiplied by the
factor:

$$k_\nu (P.I.) = k_\omega \frac{n_i n_j}{c\rho} (1 - e^{-h\nu/kT}) \, cm^2 \, g^{-1}$$

where k_ω is the monochromatic absorption coefficient, n_i, n_j are the number densities of the colliding particles, c is the speed of light, ρ is the total mass density, and the factor in parenthesis is the stimulated emission.

The numerical values of the Rosseland mean opacity, k_R are plotted in figure 1 as a function of the temperature, at four different densities. The effect of the molecular absorption can be readily appreciated: as the density increases the dissociation of the molecules requires higher temperatures and the equilibrium abundance favors the molecular component at temperatures below 4000K. Since for H_2-H_2 collisions the pressure-induced absorption is proportional to the square of the molecular number density, it is clear that its role is expected to become more and more important as the density inceeases. Indeed, from figure 1 we see

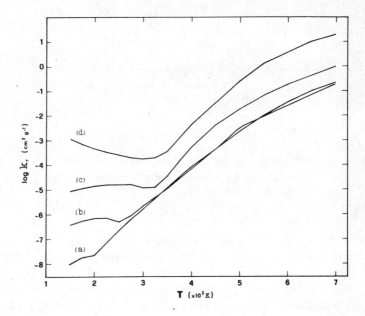

Figure 1: Run of the Rosseland mean opacity vs. temperature at four different densities:
(a)=10^{-12} ; (b)=10^{-9} ; (c)=10^{-7} ; (d)=10^{-4}.

how effectively the curve of k_R are leveled off after a rapid decrease and how, in the case of very high density, the Rosseland mean opacity still rises at the lowest temperatures.

But the net effect of the molecular absorption is better appreciated in figure 2, where we compare our results to those of CT, again for the Standard Mixture.

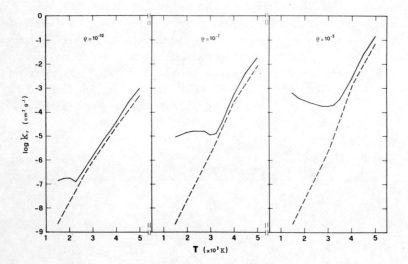

Figure 2: Comparison of the Rosseland mean opacities computed here (solid lines) and CT's (dashed lines) at three different densities.

From CT we notice that at T<5000K the decrease of k_R is very steep and that the curves have the same slope independently of the density. On the other hand, the inclusion of the molecules has the dramatic effect of increasing the absolute values of k_R over k_R(CT) by orders of magnitude. Also, as expected from simple equilibrium abundance considerations, the opacity is a strong function of the density, and its departure from CT's values becomes more evident.

Given the large effects shown in the figures, it is clear that the determination of the physical quantities which require an estimate of the absorption coefficient is strongly affected by the inclusion of molecular processes, which can lead to serious undestimates if not taken properly into account. A detailed study of the impact of a better treatment of the opacity in the case of the evolution of protostars of zero metal composition is presented in another paper (Stahler et al. 1984).

This work has been done in collaboration with Dr. S. Stahler of the Center for Astrophysics, Cambridge (USA).

REFERENCES

Alexander, D.R., Johnson, H.R., and Rypma, R.L. 1983, Ap. J. 272, pp. 773-780.

Audouze, J. 1982, "Astrophysical Cosmology", ed. H.A. Bruck, G.V. Coyne, and M.S. Longair, Pontificia Academia Scientiarum, pp. 395-422.

Castellani, V., Chieffi, A., and Tornambè, A. 1983, Ap. J. 272, pp. 249-255.

Chiosi, C. 1983, Proceedings of the Frascati Workshop 1982 on "The First Stellar Generations", Mem. S. A. It. 54, pp.251-275.

Cox, A.N., and Tabor, J.E. 1976, Ap. J. Suppl. Series 31, pp.271-312.

Guenther, B., and Demarque, P. 1982, Astron. Astrophys. 118, pp.262-266.

Hubbard, W.B., and Lampe, M. 1969, Ap. J. Suppl.Series 18, pp.297-346.

Kurucz, R.K. 1970, Smithsonian Astr. Obs. Spec. Report 309.

Linsky, J.L. 1969, Ap. J. 156, pp. 989-1005.

------ 1983, private communication.

Patch, R.W. 1971, J. Quant. Specrtosc. Radiat. Transfer 11? pp. 1331-1353.

Stahler, S.W., Palla, F., and Salpeter, E.E. 1984, in preparation.

Trafton, L.M. 1964, Ap. J. 140, pp. 1340-1341.

------- 1966, Ap. J. 146, pp. 558-571.

CONTRIBUTED ORAL PRESENTATIONS

A HIGH SENSITIVITY SPECTRAL SURVEY OF ORION A IN THE 1.3mm REGION

G. A. Blake
320-47 Caltech
Pasadena, CA 91125, U.S.A.

Abstract

Results from a high resolution spectal line survey of the Orion molecular cloud conducted with the #1 10.4 m telescope of the Owens Valley Radio Observatory in the frequency ranges 216-246 and 254-264 GHz are presented. Over 500 emission lines from 28 molecules in 14 isotopic variations have been detected and identified. The great wealth of data obtained is providing new clues to the nature of both the physical and chemical properties of this well studied star formation region.

For example, it is becoming increasingly clear that a small number of molecules are responsible for an inordinate share of the observed interstellar emission lines. In our survey emission from the five terrestrially stable molecules SO_2, CH_3OH, $(CH_3)_2O$, HCO_2CH_3, and C_2H_5CN easily accounts for over half of the observed transitions, including many initially unidentified lines. Structural isomers of these abundant species have not been detected, nor have several promising interstellar candidates. Several earlier tentative detections have not been confirmed by the spectral line survey. Most importantly, emission originally ascribed to CO^+ has now been identified as $^{13}CH_3OH$, resolving one of the greatest discrepancies between theoretical chemical models of molecular clouds and the observational data. The small number of unidentified lines present implies that the important *heavy* constituents of Orion have been determined.

The high spatial resolution employed has shown the Orion cloud to be extremely clumpy, with several distinct regions contributing to the overall emission. Particularly notable are the warm, high density regions traced out by the strong emission detected from a number of vibrationally excited species. Remarkable chemical variations are found between the distinct regions, which presumably reflect the great variety of physical conditions prevailing in Orion.

Finally, the large contiguous frequency ranges covered have allowed us to estimate the contribution of integrated molecular line emission to the measured broadband millimeterwave flux. The derived molecular flux and spectral index is similar to the total observed broadband flux, arguing that estimates of the emission from dust in Orion must be revised sharply downward.

THE IMPORTANCE OF NEUTRAL-NEUTRAL REACTIONS IN THE CHEMISTRY OF CN IN DIFFUSE CLOUDS

Steven Federman
Jet Propulsion Laboratory, M.S. 183-601
4800 Oak Grove Drive
Pasadena, CA 91109, U.S.A.

Abstract

Previous observational studies of the chemistry in diffuse interstellar clouds of diatomic molecules incorporating carbon atoms revealed the importance of ion-molecule reactions. In particular, the measurements of CH and CO are reproduced quite well when ion-molecule reactions form the basis of theoretical chemical models. Although the data for C_2 is sparse, the abundance of this molecule most likely also depends on ion-molecule reactions. When the chemistry of CN is scrutinized in the same way, it is found that neutral-neutral reactions play a key role. The CN molecule probably is produced by reactions between C_2 and N, with reactions between CH and N of some importance. Thus, we stress, that in addition to the need for precise rate constants for ion-molecule (and radiation association, etc.) reactions, more work on the rate constants for neutral-neutral reactions of astrophysical importance is required.

LONG CHAIN MOLECULES IN INTERSTELLAR CLOUDS

H. W. Kroto
School of Chemistry and Molecular Sciences
University of Sussex
Brighton BN1 9Qj, U.K.

Abstract

A combined programme of laboratory synthesis/microwave study and astrophysical radio searches on molecules chemically, closely related to the cyanopoly-ynes has been initiated in an attempt to explain the implications of the unexpected abundance of the long chain species in the ISM. The way in which these and related molecules can shed light on interstellar synthesis is discussed.

DESORPTION OF MOLECULES FROM INTERSTELLAR GRAINS

A. Leger, M. Jura, A. Omont
Groupe d'Astrophysique, CERMO
B.P. 68, 38402 Saint-Martin d'Heres Cedex, France

Abstract

Desorption due to impulsive heating by X-rays and cosmic rays is reconsidered both in the classical thermal evaporation picture and in chemical explosion models, the results of whole grain heating as well as desorption by local hot spot are evaluated.

The effects of rare heavy cosmic rays dominate except for the smallest grains. Whole grain heating can prevent important depletion of the gas in moderately dense regions. Spot heating cannot prevent a substantial depletion but it is active even on very large grains and in very dense regions. Over all we find that these mechanisms can explain the observations of gaseous interstellar molecules.

MOLECULAR MANTLES ON INTERSTELLAR GRAINS

D. A. Williams
Dept. of Mathematics, UMIST
P.O. Box 88
Manchester M60 1QD, U.K.

Abstract

The observation of the 3.08 μm ice band towards several stars in Taurus is shown to require an efficient surface chemistry leading to the formation of hydrides. The evaporation of molecular mantles from warm grains may be the origin of the very high abundances of NH_3 in some localized regions, for example IRC2 in Orion. Evidence will be presented that shocks may periodically return molecular mantle material to the gas phase. This material may be returned partly as molecules, and partly dissociated into atoms.

ELECTRON CAPTURE BY MULTIPLE CHARGED IONS; POPULATION OF EXCITED STATES

R. McCarroll
Observatoire de l'Université de Bordeaux I
33270 Floirac, France

Abstract

Both the abundance and recombination spectra of certain multiply charged ions in many astrophysical plasmas - planetary and gaseous nebulae, extended atmospheres of hot stars - may be considerably modified by charge transfer processes. While the cross sections for charge transfer in the thermal - eV energy range can, in general, be reliably estimated by the quasi-molecular model, the cross sections for electron capture into a specific excited state are still subject to much uncertainty, arising from the inability of a finite basis of adiabatic (or diabatic) states to represent the translation. It is necessary to modify the basis set either by the introduction of translation factors, if a semi-classical method is used, or by the introduction of suitable reaction coordinates, if a quantum mechanical method is used to treat the collision dynamics. Examples of the problem and possible solutions will be discussed.

PHOTOABSORPTION AND PHOTODISSOCIATION OF CO IN THE 900-1200 Å REGION

F. Rostas
Observatoire de Paris-Meudon
Dept. d'Astrophysique Fondamentale
F-92190, Meudon Cedex, France

Abstract

CO is one of the most useful molecules for determining the characteristics of interstellar matter. Its chemistry in diffuse clouds is governed to a large extent by photodissociation which can occur between 912 Å (H ionization cutoff) and 1116 Å (first photodissociation threshold of CO). A recent radioastronomy study by Bally and Langer has shown anomalous relative abundances for the ^{12}CO and ^{13}CO molecules which could be explained by a self shielding model based on photodissociation in discrete levels which is expected to be isotope selective. In this context we have undertaken a complete reevaluation of the photodissociation processes in CO in the 900-1200 Å range.

The experiments undertaken involve:
-high resolution photographic absorption spectroscopy

-high resolution photoelectric measurement of absorption cross sections
-fluorescence yield measurements in the discrete absorption bands conducted in a windowless cell at the LURE synchrotron radiation facility of Orsay.

The rotational analysis of several bands which had not been previously analysed is in progress. In particular the band at 1029 Å observed by Ogawa and Ogawa can now be attributed with confidence to the $E\ ^1\Pi \leftarrow X\ ^1\Sigma$ (2-0) transition. At high resolution a number of bands exhibit a definitely diffuse character indicating a large predissociation probability. The fluorescence measurements carried out with synchrotron radiation indicate that a number of states, even though they are well resolved at high resolution, must be considerably predissociated.

PHOTODISSOCIATION OF H_2O IN THE FIRST ABSORPTION BAND IN THE
VUV AS THE PUMPMECHANISM OF THE ASTRONOMICAL OH-MASER

P. Andresen
MPI für Strömungsforschung
3400 Göttingen, F.R.G.

Abstract

The photodissociation of H_2O in the first absorption band in the VUV proceeds via the electronically excited $^1B_1$ state:

$$H_2O(^1A_1) + h\nu(140\text{-}185 \text{ nm}) \rightarrow H_2O(^1B_1) \rightarrow OH(^2\pi) + H(^2S).$$

H_2O in its electronic groundstate $^1A_1$ is excited to the $^1B_1$ state with the VUV radiation of an F_2 excimer laser at 157 nm. At this wavelength no other states of H_2O are involved and we have a rare case of direct photodissociation on a single potential surface. The products are exclusively OH and H in their electronic groundstate.

The nascent internal state distribution of OH is studied in great detail by Laser Induced Fluorescence (LIF) via the $OH(^2\pi\text{-}^2\Sigma)$ absorption band. Vibrational, rotational distributions are measured as well as the population in the electronic fine structure states (spin and Λ-doublet). The measurement are done both for rotationally cold H_2O (i.e. H_2O that is cooled in a nozzle expansion) and rotationally warm H_2O (i.e. H_2O at 300 K in a flow system). In addition, the alignment of the OH rotation plane relative to photoselected H_2O planes is determined in a polarization experiment.

Special emphasis has been given to the relative population of the Λ-doublet states in the product OH. For the rotationally cold H_2O we obtain a population inversion between Λ-doublet states for all rotational levels in the $^2\pi_{3/2}$ manifold. This population inversion is very high (∿ 20) at high rotational states. But even for the rotational groundstate a population inversion of ∿ 2 is found. The population inversion is explained qualitatively by an approximate symmetry conservation of the electronic wave function (with respect to the H-O-H plane) in a planar decay. The strong j-dependence is also understood quantitatively in terms of a simple one electron overlap model.

A very important consequence of this work is that the orientation effects of the unpaired π-electron have been treated wrong for a long time: for the upper Λ-doublet in the lowest quantum state of OH the unpaired -electron is preferentially oriented <u>perpendicular</u> to the OH-rotation plane. This is in contradiction to the assumptions of the collision theories and implies <u>that the</u>

collision theories yield anti-inversion. Collision theories cannot explain the astronomical OH-maser.

The explanation of the pumpmechanism of the astronomical OH-maser by photodissociation of H_2O is very propable because both VUV radiation and H_2O seems to be present in the regions of new born stars. The effect should also occur in comets and explain the OH-masers there. Possible pumpmechanisms for the $OH(^2\pi_{1/2})$, for the 1612 MHz-maser and the H_2O-maser will also be discussed according to the new point of view of the Λ-doublet physics of OH and planar dynamics.

LASER SPECTROSCOPY ON SMALL MOLECULES OF ASTROPHYSICAL INTEREST IN THE VUV

C. R. Vidal
MPI für Extraterrestrische Physik
8046 Garching, F.R.G.

Abstract

In recent years we have developed coherent vacuum uv sources for high resolution spectroscopy having a line width as small as 0.1 cm^{-1} and covering typically the spectral region from 100 to 200 nm (Vidal, 1980). These sources are presently tested for various applications in laser spectroscopy. The following methods have already been demonstrated:

1. Excitation spectroscopy of the NO molecule in the spectral range of 180-200 nm: The resolution exceeds that of existing absorption spectra.

2. Frequency selective excitation spectroscopy of the CO molecule in the spectral range of 140-150 nm: Intercombination bands which are strongly overlapped by the A-X system, have been measured and analyzed for the first time.

3. Fluorescence spectroscopy of the NO molecule: Pumping individual

levels of the $A^2\Sigma$ state the spectrally resolved fluorescence spectra provide Franck Condon factors. The J-dependence has been investigated for intermediate Hund coupling cases. The measurements are in satisfactory agreement with theoretical predictions (Earls, 1935).

4. Double resonance spectroscopy of the CO molecule: Pumping individual lines of the $A^1\Pi - X^1\Sigma^+$ system in the vacuum uv a second visible laser pumps the CO molecule to the $B^1\Sigma^+$, $C^1\Sigma^+$, $E^1\Pi$ or the $c^3\Pi$ state. Looking at the visible and infrared fluorescence double resonance spectroscopy between excited states has been carried out revealing new information on the predissociation and on the assignment of levels with large total angular momentum J.

References

Earls, L.T.: 1935, Phys. Rev. 48, 423.
Vidal, C.R.: 1980, Appl. Opt. 19, 3897.

EXPERIMENTAL MEASUREMENTS OF THE BRANCHING RATIO FOR DISSOCIATIVE RECOMBINATION PROCESSES

J. B. A. Mitchell
Dept. of Physics, University of Western Ontario
London, Ontario
Canada N6A 3K7

Abstract

When polyatomic ions undergo dissociative recombination, they can decay into a variety of exit channels. The relative importance of these channels is a vital parameter for interstellar chemistry modelling. Until recently however, very little information has been available concerning this topic.

A new technique has been developed using the MEIBE I merged beam apparatus at UWO which allows the cross sections for recombination into specific exit channels to be determined. Preliminary studies have been made for H_3^+ recombination (Mitchell et al. 1983) and work is continuing on this ion and its isotopic variants. These measurements and future plans will be discussed.

Mitchell, J.B.A., Forand, J.L., Ng, C.T., Levac, D.P., Mitchell, R.E., Mul, P.M., Claeys, W., Sen, A., and McGowan, J.Wm., Phys. Rev. Letters 51, 885, 1983.

ASTROPHYSICALLY IMPORTANT CHARGE TRANSFER REACTIONS, RECENT THEORETICAL RESULTS

T. G. Heil
Dept. of Physics, University of Georgia
Athens, GA 30606, U.S.A.

Abstract

The survey calculations of low energy charge transfer reactions between doubly and trebly charged first row ions and hydrogen or helium atoms, done in collaboration with S. Butler and D. Dalgarno, are being extended.

The agreement of the original calculations with available experimental results is good for trebly charged ion reactions but discrepancies are apparent for certain reactions involving doubly charged ions, particularly C^{2+}, N^{2+} and O^{2+}. Further analysis of these reactions, including more accurate molecular structure calculations and the consideration of the role of metastables in experimental measurements, has led to refined estimates of these reaction rates.

In addition, calculations have been done on reactions involving certain second and third row ions such as iron and aluminum.

Radiative charge transfer processes are also important in low temperature astrophysical plasmas. The detailed emission spectrum of this type of reaction is being calculated for possible use in the interpretation of observational data.

POTENTIAL SURFACE OF C_2-H_2

E. Roueff
Observatoire de Meudon
Dept. d'Astrophysique Fondamentale
F-92190 Meudon, France

Abstract

Calculations have been undertaken of the potential surface of the C_2-H_2 pair. The aim of this work is to calculate subsequently the rotational excitation of the C_2 molecule by H_2 in order to interpret the observations of several rotational levels of this molecule.

THEORETICAL PREDICTIONS OF ATOMIC AND MOLECULAR PROPERTIES
FOR THE CONSTRUCTION OF EMPIRICAL SCATTERING POTENTIALS

A. J. Sadlej
Chemical Centre, University of Lund
Box 740, S-22007 Lund 7, Sweden

Abstract

The analysis and interpretation of atomic and/or molecular scattering data requires the knowledge of fairly accurate interaction potentials. Reliable non-empirical calculations of multidimensional energy hypersurfaces for interacting species are currently limited to relatively small systems and no major steps forward are expected to be made in this respect in the nearest future. For this reason a considerable attention is given to the construction of semi-empirical interaction potentials based on either experimental or theoretical data for isolated subsystems.

For most atoms and molecules the relevant data, i.e., their multipole moments and multipole polarizabilities, are rather difficult to measure. Moreover, the accuracy of measured electric properties is usually quite low. Hence, the ab initio methods of quantum chemistry and molecular physics become frequently the only source of these parameters.

The accuracy of theoretically calculated values of atomic and molecular electric properties is also limited by a number of factors. According to the present experience the most important of them appears to be the appropriate choice of the basis set functions. Different proposals of the choice of basis set functions for calculations of atomic and molecular properties will be discussed and exemplified.

Another factor which limits the accuracy of the calculated data is the amount of correlation effects included in the given wave function. The correlationless calculations can only provide an estimate of the property value which may differ from the exact result by more than 10%. The major portion of the electron correlation contribution to atomic and molecular electric properties can be accounted for either by using the methods based on the configuration interaction approach or the perturbation techniques. The relative advantages and disadvantages of these methods will be compared and illustrated.

ROVIBRATIONAL EXCITATION OF $^{12}C\,^{16}O$ BY H_2

D. R. Flower
Physics Dept., Durham University
South Road
Durham DH1 3LE, U.K.

Abstract

Cross-sections for the excitation of $^{12}C\,^{16}O$ by H_2 to the $v = 1$ and $v = 2$ vibrational manifolds have been calculated on the basis of two ab initio interaction potentials. The calculations treat the vibrational motion of the CO within the quantum mechanical coupled channels formalism. The infinite order sudden (IOS) and coupled states approximations are applied to the rotational motions of the CO and H_2 molecules, respectively. Comparison of the results of these calculations with laboratory measurements leads to conclusions regarding the accuracy of both the ab initio potential energy surfaces and the dynamical approximations.

The extension of this work to treat the excitation of $^{12}C\,^{16}O$ by HD is currently being considered, as this would enable further comparisons with experimental data to be made.

LIST OF PARTICIPANTS

Adams, N. G.
Dept. of Space Research
University of Birmingham
P.O. Box 363
EDGBASTON
Birmingham B15 2TT
England

Andresen, P.
MPI für Strömungsforschung
Böttingerstr. 4-8
D-3400 Göttingen
FRG

Bergh, C., De
Observatoire de Paris-Meudon
F-92190 Meudon Cedex
France

Billing, G. D.
Dept. of Chemistry
Panum Institute
Blegdamsvej 3
DK-2200 Copenhagen N
Denmark

Black, J. H.
Steward Observatory
University of Arizona
Tucson, AZ 85721
USA

Blake, G. A.
320-47 Caltech
Pasadena, CA 91125
USA

Böhringer, H.
MPI für Kernphysik
Saupfergeckweg 1
D-6900 Heidelberg
FRG

Buck, U.
MPI für Strömungsforschung
Böttingerstr. 4-8
D-3400 Göttingen
FRG

Bunker, P. R.
Herzberg Institute of Astrophysics
National Research Council of Canada
Ottawa, Ontario
Canada, K1A OR6

Chandra, S.
Institut für Theoretische Physik
Universität Frankfurt
Robert-Mayer-Str. 10
D-6000 Frankfurt/Main
FRG

Dalgarno, A.
Harvard-Smithsonian Center for Astrophysics
60 Garden Street
Cambridge, Massachusetts 02138
USA

Destombes, J. L.
Laboratoire de
Spectroscopie Hertzienne
Universite de Lille I
F-59655 Villeneuve d'Ascq Cedex
France

Diercksen, G. H. F.
MPI für Astrophysik
Karl-Schwarzschild-Str. 1
D-8046 Garching
FRG

Dishoeck, E. F. van
Harvard-Smithsonian Center for Astrophysics
60 Garden Street
Cambridge, Massachusetts 02138
USA

LIST OF PARTICIPANTS

Draine, B.T.
Peyton Hall
Princeton University Observatory
Princeton, NJ 08544
USA

Erman, P.
Research Institute for Physics
S-104 05 Stockholm 50
Sweden

Fatyga, B.W.
Dept. of Chemistry
Indiana University
Bloomington, IN 47405
USA

Federman, S.
Jet Propulsion Laboratory
M/S 183-601
4800 Oak Grove Drive
Pasadena, CA 91109
USA

Ferguson, E.E.
Aeronomy Laboratory, NOAA
325 Broadway
Boulder, Colorado 80303
USA

Flower, D.
Department of Physics
University of Durham
South Road
Durham DH1 3LE
England

Gargaud, M.
Observatoire de l'Universite de Bordeaux I
B.P. 21
F-33270 Floirac
France

Graff, M.M.
Harvard-Smithsonian Center for Astrophysics
60 Garden Street
Cambridge, Massachusetts 02138
USA

Grüner, N. E.
MPI für Astrophysik
Karl-Schwarzschild-Str. 1
D-8046 Garching
FRG

Guelin, M.
IRAM
Voie 10
Domaine Universitaire de Grenoble
F-38406 St. Martin d'Heres Cedex
France

Heil, T. G.
Department of Physics
University of Georgia
Athens, GA 30606
USA

Herbst, E.
Department of Physics
Duke University
Durham, North Carolina 27706
USA

Herzberg, G.
Herzberg Institute of Astrophysics
National Research Council of Canada
Ottawa, Ontario
Canada, K1A OR6

Hjalmarsson, A.
Onsala Space Observatory
S-43900 Onsala
Sweden

Huebner, W. F.
T-4, MS B 212
Los Alamos National Laboratory
Los Alamos, NM 87545
USA

Joerg, H.
Lehrstuhl für Theoretische Chemie
Technische Universität München
Lichtenbergstr. 4
D-8046 Garching
FRG

LIST OF PARTICIPANTS

Kegel, W. H.
Institut für Theoretische Physik
Universität Frankfurt
Robert-Mayer-Str. 10
D-6000 Frankfurt/Main
FRG

Kraemer, W. P.
MPI für Astrophysik
Karl-Schwarzschild-Str. 1
D-8046 Garching
FRG

Kroto, H. W.
School of Chemistry and Molecular Sciences
University of Sussex
Brighton BN1 9QJ
England

Langhoff, P. W.
Dept. of Chemistry
Indiana University
Bloomington, IN 47405
USA

Leach, S.
Laboratoire de Photophysique Moleculaire
Batiment 213
Universite Paris-Sud
F-91405 Orsay Cedex
France

Li Shou-zhong
MPI für Astrophysik
Karl-Schwarzschild-Str. 1
D-8046 Garching
FRG

Lindinger, W.
Institut für Experimentalphysik
Karl-Schönherr-Str. 3
A-6020 Innsbruck
Austria

Massi, M.
Osservatorio Astrofisico di Arcetri
Largo E. Fermi, 5
I-50125 Firenze
Italy

McCarrol, R.
Observatoire de l'Universite de Bordeaux I
B.P. 21
F-33270 Floirac
France

Millar, T.J.
Department of Mathematics
UMIST
P.O. Box 88
Manchester M60 1QD
England

Oddershede, J.
Dept. of Chemistry
Odense University
Campusvej 55
DK-5230 Odense M
Denmark

Omont, A.
Groupe d'Astrophysique
CERMO
B.P. 68
F-38402 Saint-Martin d'Heres Cedex
France

Palla, F.
Osservatorio Astrofisico di Arcetri
Largo E. Fermi, 5
I-50125 Firenze
Italy

Prasad, S.S.
Jet Propulsion Laboratory
M/S 183-601
4800 Oak Grove Drive
Pasadena, CA 91109
USA

LIST OF PARTICIPANTS

Rostas, F.
Observatoire de Paris-Meudon
Dept. d'Astrophysique Fondamentale
F-92190 Meudon Cedex
France

Rowe, B.
Laboratoire d'Aerothermique du Centre National
de la Recherche Scientifique
4 ter, route des Gardes
F-92190 Meudon
France

Sadlej, A. J.
Chemical Centre
University of Lund
Box 740
S-22007 Lund 7
Sweden

Saykally, R. J.
Dept. of Chemistry
University of California
Berkeley, CA 94720
USA

Schäfer, J.
MPI für Astrophysik
Karl-Schwarzschild-Str. 1
D-8046 Garching
FRG

Scheingraber, H.
MPI für Extraterrestrische Physik
D-8046 Garching
FRG

Scoville, N. Z.
Astronomy Department
Caltech 105-24
Pasadena, CA 91125
USA

Sharp, C. M.
MPI für Astrophysik
Karl-Schwarzschild-Str. 1
D-8046 Garching
FRG

Smith, D.
Dept. of Space Research
University of Birmingham
P.O. Box 353
EDGBASTON
Birmingham B15 2TT
England

Solomon, P.
Dept. of Earth and Space Sciences
State University of New York
Stony Brook, N.Y. 11794
USA

Valiron, P.
Groupe d'Astrophysique
CERMO
B.P. 68
F-38402 Saint-Martin d'Heres Cedex
France

Vidal, C.R.
MPI für Extraterrestrische Physik
D-8046 Garching
FRG

Watson, W.D.
Department of Physics
University of Illinois
1110 W. Green Street
Urbana, IL 61801
USA

Williams, D.A.
Department of Mathematics
UMIST
P.O. Box 88
Manchester M60 1QD
England

Wilson, T.L.
MPI für Radioastronomie
Auf dem Hügel 69
D-5300 Bonn 1
FRG

Winnewisser, G.
I. Physikalisches Institut
Zülpicher Str. 77
Universitätsstr. 14
D-5000 Köln 41
FRG

Winnewisser, M.
Physikalisch-Chemisches Institut
Justus-Liebig-Universität Giessen
Heinrich-Buff-Ring 58
D-6300 Giessen
FRG

Winnewisser, B.
Physikalisch-Chemisches Institut
Justus-Liebig-Universität Giessen
Heinrich-Buff-Ring 58
D-6300 Giessen
FRG

Ziurys, L. M.
Dept. of Chemistry
University of California
Berkeley, CA 94720
USA

SUBJECT INDEX

Abundance	421
Abundance, cosmic	240
Abundance, deuterium	685
Abundance, isotopic	164
Abundance, molecular	154, 156
Abundance, OH	685
Adsorption, collision-induced	334
Approximation, centrifugal sudden (CS)	445
Approximation, Koopmans	567
Association, radiative	11, 239, 257, 462
	626, 645
Association, ternary	462
Atmosphere, planetary	19
Auger effect	559, 567
Autoionization	556, 567
Background, cosmic	13
Background, microwaves	51, 52
Band system, Lyman-Birge-Hopfield	535
Barrier, activation energy	243
Bender, nonrigid	493
BN-KL cluster	305
Calculation, ab-initio	137, 139, 492
Calculation, self-consistent-field	435
Calculation, semi-detailed	249
Calculation, static-exchange	561
Calculatios, close coupling	441
Carbon chains	157, 164
Cation, doubly charged molecular	354
Center, galactic	193
Charge transfer, positive ion	322
Circulation currents	609
Clouds, dark	613
Clouds, diffuse interstellar	256
Clouds, interstellar	7, 16, 403, 491,
	649, 684
Clouds, molecular	177, 295, 390, 395
Clouds, pregalactic	7
Coefficient, dissociative recombination	657
Coefficient, ion-molecule reaction rate	631, 637, 638
Coefficient, rate	649
Coefficient, second virial	510
Coefficient, Einstein	535

Coefficient, Langevin rate 634
Coincidence techniques 358
Collapse, gravitational 604
Collision, H2-H2 690
Comets 15, 18, 491, 685
Comparison with observation 271
Complex 242
Confusion 148, 149
Confusion limit 146
Constant equilibrium 664
Constant, radiative association rate 473
Contraction, gravitational 250
Coolants 165
Cooling 88, 163
Cooling, interstellar cloud 49
Cooling, radiative 318
Core, central 605
Core, density 605
Cosmic ray bombardment 239
Coupled states 517
Cross section expressions 559
Cross section sums 562
Cross section values 568
Cross section, channel 563
Cross section, elastic differential 509
Cross section, partial 551, 554, 560
Cross section, partial photoionization 552
Cross section, partial-channel photoionization 566
Cross section, photon attenuation 552
Cross section, photoionization 551
Cross section, state-resolved differential 434
Cross section, state-to-state differential 434
Cross section, total differential 514
Cross section, total ionization 555
Cross section, K-edge photoabsorption 567
Cyanopolyenes 11
Cyanopolyines 381, 613
Cyanopolyynes 246, 398
Damping function 444
Daser 86
Deexitation, radiative electronic state 322
Densities, CH column 593
Desorption 609
Detection, acoustic 382
Deuterium fraction 105
Diffraction oscillations 435
Dipole moment 493
Dipole oscillator-strength density 559
Dispersion coefficients 435
Dissociation 314

Dissociation, collision-induced 282, 288
Dissociation, collisional 301
Dissociation, shock 282
Distance scales, cosmological 166
Doublet, regular Lambda- 117
Doubling, Lambda- 56, 129
Doubling, Rho- 53, 54
Dust 676
Dust particles 238
Effects, radiative transfer 679
Electrons, photoejected 555, 559
Energy change, Gibbs 665
Energy sinks 315
Energy sources 315
Energy transfer, rotational 433
Energy transfer, rotational-translational 498
Energy, zero-point- 461
Equation, generalized Saha 665
Equation, kinetic 240
Evolution, hydrodynamical-chemical 604
Evolutionary perspective 603
Excess photon energy 316
Excitation 53, 88
Excitation, collisional 76, 299, 593
Excitation, radiative 593
Excitation, resonant 675
Factor, anisotropy 560
Factor, Franck-Condon 561
Field, magnetic 15, 281, 284, 288
First flight current density 318
First flight flux 318
Fixations 244
Flow, multi-fluid 314
Fluorescence 228, 559, 567
Fluorescence, laser induced 440
Force, pressure gradient 605
Formation of H2 260
Fractionation, chemical 143, 652
Fractionation, isotopic 273, 395
Fragmention 559
Free fall 606
Frequencies, rest 596
Frequency multiplier 378, 381
Function, continuum 561
Function, dissociation 664
Function, numerical scattering wave 510
Function, pair correlation 510
Function, scattering 561
FALP (flowing afterglow / Langmuir probe) apparatus 657
Galaxies 159

Gas, shock-heated	284
Gas, shocked	15
Gasphase reaction	238, 239
Grain adsorption	241
Grains	608
Gunn-diode	392
High-energy limits	561
Hole states, electronic	558
Hydrides of "heavy" elements	263
Hydrides of nitrogen	267
Hydrides of oxygen	266
Hydrocarbons	237, 244
Hypothesis, young cloud	250
Infrared	215
Infrared astronomy	398
Infrared Laser techniques	403
Integration, numerical	493
Intensity borrowing	536
Interactions, H2-He	690
Interchange, positive ion-atom	322
Inversion, potential	384
Invertor, nonrigid	493
Ion drift tube	631, 639, 649
Ion sources	358
Ion trap	631
Ionization	163, 314
Ionization, collisional	301
Ionization, dissociative	559
Ionization, photodissociative	321
Ionization, soft X-ray	569
Ionization, UV-	569
Ions	164
Ions, deuterated	11
Ions, fragments	551
Ions, kinetically excited	458
Ions, molecular	135, 353, 403
Ions, parent	566
Isotope exchange	453, 460
Isotope fractionation	461
Isotopic refrigeration	464, 485
ISC, shocked	459
Kinetics, chemical	314
Limit, large J-	517
Line searches	592
Line, adsorption	682
Line, hyperfine	592
Line, optical absorption	215
Line, undefined	137
Lines, U-	138
LKHa 101	209

L183	613
Mapping	395
Martian dayglow	365
Maser	59, 93, 95, 284
Maser, interstellar	15, 187, 592
Maser, CH	117
Maser, H2O	673
Maser, SiO	110
Mass spectrometer	439
Mass-loss	141
Mass, multiple reduced	666
Material, shocked	281, 286
Mean free path	315
Method, polarization propagator	534, 538
Method, quantum chemical	682, 683
Millimeter wave region	375
Model	248
Model of interstellar clouds	685
Model, endothermic trapping	481
Model, static	608
Model, Gwin-Townes	591
Molecule formation	613
Molecules, carbon chain	249
Molecules, complex	8, 237, 248, 250, 288
Molecules, deuterated	11, 284
Molecules, interstellar	8, 13
Molecules, new	152, 153
Molecules, new interstellar	131, 165
Molecules, quasilinear	491
Molecules, quasiplanar	491
Molecules, synthesis	639
Motion, proper	98, 166
MWC 297	209
Nozzle beams	439
NMR relaxation time for H2-He	503
Objects, pregalactic	7
Objects, protostellar	201
Objects, quasi-interstellar	229
Objects, Becklin-Neugebauer	202
Opacities	687
Orbitals, innner valence	558
Orbitals, outer valence	558
Orion	182, 678
Orion Molecular Cloud 1	305
Oscillatorstrength	422
Photochemisty	238
Photodissociation	13, 55, 158, 222, 258, 551, 625
Photoionization	551

Photons, X-ray 569
Planets, outer 331
Planets, Iovians 19
Population inversion 76, 101
Populations, rotational 681
Potential, ionization 552, 557
Potential, non-spherical intermolecular 433
Potential, square well 671
Precursor 281, 289
Precursor, magnetic 296
Predissociation 421, 564
Predissociation, inverse 426
Problem, rotation-vibration 493
Problem, CH+ 263
Problem, NH3 456
Procedure, Feshbach-Fano 563
Processes, fragmentation 569
Processes, ionization 569
Processes, photodissociation 683
Quadrupole interactions 387
Radiation field, interstellar 553
Radiation transfer 332
Radiotelescopes 390
Radius, critical 318
Rainbow, rotational 437
Raman broadening and shift 500
Rate, reaction 283
Ratios, branching 329
Ratios, isotopic 143
Rays, cosmic 7, 263
Rays, X- 263
Reactions, association 638, 640
Reactions, chemical 302
Reactions, chemical exchange 282
Reactions, condensation 245
Reactions, endoentropic 654
Reactions, hydrogenation 243
Reactions, insertion 244
Reactions, ion-electron dissociative recombination 240
Reactions, ion-molecule 135, 239, 258
453, 471
Reactions, isotope exchange 649
Reactions, neutral-neutral 258
Reactions, on dust grains 257
Reactions, radiative associations 241
Reactions, ternary association 243
Reactions, Penning 635
Receiver 392
Recombination, dissociation 8, 11, 287, 466
Recombination, dissociative electron 259, 322

SUBJECT INDEX

Regions, shocked	15, 281
Relaxation cross sections	499, 506
Relaxation, intramolecular	358
Relaxation, vibrational	474
Resonance autoionization	564
Resonance phenoma	563
Resonance, shape	564, 565, 567
Resonance, Feshbach	556, 564
Reversibility, microscopic	242
Rotation, quadruple transition	201
Sequences, chemical	8, 13
Shells, circumstellar	15
Shielding, self-	261
Shock	281, 282, 286, 385
Shock front	281
Shock model	250
Shock wave	295
Shocks, dissociative	289
Shocks, C-type	295
Shocks, J-type	296
Slip, magnetic-ion	295
Sources, soft X-ray	569
Spectra of ions, rotational	376
Spectra, differential energy	434
Spectra, infrared and microwave of small molecules	491
Spectra, rotational	411
Spectrometer, acousto-optical	392
Spectroscopy, electron	557
Spectroscopy, millimeter wave	621
Spectroscopy, molecular ion	355
Spectroscopy, molecular	331
Spectroscopy, optical	533
Spectroscopy, photofragment	625
Spectroscopy, time resolved	422
Spectroscopy, velocity-modulation Laser	406
Spectroscopy, Laser magnetic resonance rotational	407
Spectrum, solar	553
Spin statistics	395
Spin-rotation interaction	387
Splitting, hyperfine	410
Stabilization, radiative	282, 287
Star formation	201
Star, nascent	609
Star, B-type	569
Star, O-type	569
States, inner-valence ionic	566
States, parent-ionic	559
Steady-state	240
Submillimeter wave region	375
Sum rules	561

Swarm experiments	649
Synthesis, interstellar molecular	649
Technique, CRESU	631
Technique, High Frequency Deflection	421
Technique, SIFT	631
Telescope, 10-m submillimeter	194
Theory, perturbation	493
Theory, variation	493
Thread, common	608
Threshold effect	564
Threshold law	561
Threshold law, Wigner	562
Threshold, K-edge	558, 567
Threshold, L-edge	558
Time dependence, chemical	240, 250
Titan	20, 331
Transition, dipole forbidden	535
Transition, moments	535
Transition, probability	424
Transition, radiative	533, 539
Transition, spin forbidden	536
Transition, transfer	428
Transition, vibrational	201
Transition, Lambda-doubling	592
Triton	331
TMC-1	398, 613
Ultraviolett	215
UOA 27	209
UV fields, internal	241
UV radiation	556
UV stars	558
VLBI	66, 97
VT-SIFDT	455, 457
VT-SIFT	454
W 51	591
Wave approximation, distorted	436
Zeemann splitting	63, 64, 65
Zeta Ophuichi	274

MOLECULE INDEX

Ar	347, 523, 524
B	687
Be	687
C	224, 240, 244, 249, 250, 251, 255, 256, 258, 261, 262, 264, 267, 268, 269, 270, 288, 289, 336, 465, 477, 614, 618, 662
C+	9, 10, 13, 15, 220, 221, 224, 240, 241, 244, 249, 250, 251, 263, 270, 273, 287, 288, 289, 459, 460, 485, 488, 614, 618
Ca+	485
CD	19
CD2H+	486
CD3+	464, 485, 486
CD3H2+	464, 485
CH	9, 10, 13, 16, 50, 76, 117, 215, 219, 221, 227, 228, 230, 256, 264, 268, 269, 282, 287, 289, 555, 591, 593, 616, 621, 669
CH+	13, 15, 215, 219, 220, 244, 256, 263, 264, 265, 271, 273, 287, 288, 459, 473, 488, 618, 625, 650, 651, 669, 670
CH2	9, 10, 127, 264, 269, 669
CH2+	9, 10, 244, 264, 287, 473, 476
CH2CHCHO	376
CH2CHCN	376, 616
CH2CHNC	376
CH2CO	248
CH2D+	276, 460, 461, 464, 465
CH3	244, 246, 248, 288, 348, 477, 669
CH3+	9, 10, 11, 241, 243, 244, 246, 248, 276, 463, 473, 476, 478, 479, 485, 489
CH3CHO	250
CH3CH2CN	194
CH3CH2OH	145

MOLECULE INDEX

CH3CN	248, 250, 348, 464, 616
CH3CNH+	248
CH3CO+	248
CH3C2H	382, 615, 616
CH3C3N	7, 152, 616
CH3C4H	152, 382, 395, 398
CH3D	334, 337, 340, 343, 344
CH3HCN+	463, 464
CH3H2O+	463, 464
CH3NH2	250, 348
CH3O+	479
CH3OCH3	145, 187, 248, 250
CH3OCH4+	248
CH3OH	16, 142, 143, 187, 189, 190, 191, 192, 248, 250, 464
CH3OH2+	248
CH3OOCH	145
CH3OOH	136
CH3O2+	478, 633, 634
CH4	10, 11, 18, 20, 136, 244, 261, 288, 289, 334, 336, 337, 340, 341, 343, 344, 345, 347, 454, 476, 477, 479, 615, 631, 633, 634, 635, 636, 638, 669, 670
CH4+	652, 653, 654
CH4D+	464, 465
CH5+	9, 10, 244, 462, 463, 473, 476, 477, 485, 652, 653, 654, 657, 658
CI	179, 193, 240, 250, 615, 616
CN	13, 14, 50, 54, 215, 219, 221, 227, 228, 230, 256, 268, 269, 271, 273, 289, 669
CN2H2	385
CO	9, 12, 13, 15, 19, 178, 181, 188, 191, 192, 193, 215, 220, 222, 223, 224, 225, 229, 230, 240, 251, 256, 260, 261, 268, 269, 270, 273, 274, 276, 282, 283, 284, 286, 288, 289, 334, 343, 379, 454, 458, 460, 461, 463, 477, 523, 524, 564, 568, 616, 631, 635, 636, 669
CO+	12, 19, 269, 270, 286, 481, 566, 567, 621, 650, 651
COH+	650
CO2	18, 19, 222, 342, 463, 480, 523, 524, 568
CO2+	19, 364, 650

CO2H+	650
CS	180, 184, 286, 568
CS+	215, 230, 361
CS2	568
C2	223, 215, 219, 223, 227, 229, 230, 268, 269, 273, 274, 555, 681
C2+	269, 669
C2−	668
C2H	244, 245, 616
C2H+	269
C2H2	245, 334, 341, 342, 458, 464, 662, 669
C2H2+	244, 245, 269, 458, 464, 482, 483
C2H3	245, 464
C2H3+	244, 245, 482, 484
C2H4	245
C2H4+	245, 464, 482, 483
C2H5+	245, 248, 482
C2H5OH	250
C2H5OH2	248
C2H5OH2+	248
C2H6	20, 334, 341, 342, 343, 344
C2N2	342
C3	219, 223, 229, 230, 668, 669, 670
C3+	245
C3H	139, 464, 616, 618
C3H+	464, 614
C3H2	464
C3H2+	458, 482, 483, 484
C3H3+	245, 246, 458, 464, 482, 483, 614
C3H4	249, 334, 342, 458
C3H4+	482
C3H5+	482
C3H7+	482
C3H8	20, 334, 342
C3H9+	20
C3N	152, 245, 246, 249, 616
C3O	152
C4	245
C4H	245, 249, 250, 616
C4H2	342
C4H2+	245, 615
C4H3	245
C4H3+	245
C4N	152, 246
C4N+	246

MOLECULE INDEX

C5H	616
C5H+	246
C5H2+	246
C5N	246
D	7, 11, 262, 263, 266, 275, 282, 284, 337, 344, 345, 462, 685
D+	5, 7, 263, 460
DCO+	276, 462, 652
DNC	276
D2	486
Fe+	485
GeH4	334
H	4, 6, 11, 12, 17, 20, 220, 226, 244, 256, 258, 259, 260, 261, 263, 264, 265, 266, 269, 270, 271, 276, 282, 283, 284, 285, 286, 287, 288, 335, 456, 458, 459, 465, 481, 488, 614, 649, 650, 662, 687, 688
H+	4, 5, 6, 7, 12, 260,
H+	263, 269, 481
H−	4, 5, 260, 555, 556, 561,
H−	562
He	19, 240, 256, 334, 335, 336, 344, 348, 458, 463, 473, 474, 482, 485, 522, 524, 555, 556, 561, 631, 633, 635, 687
He.CO	514
He.H2	522
He+	13, 635, 636, 641
He−	689
HeI	689
HCl	683
HCl+	409
HCN	180, 187, 241, 244, 269, 276, 289, 334, 342, 463, 616, 662, 669
HCND+	276
HCNH+	246
HCO	288, 478
HCO+	10, 12, 135, 180, 188, 248, 260, 269, 274, 286, 289, 458, 461, 462, 477, 479, 480, 482, 616, 650, 652, 657, 658
HCOOCH3	136, 395
HCOOH	145, 248, 478
HCOOH2+	248
HCO2+	621
HCP	348

MOLECULE INDEX

HCS+	473
HC2I	379
HC3N	246, 250, 342, 343, 381, 616
HC5N	16, 223, 246, 250, 381
HC5N+	246
HD	5, 11, 13, 15, 221, 260, 262, 263, 266, 274, 275, 276, 282, 284, 334, 337, 461, 462, 486, 555, 652
HD+	11
HDO	104, 284
HF	523
HF+	410
HHe	688
HI	689
HII	3, 187, 191, 193, 287
HNC	237, 616
HNCO	376
HN2+	473
HOC+	137, 473
HOSi+	473
HO2	285
H2	4, 5, 6, 7, 9, 11, 12, 15, 16, 18, 177, 181, 182, 184, 185, 186, 188, 191, 192, 193, 201, 215, 217, 220, 221, 223, 224, 225, 226, 227, 228, 229, 230, 238, 239, 240, 241, 243, 244, 251, 256, 258, 260, 261, 262, 263, 264, 265, 271, 274, 275, 276, 282, 283, 285, 287, 288, 295, 334, 335, 336, 337, 341, 342, 344, 456, 457, 458, 459, 460, 463, 466, 475, 477, 481, 485, 486, 488, 510, 522, 524, 555, 556, 561, 614, 635, 636, 637, 641, 644, 649, 650, 651, 681, 688, 689, 690
H2.He	690
H2.H2	441, 690
H2+	4, 5, 19, 229, 239, 260
H2-	689
H2CNCN	389
H2CN2	387
H2CO	182, 184, 186, 187, 194, 244, 288, 478, 480, 616
H2CO+	10
H2C3N+	245, 246
H2D+	11, 276, 621, 460, 461

H2He	334, 335, 336
H2N2	342
H2O	9, 12, 15, 16, 17, 18, 19, 95, 186, 187, 190, 191, 221, 222, 244, 261, 266, 283, 284, 287, 288, 289, 334, 341, 463, 465, 480, 568, 662, 669, 674, 677
H2O+	12, 17, 221, 222, 230, 240, 266
H2S	285, 289, 348, 454, 459, 460
H2S+	362, 459, 460
H3	229
H3+	11, 12, 19, 229, 240, 267, 268, 276, 278, 458, 460, 466, 477, 614, 618, 657, 658
H3CO+	10
H3O+	12, 17, 222, 223, 240, 266
H3S+	459, 460, 465
H4C3N+	246
H4C5N+	246
H5+	20
K+	485
Li	687
Li+	485
Mg	485
MgH	230
MgO	230
N	19, 240, 251, 256, 258, 261, 262, 264, 267, 268, 271, 289, 337, 341, 465, 614, 662
N+	457, 635, 636, 637
Na+	465, 485
NaH	230
NaOH	465
Ne	454, 523, 524
Na+.H2	465
NH	230, 271, 555, 669
NH+	457, 635, 636, 637
NH2+	267, 457, 614, 615
NH2D	384, 395
NH3	20, 72, 130, 184, 185, 186, 184, 185, 186, 188, 190, 192, 193, 194, 261, 289, 334, 337, 340, 341, 454, 456, 457, 463, 522, 617
NH3.H2O	340
NH3+	456, 457, 636, 644
NH4+	456, 457, 462
NH4SH	340

MOLECULE INDEX

NO	268, 270, 271, 484, 568
NO+	19, 270, 475, 476, 484
NO+.N2	475
NO2	480
NS	568
N2	19, 260, 261, 268, 270, 271, 276, 289, 342, 343, 345, 347, 454, 456, 461, 474, 475, 476, 484, 523, 555, 564, 568, 631, 633, 635
N2+	457, 474, 566, 567, 631, 633, 636, 637
N2D+	276, 462, 658
N2H+	187, 260, 458, 461, 462, 616,
N2H+	644, 657, 658
N2H4	348
N2O	523
N4+	474, 633
O	17, 19, 222, 240, 251, 256, 261, 262, 263, 264, 266, 267, 268, 270, 282, 283, 289, 343, 465, 479, 480, 568, 614, 662, 685
O+	19, 263, 265, 270, 456, 635
OCS	286
OC3O	382
OC3S	382
OD	684
OH	13, 15, 16, 17, 19, 55, 61, 74, 76, 186, 187, 188, 190, 191, 221, 222, 230, 239, 268, 269, 270, 273, 283, 285, 286, 287, 288, 289, 295, 568, 683, 685
OH+	12, 17, 240, 668, 669, 670
OH−	414
O2	221, 222, 225, 261, 268, 270, 285, 289, 454, 458, 463, 479, 555, 631, 633, 634, 635
O2+	19, 456, 475, 479, 484, 633, 634
O2+.N2	484
O2H+	458
O4+	474, 475, 633
PH3	334, 340
S	285, 286, 289, 465, 568, 662
S+	459
Si	240, 286, 485
Si+	485
SiC2	229, 538

SiH				56
SiH+				230
SiO	16,	110,	191,	192
SiO2				485
SiO2+				485
SiS				16
SC3S				382
SH		230,	285,	568
SH+	230,	361,	459,	465
SO		192,	285,	669
SO2			192,	285
S2				555
Ti				662